Günter Gottstein

Physical Foundations of Materials Science

Springer

Berlin
Heidelberg
New York
Hong Kong
London
Milan
Paris
Tokyo

Günter Gottstein

Physical Foundations of Materials Science

With 472 Figures

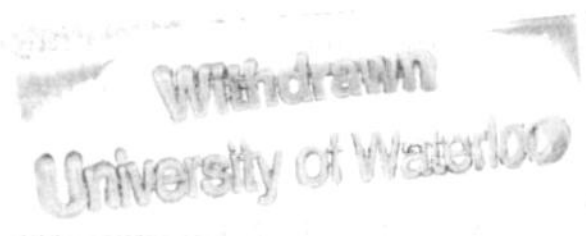

Springer

Professor Dr. Günter Gottstein
RWTH Aachen
Institut für Metallkunde und Metallphysik
Kopernikusstr. 14
52074 Aachen
Germany

Cataloging-in-Publication Data applied for
Bibliographic information published by Die Deutsche Bibliothek
Die Deutsche Bibliothek lists this publication in the Deutsche Nationalbibliografie; detailed bibliographic data is available in the Internet at <http://dnb.dd.de>

ISBN 3-540-40139-3 Springer-Verlag Berlin Heidelberg New York

Springer-Verlag is a part of Springer Science+Business Media

springeronline.com

Printed in Germany

Typesetting: Digital data supplied by author
Cover-Design: Design & Production, Heidelberg
Printed on acid-free paper 62/3020 Rw 5 4 3 2 1 0

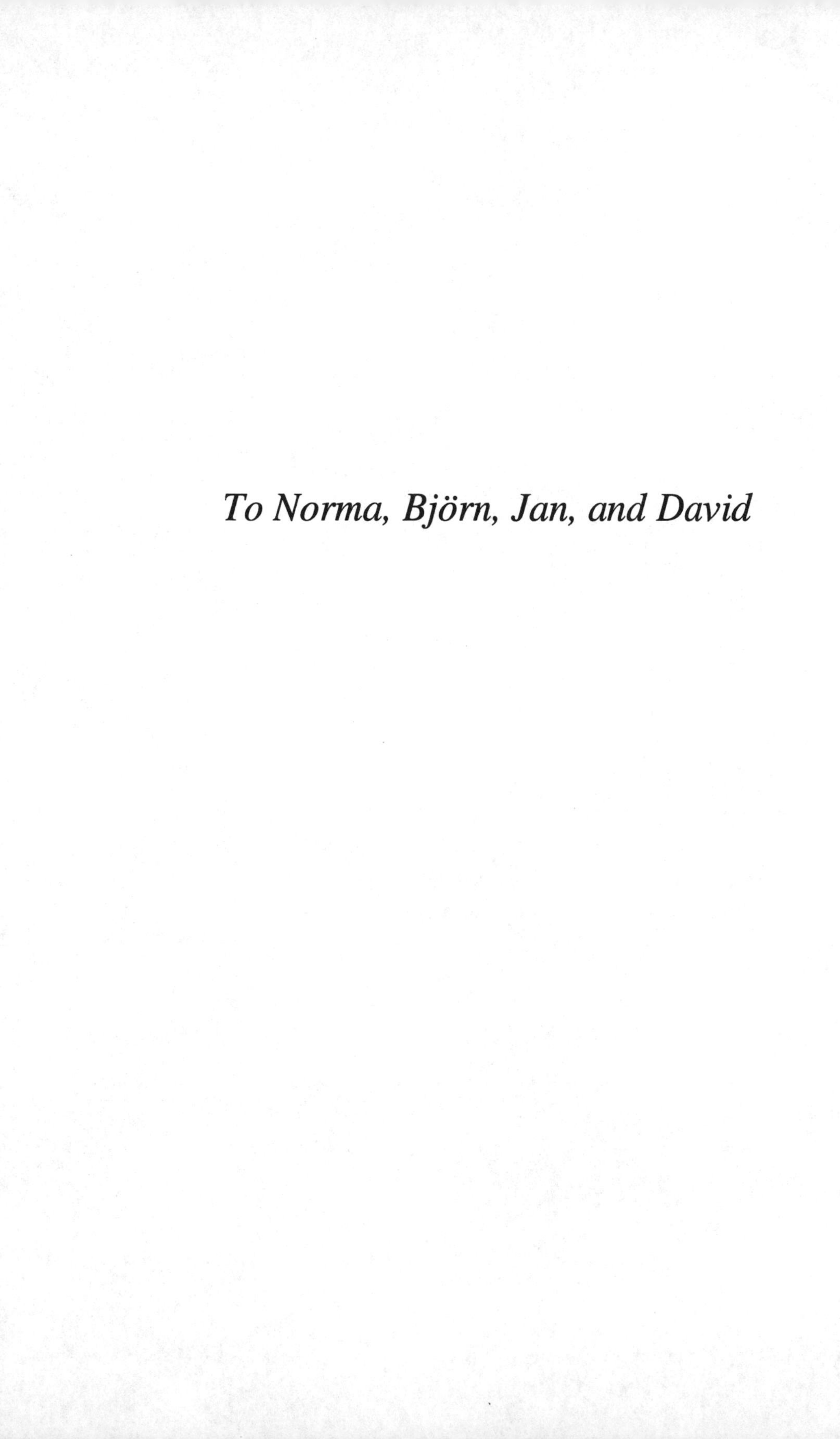

To Norma, Björn, Jan, and David

Acknowledgement

To begin with I would like to express my appreciation for the advice, recommendation, and encouragement that I received from many colleagues abroad who urged me to translate my German textbook into English. Most of all, I am indebted to Mrs. Irene Zeferer, who was in charge of typescript and layout of the text, for her dedicated personal engagement in the project. Mrs. Barbara Eigelshoven assisted with quality enhancement of the figures and in securing top quality micrograph reproduction. Dipl.-Ing Petar Mijatovic lent a hand in circumnavigating unanticipated problems with the desk top publishing system. Also, he regenerated all the tables in a more aesthetically pleasing format. The invaluable advice of Profs. Tony Rollett, CMU, and Martin Glicksman, RPI, on the correct use of language and grammar was highly appreciated. Finally, I want to thank all those who assisted in proof reading the manuscript, in particular my doctoral students.

Preface

This book is a translation of a German textbook on Materials Science. It originates from a set of handouts at the RWTH Aachen university for students of materials engineering and of metal physics and was developed over the years to a compact manuscript. From the core of a physical metallurgy text it was extended to a broader coverage of materials in the frame of established scientific concepts.

The text aims at providing the physical fundamentals to understand materials behavior and at preparing the reader for more advanced litarature on the subject. On the other hand the book is designed not to follow the common scheme of traditional introductory Materials Science texts which primarily introduce into the phenomenology of materials on an elementary level. Rather this book tries to bridge the scope from atomistic mechanisms to engineering properties of materials not staying away from mathematics where necessary. The manuscript does not pretend to give a comprehensive coverage of materials science, and as a textbook it has to find a compromise between comprehensive and in depth treatment of the subject. Such compromise is a matter of personal preference and taste. This is particularly true for the chapter on "physical properties" which is designed for materials engineering students who usually are less familiar with the basics of solid state physics.

The text builds on the classical German text of Masing "Einführung in die Metallkunde" (introduction to physical metallurgy) which represents the approach of the Göttingen school of physical metallurgy in the line of Tammann, Masing, Haasen, and Lücke. It is typical for this approach to develop a deeper understanding of the subject and to reduce complex phenomena to their essential physical mechanisms for refined analysis and prediction. Since understanding is essentially based on visualization the text provides an abundance of figures to guide the reader through the seemingly confusing but fascinating world of materials.

Contents

Introduction

The development of new materials is considered a key technology on an international level. The capability of production, fabrication, and application of high performance materials is a prerequisite for novel, and internationally competitive products and processes and a crucial element for more efficient use of resources and for environmental protection.

The development of novel materials and processes requires a deep knowledge of the physical foundations of materials as a tool for systematic tayloring of materials properties. These physical foundations are the roots of material science, and they are the key elements of this text book. The term material science is relatively young and not very precisely defined. Sometimes it is understood as an extension of physical metallurgy to non-metallic materials. Natural scientists frequently associate material science only with novel or even exotic functional materials.

In the following we will define materials science as our knowledge of the relation between microscopic structure and macroscopic properties of engineering materials. It combines the broad spectrum of commercially utilized solids ranging from metals to ceramics, glasses, and plastics to composites and electronic materials.

The most important group of engineering materials, both with respect to production volume and variety of applications as well as tradition and systematic development, are metals. Their excellent combination of formability and strength have made them prime materials for structural applications and their excellent electrical and thermal conductivity has rendered them indispensible for electrical engineering. Metals have determined the history and development of materials over several thousand years, even coined the names of historical periods like the bronze age which dates back to 3000BC. The need for low cost mass products and components for extreme service conditions in our industrial age have made high performance ceramics, plastics, and eventually composites highly compatitive materials.

The materials science of ceramics, plastics, and electronic materials is relatively young compared to metallurgy. The physical foundations and theoretical concepts of metals, ceramics, semi-conductors, and plastics, however, share a large common frame, which is essentially derived from the foundations of physical metallurgy. In this respect, metallurgy is the mother of materials science as evident from the extensive research in this class of material over many centuries. Despite its very long tradition, metallurgy is not a classical scientific discipline itself. Up to medieval ages knowledge of the extraction and fabrication of metals had been considered a national secret asset and had been only traded orally from generation to generation. Only in the middleages a German metallurgist of name Bauer (engl. farmer, in latin "Agricola") wrote down the recipes of metal fabrication in his famous book "De Re Metallica" [0.1]. The book reads like a mystic instruction to metal fabrication, there is mention of bull blood and clear nights of full moon, harmful creatures like cobolds and nickels (therefore the terms cobalt and nickel) all of which had practical relevance, e.g. for making a steel harder, and which we explain nowadays on a scientific basis. As a matter of fact metallurgy was a discipline of alchemie in medieval times and comprised a colorful mixture of empirical recipies and superstition.

With increasing scientific character of more recent centuries metallurgy became a displine of chemistry, where it remained even up to now at many universities. The rapid development of the understanding of materials properties, in particular due to the discovery of X-rays and their application to crystallography, revealed that the properties of metals were not determined by the gross chemical composition, in contrast to common belief at that time. This made metallurgy to become a discipline of physical chemistry. The development of the atomistic foundations of our understanding of mechanical and electronic properties of metallic materials in the frame work of dislocation theory and electron theory of metals finally shifted the focus of metallurgy to physics at the beginning of the 20th century. Eventually it engendered the discipline of metal physics, which has dominantly influenced the science of metals in the past 50 years. In fact, our current understanding of metallic and non-metallic materials on the basis of atomistic models has essentially been developed only in the past 80 years of research in metal physics. The objective of this research has been to describe the properties of a material on the basis of atomistic physical models, which can be formulated in terms of equations of state. This allows for a prediction of materials behavior on a theoretical basis and can be utilized to cut down on the costly and lengthy experimental investigations and testing of materials behavior.

In the sixties and seventies of the past century it became obvious that the urgent demand for high performance materials and competitive mass products had to include the development of non-metallic materials, for instance ceramics for high temperature components and plastics for a weight savings in automobiles and airplanes. Materials research revealed soon, however, that the foundations of physical metallurgy within limits can be readily applied to other materials, in particular to crystalline solids. Crystallography, constitution, diffusion, phase transformations, physical properties, and so on are the foundations of the understanding of all kinds of engineering materials.

Of course, there are also specific differences. For instance, dislocation theory which is indispensible for an understanding of plastic deformation of metals is of less importance for brittle ceramics, but it teaches the reasons for the brittleness and, therefore, offers respectives for counter actions. For polymers which are usually non-crystalline, an appropriate dislocation concept of their mechanical properties is still too complicated to be useful, and the deformation behavior of plastics is, therefore, currently restricted to phenomenological tribological models.

This development generated the strong belief that it is possible to derive a comprehensive description comprising the different classes of materials and that the future world be multi-material. This worldwide trend in the seventies of the past century caused the classical independent disciplines of metallurgy, ceramics, and plastics to merge to a new discipline "material science and engineering", encompasing both the science and engineering aspects of materials, and which has become our modern powerhouse of materials research and development.

1

Microstructure

When we buy a commercial product what we first note is its function and its appearence as, for example, the precious look of a noble metal, the engine of an automobile, the rope of a bridge, the wire of an electrical cable, the heat absorbing panes of a modern building, or the decorative ceramic and metallic parts of a modern bathroom. However, the usefulness of these items for their purpose and their life cycle will be determined by the properties of the material from which they were manufactured. Without any doubt we trust in the strength of the rope that suspends a large bridge, in the impact resistance of a ceramic hot plate in our kitchen or in the reliability of the little metallic buckets which provide the thrust of the turbine engine of an airplane at temperatures above 1000 °C. The properties of advanced materials are not so much affected by their overall chemical composition but rather by the specific arrangement of their constituents which we usually can not discriminate with our bare eye. The arrangement of the constituents of a material, i.e. the spatial distribution of elements, phases, orientations, and defects are subsumed under the term microstructure.

Castings or hot dip galvanized sheet reveal already to the bare eye that they are composed of small blocks that completely fill space. These blocks are referred to as grains or crystallites, if the nature of the material is crystalline, like metals, minerals or ceramics. Usually, however, the grain structure of a material is too small to be discerned by the bare eye. By careful surface treatment with grinding, polishing, and chemical etching the crystallites can be made visible under the optical microscope (Fig. 1.1). The corresponding image is referred to as a micrograph. Metallographic observation by optical microscopy is still an important stage of materials characterization and is referred to as optical metallography. The microstructure of the material as seen under the optical microscope is comprised of the grain structure of the material and its macroscopic chemically distinct phases.

The microstructure as revealed under the optical microscope is only a very rough characterization of the material. At higher magnification in the electron microscope it can be recognized that the macroscopically seemingly homoge-

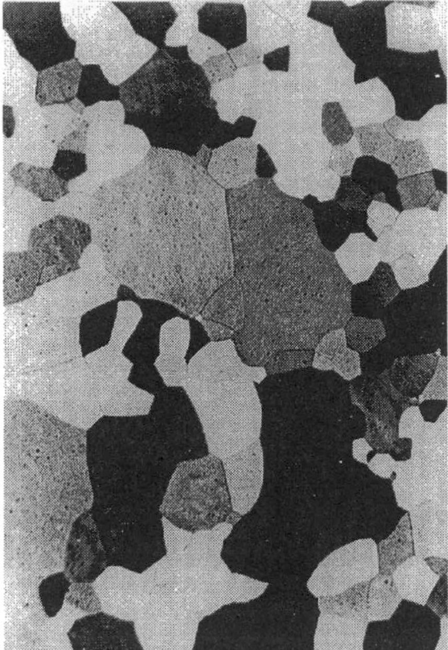

Fig. 1.1. Microstructure of recrystallized aluminium (a) and α-brass (b). The typical straight grain boundaries (twin boundaries) observed in brass are not present in aluminum and give the two structures a totally different appearance.

neous and perfect material itself contains a microstructure, i.e. microstructural defects, in particular dislocations (see Chapter 3), which are arranged in particular patterns, in addition to stacking faults, and, in most commercial materials, finely dispersed second phases (Fig. 1.2). Special materials also reveal additional microstructural elements, such as domain boundaries in magnetic materials, or antiphase boundaries in ordered solid solutions. The chemical composition can fluctuate over small distances as revealed by chemical microanalysis. Advanced materials can have a grain size in the submicrometer range (1 μm = 10^{-6}m) or even nanometer range (1 nm = 10^{-9}m) which can be resolved only under high magnification in the electron microscope. Therefore, imaging and chemical microanalysis in the electron microscope or with the electron probe microanalyzer have become standard techniques of modern metallography, i.e. advanced microstructural characterization.

Microstructural characterization is in itself interesting but becomes significant only in its relation to macroscopic properties and, therefore, requires a quantitative representation. The most fundamental microstructural information is the characteristic length scale, i.e. the grain size (grain diameter). One never, however, observes a uniform grain size, rather crystalline solids are composed of grains of different size, which is represented by the grain size dis-

Fig. 1.2. Microstructure of an engineering material (Al-alloy 2014), as it appears in an electron microscope image. Particles of a second phase and one dimensional crystal defects (dislocations) are visible. Also a grain boundary.

tribution. The characteristic value (1st moment, in the statistical sense) of the grain size distribution is the average grain size. Accurate values of grain size require advanced stereological methods. The most convenient and thus, most common is its derivation from optical micrographs, for instance by counting the intersections of grain boundaries with overlaid special geometrical curves in the micrograph, in the simplest case with a straight line or with spirals. Fortunately stereology proves that the average grain size obtained as the average intersection distance from a two dimensional micrograph also corresponds to the three dimensional average grain size within a factor of the order of one. More detailed information on the microstructure can be obtained from the grain size distribution. The grain size distribution function is defined as the statistical frequency of a specific grain size. Evidently, a mathematically exact given grain size may not exist, therefore, it is common and reasonable to count the frequency of grains with sizes that fall within a predefined grain size interval. For instance, all grains with a grain size between 0 and 10 μm, from 10 to 20 μm, and so on, are summed up. The representation of the frequency of a particular characteristic based on intervals of a measurable feature is referred to as histogram (Fig. 1.3).

A measured histogram of the grain size distribution is not symmetrical with regard to the average value, i.e. the most frequent value (median value) or the grain size D_m where the grain size distribution has its maximum, is

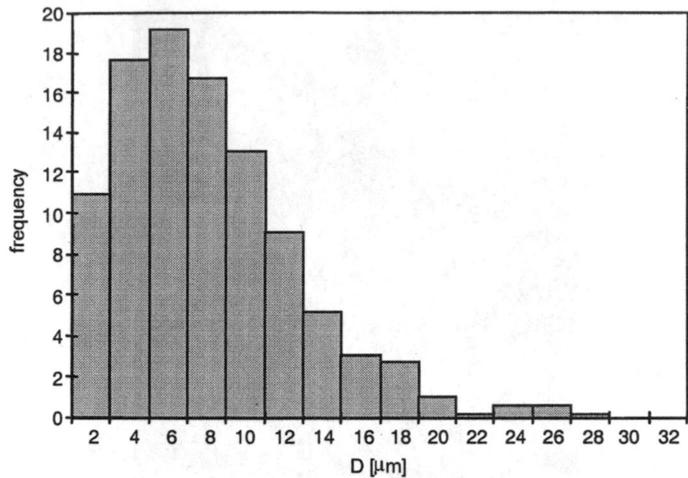

Fig. 1.3. Histogram of grain size distribution in recrystallized Fe-17%Cr commercial sheet metal (70% rolled, annealed 250 min. at 1050°C). The distribution is not symmetrical.

not identical with the average value D_0. Therefore, grain size distributions are not represented by a normal distribution (Gauss distribution or bell curve) in statistics, which is plotted in Fig. 1.4 and which can be expressed mathematically for a single observable x

$$w(x)\,dx = \frac{1}{\sqrt{2\pi}\sigma} \cdot \exp\left(-\frac{1}{2}\left(\frac{x-x_0}{\sigma}\right)^2\right) dx \tag{1.1}$$

Here $w(x)dx$ is the probability of finding the measured value in the interval $[x, x+dx]$; x_0 is the average and σ the standard deviation , which is a measure of the width of the distribution. Grain size distributions can be converted to a symmetrical shape, if the frequency is plotted against the logarithm of the grain size, lnD (Fig. 1.5). Such a distribution is termed a log-normal distribution and can be mathematically represented for a single observable x

$$w(x)\,dx = \frac{1}{\sqrt{2\pi}\sigma} \cdot \frac{1}{x} \exp\left(-\frac{1}{2}\left(\frac{\ln(x/x_m)}{\sigma}\right)^2\right) dx \tag{1.2}$$

Applied to the grain size, i.e. $x = D$, the average value of the log-normal grain size distribution is $(lnD)_0 = lnD_m$. The reason why grain size distributions follow a log-normal distribution is not known and all attempts to derive the log-normal distribution from first principles have been dissatisfactory so far.

For log-normal distributions there is a difference between the average size x_0 and the most frequent size x_m:

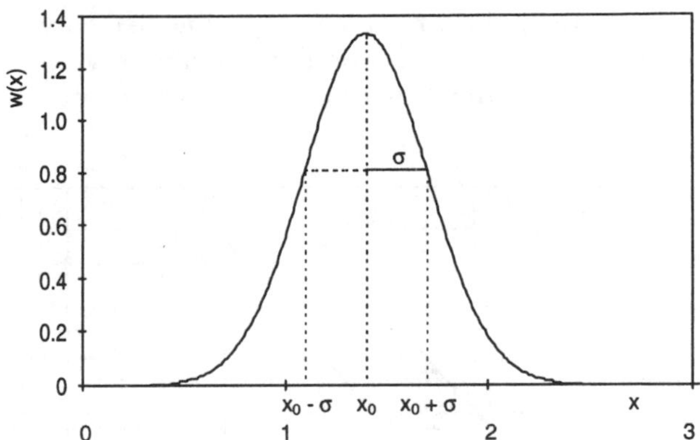

Fig. 1.4. Ideal normal distribution curve as given by Eq.(1.1) with median $x_0 = 1.4$ and standard deviation $\sigma = 0.3$.

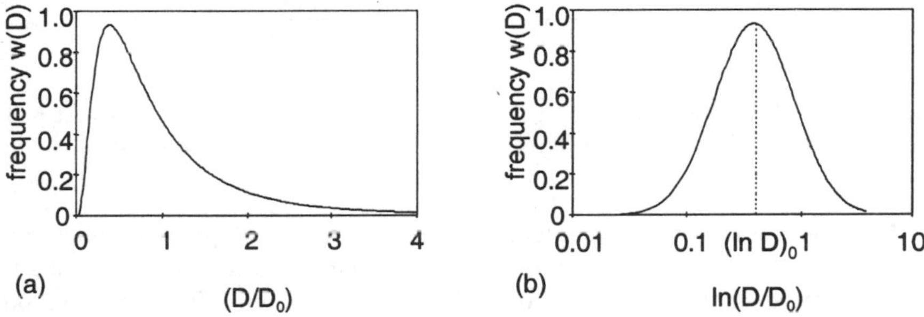

Fig. 1.5. (a) Logarithmic normal distribution as given by Eq.(1.2) plotted linearly (normalized to median grain size). The maximum is not at D_0 (i.e. $D/D_0 = 1$). (b) Plot of the distribution in (a) over the logarithm of grain size. The distribution is symmetrical, hence $(\ln D)_0 = (\ln D)_{\max}$.

$$x_m = \exp\left(\ln x_0 - \sigma^2\right) \tag{1.3}$$

Obviously, besides the average value of a distribution also its width (scattering width) is important. Commonly it is represented by the standard deviation, i.e. the deviation from the average value where the frequency has decreased to $1/e$ (e - Euler's number) of the maximum value. In a normal distribution 68,3% of all measured values are found within the standard deviation from the mean value.

For very broad distributions it is not always obvious whether they consist of a single distribution or of the superposition of several distributions. In such cases it is helpful to plot the integral of the distribution (frequency sum) on a so-called probability plot. For an exact normal distribution this plot will yield a straight line. Deviations from a straight line like curved segments or kinks indicate the superposition of several distributions which, however, can be deconvoluted by the appropriate mathematical procedures (Fig. 1.6).

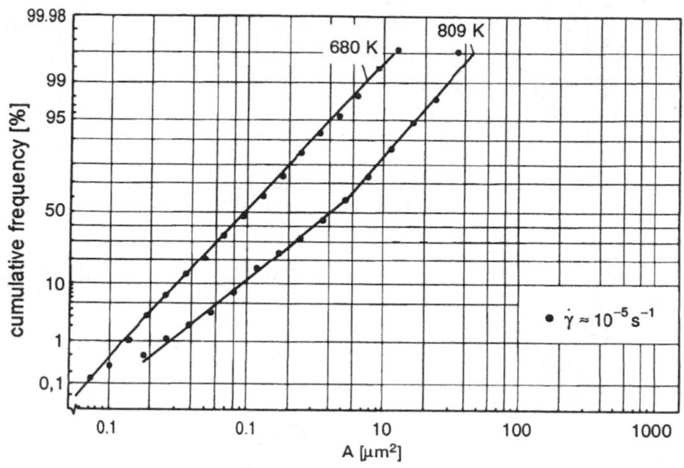

Fig. 1.6. Cumulative frequency of (sub-)grain sizes measured in copper. At 680K there is only one distribution, whereas at 809K two distributions coexist.

Commercial materials are usually comprised of several phases . Therefore, information on the phase distribution is necessary besides the grain size for microstructure characterization. In addition to the volume fraction of second phases their spatial arrangement and the size distribution of each phase is of importance. We distinguish the cases where the secondary phase has a volume fraction comparable to the parent phase (usually referred to as matrix) or where it represents only a small fraction of the total volume. In particular the mechanical properties will show that there is not only a quantitative difference between these cases since a small volume fraction of a second phase primilary influences the properties of the parent phase through its particle size and spacing, while for a massive second phase the properties of the material are affected by the properties of both constituents.

Depending on phase morphology we distinguish typical microstructures. If both phases have a similar arrangement but are separated from each other, we call this a duplex structure (Fig. 1.7a). Where the phase boundaries are parallel to preferred crystallographic planes, which often appears macroscopically as a "basket weave" pattern, the microstructure is referred to as a Widmannstätten structure (Fig. 1.7b). Martensitic microstructures (see Chapter 8) typically appear plate-like or lenticular (Fig. 1.7c). In contrast the structure of Bainite appears to be composed of very fine feathers. Frequently second phases predominantly appear on grain boundaries or at triple junctions, for instance in case of discontinuous precipitation (see Chapters 7 and 9). If in a

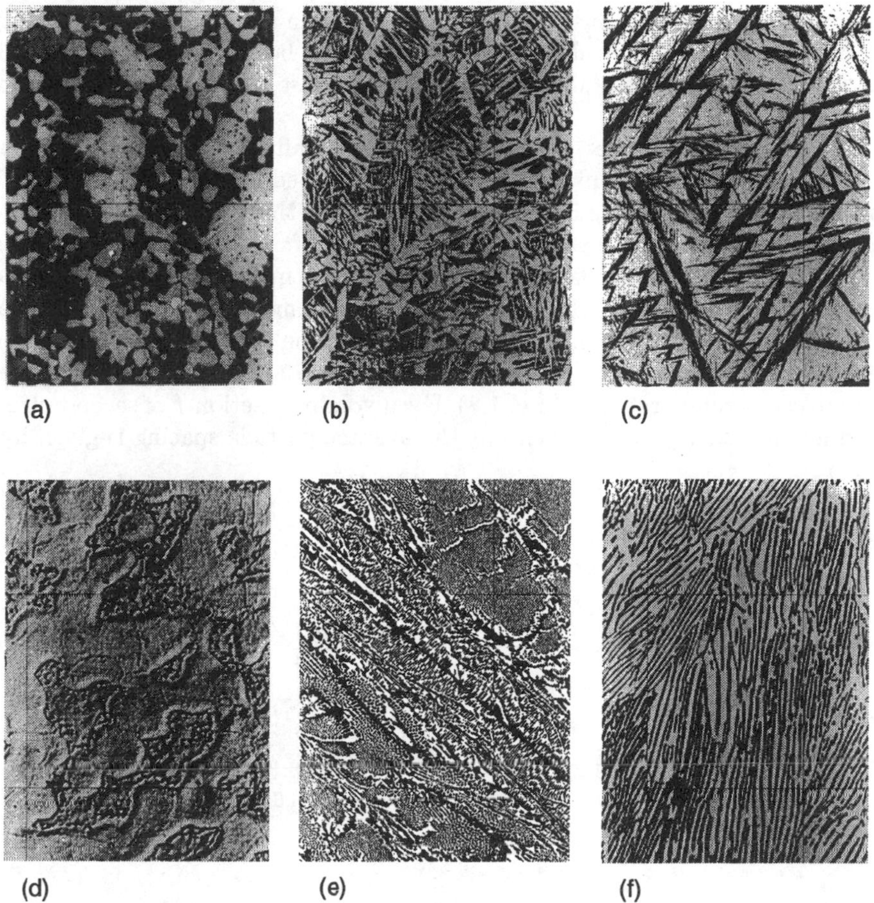

Fig. 1.7. Typical microstructures of metallic materials [1.1]. (a) Duplex microstructure of austenite (light) and ferrite (dark). Material: X2 CrNiMo N 2253; (b) Widmannstätten microstructure in C35 cast steel; (c) Martensitic microstructure (martensite plate) in C150; (d) Dual phase microstructure of steel. Ferritic islands are visible in an austenitic matrix; (e) Eutectic microstructure in white iron (carbon content 4,3%); (f) Eutectoid microstructure (pearlite) in C80 steel.

micrograph this is a dominant feature we refer to the microstructure as a dual phase structure. Highly distinctive microstructures are obtained in solidified alloys. Important examples are eutectic microstructures, where both phases appear as lamellae next to each other (Fig. 1.7e). This will be of concern in chapter 8. With some practice typical microstructures are readily recognized and reveal information on the state of the material. In Germany special schools (for instance the Lette school in Berlin) train their students in metallography to generate micrographs of high quality. The right grinding, polishing, or etching is different for each material and requires much experience and ingenuity and still has a touch of alchemy. Nevertheless, a basic knowledge of metallography remains a helpful tool for the metallurgist and material scientist of today.

If the volume fraction of second phase is small the particles cannot be observed in the optical microscope, but are revealed only under high magnification in the electron microscope (Fig. 1.2). In this case the physical and chemical properties of the second phase are only of concern as a perturbation of the matrix. This is of particular importance for mechanical properties and for recrystallization phenomena (see Chapters 6 and 7). For these particles their size and their spacing is most important. The particle size distribution usually does not exhibit a log-normal distribution and is subject to changes due to coarsening processes (Fig. 1.8). For a volume fraction f of second phase and for equiaxed particles of size d_0 the average particle spacing is given by

$$R = d_0/\sqrt{f} \tag{1.4}$$

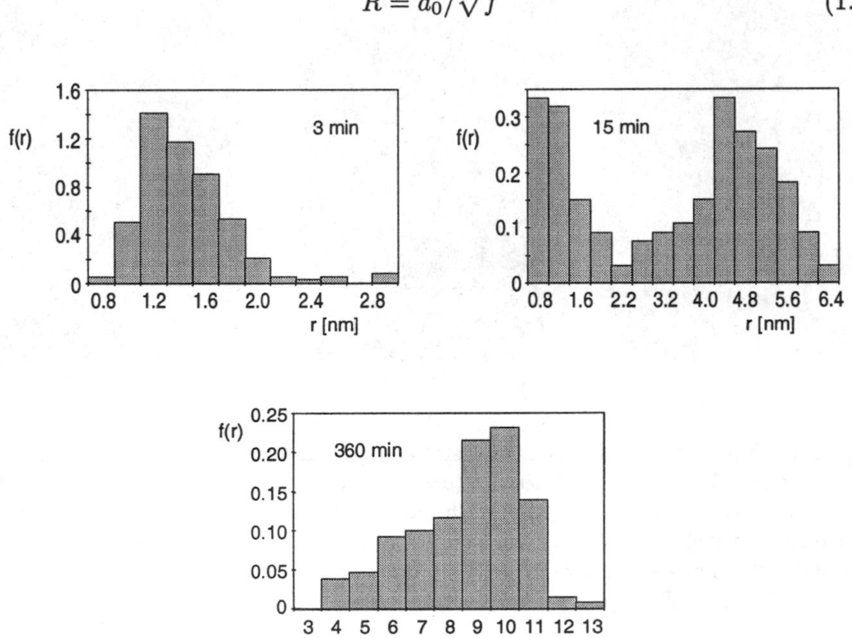

Fig. 1.8. Particle size distribution of the metastable δ'-phase of Al-7at.%Li annealed and aged at 190°C (after [1.2]).

For non-equiaxed morphologies respective geometrical factors have to be introduced. There are occasions where the spacing of particles along specific planes or directions is of importance. We will address these issues in their respective context.

2

Atomic Structure of Solids

2.1 Atomic Bonding

The structural elements of matter are the atoms, which consist of a nucleus and the atomic shell. The properties of solids are essentially determined by the electron shell structure. According to the Bohr model of an atom, the electrons occupy specific orbitals (Fig. 2.1) the configuration of which, i.e. number of electrons and their spatial arrangement, follows the laws of quantum mechanics. The most important electrons for the properties of a solid are the electrons in the outermost orbital, because they determine the interaction with other atoms. The dominant principle of atomic interaction is the tendency of an atom to have its outermost shell filled with eight electrons, i.e. the noble gas configuration. This simple principle is the foundation of chemical bonding. If an atom has already a complete outer shell with eight electrons, like the noble gases, then its tendency to interact with other atoms, i.e. for chemical bonding or even for solidification is very small. Helium has to be cooled to 0.1 K to make the interaction forces between the atoms sufficiently large compared to thermal vibrations to generate a solid. All elements which do not have a noble gas configuration have the tendency (since associated with an energy gain) to accept, to donate, or to share the outermost electrons, also referred to as valence electrons, when in contact with other atoms. From these principles we obtain the fundamental types of atomic bonding (Fig. 2.2):

i) Heteropolar or ionic bond (a): The number of valence electrons of the partners adds up to eight. The partner with smaller valency donates its valence electrons to the other partner with higher valence. Both elements attain a noble gas configurations, but the atoms lose their charge neutrality. For instance: Na^+Cl^-; the sodium atom donates its only valence electron to the chlorine atom with seven valence electrons. It is also possible that more than two atoms are involved in the formation of a molecule, for instance $Ca^{2+}(F^-)_2$, where each of the three atoms attains the noble gas configuration.

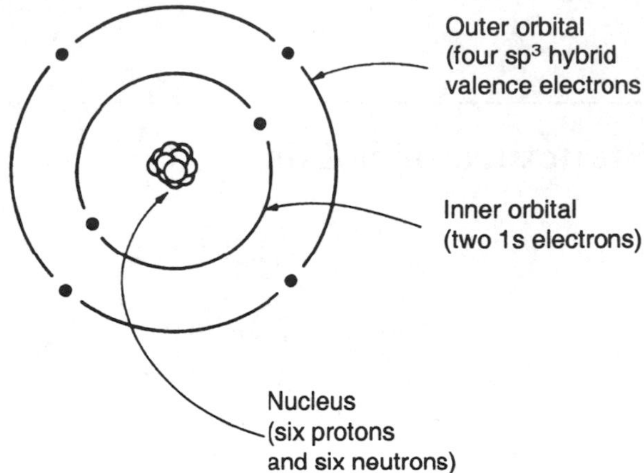

Fig. 2.1. Diagrammatic illustration of the electronic configuration of the ^{12}C atom after Bohr's atomic model.

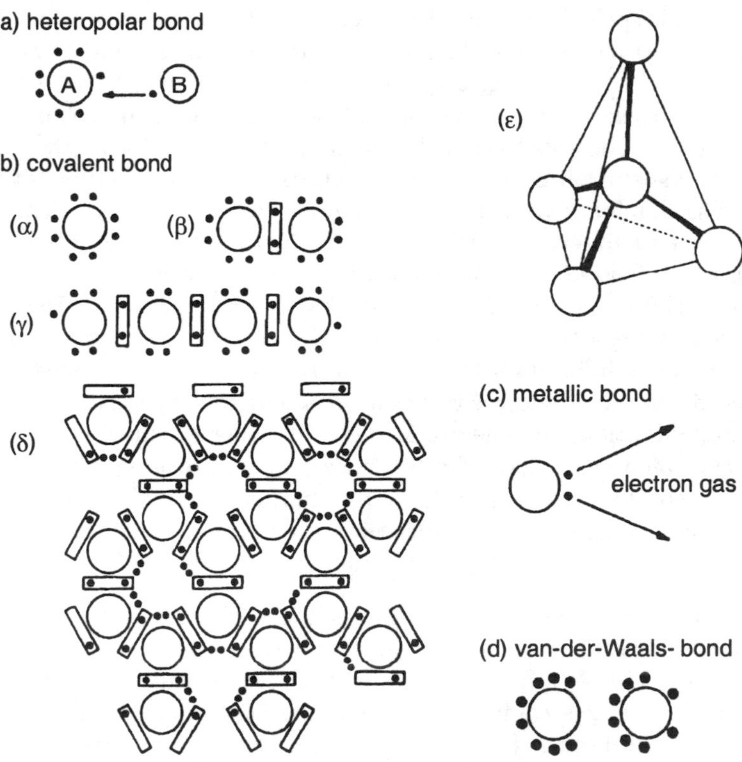

Fig. 2.2. Basic types of chemical bond. When atoms of the same kind are bound covalently, specific arrangements occur.

ii) Covalent or electron pair bond (b): If the noble gas configuration cannot be established by exchange of electrons since the sum of valence electrons does not add to eight, a stable arrangement of atoms in a molecule can be obtained by the formation of electron pairs (α). For instance, two chlorine atoms with valency seven can form a stable chlorine molecule Cl_2 by the formation of an electron pair (β), which is shared by the two Cl atoms, and both atoms attain the noble gas configuration. For atoms with valency six, two electron pairs have to be formed per atom. This leads to the formation of long chain molecules (γ) as for example in sulphur. For valency five, three electron pairs are necessary per atom, which can only be realized in a two-dimensional arrangement (δ), for instance for arsenic layers. If the valency is four, a three-dimensional arrangement is needed to have four electron pairs per atom properly established (ε). Examples are the semiconductors silicon and germanium as well as carbon in the diamond structure.

iii) Metallic bond (c): If the number of valence electrons is less than four a noble gas configuration cannot be established by electron pairs in a three-dimensional lattice. In this case the atoms prefer to donate their valence electrons to a common electron gas (Fig. 2.3) so that the ionic cores attain the noble gas configuration and the electrons in the electron gas are not associated with a particular atom. This is the most frequent type of bond among the elements, because about three quarters of all natural elements are metals. While in a covalent bond the valence electrons are localized at the atom - or in terms of quantum mechanics, are preferentially located at that atom - in the metallic bond the electrons are not localized and practically belong to all atoms at the same time. The weak localization of the electrons in the metallic bond is the reason for its weakness in comparison to the other types of bonds. In turn, this causes the high mobility of defects in metals and, therefore, their excellent formability, which has made metals the dominant structural materials.

iv) Van-der-Waals bond (d): There is another type of bond which is not associated with the exchange of electrons, namely the so-called Van-der-Waals bond. It is caused by a dipolar interaction of atoms. This dipole interaction is caused by the fact that the center of gravity of charge of the nucleus is not identical with that of the electron shell. This causes a dipole moment of the atoms which invariably generates an attractive interaction with other atomic dipoles (Fig. 2.4). This attractive force is the reason for bonding in noble gas molecules and the interaction of far apart atoms, when no exchange of electrons can take place.

The formation of a molecule and equivalently the formation of a crystalline phase can be understood in the following way. When the atoms are far apart they interact by Van-der-Waals forces. Decreasing their spacing causes the electron shells to overlap, and the electronic interaction leads to electron transfer processes, which cause bonding. The atoms are forced even closer together, the electron shells will overlap and eventually react with strong re-

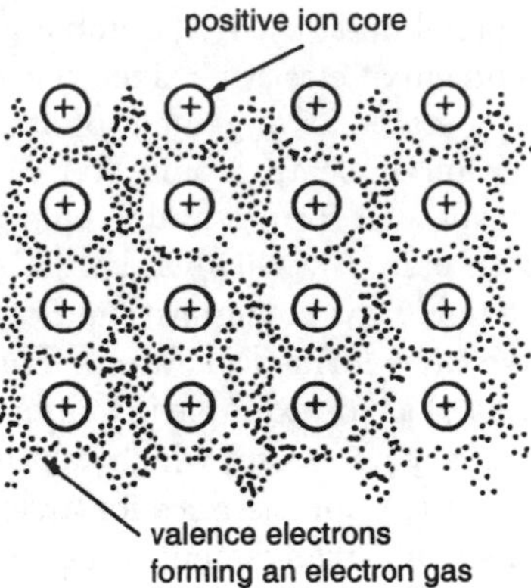

Fig. 2.3. Principle of the metallic bond. Ion cores are surrounded by the electron gas formed by the valence electrons.

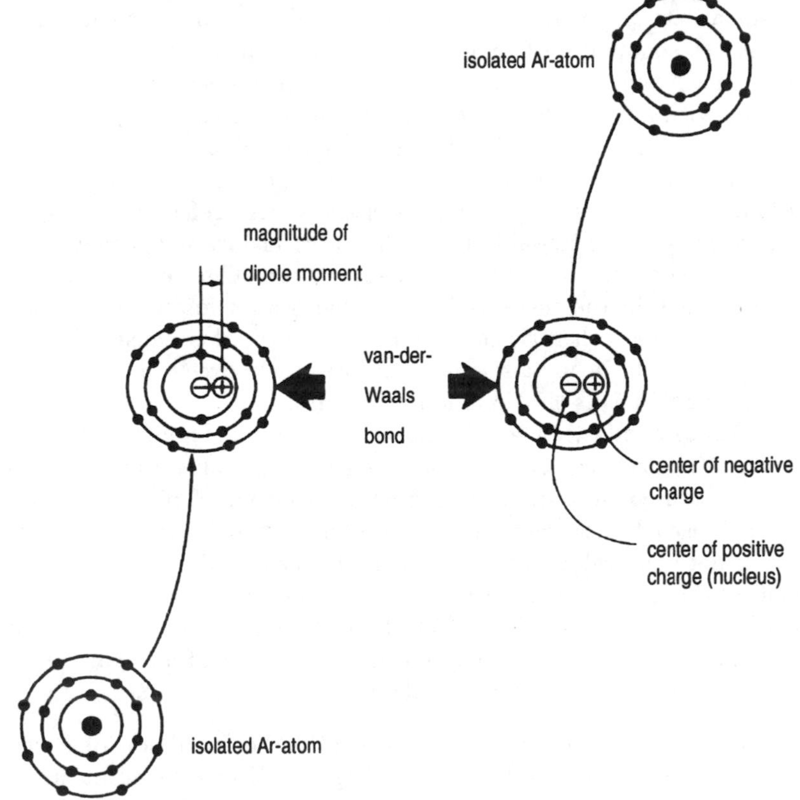

Fig. 2.4. Schematic representation of how the van-der-Waals bond is formed by interaction of induced dipole moments.

pulsion because of the Pauli principle, which will be explained in more detail in Chapter 10. The force-spacing curve of two approaching atoms typically has a shape as depicted in Fig. 2.5a, which is connected to the associated potential energy (Fig. 2.5b) by integration. The spacing at which the force between the atoms vanishes, i.e. repulsive and attractive forces compensate each other, is the equilibrium spacing a_0 (Fig. 2.5a). Generalizing this consideration from two atoms to many atoms yields a periodic arrangement of the atoms of an element as a crystalline solid, where the equilibrium spacing is the distance between the neighboring (touching) atoms.

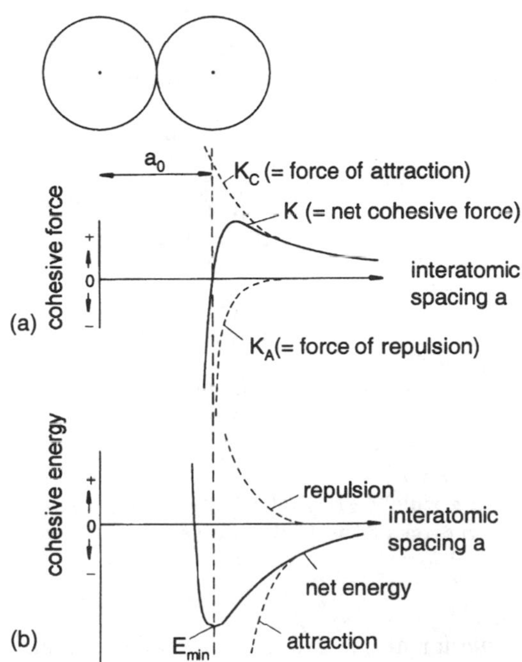

Fig. 2.5. Cohesive force and cohesive energy of a pair of atoms plotted as a function of atomic distance.

The simplest arrangement in solids is found in metals. Their bonding is virtually independent of spatial orientation so that metal atoms can be considered to be hard spheres, that are (most) densely packed to maximize atomic attraction. This hard sphere model of a metallic solid is very simple but for many problems a very helpful model. Accordingly, we expect that metallic solids have an atomic arrangement of close-packed spheres, which corresponds to the packing in (stacks of) layers with a hexagonal atomic arrangement in each layer.

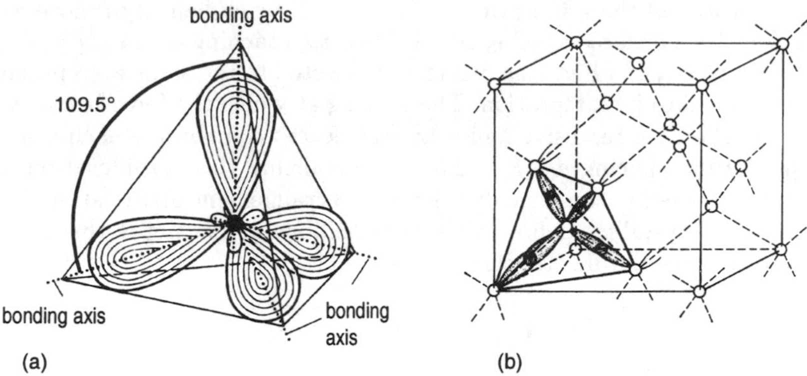

Fig. 2.6. The tetrahedral orbitals of the valence electrons of ^{12}C atoms cause their tetrahedral arrangement (a) and result in the formation of the diamond lattice (b).

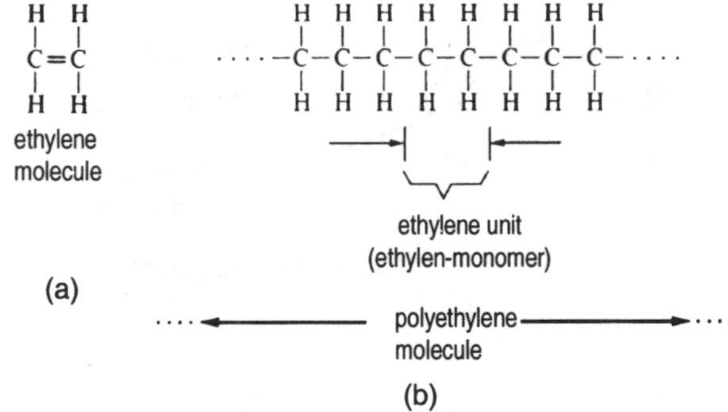

Fig. 2.7. (a) Ethylene molecule (C_2H_4) with double bond; (b) polyethylene molecule (C_2H_4) formed by conversion of the C=C double bond to two C-C single bonds (polymerization).

In fact, this is true for about 2/3 of all metallic elements. However, less densely packed structures are also observed, if other electronic effects play a role, because bonds in most cases are of mixed type, e.g. metallic and covalent. We will introduce the kinds of crystal structures formed by these arrangements in the next chapter.

The covalent bond strongly depends on orientation, since the electrons which form pairs have to adjust their orbitals such that the center of gravity of charge remains in the center of an atom. For instance, for carbon and, correspondingly for other elements of valency four, the paired electrons are arranged in the corners of an equilateral tetrahedron for reasons of symmetry, i.e. the interbond angles are all equal to the tetrahedral angle of 109.5° (Fig. 2.6). (In these directions the bonds will be formed.) A crystal which is composed of such atoms has to arrange the atoms in such a way that the tetrahedral environment is established for each atom. This can actually be accomplished in the diamond crystal structure (Fig. 2.6b) - diamond is pure crystalline carbon - as will be considered below. Since the directionality of bonding dominates the atomic arrangement, the packing density plays a less important role for bonding. If the partner atoms are unlike elements, as in ethylene C_2H_4 then the electronic structure gets distorted by the hydrogen atoms which results in a linear arrangement of molecules or "mers", i.e. the formation of polyethylene $(C_2H_4)_n$ (Fig. 2.7).

The ionic bond does not show a spatial dependency, but it occurs between unlike atoms. The repulsive action of similar electrical charges favors the formation of special structures, which optimize the contact of unlike atoms and prevents an overlapping of like atoms or ions (Fig. 2.8). The respective arrangement, i.e. the number of possible next neighbors (coordination number) depends on the ratio of the size of the atoms. If all atoms are equally large one can obtain the most dense packing with twelve nearest neighbors (Fig. 2.8b). If the ratio of atomic radii (assuming a spherical shape of the atoms) is smaller than one, the coordination number changes discontinuously at specific ratios, when atomic shells begin to overlap. For $r/R < 0.155$ only the formation of two-dimensional chains is possible.

Table 2.1. Nature of chemical bond of the four most important types of engineering materials.

Material	Nature of bond	Example
metals	metallic	iron (Fe) and alloys of iron
ceramics and glasses	ionic/covalent	silica (SiO_2): crystalline and noncrystalline
polymers	covalent and Van-der-Waals	polyethylene $(C_2H_4)_n$
semiconductors	covalent and covalent/ionic	silicon (Si) and cadmium sulfide (CdS)

The bonds in a solid are usually mixed bonds, but one or the other type of bond will dominate. If we classify the solids according to the type of material, they can be associated with specific types of bonds according to Fig. 2.9 or Table 2.1. Metals have metallic bonds with small covalent or ionic contributions. Compounds of metallic elements (intermetallic compounds, see Chapter 4) often exhibit a high fraction of covalent bonding, which has drastic consequences, for instance with regard to formability. In ceramics and polymers usually mixed types of bonds dominate. For instance, in polymers, covalent bonds act along the chains while Van-der-Waals bonds cause attractive interaction between chains which can be sufficiently strong to result in crystalline forms of polymers (Chapter 8).

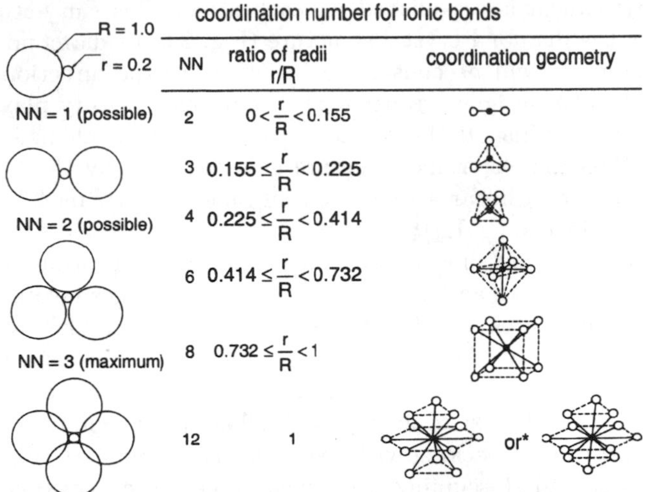

R = 1.0
r = 0.2
NN = 1 (possible)
NN = 2 (possible)
NN = 3 (maximum)
coordination number for ionic bonds
NN
ratio of radii r/R
coordination geometry
2 0<r/R<0.155
3 0.155≤r/R<0.225
4 0.225≤r/R<0.414
6 0.414≤r/R<0.732
8 0.732≤r/R<1
12 1
or*

NN = 4 (unstable) * the geometry on the left is for the hexagonal closepacked (hcp) structure, that on the right for the face-centered cubic (fcc) structure.

(a) (b)

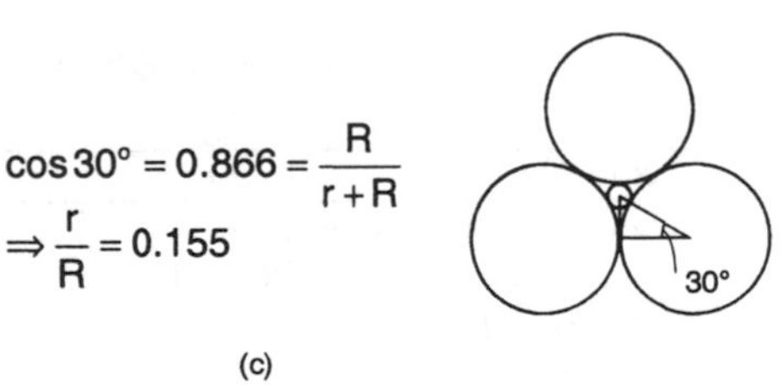

Fig. 2.8. Number of nearest neighbors NN (coordination number) as a function of the relative size of the atoms. (a) For an atomic radii ratio of $r/R = 0.2$ the greatest possible number of nearest neighbors is three; (b) coordination number for different atomic radii ratios and the resulting coordination geometry; (c) the smallest ratio of radii resulting in a coordination number of three is 0.155.

2.2 Crystal Structure

2.2.1 Crystal Systems and Space Lattice

Metallic and ceramic materials are usually crystalline. Also polymers can partially crystallize, which will be addressed in more detail in Chapter 8. By definition glasses are not crystalline.

A crystalline structure means in a physical sense, a strictly periodic arrangement of atoms. However, long before the atomistic structure of solids was known, crystals of minerals fascinated man and became a subject of scientific interest. The prominent feature of crystalline minerals is their external ap-

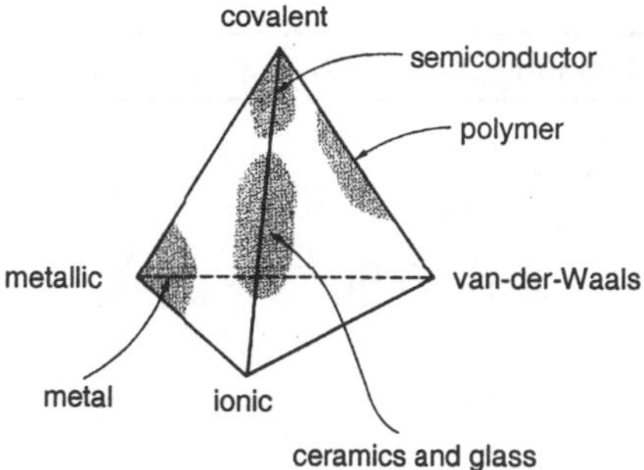

Fig. 2.9. Schematic illustration of the contribution of the different bond types in the most important commercial materials.

pearence with planar facets, which are characteristic of each mineral. According to the geometry of the facets, the crystallographers were able to group the crystals in terms of their shapes and symmetries into 32 classes (also referred to as point groups) , which could be subdivided into seven crystal systems. These seven crystal systems can be defined by the macroscopic orientation of the crystal surfaces and their lines of intersection in an appropriately chosen crystal coordinate system (Fig. 2.10). If there is no apparent symmetry, the crystal structure is called triclinic, and the angles between the crystal axes and their respective lengths are all different. The highest symmetry is represented by cubic crystals, where all crystal axes are equally long and are arranged under a mutual angle of 90°.

If a solid is considered to consist of atoms, the arrangement of these atoms must conform with the observed crystal symmetries. For this Bravais introduced the concept of the space lattice, a mathematical spatial point pattern, the lattice, and for a physical representation, each lattice point has to be associated with an atom or a molecule. The point lattice must be strictly periodic and, therefore, its characterization can be reduced to a unit cell, such that an assembly of the unit cells generates the space lattice. Bravais was able to show that there are only 14 different point lattices (Fig. 2.11). Besides the primitive structures, where lattice points coincide only with the corners of the unit cell, there can be a point in the center of the cell (body-centered) or on the centers of adjacent faces (face-centered) because of crystal symmetry. Body-centered and face-centered arrangements cannot be realized for all crystal classes without loss of symmetry. For instance, a face-centered version of the tetragonal lattice with lattice points on the center of the faces on the basal planes (Fig. 2.12) corresponds exactly to a primitive tetragonal lattice with the base length $a' = a\sqrt{2}$ instead of a.

The lattice points can simply represent single atoms, but can also be associated with groups of atoms or molecules. The number of possible periodic arrangements of atoms in space which can be distinguished and which comply with the constraints of symmetry is limited. There are 230 possible different

system	axis lengths and angles	symmetry of the unit cell
cubic	$a=b=c$, $\alpha=\beta=\gamma=90°$	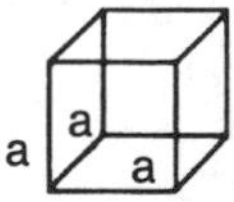
tetragonal	$a=b\neq c$, $\alpha=\beta=\gamma=90°$	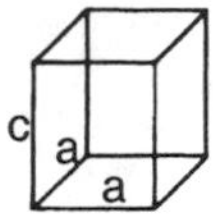
orthorhombic	$a\neq b\neq c$, $\alpha=\beta=\gamma=90°$	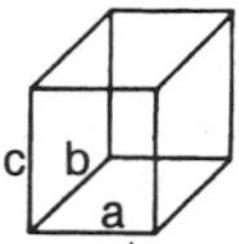
rhombohedral (trigonal)	$a=b=c$, $\alpha=\beta=\gamma\neq 90°$	
hexagonal	$a=b\neq c$, $\alpha=\beta=90°$, $\gamma=120°$	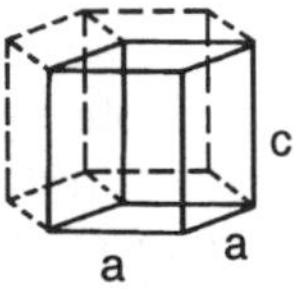
monoclinic	$a\neq b\neq c$, $\alpha=\gamma=90°\neq\beta$	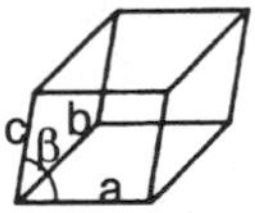
triclinic	$a\neq b\neq c$, $\alpha\neq\beta\neq\gamma$	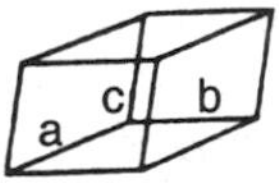

Fig. 2.10. Definition of the seven crystal systems.

arrangements, to one of which a crystal has to belong. These arrangements are referred to as space groups and are distinguished from the point groups of crystal classes since the symmetry relations do not refer to a particular point, i.e. the origin of the selected coordinate system, but to each point of the space lattice.

The different space groups have been tabulated and are commonly denoted with symbols according to their symmetries, for instance by the Schoenflies or Hermann-Mauguin symbols. In materials science the simplified terminology of the structure reports (Strukturberichte) have become widely accepted. These structure reports periodically publish the structure of new substances.

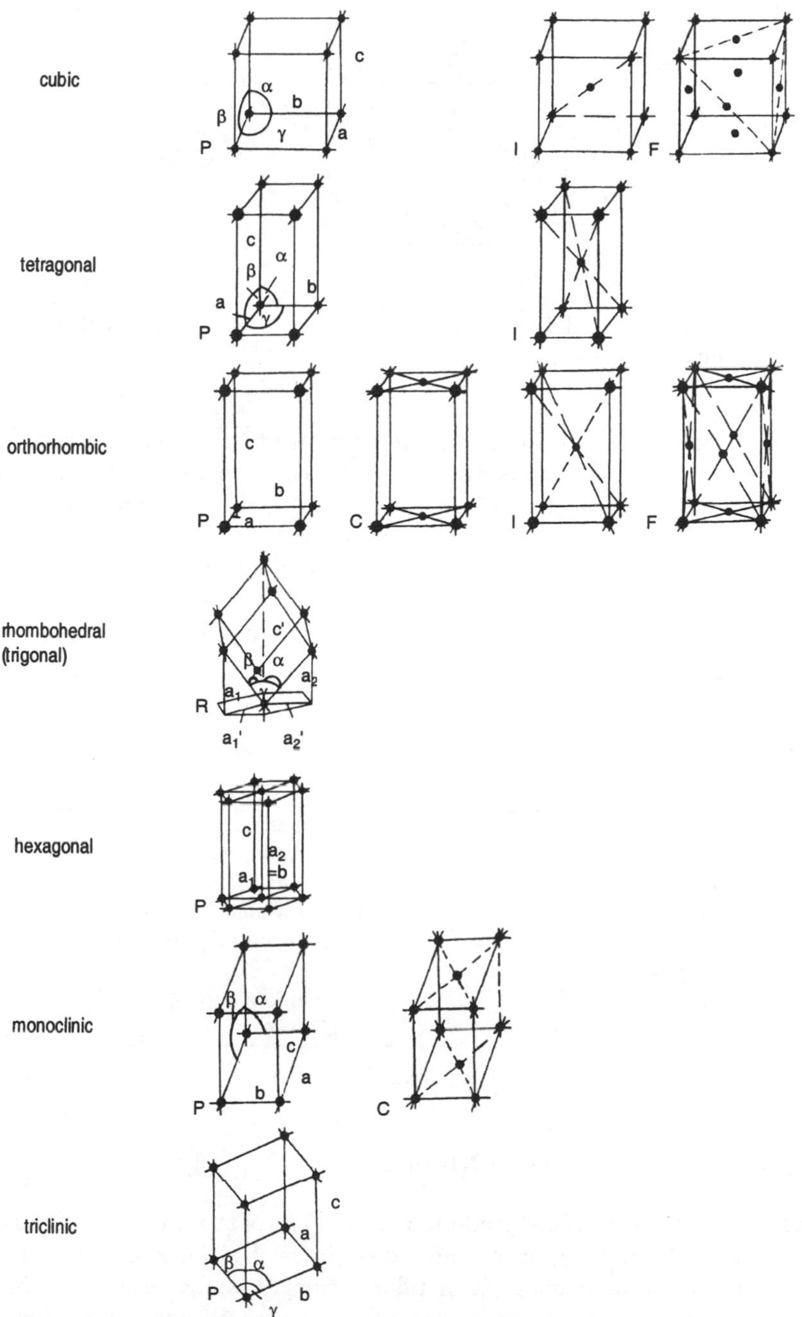

Fig. 2.11. The conventional unit cells of the 14 different Bravais lattices.

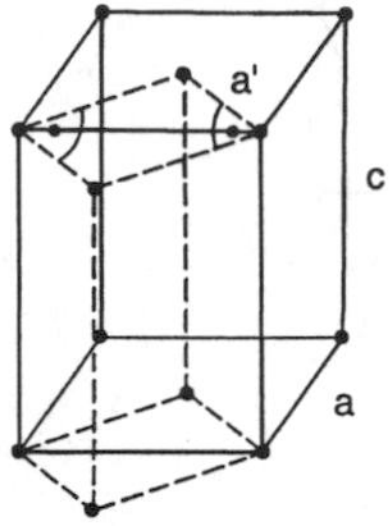

Fig. 2.12. A face centered-tetragonal cell and a simple tetragonal structure (dashed lines) are equivalent.

The editors of the report found it more convenient to introduce a simple notation for frequently found crystal structures, consisting of a letter according to substance and chemical composition (Table 2.2) and a running number, for instance for a cubic face-centered (fcc) element, the notation is $A1$.

Finally we note, that the crystal structure defines the atomic arrangement of atoms, which is not necessarily described by the crystal lattice. Frequently, these two terms are used synonymously, which can lead to confusion in the case of alloys.

Table 2.2. Structure report designations: Simple nomenclature for the common types of crystal structure.

A-type	elements
B-type	AB-compounds
C-type	AB_2-compounds
D-type	A_mB_n-compounds
E....K-type	more complicated compounds
L-type	alloys
O-type	organic compounds
S-type	silicates

2.2.2 Crystal Structures of Metals

Metallic elements crystallize predominantly in one of the three crystal lattices, body-centered cubic (bcc), face-centered cubic (fcc), and hexagonal (hcp) with about equal fractions among all metallic elements. Many properties, in particular the mechanical properties, depend on the crystal structure. Therefore, these three important types of crystal structures will be addressed in more detail.

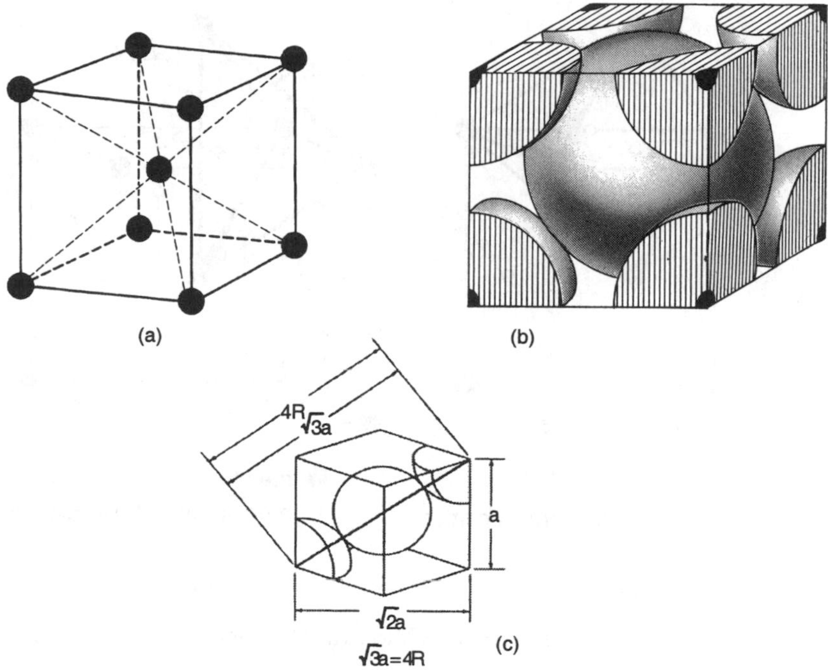

Fig. 2.13. Body-centered cubic structure. (a) Unit cell of the bcc point lattice; (b) unit cell composed of spheres; (c) atoms are in contact along the body diagonal.

In the bcc lattice, the atoms are located on the corners and in the center of a cube (Fig. 2.13). This arrangement can also be visualized as two interpenetrating cubic primitive lattices. If the atoms are considered as hard spheres, they touch along the space diagonal. The atomic spacing b is the distance between the centers of the next neighbor atoms, i.e. along the most densely packed directions. It is identical with the atomic diameter or twice the atomic radius R. Correspondingly, (Fig. 2.13c) we obtain for the bcc lattice

$$R = \frac{a}{4}\sqrt{3} \tag{2.1a}$$

$$b = 2R = \frac{a}{2}\sqrt{3} \tag{2.1b}$$

where a is the lattice parameter.

The spheres do not completely fill space. Between the spheres there remains free space also referred to as interstices. A unit cell of the bcc lattice contains two atoms, namely the atom in the center of the cube and 1/8 of each of the eight atoms on the corners of the cube, since they have to be shared among eight unit cells (Fig. 2.13b). The packing density is the ratio of the volume of the two spheres and the volume of the cube

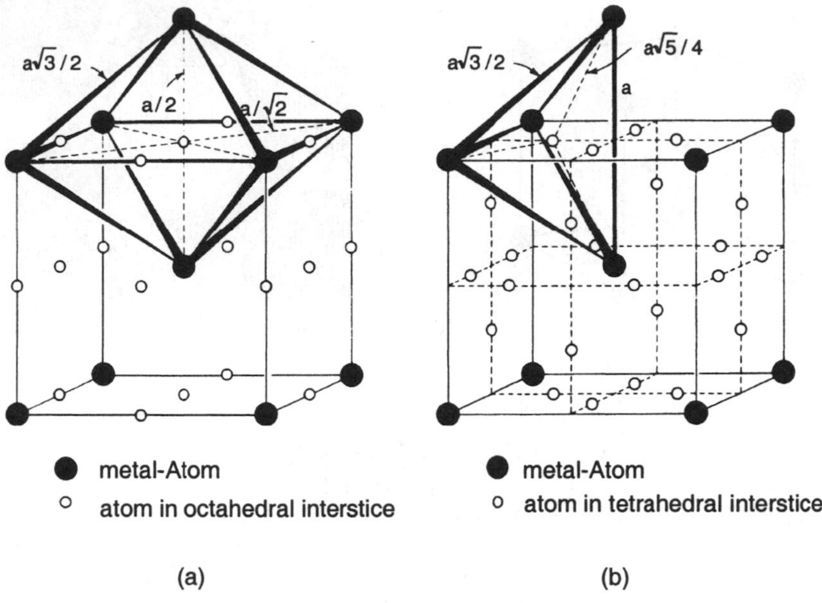

Fig. 2.14. Interstitial sites in the bcc lattice. (a) Octahedral interstices; (b) tetrahedral interstices.

$$V_f^{bcc} = \frac{2 \cdot \frac{4}{3}\pi R^3}{a^3} = \frac{\frac{8}{3}\pi\left(\frac{a}{4}\sqrt{3}\right)^3}{a^3} = \frac{\pi\sqrt{3}}{8} = 68\% \tag{2.2}$$

There are two types of interstitial sites, namely octahedral interstitial sites and tetrahedral interstitial sites. These terms denote the geometrical arrangement of the atoms enclosing an interstitial site (Fig. 2.14). In the bcc lattice the centers of the octahedral interstitial sites (Fig. 2.14a) are the centers of the faces and the centers of the edges of the unit cell. There are six octahedral interstitial sites per unit cell, i.e. three times as many as atoms. A tetrahedral interstitial site is enclosed by a tetrahedron cornered by two corner atoms and two atoms on adjacent faces of a unit cell. Their centers are located on the cube faces with the coordinates [0, 1/2, 1/4] and all crystallographically equivalent sites (Fig. 2.14b). There are four tetrahedral interstitial sites on each cube face, that is in total 24/2 tetrahedral sites per unit cell, i.e. six times as many as atoms and two times as many as octahedral interstitial sites. The size of an interstitial site is defined by the radius of a sphere which exactly fits into the gap. Correspondingly, one obtains for the size of the

$$\text{octahedral gap} \qquad \frac{r}{R} = 0.155 \tag{2.3a}$$

$$\text{tetrahedral gap} \qquad \frac{r}{R} = 0.291 \tag{2.3b}$$

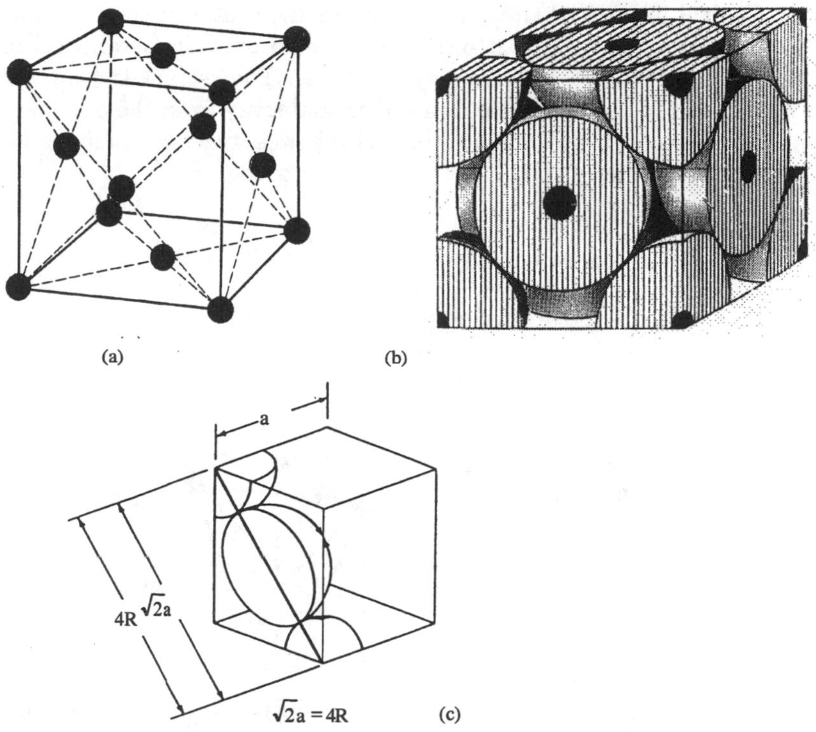

Fig. 2.15. Face-centered cubic structure. (a) Unit cell of the fcc point lattice; (b) unit cell composed of spheres; (c) atoms are in contact along the face diagonal.

Interstitial sites are very important in the context of solid solutions and their properties, and therefore, will be addressed in more detail in Chapter 4.

The face-centered cubic lattice has atoms on all corners and on the centers of the cube faces (Fig. 2.15), i.e. four atoms per unit cell, corresponding to four interpenetrating simple cubic lattices. Atoms as hard spheres touch along the face diagonals. Correspondingly, the radius and the spacing of the atoms are related to the lattice parameter

$$R = \frac{a}{4}\sqrt{2} \tag{2.4a}$$

$$b = \frac{a}{2}\sqrt{2} \tag{2.4b}$$

With this we obtain the packing density

$$V_f^{fcc} = \frac{4\pi \cdot \frac{4}{3}\left[\frac{a}{4}\sqrt{2}\right]^3}{a^3} = \pi\frac{\sqrt{2}}{6} = 74\% \tag{2.5}$$

The octahedral interstitial sites are located in the center of the cube and on the center of the edges (Fig. 2.16a). Hence, there are four octahedral interstitial sites per unit cell, i.e. as many as atoms. The tetrahedral interstitial sites are located at 1/4 along the space diagonal away from the corners (Fig. 2.16b). There are eight tetrahedral interstitial sites, twice as many as atoms or octahedral interstitial sites. Their size amounts to

$$\text{octahedral gap} \qquad \frac{r}{R} = 0.41 \tag{2.6a}$$

$$\text{tetrahedral gap} \qquad \frac{r}{R} = 0.22 \tag{2.6b}$$

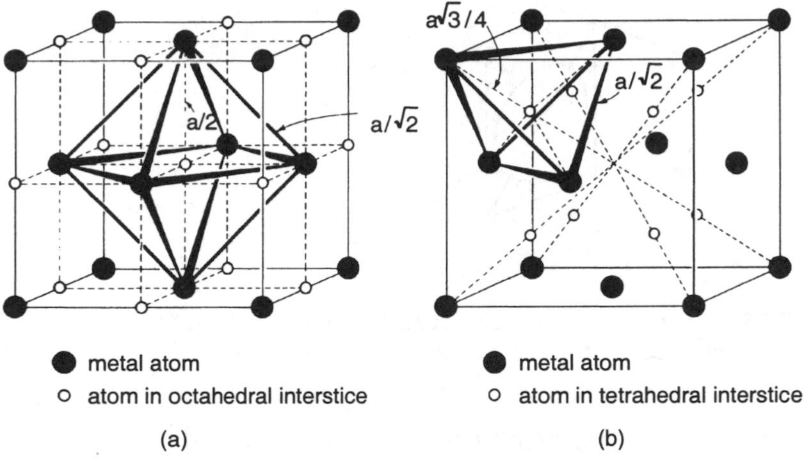

Fig. 2.16. Interstitial sites in the fcc lattice. (a) Octahedral interstice; (b) tetrahedral interstice.

In comparison to the bcc lattice the fcc lattice has fewer but larger interstitial sites. This has important consequences for the structure of alloys (Chapter 4).

The hexagonal lattice consists of layers of hexagonal lattice point arrangements. The c axis has a length different from the a axes (Fig. 2.17). The unit cell is shaded in Fig. 2.17a. It contains two atoms. In order to reveal the hexagonal symmetry, commonly the arrangement of three unit cells is combined to a structural unit of the hexagonal lattice.

If the structure is composed of equally large hard spheres, they touch each other in the hexagonal basal plane and in adjacent layers. The ratio of the length of c and a axes then yields

$$\frac{c}{a} = \sqrt{\frac{8}{3}} = 1.63 \tag{2.7}$$

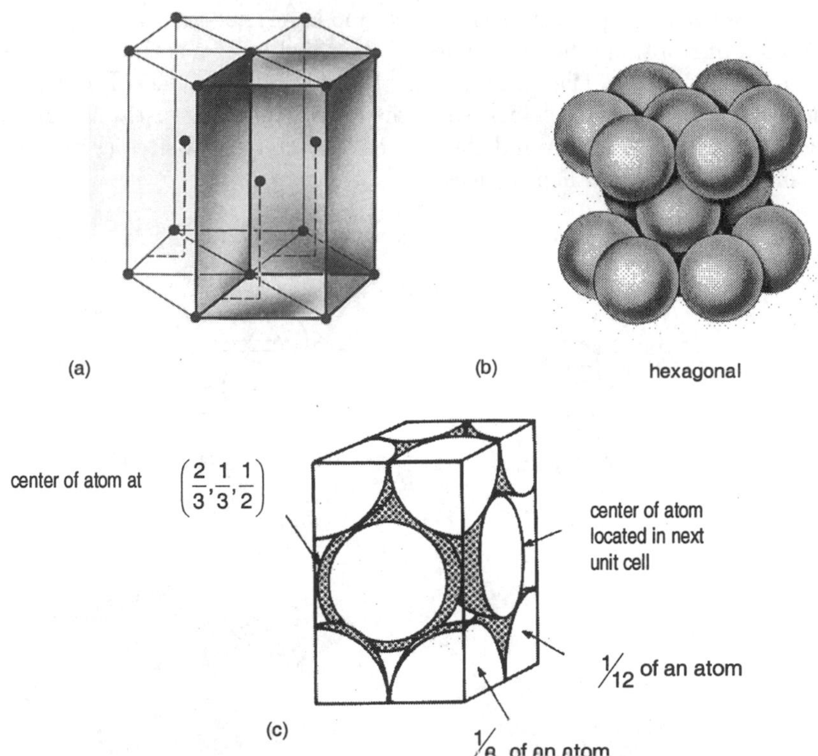

Fig. 2.17. Hexagonal structure. (a) (Triple) unit cell of the hexagonal point lattice; (b) spheres arranged to form the structure; (c) unit cell composed of spheres.

Table 2.3. c/a ratio of several elements with hexagonal structure.

	Cd	Zn	Mg	Co	Zr	Ti	Be
c/a	1.88	1.86	1.62	1.62	1.59	1.58	1.57

This structure is called hexagonal close-packed (hcp). This ideal c/a ratio is found, for instance, in magnesium, but for many other hexagonal metals the ratio deviates substantially from the ideal ratio to larger or smaller values (Table 2.3).

The hexagonal close-packed lattice is very similar to the fcc lattice (Fig. 2.18) and both are close-packed structures. In the fcc lattice the plane which contains the three face diagonals also has a hexagonal structure and the fcc lattice can be visualized as being composed of layers of such hexagonal planes. The difference between the fcc and hcp lattice lies in the difference of the stacking sequence of the layers. A hexagonal layer has two hollows - referred to as B and C in Fig. 2.18 - where the atoms of the next layer can be placed. If in the third layer the same position is picked as in the first layer and in the fourth layer the same position as in the second layer, i.e. the stacking sequence ...ABAB... one obtains the hcp lattice. If the third layer has the atoms placed on top of position C of the first layer, i.e. the stacking sequence ...ABCABC... an fcc lattice is generated. Because of the similar packing of the fcc and hcp lattice the volume fraction and the size of the interstitial sites (Fig. 2.19) is the same in the fcc and hcp lattices.

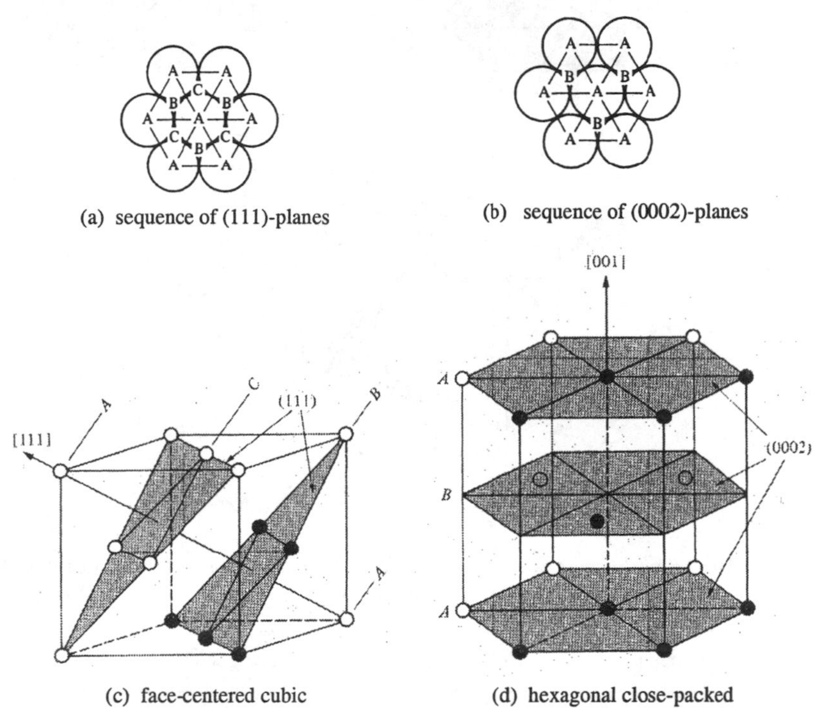

(a) sequence of (111)-planes (b) sequence of (0002)-planes

(c) face-centered cubic (d) hexagonal close-packed

Fig. 2.18. Comparison of two close-packed structures: fcc and hcp. The (111)-plane of the fcc lattice corresponds to the (0002)-plane of the hcp lattice. The only difference between the two structures is the stacking sequence of these two planes.

2.2.3 Crystal Structure of Ceramic Materials

Ceramic materials are primarily bonded by ceramic materials of metals with nonmetals, in particular with oxygen (oxides), nitrogen (nitrides), and carbon (carbides). As shown in Section 2.1 the crystal structure depends on several factors, in particular on composition and size of the atoms, which determines the coordination. Correspondingly, there is a large number of crystal structures in ceramic materials. We confine ourselves here to the simplest structures, in particular to cubic crystal symmetry. The CsCl structure (Fig.

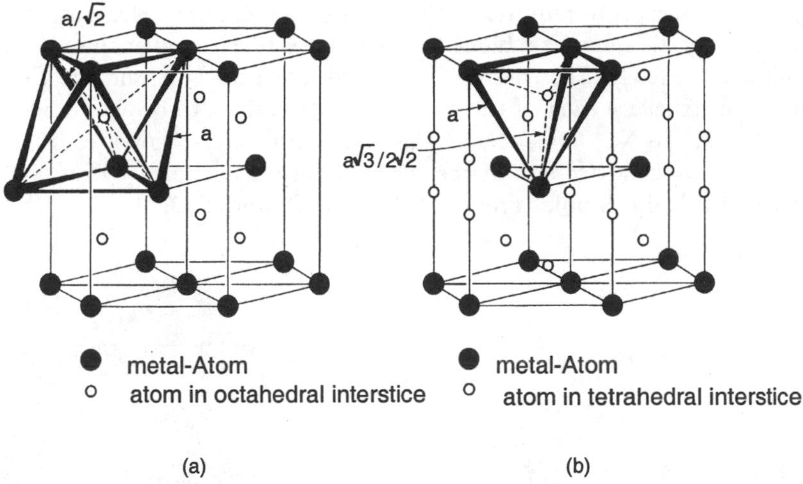

Fig. 2.19. Interstitial sites in the hcp lattice. (a) octahedral interstice; (b) tetrahedral interstice.

2.20) requires an identical number of Cs^+ and Cl^- ions to maintain charge neutrality. This is accomplished most easily by two atoms per unit cell. The structure, therefore, consists of the Cs^+ ion in the center of the cube and the Cl^- ions on the corners of the cube. Despite its similarity to the bcc lattice the crystal structure is only simple cubic, because the center of the cube is occupied by an atom of a different element. Each simple cubic lattice site can be associated with a CsCl molecule.

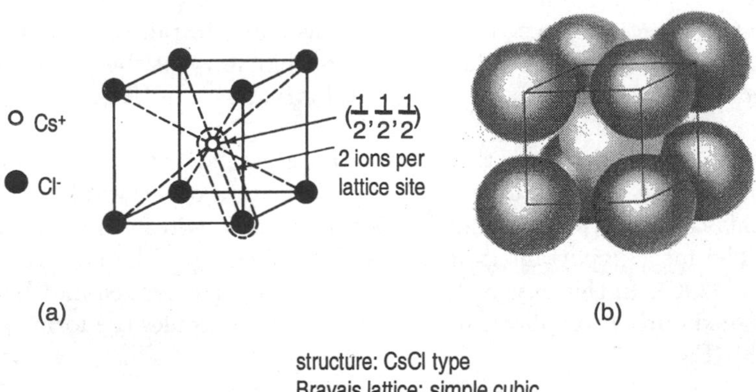

Fig. 2.20. Unit cell of cesium chloride (CsCl). (a) Position of the atoms in the lattice; (b) hard sphere model.

The CsCl structure requires both atoms of the different elements to have approximately the same size. If one of the atoms is much smaller than the other one obtains the fcc NaCl structure (Fig. 2.21). Both the Na^+ and the Cl^- ions form an fcc structure each. If the structure is visualized to be assembled from hard spheres, the Na^+ ions occupy the octahedral interstitial sites of the fcc Cl^- lattice. Each lattice site of the fcc structure is occupied by an Na^+Cl^- molecule. Typical examples are MgO, CaO, FeO, and NiO.

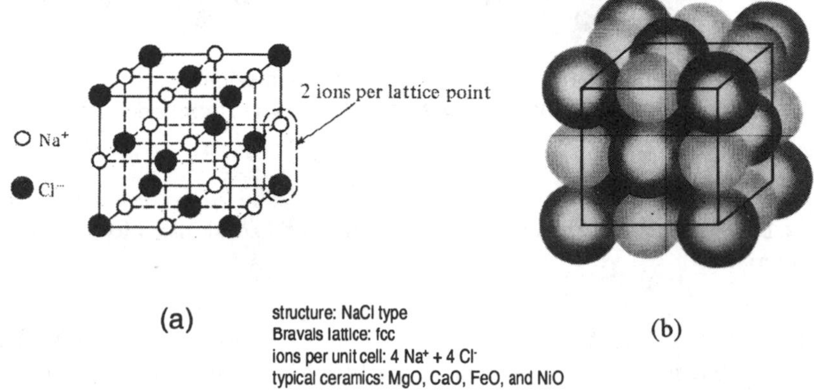

Fig. 2.21. Unit cell of sodium chloride (NaCl). (a) Position of the atoms in the lattice; (b) hard sphere model.

The Na^+Cl^- structure is not feasible for compounds of ions with different valency, for instance $Ca^{2+}F_2^-$ (Fig. 2.22). In this case the Ca^{2+} ions form an fcc lattice, but the F^- ions are located on the tetrahedral interstitial sites of the fcc Ca^{2+} lattice. Since there are twice as many tetrahedral interstitial sites as lattice sites in the fcc lattice the stoichiometry of the composition is satisfied. Typical examples for this crystal structure are UO_2, ThO_2, and TeO_2.

The strong dependency of the bonding on crystallographic direction for covalent bonds can result in a barrier to the formation of a crystalline structure. In this case amorphous solids, for instant glasses will form rather than crystals. The most important examples are the silicates with the structural complex $Si^{4+}O_2^{2-}$. In this case a strictly periodic arrangement cannot be easily established during solidification, and only chain molecules are formed (see Chapter 8) (Fig. 2.23).

2.2.4 Crystal Structure of Polymers

The chain structure of polymers also has the tendency to generate specially ordered structures due to the property of the hydrogen atoms to form so-called

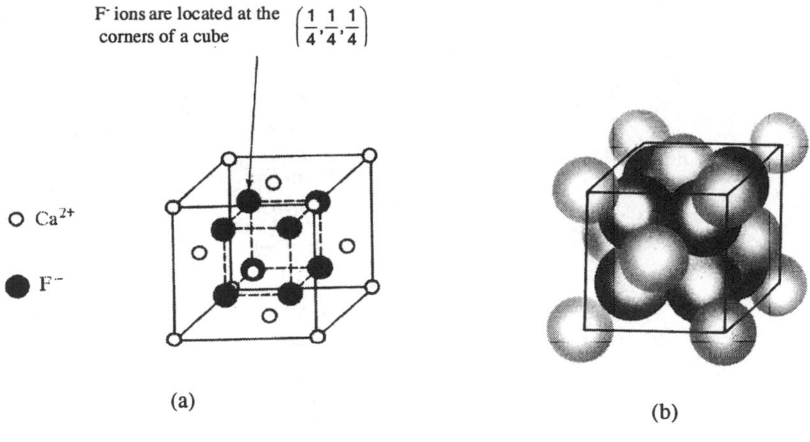

Fig. 2.22. Unit cell of fluorite (CaF_2). (a) Position of the atoms in the lattice; (b) hard sphere model.

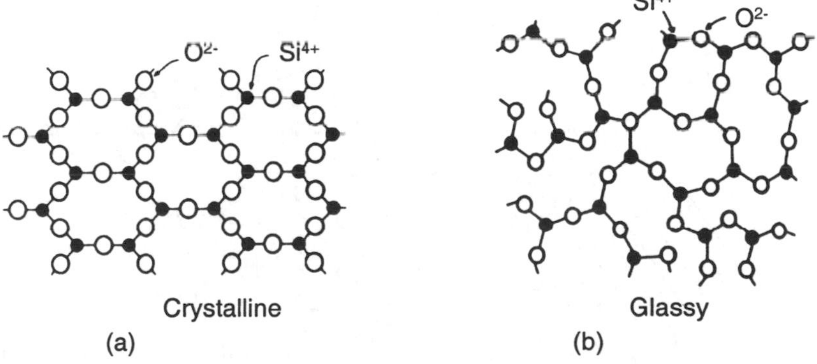

Fig. 2.23. (a) Structure of crystalline SiO_2; (b) glassy.

hydrogen bonds owing to the Van-der-Waals interaction (Fig. 2.24). This can be obtained by folding of the polymer chains to establish periodic molecular arrangements which can be described by a space lattice. Usually the unit cell of such a lattice contains a large number of atoms (50 and more), and, correspondingly, the symmetry of the space lattice is very low. The crystal structure of polyhexamethylene apidamide (better known as nylon 66), for example (Fig. 2.25) is triclinic, i.e. the lowest crystal symmetry.

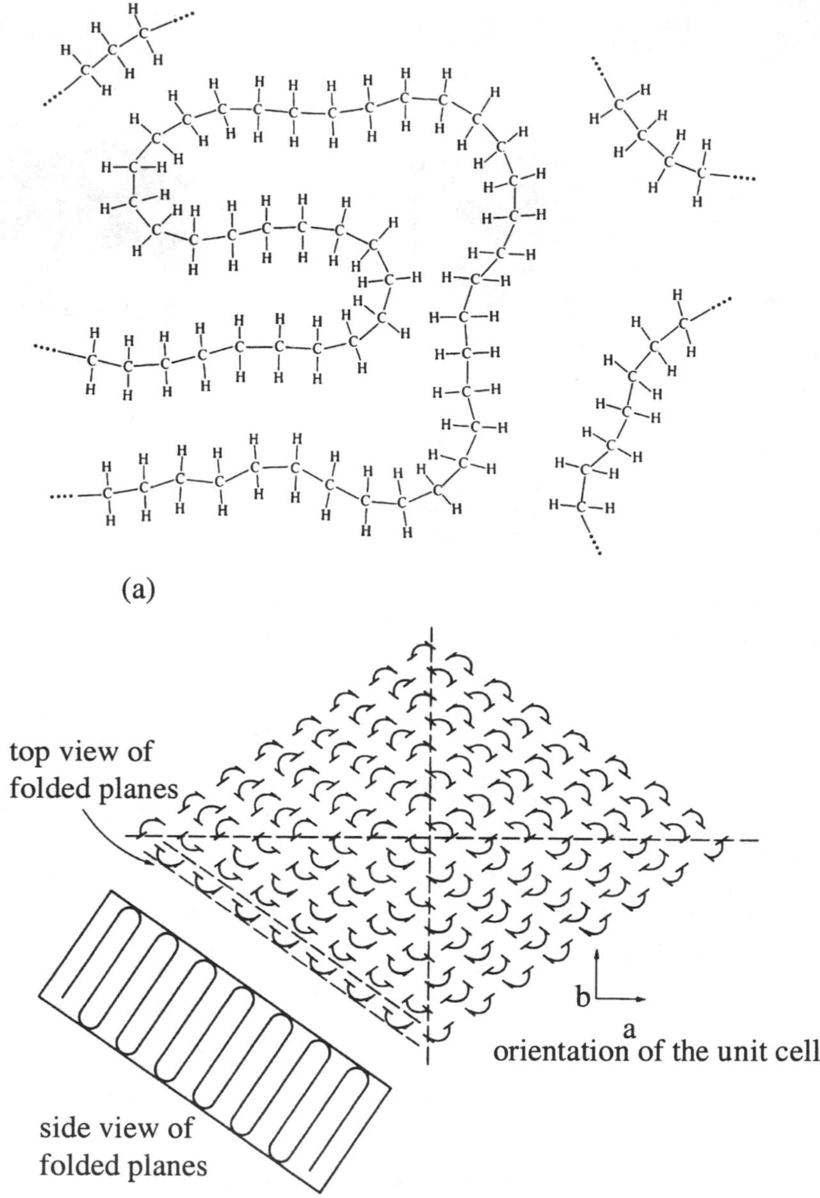

Fig. 2.24. (a) Schematic representation of the chain structure of solid polyethylene. (b) Folded polymer chains in the planes of crystalline polyethylene.

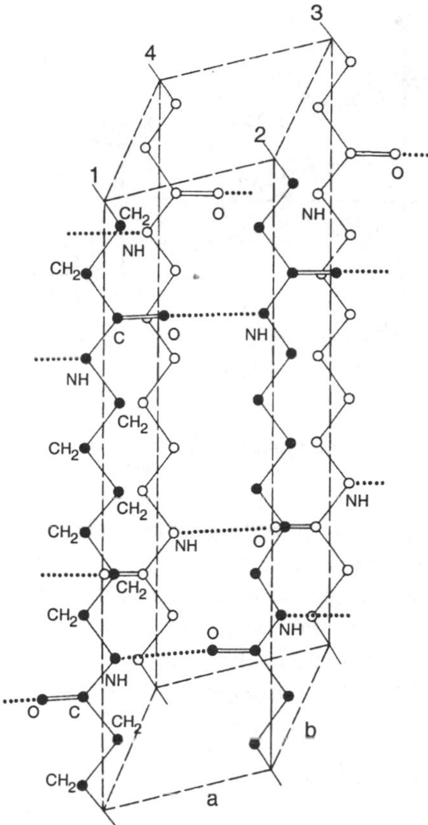

Fig. 2.25. Unit cell of polyhexamethylenapidamide (nylon 66). The nylon molecules are arranged in a trinclinic unit cell.

2.3 Indexing Crystallographic Planes and Directions

For quantitative crystallography it is necessary to uniquely characterize planes and directions of the crystal lattice and the positions of the atoms in the unit cell. The positions of the atoms in the unit cell are characterized by their coordinates with respect to the origin of the coordinate system that defines the lattice (Fig. 2.26, internal coordinates). The scale is given by the lattice constants in the directions of the coordinate axes. Atoms within the unit cell, therefore, have coordinates with components less than one, for instance (1/2, 1/2, 1/2), for the center of the cube in cubic crystals. Positions of atoms in other unit cells can be obtained by addition of the internal coordinates to the translation vector, which connects the origin of the coordinate system to the respective origin of the unit cell under consideration.

For a quantitative characterization of crystallographic planes and directions Miller indices are used. They are defined as follows. The intersections

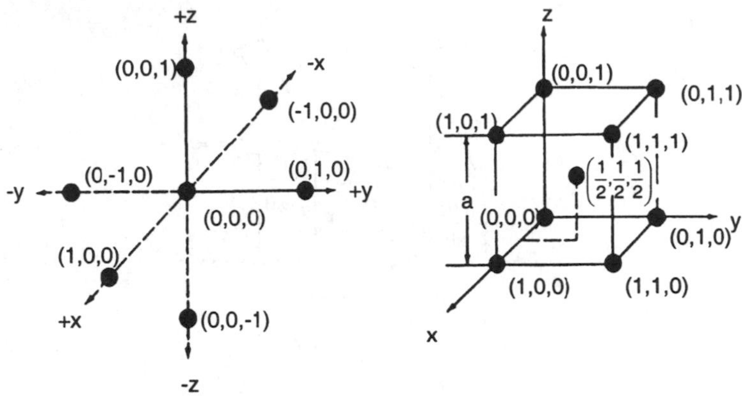

Fig. 2.26. How to specify the positions of atoms in the conventional unit cell.

of a chosen plane with the coordinate axes are determined in multiples of the lattice parameter along those axes (Fig. 2.27). For instance, if a lattice has the axes **a**, **b**, and **c** which are potentially of different length, then the intercepts with the crystal axes may be (ma, nb, qc). From the reciprocal values $(1/m, 1/n, 1/q)$, one obtains the Miller indices by multiplication with a factor r such that m, n, and q become integers with no common divisor, i.e.

$$r \cdot \left(\frac{1}{m}, \frac{1}{n}, \frac{1}{q} \right) = (hkl) \tag{2.8}$$

where h, k, and ℓ, are integers.

For instance (Fig. 2.27d), in a cubic lattice, a plane with the intersections (multiples of the lattice parameter) (1, 2/3, 1/3), has the Miller indices

$$2 \cdot \left[\frac{1}{1}, \frac{3}{2}, \frac{3}{1} \right] = (236) \tag{2.9}$$

Although mathematically slightly incorrect, we use the number infinity for a missing (finite) point of intersection. For example, the cube face in Fig. 2.27a intersects neither the y axis nor the z axis, so the intersections are $(1, \infty, \infty)$. Thus, we obtain the Miller indices

$$1 \cdot \left[\frac{1}{1}, \frac{1}{\infty}, \frac{1}{\infty} \right] = (100) \tag{2.10}$$

We recognize that only in the cubic case the Miller indices of a crystallographic plane are identical to the components of a vector perpendicular to that plane (plane normal).

The Miller indices of directions are just the vector components of the respective direction, modified to become integer numbers. For instance, the

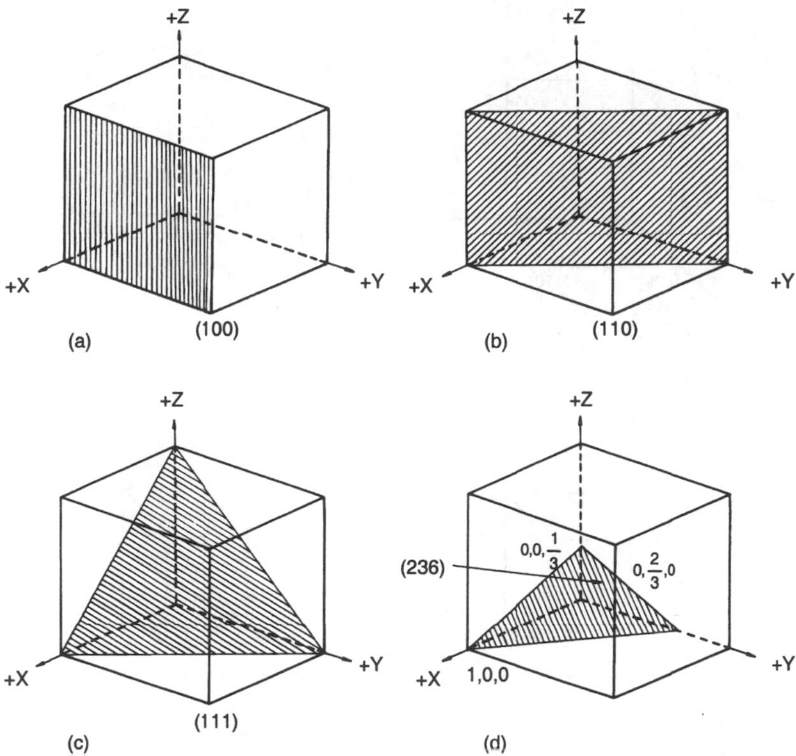

Fig. 2.27. Several lattice planes, their orientations and Miller indices in cubic crystals. (a) (100) plane; (b) (110) plane; (c) (111) plane; (d) the (236) plane.

vector with the components [1/2, 1/2, 1] corresponds to the Miller indices [112] of the respective direction (Fig. 2.28b).

The symmetry of the cubic lattice causes the atomic arrangement of some planes and directions to become indistinguishable. This is reflected in equivalent directions having Miller indices with exchanged components or opposite sign. Only if the coordinate system is uniquely specified can the crystallographically equivalent planes and directions be discriminated. For instance, a rotation by 90° around a cube axis generates an arrangement of lattice points identical with the unrotated arrangement. However, if the coordinate system would have been defined first, then, for instance, the [100] direction would have been transferred into the [010] direction by a rotation about the z axis [001]. For physical properties only the atomic arrangement plays a role, while the definition of the coordinate system is quite arbitrary. Therefore, all crystallographically equivalent planes and directions are identified by braces { } for planes and angular brackets ⟨ ⟩ for directions. In contrast, specific planes and directions are denoted by parentheses ( ), and square brackets [ ], respectively. For instance, the family of planes {111} comprises the planes

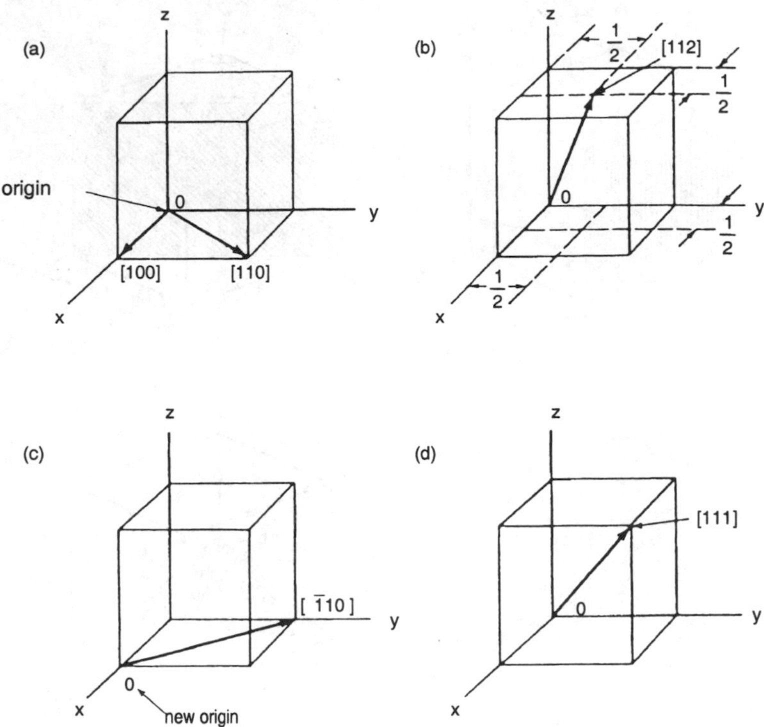

Fig. 2.28. Miller indices of several directions in cubic crystals.

$(111), (\bar{1}11), (1\bar{1}1), (\bar{1}\bar{1}1)$, and all planes with normals of opposite sign, for instance $(\bar{1}\bar{1}\bar{1})$ instead of (111), although these do not represent different physical planes. Analogously, a family of $\langle 111 \rangle$ directions is composed of the directions $[111], [\bar{1}11]$, and so on (see above) and also the antiparallel directions if the orientation sense is to be discriminated. In the case of cubic crystal symmetry a plane $\{hkl\}$ (with $h \neq k \neq l$) represents 24 different planes, or 48 planes when including opposite signs. For crystal lattices with lower symmetry there is a smaller variety of equivalent planes and directions, since in this case some or all of the crystal axes can be distinguished.

The Miller indices can be defined for any crystal system. For the hexagonal lattice this causes the disadvantage that the hexagonal symmetry in the basal plane is not apparent in the Miller indices if two axes, for instance a_1 and a_2 are selected from the three equivalent axes (a_1, a_2, a_3) (Fig. 2.29). Therefore, for hexagonal crystal structures Miller-Bravais indices are used which consist of four components $(hkil)$ with the constraint

$$h + k + i = 0 \tag{2.11}$$

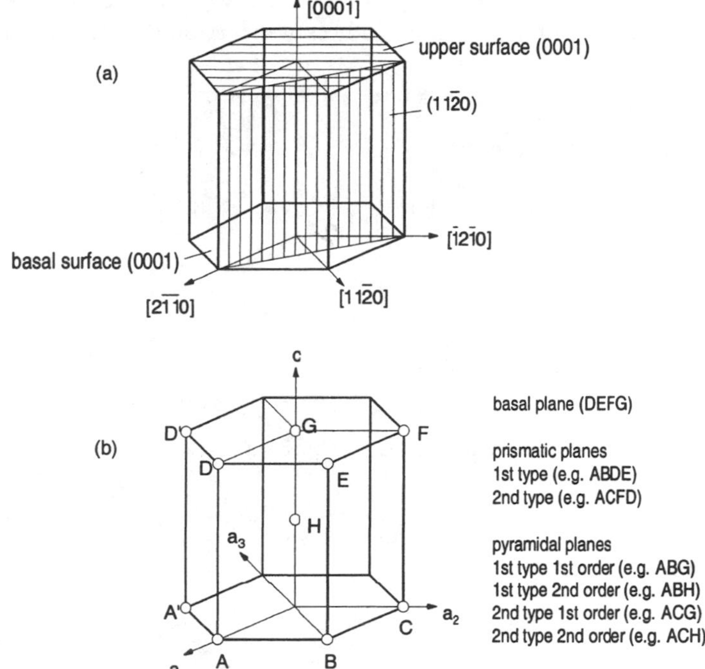

Fig. 2.29. Indices and orientations of planes and directions in the hexagonal lattice; (a) illustration of the indices of some directions in the basal plane after Miller and Miller-Bravais; (b) orientation of the basal plane and prismatic and pyramidal planes.

The indices h, k, i, and l again are the integer numbers of the reciprocal axial intercepts of the four crystal axes $\mathbf{a_1}, \mathbf{a_2}, \mathbf{a_3}$, and $\mathbf{c}$.

The Miller-Bravais indices can be transformed into Miller indices with the axes $\mathbf{a_1}, \mathbf{a_2}$, and $\mathbf{c}$ and vice versa. If we denote the Miller indices with capital letters to distinguish them from Miller-Bravais indices with small letters we obtain for planes

$$(HKL) \rightarrow (hkil) = (H, K, -(H+K), L) \tag{2.12}$$

The advantage of the Miller-Bravais indices is that crystallographically equivalent planes are reflected by crystallographically equivalent indices. For instance, in Fig. 2.29b the prism planes ADEB and A'D'DA are crystallographically equivalent, however, they are represented by different Miller indices, namely (100) and (1$\bar{1}$0), respectively. In contrast, the Miller-Bravais indices are (10$\bar{1}$0) and (1$\bar{1}$00). Sometimes, the notation $(hk \cdot l)$ can be found in the literature for Miller-Bravais indices. This, however, conceals the obtained crystallographic equivalence by the Miller-Bravais indices.

The transformation from Miller to Miller-Bravais indices for directions is slightly more complicated. In Miller indices a direction is given by the vector

$$\mathbf{r}_{UVW} = U\mathbf{a}_1 + V\mathbf{a}_2 + W\mathbf{c} \tag{2.13}$$

in Miller-Bravais indices, however

$$\mathbf{r}_{uvtw} = u\mathbf{a}_1 + v\mathbf{a}_3 + t\mathbf{a}_3 + w\mathbf{c} \tag{2.14}$$

With

$$\mathbf{a}_1 + \mathbf{a}_2 + \mathbf{a}_3 = 0 \tag{2.15}$$

$$\mathbf{r}_{uvtw} = u\mathbf{a}_1 + v\mathbf{a}_2 + t\left(-\mathbf{a}_1 - \mathbf{a}_2\right) + w\mathbf{c} \tag{2.16}$$

By comparison of the coefficient of Eqs. (2.13) and (2.16) one obtains

$$\begin{aligned} U &= u - t \\ V &= v - t \\ W &= w \end{aligned} \tag{2.17}$$

and because of symmetry

$$u + v + t = 0 \tag{2.18}$$

we obtain from Eqs. (2.17), (2.18), and inserting,

$$\begin{aligned} U &= 2u + v \\ V &= 2v + u \end{aligned} \tag{2.19}$$

$$\begin{aligned} u &= \frac{1}{3}\left(2U - V\right) \\ v &= \frac{1}{3}\left(2V - U\right) \\ t &= -\frac{1}{3}\left(U + V\right) \\ w &= W \end{aligned} \tag{2.20}$$

To obtain integer indices one has to multiply by three, so that

$$[UVW] \rightarrow [uvtw] = [2U - V, 2V - U, -\left(U + V\right), 3W] \tag{2.21}$$

2.4 Representation of Orientations: Stereographic Projection

For representation of orientations, i.e. the orientation of the unit cell in space, the orientation sphere is most useful, on the surface of which each point represents the intersection with the normal of a crystallographic plane. However, the 3D orientation sphere is poorly suited for reproduction on paper, where only 2D representations are possible. Hence, the orientation sphere must be projected on to a plane. Among several mathematically possible projections the stereographic projection (Fig. 2.30) has become the standard tool for representation of orientations. For this consider the unit cell to be located at the center of a sphere (reference sphere). The normal of a plane E intersects the surface of the sphere at point P. The connection of the point P with the center of projection, the south pole of the sphere, intersects the equatorial plane [1] at P'. The point P' is referred to as pole of the plane E. The pole is identified by two angles, α and β, as defined in Fig. 2.30. Such planes, the normals of which intersect the southern hemisphere, are not included in this projection, because their poles would be located outside of the equator. However, these planes are also uniquely represented by the normals of opposite sense.

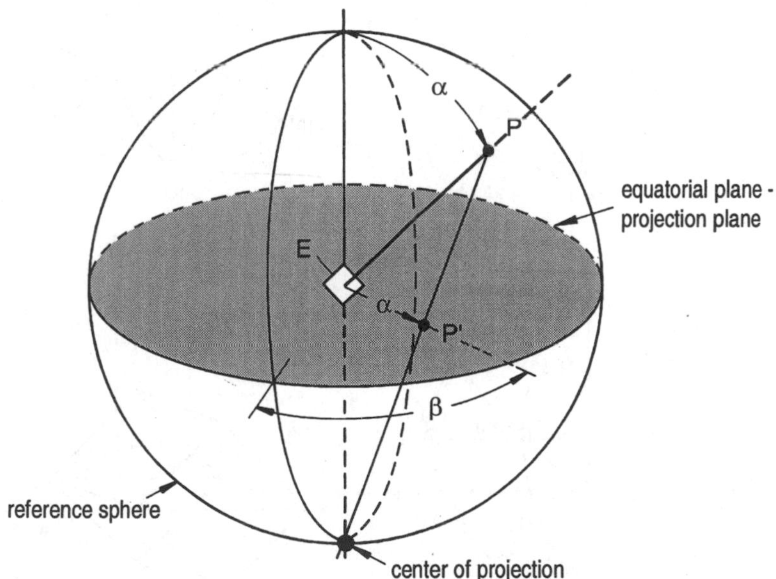

Fig. 2.30. Illustration of the principle of stereographic projection: E - crystallographic plane, P' pole of E in stereographic projection.

[1] Frequently, the tangential plane to the north pole of the sphere is used instead of the equatorial plane. The result is the same, except for a different magnification.

As an example, consider an orientation, where the [001] axis intersects the reference sphere in the north pole. Correspondingly, the pole of the (001) plane is located in the center of the projection. This defines the (001) projection. If all {100}, {110}, and {111} planes are included in the projection, the projection of the northern hemisphere comprises 24 stereographic triangles, each of which is cornered by a {100}, {110}, and {111} pole. These 24 triangles reflect the 24-fold cubic symmetry. For each pole in any triangle there is a crystallographically equivalent pole (i.e. indices permutated and/or of opposite sign, however, limited to the northern hemisphere because of $l \geq 0$) in another triangle, e.g. (123) and ($2\bar{1}3$). For cubic crystal symmetry, a single triangle, the stereographic standard triangle, is sufficient for representation of a family of planes or directions. Usually, the triangle (001) – (011) – ($\bar{1}11$) is chosen (Fig. 2.32), but any other triangle can also serve as a standard triangle.

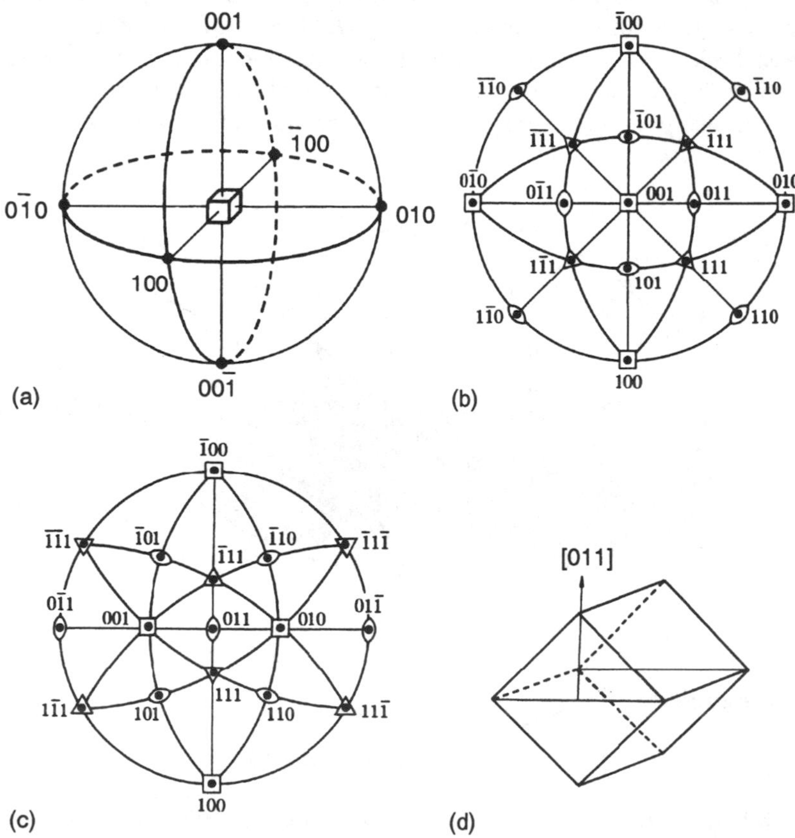

Fig. 2.31. Spatial orientation of the {100} planes in (001) projection, (b) standard projection of a cubic crystal in (001) and (c) (110) orientation and three-dimensional position of the (110) plane in (011) projection (d).

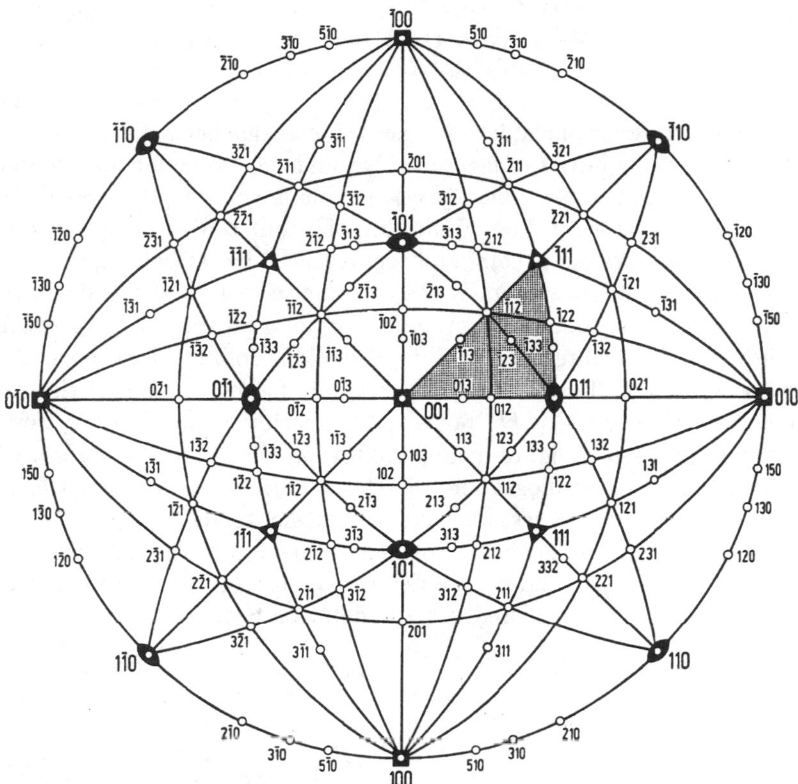

Fig. 2.32. Standard (001)-projection of some low index planes and zones in a cubic lattice. The standard triangle is shaded.

The (011) projection (Fig. 2.31c) is obtained if the (011) pole is located in the center of the projection (cube on edge). The (011) projection can be derived from the (001) projection by rotation of the (011) pole into the center of the stereographic projection. For convenience, a Wulff net can be used for rotations in stereographic projections. The Wulff net is the stereographic projection of circles on the reference sphere, where the different circles have a constant angular distance (typically 2°). With a computer at hand it is even easier to use matrix algebra, utilizing the rotation matrix **A**. The rotated direction (vector $\mathbf{r}'$) is obtained from the original direction **r** (vector **r**) by

$$\mathbf{r}' = \mathbf{A}\mathbf{r}$$

If the rotation axis is $\mathbf{a} = (a_1, a_2, a_3)$ where $|\mathbf{a}| = 1$, and the angle of rotation φ, then the rotation matrix $\mathbf{A}(\mathbf{a}, \varphi)$ is as follows:

$$\mathbf{A}(\mathbf{a},\varphi)=\begin{bmatrix} \left(1-a_1^2\right)\cos\varphi+a_1^2 & a_1a_2(1-\cos\varphi)+a_3\sin\varphi & a_1a_3(1-\cos\varphi)-a_2\sin\varphi \\ a_1a_2(1-\cos\varphi)-a_3\sin\varphi & \left(1-a_2^2\right)\cos\varphi+a_2^2 & a_2a_3(1-\cos\varphi)+a_1\sin\varphi \\ a_2a_3(1-\cos\varphi)+a_2\sin\varphi & a_2a_3(1-\cos\varphi)-a_1\sin\varphi & \left(1-a_3^2\right)\cos\varphi+a_3^2 \end{bmatrix} \tag{2.22}$$

Because of crystallographically fixed angular relations between the planes and directions in a cubic lattice, a few poles are sufficient to define an orientation. More specifically, only two poles $\{hkl\}$, for instance two $\{100\}$ poles or two $\{111\}$ poles are necessary. The orientation of a unit cell with regard to a reference coordinate system is commonly represented by its pole figure. Usually the reference coordinate system is the coordinate system of the macroscopic specimen (specimen coordinate system), for instance of a rolling specimen with the orthogonal directions: rolling direction, rolling plane normal, and transverse direction (Fig. 2.33). An $\{hkl\}$ pole figure consists of the stereographic projection of the $\{hkl\}$ poles in the specimen coordinate system, for instance the $\{100\}$ pole figure in Fig. 2.33. From the angles α_i and β_i of the $\{100\}$ poles those crystallographic directions can be determined which are parallel to the specimen axes, for instance parallel to the sheet plane normal and the rolling direction of the rolling sheet. Pole figures are important tools in materials engineering and can be determined directly by X-ray or neutron diffraction, as will be considered in more detail in Section 2.5.

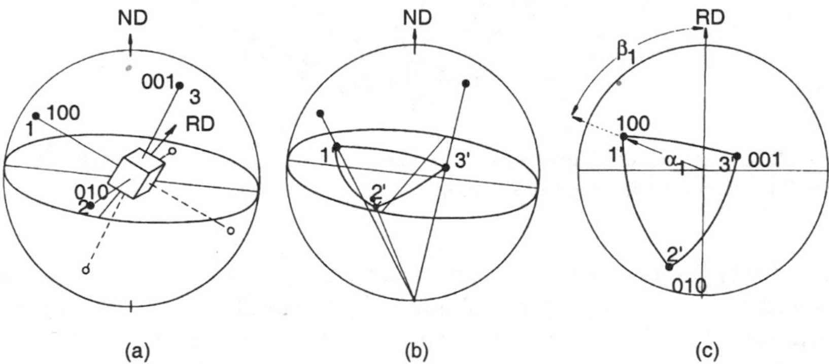

Fig. 2.33. Representation of one orientation by the $\{100\}$ axes in stereographic projection. (a) Position of the crystal at the center of the orientation sphere; (b) stereographic projection of the cubic axes; (c) $\{100\}$ pole figure showing orientation and definition of the angles α_1 and β_1 associated with pole 1.

While the pole figures denote the orientation of a crystal axis in the specimen coordinate system, the inverse pole figure represents the orientation of a specimen axis with regard to the crystal coordinate system (Fig. 2.34). If a single specimen axis is relevant only, for instance the axis of a wire or of a tensile specimen, representation can be reduced to the standard triangle, in which the orientation of the respective sample axis is noted (Fig. 2.34b).

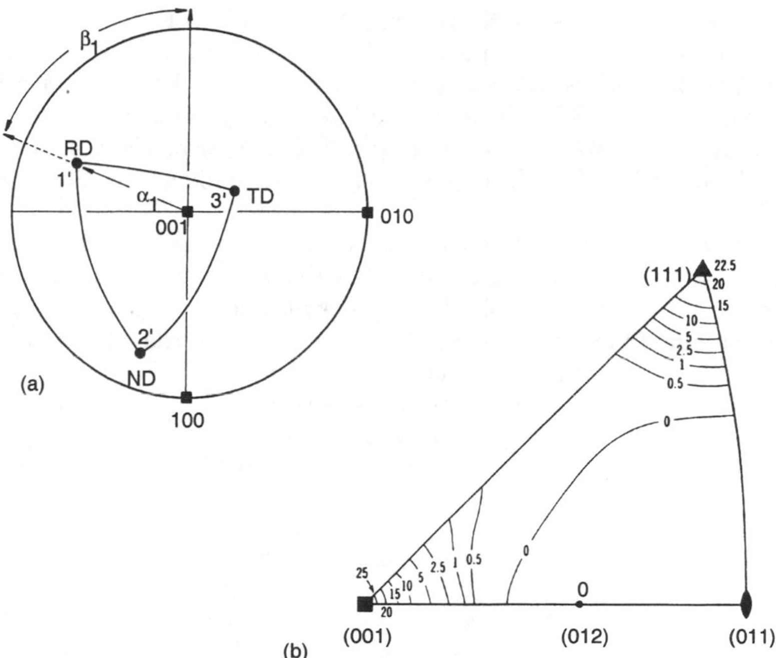

Fig. 2.34. (a) Diagrammatic illustration of the inverse pole figure of a sample of rolled material. (b) Inverse pole figure of extruded aluminum wire (sample orientation = wire axis). The lines indicate contours of equal measured intensity (after 2. [2.1])

2.5 Experimental Crystallographic Methods

2.5.1 Bragg's law

The theoretical foundation for the investigation of crystals by X-rays, neutrons, or electrons is Bragg's law. It describes the diffraction (i.e. the coherent elastic scattering) of waves by the crystal lattice and reads

$$n\lambda = 2d\sin\Theta \tag{2.23}$$

where λ - wavelength; d - spacing of reflecting planes; Θ - angle of incidence and reflection; n - order of diffraction.

Correspondingly, if an X-ray of wavelength λ hits a set of lattice planes of spacing d under an angle Θ, the radiation will be reflected. More specifically, reflection occurs only if this condition is met. The reflection of electromagnetic radiation in crystals is due to the interaction of radiation with the electron shell of atoms, which is very complex in detail. To understand Bragg's equation, it is helpful and sufficient to consider the lattice planes as semi transparent mirrors for X-rays (Fig. 2.35). Only if the diffracted radiation of parallel planes are in phase, will the reflected intensity be non-zero (Fig. 2.35b). If there is any phase difference between diffracted rays from parallel planes (Fig. 2.35a) the diffracted intensity will vanish because of the many lattice planes which contribute to diffraction, since there is always a diffracted wave which has the opposite phase (phase difference $180° = \pi$) to any considered wave (Fig. 2.36).

Two waves which are diffracted by next neighbor parallel planes can only be in phase if the difference of the path lengths is an integer multiple of the wavelength. This is the case only if Bragg's law is obeyed (Fig. 2.35). Usually, only diffraction of first order ($n = 1$) is considered, i.e. if the diffracted waves of next neighbor lattice planes differ just by one wavelength.

The lattice plane spacing d depends on the crystal structure and on the Miller indices $\{hkl\}$ of the crystallographic planes.
One obtains for

$$\text{cubic structure} \qquad d = \frac{a}{\sqrt{h^2 + k^2 + l^2}} \tag{2.24}$$

$$\text{hexagonal structure} \qquad d = \frac{a}{\sqrt{\frac{4}{3}\left(h^2 + k^2 + h \cdot k\right) + l^2 / \left(c/a\right)^2}} \tag{2.25}$$

where a, respectively a and c are the lattice parameters. Bragg's equation has a real solution only for

$$\frac{n\lambda}{2d} \leq 1 \tag{2.26}$$

because of $\sin x \leq 1$. Since the lattice plane spacing according to Eqs. (2.25) and (2.26) is smaller than the lattice parameter, a crystal lattice can diffract radiation only if the wavelength is of equal magnitude or smaller than the lattice parameter. This is only the case for hard X-rays or matter waves from electrons or neutrons. The wavelength of X-rays of conventional X-ray tubes is of the order of about 0.1 nm. High energy electrons, for instance in the electron microscope with an acceleration voltage of $U = 100$ kV have a wavelength $\lambda = 0.0037$nm.

2.5.2 X-ray Methods

X-rays are generated by irradiation of a material with electrons. The structure of an X-ray tube is shown in Fig. 2.37. Electrons that are emitted by thermal emission from a cathode, e.g. hot wire, are accelerated by a high voltage (about 20 - 30 kV) and hit a target (anode) which consists of a pure element. The strong deceleration of the electrons in the anode target generates Bremsstrahlung, an X radiation with a broad wave length spectrum, the so-called continuous radiation (Fig. 2.38). In addition, sharp peaks of characteristic radiation are observed which are generated by excitation of electrons

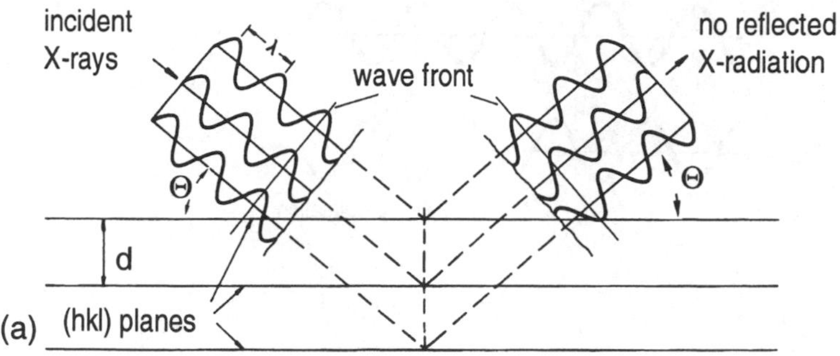

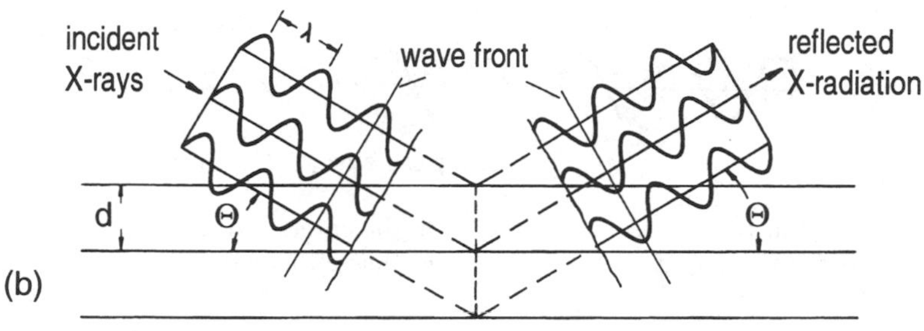

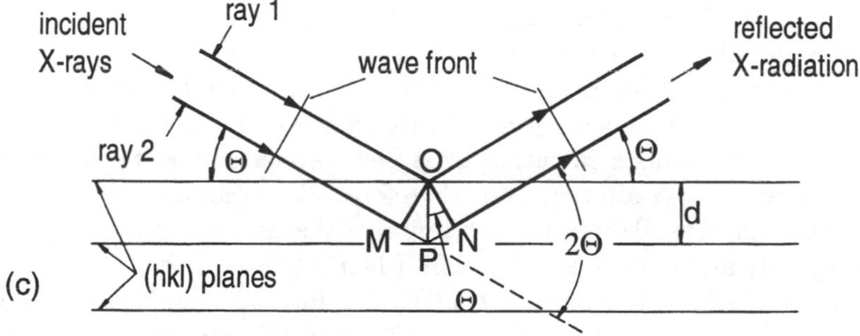

Fig. 2.35. X-ray diffraction by a crystal lattice can be regarded as reflection from a semi-inflecting mirror. Only if rays diffracted by parallel planes are in phase (b) (equal phase length in every plane perpendicular to the direction of propagation), they will not cancel each other out (as illustrated in figure 2.36) and reflection will occur. Equal phase length will result, only if the path difference from successive planes (MPN in (c)) is an integral number of wavelength. This condition leads to the Bragg equation.

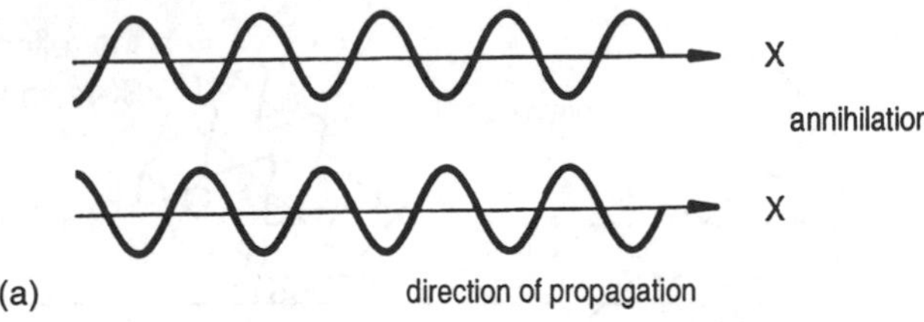

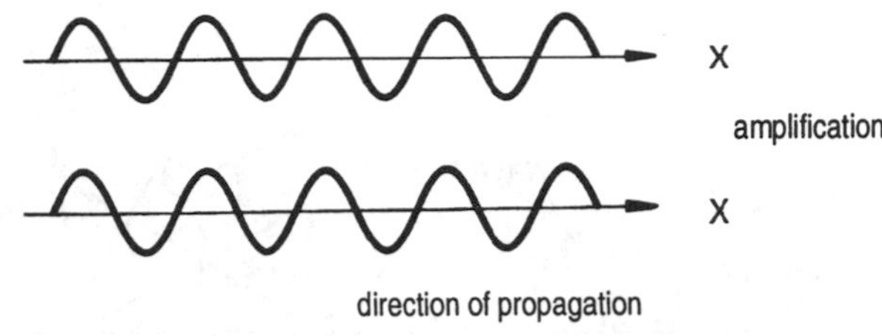

Fig. 2.36. (a) Due to a phase difference of π the interference of the two waves results in annihilation. (b) If both waves are in phase, this results in amplification of their amplitudes.

in the atoms of the target material. Their wavelength depends only on the target material. X-rays of a single wavelength are called monochromatic X-rays. They are obtained from the typical spectrum of an X-ray tube by filter absorption of the continuous spectrum below the wavelength of the characteristic radiation. For instance, a typical filter material for copper X-ray tubes is nickel. The continuous radiation at larger wavelengths is weak compared to the characteristic radiation and, therefore, can practically be neglected. Monochromator crystals can further improve the quality of monochromatic X-rays, usually at the expense of a diminished intensity, however.

For crystallographic structure analysis, or for chemical analysis powder X-ray diffractometry is employed (Fig. 2.39a). In this method a powder of a substance or a polycrystal with random orientation distribution is illuminated with monochromatic X-rays. An X-ray detector moves about the specimen, such that an angular range $0 \leq 2\Theta \leq 2\Theta_{max} \cong 120°$ is swept. By this a diffractogram is obtained with diffracted intensities at specific values of 2Θ (Fig. 2.39b). A reflection is obtained only if in the crystals of the investigated substance there is a crystallographic plane that obeys Bragg's law for the selected wave length. The angles of the obtained reflections for a given wavelength are a fingerprint of the investigated crystal structure. For an accurate analysis a structure factor has to be considered which can be zero for certain crystallographic planes. For instance, in fcc crystals there is a reflection only if all Miller indices are even or all Miller indices are odd. Therefore, there is no

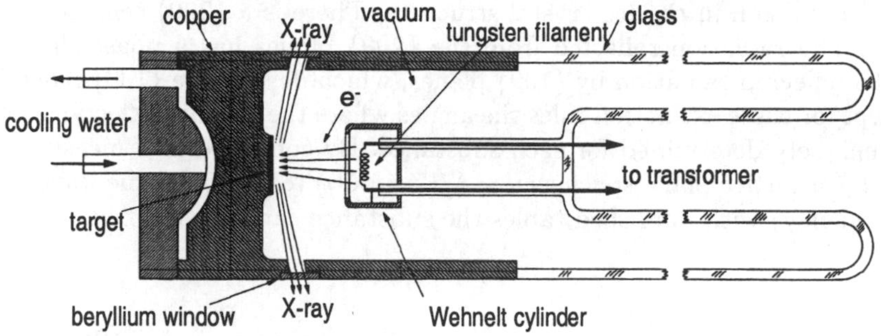

Fig. 2.37. Cross section of an X-ray tube.

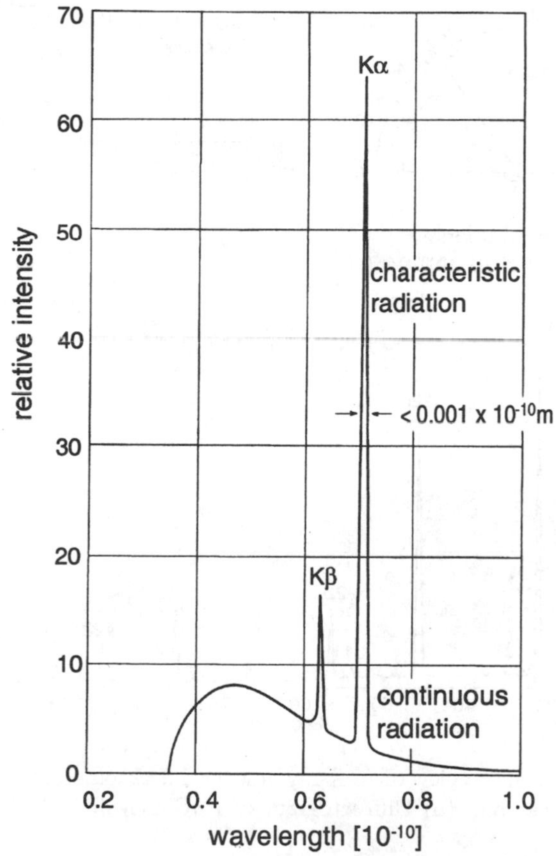

Fig. 2.38. Schematic X-ray spectrum of Mo at 35kV.

{100} reflection in the fcc crystal structure. There is a {200} reflection (Fig. 2.40). The radiation reflected from the {200} planes has a phase shift of π to the reflected radiation by {100} planes, which deletes the {100} reflection. Except for these extinction rules the angles where the specific reflections occur are uniquely determined for each substance. By comparison of measured 2Θ angles (or lattice plane spacings $d = \lambda/(2 \sin \Theta)$) to eliminate the wavelength dependency) with published tables the substance can be identified.

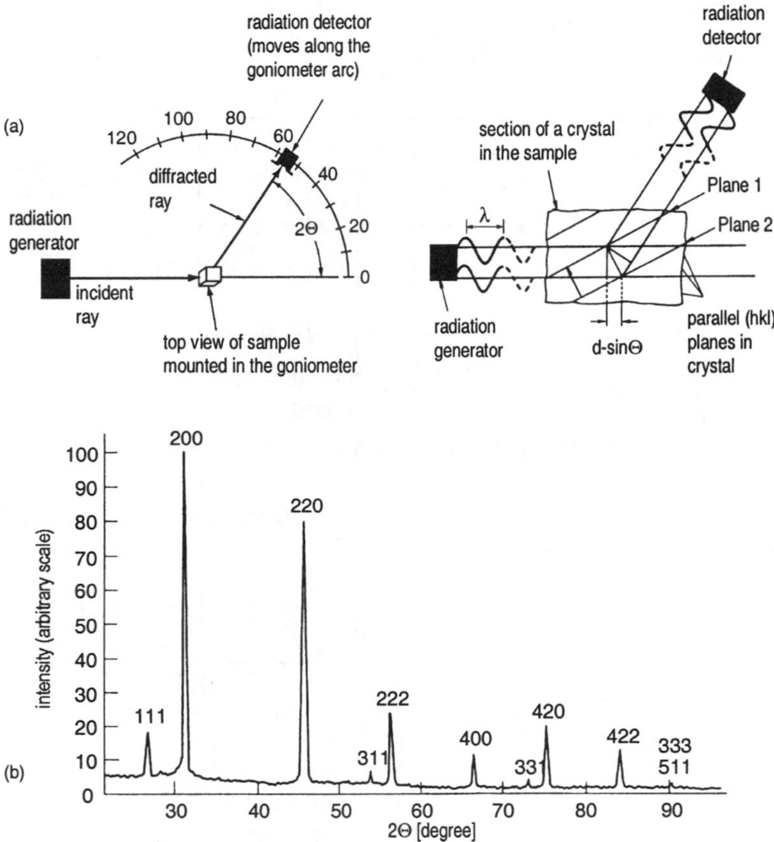

Fig. 2.39. (a) Schematic view of an X-ray powder diffractometer and the diffraction geometry in the sample. (b) Diffractogram of NaCl powder; Cu-K_α-radiation, Ni filter.

Instead of an X-ray detector also a film can be used as a detecting medium (Debye-Scherrer method). Typically, the film is arranged cylindrically around the specimen. The reflected X-ray intensities of a powder or polycrystal lie on a cone of angle 2Θ (Fig. 2.41) about the incident beam. A powder specimen

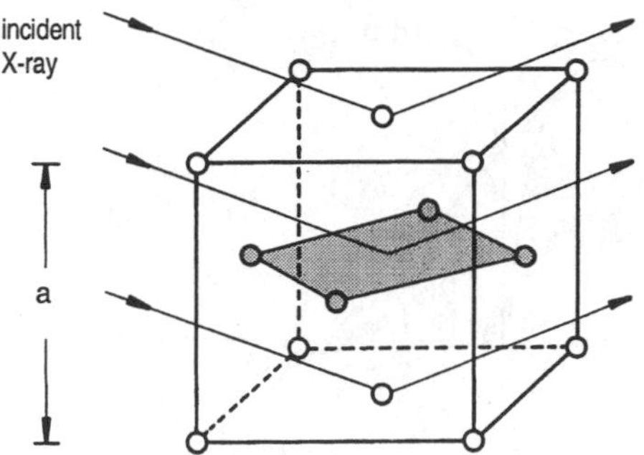

Fig. 2.40. Schematic illustration of (100) - (200) annihilation in a fcc lattice.

of crystallites with random orientation distribution contains virtually every orientation, in particular all orientations that are related by a rotation about the direction of the incident beam. If for one crystal Bragg's equation holds for an angle 2Θ, it invariably holds for all related crystals which results in reflections along the surface of a cone. The intersection of the cone with a planar film perpendicular to the direction of the beam is a circle, the so-called Debye-Scherrer ring (Fig. 2.41a). A cylindrical film around the specimen contains small sections of these circles (Fig. 2.41b). Since the angle 2Θ can be measured very accurately by the Debye-Scherrer method, it has been used for high precision measurements of the lattice constant.

If a single crystal is illuminated with monochromatic X-rays, usually no reflected intensity is detected, except if the single crystal is oriented in such a way that it meets the geometry of Bragg's equation. However, if instead of monochromatic X-rays so-called white X-rays, i.e. the whole continuous spectrum is used for illumination, then there is a wave length available for each crystallographic plane that can satisfy Bragg's law (Laue-method). A planar film mounted between single crystal and X-ray source will show a spot pattern of back diffracted X-rays (Fig. 2.42). From the arrangement of reflections on such a Laue pattern the orientation of a single crystal can be determined if the structure of the material is known. For thin specimens the diffracted intensity can also be determined in transmission. All crystallographic planes that are related by a rotation about a common axis (zone axis) form a crystallographic zone. The reflections of the planes of a zone fall on a hyperbola in the Laue pattern of a planar film (Fig. 2.42a). Therefore, a Laue pattern contains a multitude of intersecting hyperbolas (Fig. 2.42c). By identification of the zones the crystallographic directions parallel to the incident beam or to any other specimen direction can be obtained and is usually represented in an inverse pole figure.

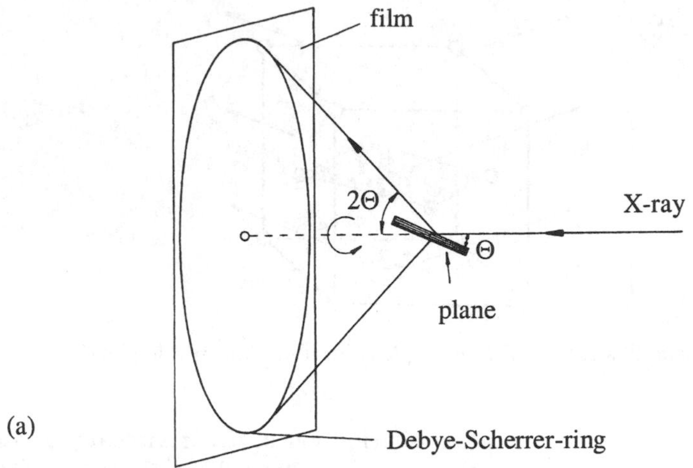

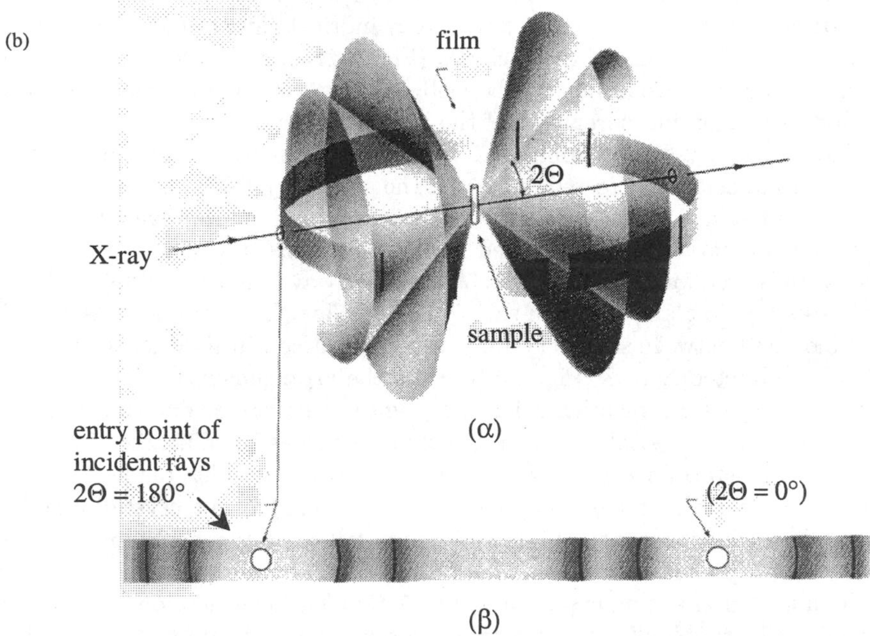

Fig. 2.41. (a) Schematic representation of a Debye-Scherrer image on plane film. (b) Rotation of the crystal about the radiation incidence axis results in reflected radiation forming a cone with an acceptance angle of 4Θ.

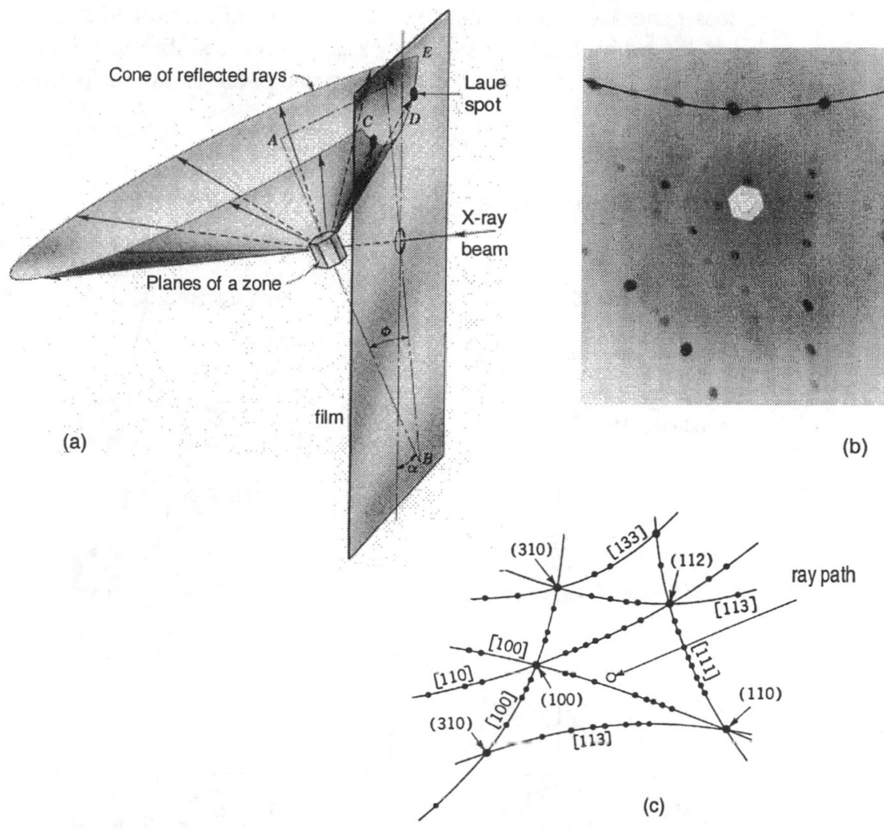

Fig. 2.42. (a) Laue-back-scattering-image; (a) Reflections from one zone are along a hyperbola on the film; (b) sample: aluminum (tungsten radiation 30kV, 19mA); (c) sample: α-iron; zones and major reflections are indexed.

2.5.3 Electron Microscopy

Instead of X-rays also electrons can be used for structure and orientation determination. Like all elementary particles, electrons (in principle all matter) can behave like a wave and, therefore, can be diffracted by a crystal lattice. Electron diffraction is the key for imaging in the Transmission Electron Microscope (TEM). Imaging in TEM by electrons is quite analogous to optical microscopy with visible light (Fig. 2.43). If a monochromatic electron beam passes through a thin crystal volume electron diffraction occurs (Fig. 2.44). Typical spot patterns are obtained for a single crystal, from which the lattice parameter and the orientation can be determined. If the electron beam traverses/crosses many grains in a polycrystal, diffraction rings are obtained, in analogy to the Debye-Scherrer rings of X-ray diffraction. An amorphous substance or a glass generates diffuse ring patterns. The diffraction patterns in TEM are also governed by Bragg's law (Eq. (2.23)). Electron diffraction is the basis for TEM imaging, for instance to reveal crystal defects, like dislocations (see Chapter 3).

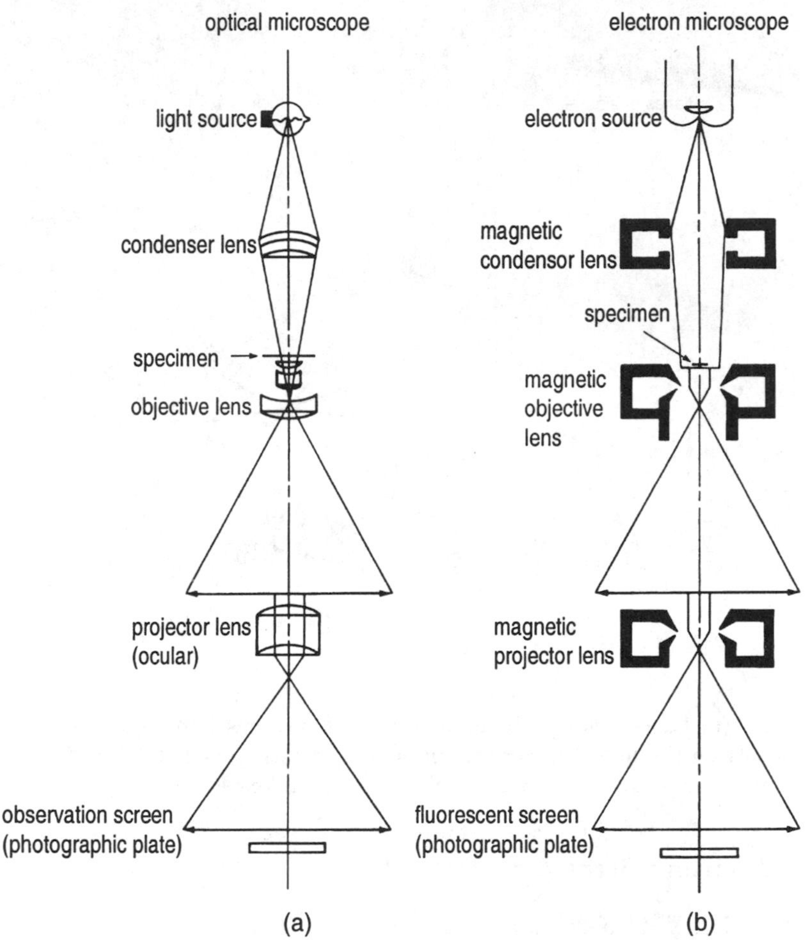

Fig. 2.43. Comparison of layout and ray path of a light microscope (a) and a transmission-electron-microscope (b).

2.5.4 Crystallographic Textures

The orientations in a polycrystal are not necessarily distributed randomly, as in a powder. Generally, the opposite is closer to reality. Due to materials processing such as forming or heat treatment, most metallic materials have

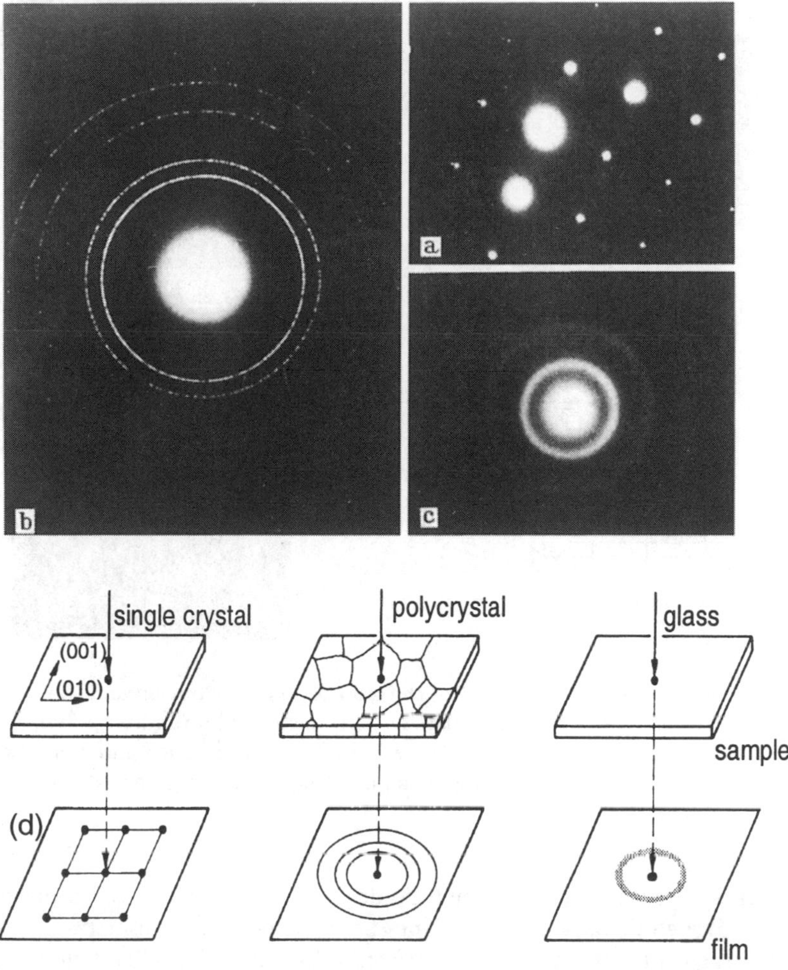

Fig. 2.44. Diffraction patterns of X-rays and electron radiation in (a) monocrystalline, (b) polycrystalline and (c) amorphous material, as illustrated schematically in (d) [2.2].

an orientation distribution where certain orientations are preferred. The orientation distribution is referred to as crystallographic texture. A non-random texture causes a non-homogeneous distribution along Debye-Scherrer rings (Fig. 2.45).

For quantitative determination of texture traditionally an X-ray texture goniometer is employed (Fig. 2.46). In this device monochromatic X-rays are used and X-ray source and counter are arranged in a fixed geometry, depending only on the Bragg angle of the investigated crystallographic plane. The

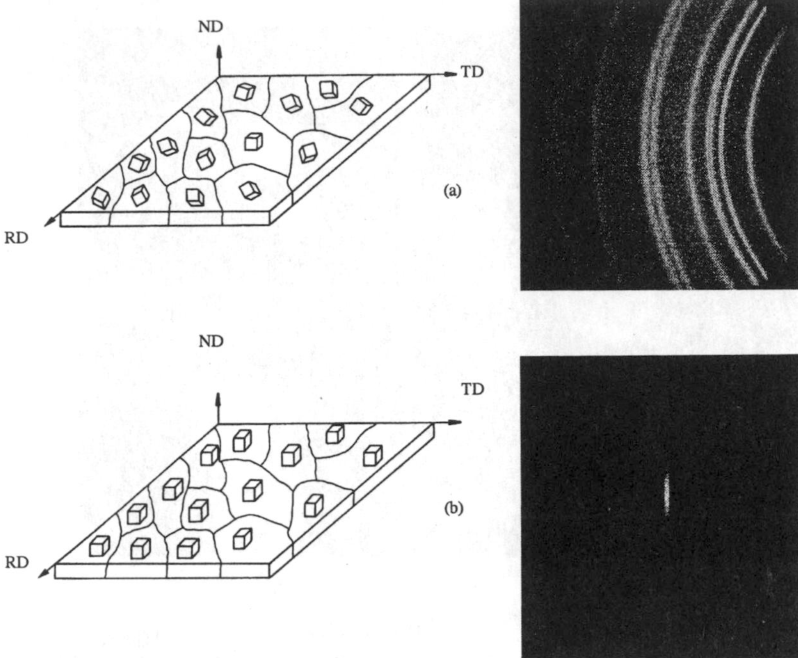

Fig. 2.45. Schematic representation of the orientation distribution and the resulting Debye-Scherrer images (a) random distribution, pattern of a random polypropylene polycrystal, (b) preferred orientation (here: cube texture) in rolled and recrystallized Al. Only a small section of the ring remains. (Debye-Scherrer images obtained by 2D X-ray detector [2.3])

specimen is mounted on a specimen holder which can be rotated around two mutually perpendicular axes to orient the specimen in practically any position with regard to the incident X-ray beam. Because of the fixed geometry of source and counter, diffraction occurs only if a particular family of crystallographic planes $\{hkl\}$, for instance $\{111\}$, obeys Bragg's law. The movement of the specimen unveils the spatial orientation of the respective poles $\{hkl\}$. In the stereographic projection the measured intensity distribution generates the $\{hkl\}$ pole figure. If in a cubic single crystal the cube faces are parallel to the specimen axes, for instance in a rolling specimen, one obtains the $\{200\}$ pole figures given in Fig. 2.47. In a polycrystal the respective distribution of $\{hkl\}$ poles in the specimen is measured, for instance after heavy rolling deformation of copper or brass (Fig. 2.48). This distribution of $\{hkl\}$ poles does not allow one, however, to identify the respective orientations of the crystals, because an orientation is given by three $\{100\}$ poles or four $\{111\}$ poles, the association of which is not known in the pole figure of a polycrystal, in contrast to a single crystal. From the measurement of several pole figures and respective

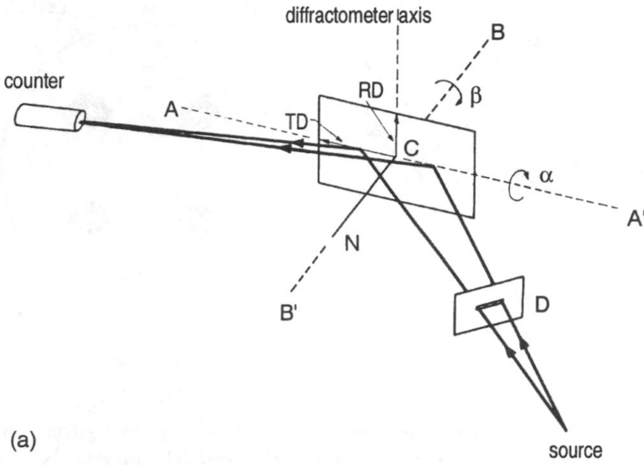

Fig. 2.46. (a) Ray path and sample rotation in an X-ray-texture-goniometer (here: Schulz-reflection-method). (b) X-ray-texture-goniometer at IMM.

computational methods, the orientation distribution function (ODF) can be determined. It is represented in a space, where each orientation is given by a point. Such a space is called orientation space. The orientation of a crystal can be described in terms of the orientation of its unit cell, i.e. the orientation of its crystal coordinate system with regard to the specimen coordinate system. To make the crystal coordinate system coincide with the specimen coordinate system requires a pure rotation. A rotation can be expressed by a rotation axis and angle as in Eq. 2.22 but also by three angles, for instance the three Euler angles $\varphi_1, \Phi, \varphi_2$ as is common in engineering practice. One

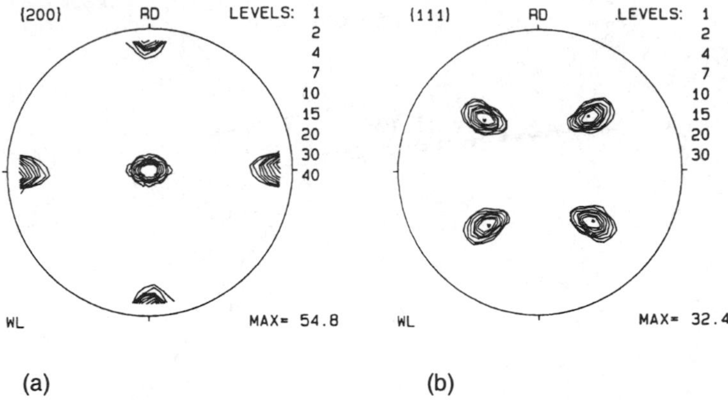

Fig. 2.47. Representation of the cube orientation in pole figures. (a) The {200} axes normal to the lattice planes occur most frequently in RD, ND, and TD. (b) Corresponding orientation of {111} poles. Compare (001) projection (figure 2.31b).

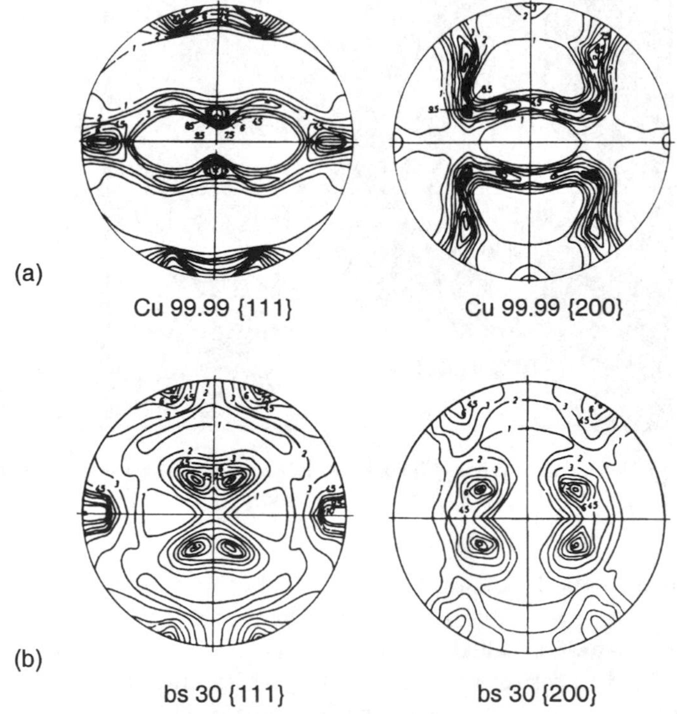

Fig. 2.48. Experimentally determined {111} and {200} pole figures of rolled material. (a) Copper 99.99% pure (rolled at room temperature); (c) copper with 30%Zn (α brass) (rolled at room temperature).

possible - and most commonly used - orientation space is, therefore, defined by three orthonormal coordinate axes, each of which corresponds to one Euler angle (Euler space). The orientation distribution function (ODF) of the pole figure of rolled copper as shown in Fig. 2.48 is represented in Euler space in Fig. 2.49a. For a two-dimensional representation of the three-dimensional distribution, it is typical to plot sections through Euler space at intervals of 5° parallel to the angle φ_2 (Fig. 2.49b).

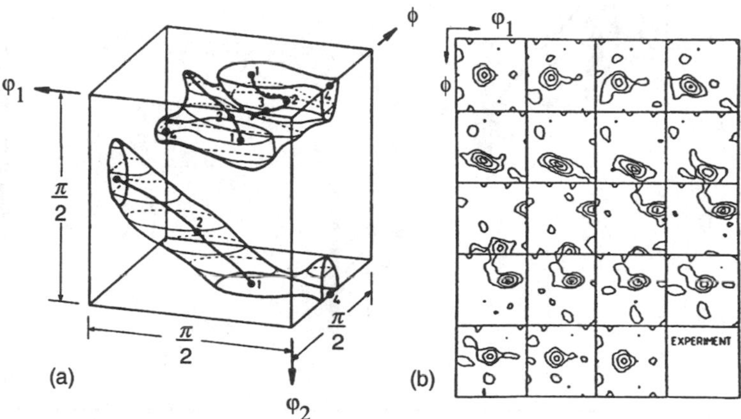

Fig. 2.49. 3-dimensional representation of the orientation distribution of rolled high purity copper (figure 2.48a) in Euler space. Unlike in the 2-dimensional pole figure, here an orientation is represented unambiguously in three coordinates (by three Euler angles). (b) Illustration of the orientation distribution by sections perpendicular to the angle φ_2 at 5° intervals.

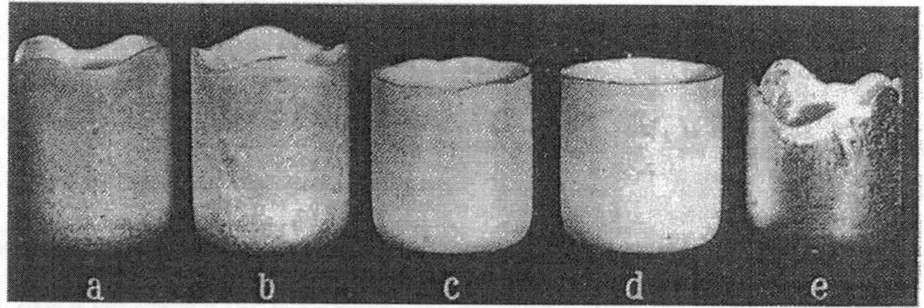

Fig. 2.50. Earing in deep drawn highest grade aluminum plate. (a) Ears at ± 45°; (b) ears at 0° and 90°; (c) eight ears; (d) specimen without ears; (e) ear generation due to large grain size (not a texture effect).

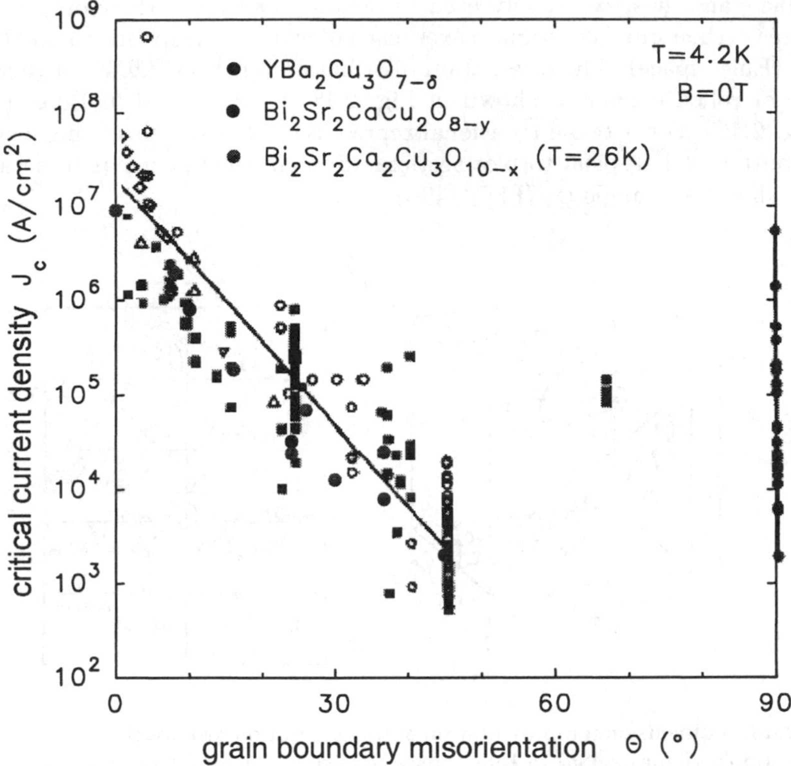

Fig. 2.51. Critical current of high temperature superconductors vs. misorientation across grain boundaries. Only for a misorientation below 5° the critical current is sufficiently high for power applications. Such low misorientations require a very sharp texture [2.4].

In modern materials textures become increasingly important because it causes anisotropy. An example is the earing of textured sheet during sheet forming, in particular deep drawing (Fig. 2.50), for example during forming of automotive sheet or for the production of beverage cans. The formation of ears during deep drawing is a consequence of texture. It leads to inhomogeneous sheet thickness, requires an additional production step to have them removed, and in particular interrupts high speed production. Therefore, in this case texture is very much undesired. For other applications a sharp texture can be highly desirable, for instance for sheet in cores of electrical transformers to minimize the losses due to changing magnetization (see Chapter 10). Sharp textures are also necessary for the production of superconducting tapes from high temperature superconductors (HTSC). Owing to the fact that the critical current to maintain the superconducting property is seriously impeded by high angle grain boundaries, a very sharp texture is needed for high current

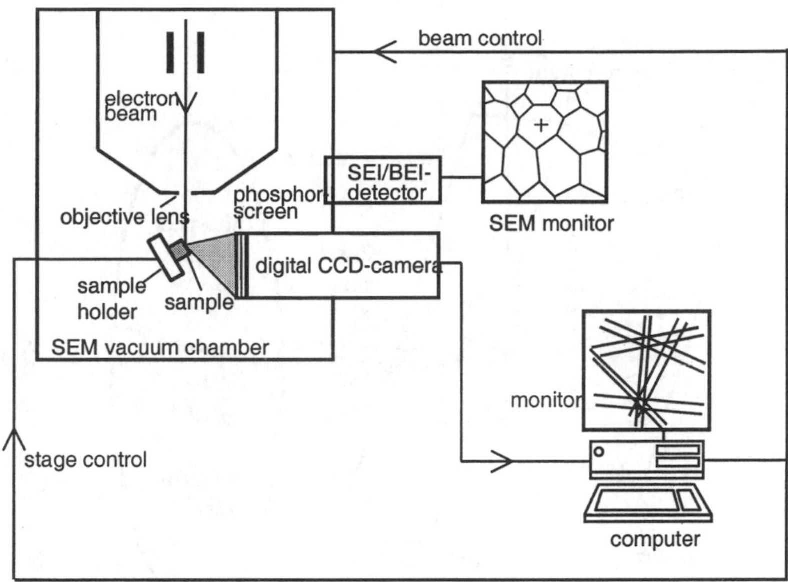

Fig. 2.52. SEM configuration for orientation mapping by EBSD. The specimen surface is scanned, and the orientation is automatically determined at every beam location.

engineering applications of high temperature superconductors (Fig. 2.51).

X-ray texture goniometers measure the average intensity of many grains, typically 10^5 grains in commercial materials. The corresponding texture is, therefore, referred to as macrotexture. In contrast the microtexture is obtained by measurement of the orientation of single grains in a polycrystal. This requires orientation measurement of very small areas which can be accomplished by electron back scatter diffraction (EBSD) in an SEM (Fig. 2.52). Modern SEM and advanced EBSD equipment and software are capable of measuring orientations of surface areas as small as 50nm at sampling rates in excess of 10000 orientation measurements per hour. Correspondingly, orientation maps (Fig. 2.53) can be generated by EBSD scanning of the polycrystalline microstructure, which is also referred to as orientation imaging microscopy (OIM). Besides local orientation information this also provides information on the misorientation across grain boundaries, i.e. the grain boundary character distribution. This is an important information for grain boundary engineering, e.g. for high intergranular corrosion resistance or HTSC with high critical currents as evident from Fig. 2.51.

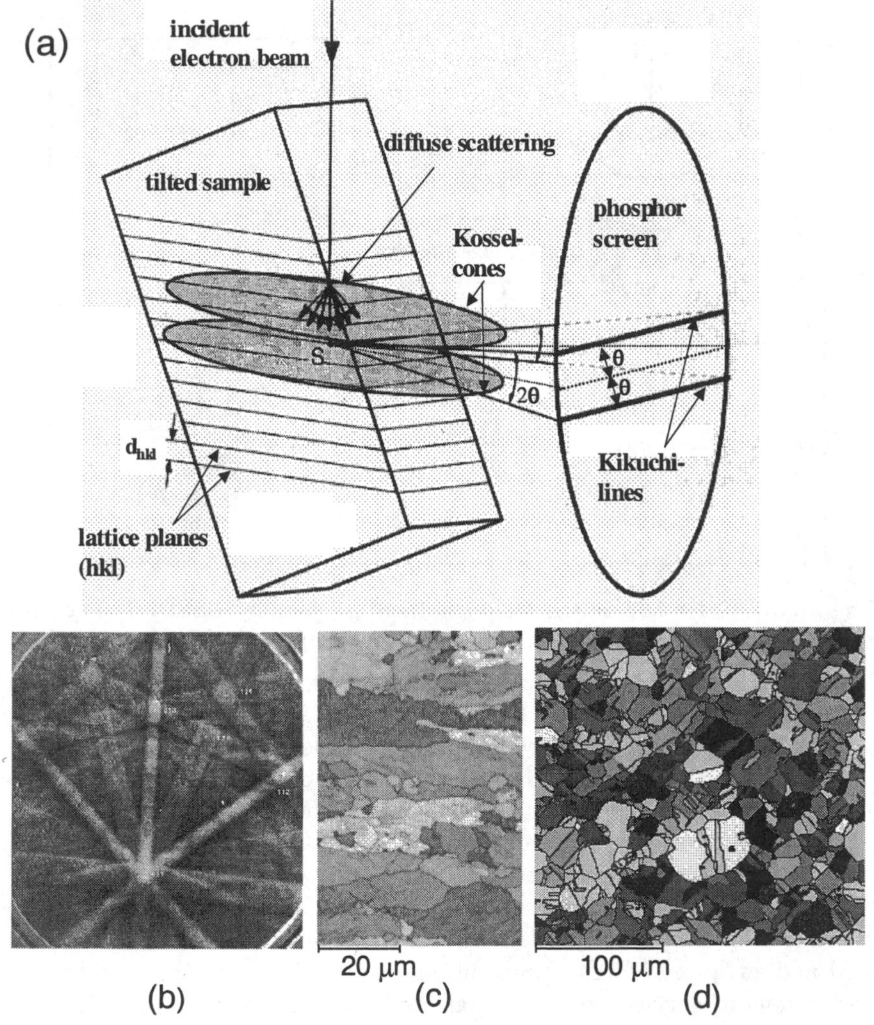

Fig. 2.53. (a) Principle of formation of Kikuchi bands by electron back scatter diffraction (EBSD). (b) EBSD pattern of an Aluminum crystal. (c) Orientation map of a 41% cold rolled and annealed (70s at 300°C) Al-4.5%Mg-0.14%Mn polycrystal. (d) Recrystallized Invar Steel (Fe-36%Ni).

3

Crystal Defects

3.1 Overview

Crystals are never free of defects. This is a fundamental consequence of equilibrium thermodynamics, as we will show in Section 3.2.2. Real crystals even have a defect structure far from thermodynamic equilibrium, due to the lack of mechanisms to establish thermodynamic equilibrium. We distinguish different kinds of crystal defects which are most easily classified according to their dimension; vacancies and interstitials (zero dimensional point defects), dislocations (one dimensional line defects), and grain and phase boundaries (two dimensional planar defects). Sometimes different phases are considered as three dimensional defects. However, these phases are constituents of the thermodynamic equilibrium and the real defect, the interface boundary, can be subsumed under the category of two dimensional defects.

It sounds odd, but it is the existence of these crystal defects which control materials properties and which, in particular, have allowed metallic materials to become the most important structural materials. For instance, plastic deformation consists of generation and motion of dislocations, diffusion controlled phase transformations need vacancies for diffusion, and recrystallization, i.e. softening during heat treatment of deformed materials, proceeds by generation and motion of grain boundaries. In this chapter we will introduce the thermodynamic foundations and the structure of crystal defects, together with their respective properties which will be addressed in their respective chapters, where they play a major role, i.e. diffusion, plasticity, recrystallization, and phase transformations in the solid state.

3.2 Point Defects

3.2.1 Types of point defects

Apart from solute atoms (impurities) which also represent disorder of the perfect crystal, there are principally two types of point defects, namely a vacant lattice site (vacancy) or the placement of an atom in an interstitial site (interstitial atoms). However, point defects can also occur in combination or in special configurations. The most important point defect is the vacancy, which is very important for diffusion (see Chapter 5). In metals a vacancy can be generated as a single defect. We can imagine the generation of a vacancy by a diffusional jump of a subsurface atom to the surface which leaves behind an empty lattice site. By subsequent diffusional jumps into nearest neighbor sites, it finally moves to the interior of the crystal so that vacancies become distributed homogeneously throughout the crystal. If an atom, however, is displaced from its regular lattice site to an interstitial site, a pair of point defects is produced, namely a vacancy and an interstitial site. This pair is referred to as a Frenkel defect. In ionic crystals a single vacancy can not occur by itself because of the constraint of charge neutrality. Therefore, either Frenkel defects or pairs of vacancies are generated (anion vacancy and cation vacancy), which are called Schottky defects. The various types of point defects are shown schematically in Fig. 3.1, however, in reality their structure is much more complicated.

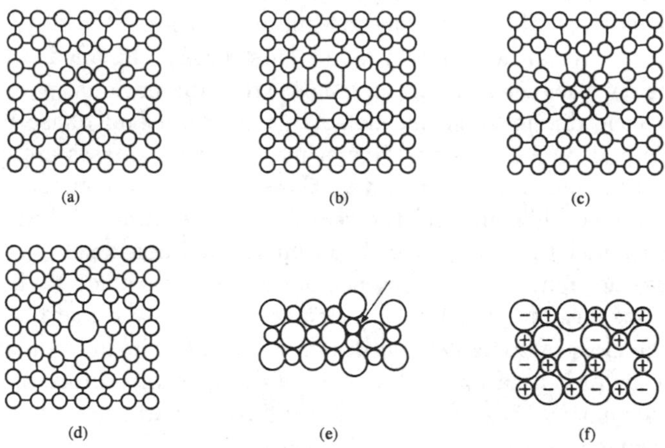

Fig. 3.1. The different types of point defects. (a) vacancy; (b) interstitial atom; (c) small substitutional atom; (d) large substitutional atom; (e) Frenkel defect; (f) Schottky defect (a pair of positive and negative ion vacancies)

In the environment of a point defect the atoms will rearrange to adapt to the perturbation of the perfect crystal. The interstitial atom usually does not exist as a single atom on an interstitial site, but rather two atoms share a lattice site, which is referred to as interstitial dumbbell (Fig. 3.2). It was also proposed that an interstitial atom can be forced into a densely packed crystallographic direction (crowdion), however, it is uncertain as to whether such a configuration really exists.

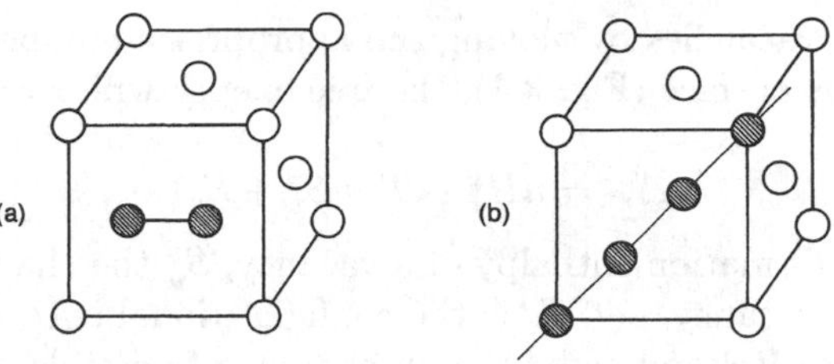

Fig. 3.2. Possible configurations of an interstitial atom: (a) ⟨100⟩ dumbbell; (b) crowdion.

3.2.2 Thermodynamics of point defects

In the following we want to determine the number n of point defects in thermodynamic equilibrium. For this we consider the example of vacancies in a crystal consisting of N sites (atoms). The ratio $n/N \equiv c^a$ is defined as atomic concentration of point defects. According to the first principle of thermodynamics

$$\delta Q = dU + pdV \tag{3.1}$$

The absorbed amount of heat δQ is converted to a change of the internal energy dU and the work of expansion pdV of the system at a given pressure p. The second principle of thermodynamics

$$dS \geq \frac{\delta Q}{T} \tag{3.2}$$

defines the entropy S, where equality holds for thermodynamic equilibrium. Combination of Eq. (3.1) and Eq. (3.2) yields

$$dU + pdV - TdS \leq 0 \tag{3.3}$$

If we define the Gibbs free energy G as

$$G = U + pV - TS \tag{3.4}$$

then

$$dG = dU - TdS - SdT + pdV + Vdp \tag{3.5}$$

At constant pressure p and constant temperature T, i.e. $dT, dp = 0$, we obtain from Eq. (3.3)

$$dG = dU - TdS + pdV \leq 0 \tag{3.6}$$

The free energy G, therefore, continuously decreases, and is at a minimum at thermodynamic equilibrium, i.e. $dG = 0$

$$dG = dU + pdV - TdS \equiv dH - TdS = 0 \tag{3.7}$$

If we generate n vacancies by moving the appropriate number of atoms from the volume to the surface (Fig. 3.3), the free energy will change by

$$\Delta G = nH_f^v - T\left(nS_v^v + S_c\right) \tag{3.8}$$

where H_f^v is the formation enthalpy of a vacancy, S_v^v the change of the vibration entropy per vacancy, and S_c is the configurational entropy. The latter is given according to Boltzmann by the number ω_n of possible configurations of n vacancies on N lattice sites

$$S_c = k \ln \omega_n \tag{3.9}$$

with the Boltzmann constant $k = 8.62 \cdot 10^{-5}\text{eV/K}$. For a single vacancy the possible number of configurations is equal to the number of lattice sites since one vacancy can be placed on any lattice site, i.e. $\omega_1 = N$. For two vacancies correspondingly $\omega_2 = N(N-1)/2$ and for n vacancies

$$\omega_n = \frac{N\left(N-1\right)\left(N-2\right)....\left(N-n+1\right)}{1 \cdot 2 \cdot 3 \cdot 4 ... n} = \frac{N!}{(N-n)!n!} \tag{3.10}$$

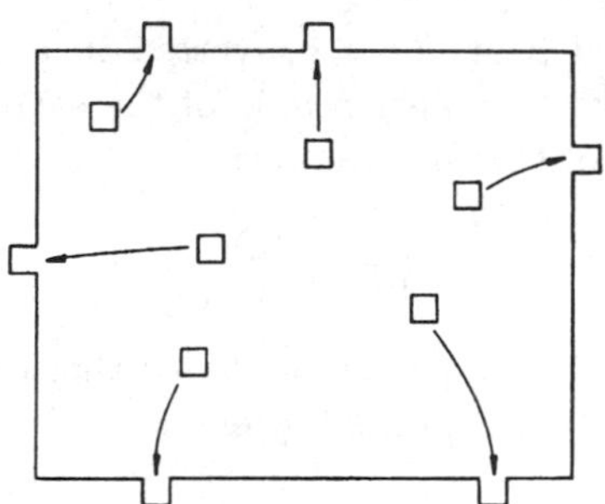

Fig. 3.3. How vacancies form: Atoms move to the surface.

The product in the denominator corrects for the cases which can not be distinguished, i.e. the exchange of two vacancies does not yield a new configuration. The function $f(x) = x!$ is mathematically difficult to handle analytically, but for $x \geq 5$ it can be approximated (Stirling's formula) by

$$\ln x! \cong x \ln x - x \tag{3.11}$$

The equilibrium number n of vacancies follows from the condition Eq. (3.7) in combination with Eq. (3.8)

$$\frac{d\left(\Delta G\right)}{dn} = H_f^v - TS_v^v - T\frac{dS_c}{dn} = 0 \tag{3.12}$$

From Eqs. (3.9) through (3.11) it follows

$$\frac{dS_c}{dn} = -k\{[\ln n + 1 - 1] - [\ln(N-n) + 1 - 1]\} = -k \ln \frac{n}{N-n} \tag{3.13}$$

Because $n \ll N$,

$$\frac{dS_c}{dn} \cong -k \ln \frac{n}{N} = -k \ln c_v^a \tag{3.14}$$

(c_v^a atomic vacancy concentration).

Inserting Eq. (3.14) into Eq. (3.12)

$$H_f^v - TS_v^v + kT \ln c_v^a = 0 \tag{3.15}$$

With the free energy of formation of a vacancy[1] $H_f^v - TS_v^v = G_f^v$, we obtain finally the equilibrium concentration of vacancies

$$c_v^a = \exp\left(-\frac{G_f^v}{kT}\right) \tag{3.16a}$$

or

$$c_v^a = \exp\left(\frac{S_v^v}{k}\right) \exp\left(-\frac{H_f^v}{kT}\right) \tag{3.16b}$$

Eq.(3.16a) holds also for any other type of point defect given its free energy of formation. The magnitude of the concentration is essentially determined by the enthalpy of formation H_f^v. Table 3.1 gives values of activation enthalpies for vacancies in some metals and the associated vacancy concentration at the melting temperature, which is about 10^{-4}, independent of material.

Table 3.1. Vacancy generation enthalpy (H_f^v), vibration entropy (S_ν^v - given in multiple of Boltzmann's constant k) and equilibrium concentration of vacancies (c_v^a) at melting temperature for selected metals.

	Au	Al	Cu	W	Cd
H_f^v [eV]	0.94	0.66	1.27	3.6	0.41
S_ν^v [k]	0.7	0.7	2.4	2.0	0.4
c_v^a [10^{-4}]	7.2	9.4	2.0	1.0	5.0

As a consequence vacancies always exist in a material in an appreciable concentration. Their existence cannot be avoided. For $T \to 0$ also $c_v^a \to 0$ in

[1] This terminology is slightly incorrect, since the free energy of vacancy formation also comprises the configuration entropy.

principle, but because of their vanishing mobility at low temperatures vacancies are unable to disappear so that a certain amount - albeit a very small concentration of vacancies - remains in the crystal. The free energy of formation of an interstitial atom is much larger (by about a factor of 3) than for the vacancy. Therefore, interstitial atoms essentially do not occur in thermodynamic equilibrium.

3.2.3 Experimental evidence of point defects

Point defects are perturbations of the ideal crystal. Therefore, they cause changes in the physical properties. Electrical resistivity is the easiest to measure. If a specimen is quenched from a high temperature T_q, the number of lattice defects at that temperature can be quenched-in and their contribution to the resistivity $\Delta\rho$ can be measured as an increase of the residual resistivity (see Chapter 10). If ρ_p is the resistivity increase per point defect, N the number of lattice sites, and if the increase of resistivity $\Delta\rho$ is proportional to the concentration of point defects c^a:

$$\Delta\rho = Nc^a\rho_p = N\rho_p \cdot \exp\left(\frac{S_f^v}{k}\right)\exp\left(-\frac{H_f^v}{kT}\right) \tag{3.17}$$

The enthalpy of formation H_f^v can be determined by measuring the increase in residual resistivity after quenching from different temperatures (Fig. 3.4). The increase of resistivity, however, does not yield information on the type of point defect, because both interstitials and vacancies will increase the electrical resistivity.

Simmons and Balluffi showed in a classical experiment that vacancies are the dominant point defects in thermal equilibrium. To prove this they concurrently measured the thermal expansion and the change of lattice parameter as function of temperature in gold. If vacancies are formed, atoms will be moved from the crystal interior to the surface as shown in Fig. 3.3 and increase the volume by one atomic volume Ω per vacancy. At the location of a vacancy the adjacent atoms will slightly relax in the free volume so that the volume per defect ΔV_D is decreased by $\Delta V_{vr}(< 0)$. The change of the total volume per defect then reads

$$\Delta V_D = \Delta V_{vr} + \Omega \tag{3.18}$$

The lattice parameter is affected only by ΔV_{vr}, because it is sensitive only to the volume of the elementary crystallographic unit cell and remains unaffected by the atoms on the surface.

For the change ΔV of the volume V_0 by n vacancies we obtain for a cube of length L_0

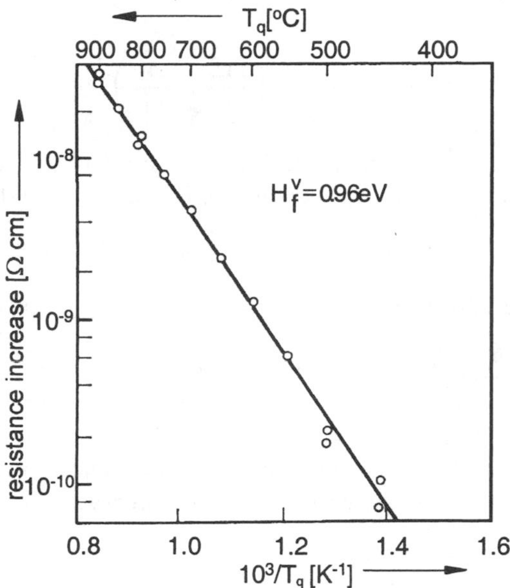

Fig. 3.4. Change in resistivity due to quenched in vacancies in Au. The formation enthalpy H_f^v can be derived from the slope of an Arrhenius plot (after [3.1].

$$\frac{\Delta V}{V_0} = \frac{n\,(\Delta V_{vr} + \Omega)}{V_0} = \frac{V - V_0}{V_0} = \frac{(L_0 + \Delta L)^3 - L_0^3}{L_0^3} \tag{3.19a}$$

$$= \frac{L_0^3 + 3L_0^2\Delta L + 3L_0\,(\Delta L)^2 + (\Delta L)^3 - L_0^3}{L_0^3} \cong 3\frac{\Delta L}{L_0}$$

neglecting higher order terms.

If the volume V_0 consists of m elementary cells, i.e. $V_0 = m \cdot V_{ec}$, then n/m is the number of vacancies per unit cell. The average volume change of a unit cell by n vacancies in the total volume is then $\Delta V_{ec} = n/m \cdot \Delta V_{vr}$, and, therefore,

$$\frac{n \cdot \Delta V_{vr}}{V_0} = \frac{m \cdot \Delta V_{ec}}{V_0} = \frac{(a_0 + \Delta a)^3}{a_0^3} \cong \frac{3\Delta a}{a_0} \tag{3.19b}$$

and

$$\frac{\Delta V}{V_0} - \frac{n \cdot \Delta V_{vr}}{V_0} = \frac{n\,(\Delta V_{vr} + \Omega)}{V_0} - \frac{n\Delta V_{vr}}{V_0} = \frac{n\Omega}{V_0} = 3\left(\frac{\Delta L}{L_0} - \frac{\Delta a}{a_0}\right) \tag{3.20a}$$

The number of lattice sites is $N = V_0/\Omega$ and, therefore, the vacancy concentration

$$c_v^a = \frac{n}{N} = 3\left(\frac{\Delta L}{L} - \frac{\Delta a}{a}\right) \tag{3.20b}$$

For vacancies $\frac{\Delta L}{L} > \frac{\Delta a}{a}$. If interstitial atoms were to be formed rather than vacancies, the observed macroscopic change of volume would be

$$\Delta V = n\left(\Delta V_{ir} - \Omega\right) < 0 \tag{3.21}$$

since an atom is removed from the surface to occupy an interstitial site and thus $\frac{\Delta L}{L} < \frac{\Delta a}{a}$.

Simmons and Balluffi measured ΔL and Δa as function of temperature simultaneously on a rod of gold. The experimental results (Fig. 3.5) prove that vacancies are formed in thermodynamic equilibrium. The enthalpy of formation of a vacancy can be determined from the measurements according to Eqs. (3.20b) and (3.16). They agree very well with results from measurements of the electrical resistivity.

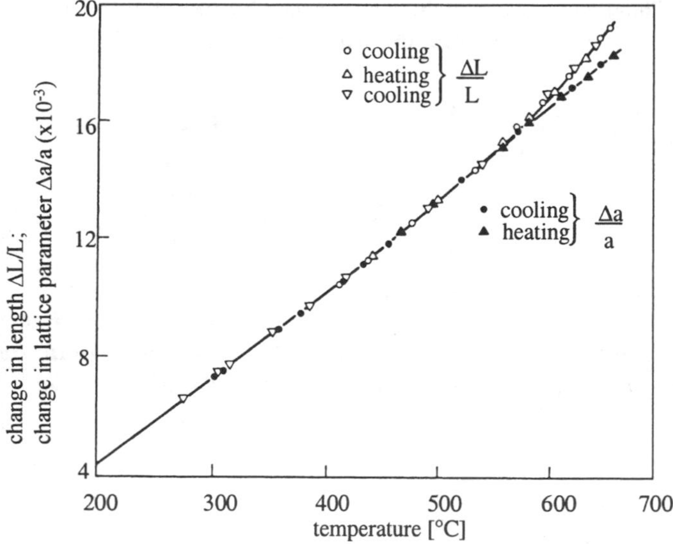

Fig. 3.5. Temperature dependence of length and lattice parameter of gold. The difference is proportional to vacancy concentration (after [3.2]).

More recently, more accurate methods to determine the vacancy concentration have become available, for instance positron annihilation. A positron has the same mass but opposite charge of an electron. If a positron meets an electron both particles decompose into γ-rays (hard X-rays). The interaction of a vacancy with a positron (Fig. 3.6) increases the life time of a positron in a material. This can be exactly measured by the emitted X-ray intensity, and, therefore, the vacancy concentration can be determined very accurately.

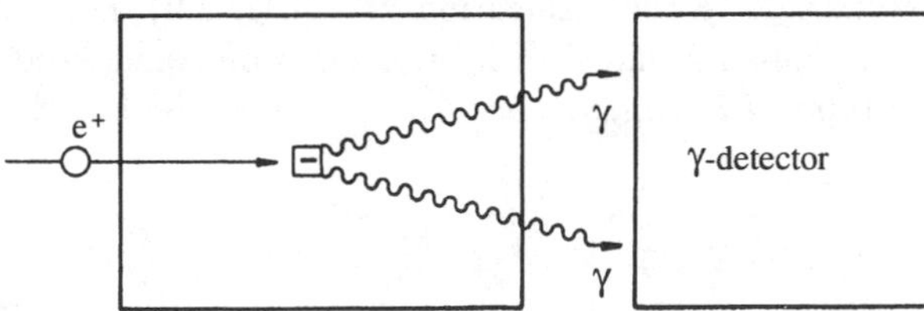

Fig. 3.6. Illustration of the principle of detecting vacancies with positrons. A vacancy has the same effect as a missing ion core and corresponds to a free negative charge, which reacts with a positron.

Point defects can also be generated by irradiation of a material with high energy elementary particles, for instance by electrons, protons, neutrons, or heavy ions. If a material is irradiated by electrons with an energy beyond a certain threshhold energy (for instance 400 kV for copper), one Frenkel pair will be produced per electron. By irradiation with high energy heavy elementary particles, so-called collision cascades will be produced with complicated defect arrangements. These defects occur in nuclear reactors and are liable to damage the material of the reactor container. The generation, properties, and control of point defects is very important in this case and, therefore, was the subject of extensive research in the 1960s when nuclear reactors were considered to be a vital prime source of energy.

Besides these structural point defects there are also special defect configurations in insulators, the so-called color centers, since they change the color of an insulator. These color centers will be addressed in Chapter 10 in the context of optical properties of solids.

3.3 Dislocations

3.3.1 Geometry of dislocations

Dislocations are perturbations of the perfect crystal along a line. The type of dislocation most easy to visualize is a line where a crystallographic plane terminates in a crystal (Fig. 3.7). This terminal line of the partial crystallographic plane in the crystal is also referred to as an edge dislocation. Alternatively, such an edge dislocation can be considered to be generated by cutting a crystal partially along a plane, displacing the two separated parts of the crystal perpendicular to the terminal line of the cut and subsequent reattaching both parts of the crystal. Another type of dislocation is obtained if both parts of the crystal are not displaced perpendicular but parallel to the terminal line of the cut. This generates a screw dislocation (Fig. 3.8). If one makes a circuit on a plane perpendicular to the line of and encircling the screw dislocation, one never returns to the starting point but rather one moves on a spiral around the dislocation line. The displacement of the separated crystallites in the plane of cut can be at any angle to the dislocation line (Fig. 3.9), i.e. neither perpendicular nor parallel. Such a mixed dislocation can be considered as composed of the two basic types, i.e. edge dislocation and screw dislocation.

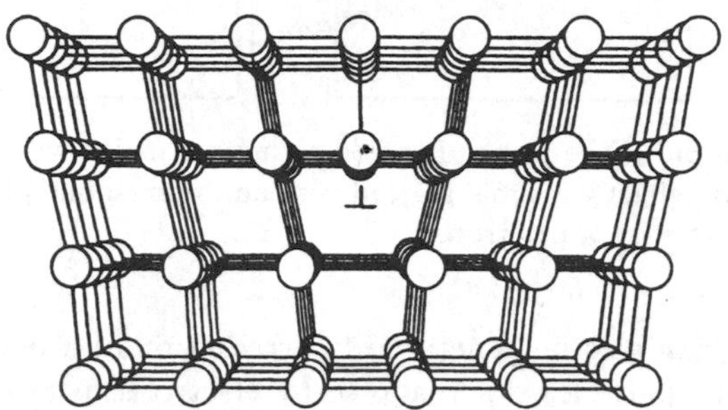

Fig. 3.7. Atomic configuration at an edge dislocation.

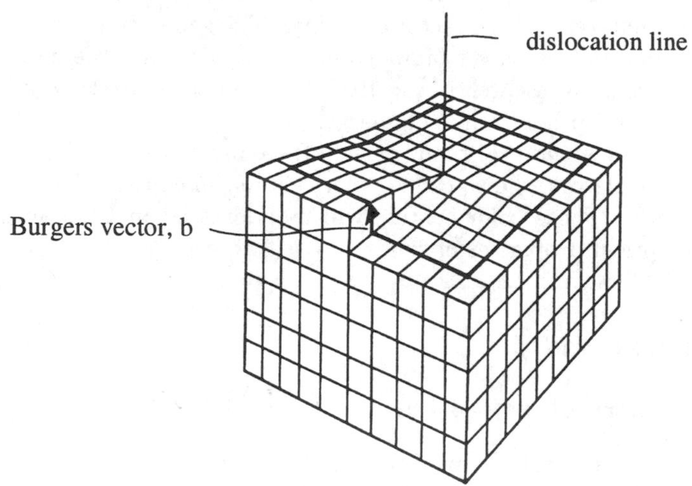

Fig. 3.8. Atomic configuration of a screw dislocation.

A dislocation is characterized by its line element **s** and its Burgers vector **b**. The line element is the unit vector tangential to the dislocation line. If the dislocation line is curved, **s** will change along the dislocation line. The Burgers vector has the length and direction of the vector by which the two parts of the crystal above and below the plane of motion of the dislocation are displaced

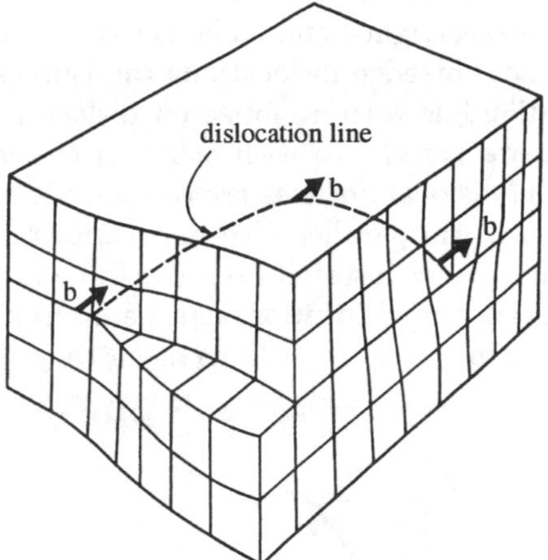

Fig. 3.9. A dislocation line that changes its character from screw dislocation to edge dislocation.

with respect to each other. It can be defined exactly by the Burgers circuit (Fig. 3.10). For this, the arrangement of lattice sites in a plane perpendicular to the dislocation line is drawn for the perfect and the dislocated crystal. After defining the direction of the dislocation line, a closed circuit around the dislocation line clockwise (right-hand screw) around the positive direction of the dislocation line (Fig. 3.10a) is created.

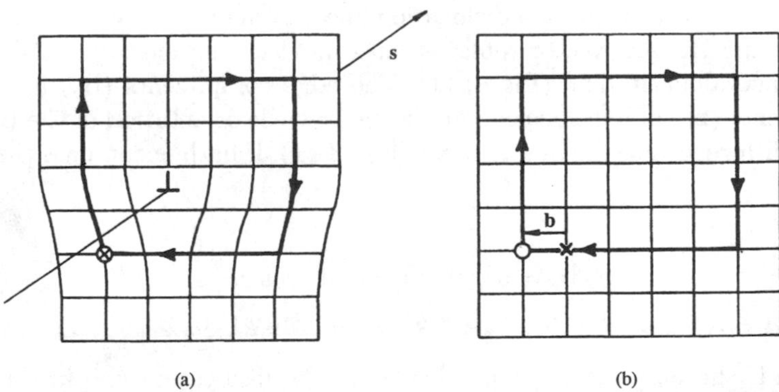

Fig. 3.10. Definition of the Burgers vector **b** by a Burgers circuit. (a) **s** indicates the direction of the dislocation line. (b) Circle and cross indicate start and finish of the circuit.

If the same circuit is then drawn in the perfect crystal (Fig. 3.10b), the start and finish points of the circuit are not identical. The difference vector between finish and start is the Burgers vector. The rule is also referred to as FS/RH rule (finish-start/right-hand). The Burgers vector does not change along a dislocation line. For edge dislocations the Burgers vector is perpendicular to the dislocation line whereas for screw dislocations the Burgers vector and line element are parallel to each other. If either Burgers vector or line element of parallel dislocation lines are of opposite sign, they are called antiparallel dislocations. Antiparallel edge dislocations can be visualized as partial crystallographic planes inserted from the top and from the bottom or vice versa. Antiparallel screw dislocations can be discriminated by the sense of their screw. If two antiparallel dislocations meet, they annihilate each other (see Chapter 7).

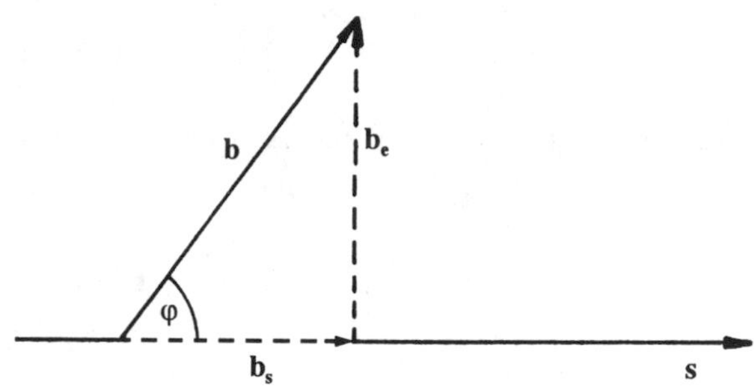

Fig. 3.11. The Burgers vector **b** can be broken down into its screw ($\mathbf{b}_s$) and edge ($\mathbf{b}_e$) components.

Since the line element of a dislocation line can change along the dislocation line, but the Burgers vector remains constant, the character of a dislocation can change along its line (Fig. 3.11). The edge components ($\mathbf{b_e}$) and screw components ($\mathbf{b_s}$) of a dislocation are defined by the orientation of the Burgers vector with regard to the line element (Fig. 3.11). If both encompass the angle φ then

$$\mathbf{b}_s = \mathbf{s} \cdot (\mathbf{b} \cdot \mathbf{s}) = (|\mathbf{b}| \cdot \cos\varphi) \cdot \mathbf{s} \tag{3.22a}$$

$$\mathbf{b}_e = \mathbf{s} \times (\mathbf{b} \times \mathbf{s}) = (|\mathbf{b}| \cdot \sin\varphi) \cdot \mathbf{n} \tag{3.22b}$$

where **n** is the unit vector perpendicular to the dislocation line in the plane defined by **s** and **b**.

It follows from the definition of a dislocation line as the terminal line of a plane of cut or of a partial plane in a crystal that a dislocation line can never end in a crystal. However, a dislocation can occur in a configuration of a closed loop in a crystal, without being in contact with the surface. During plastic deformation dislocations are mainly generated as dislocation loops (see Chapter 6). The creation of a dislocation loop can be visualized by making a cut in the interior of the crystal along a plane. For instance consider the plane defined by ABCD in Fig. 3.12a, and displace the top part of the crystal by a vector **b** with regard to the bottom part. After reattaching the two parts of the crystal, the boundary of this cut (ABCD in Fig. 3.12a) represents a dislocation loop. The dislocation loop has the same Burgers vector everywhere, namely the vector of displacement of the planes above and below the cut. The character of the dislocation is defined by the orientation of the dislocation line with respect to the Burgers vector. The loop in Fig. 3.12, for instance, consists of the edge dislocations AB and CD (Fig. 3.12b) and the screw dislocations BC and DA (Fig. 3.12c). Both edge and screw segments are antiparallel dislocations. If we move along the dislocation line, for instance clockwise, the line elements in the sections AB and CD or BC and DA are opposite, while the Burgers vector remains the same as we circumnavigate the loop.

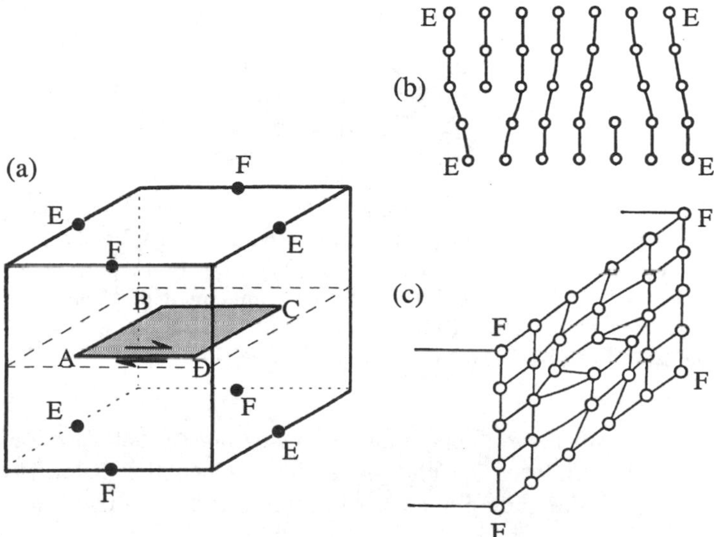

Fig. 3.12. (a) Representation of a closed dislocation loop of rectangular shape. A section is made in the ABCD plane, and the atoms on both faces of the section are shifted parallel to the plane of section. Then the surfaces are joined again. (b) Representation of the arrangement of atoms in the EEEE-plane. (c) Representation of the arrangement of atoms in the FFFF plane.

Since a closed dislocation loop has to change its direction in at least two points, and for the entire loop the Burgers vector is constant, a dislocation loop can never be composed only of screw dislocations. In contrast a dislocation loop can consist only of edge dislocations, if its Burgers vector is parallel to the plane of the loop. The corresponding dislocation loop is equivalent to an inserted or removed partial plane (Fig. 3.13).

Such dislocations are referred to as Frank dislocations or prismatic dislocation loops.

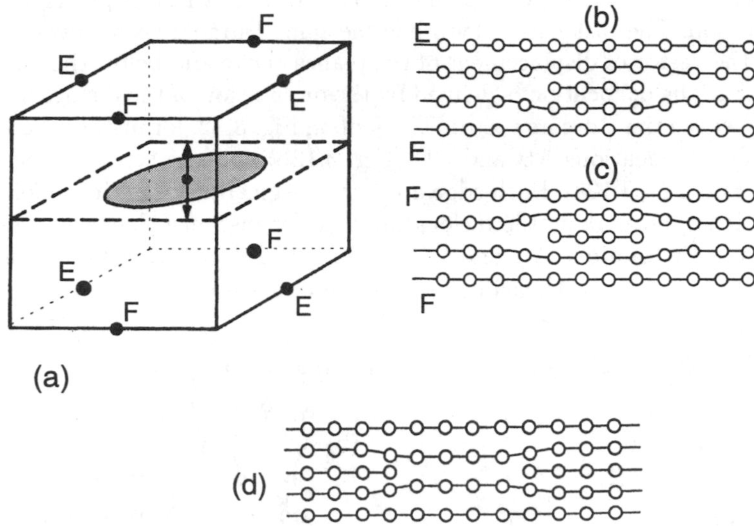

Fig. 3.13. (a) Representation of a prismatic dislocation loop. A cut has been made along the shaded plane and the surfaces of both sections have been separated. The gap has been filled with atoms. (b) Representation of the arrangement of atoms in the EEEE plane. (c) Representation of the arrangement of atoms in the FFFF plane. (d) Section through a region with a prismatic dislocation loop of opposite sign. Atoms have been removed from the plane of section.

Dislocations can move. Their motion causes plastic deformation in crystalline solids (see Chapter 6). The plane along which the dislocation line is displaced is called the glide plane (or slip plane) and its normal $\mathbf{m}$ is defined by

$$\mathbf{m} = \mathbf{s} \times \mathbf{b} \tag{3.23}$$

Accordingly, screw dislocations ($\mathbf{s}||\mathbf{b}$) do not have a defined slip plane and can change their slip plane (cross slip) (Fig. 3.14). The slip plane of prismatic dislocations changes along the dislocation line and, therefore, prismatic dislocations are immobile. Edge dislocations and mixed dislocations have a defined slip plane according to Eq. (3.23). They can leave the slip plane only by absorption or emission of point defects (climb), for instance vacancies (Fig. 3.15). The absorbed vacancies are extracted from the volume. Therefore, dislocations are sinks for vacancies. By inversion of the process vacancies can be generated at dislocations (vacancy sources). Since the generation or removal of point defects changes the volume of a crystal, climb is also referred to as non-conservative dislocation motion. The motion of dislocations and their elastic properties will be discussed in detail in Chapter 6.

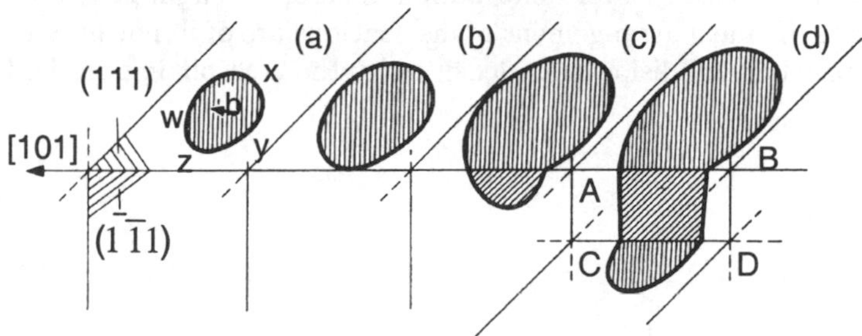

Fig. 3.14. Schematic representation of a dislocation slipping laterally. A screw dislocation at "z" can slip in both (111) and (1$\bar{1}$1). In (d) double cross slip is illustrated.

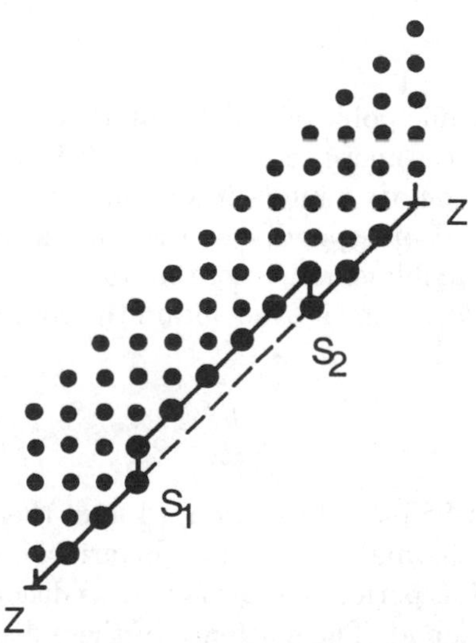

Fig. 3.15. An edge dislocation climbing by absorption of vacancies.

The number of dislocations is quantified by the dislocation density ρ. The dislocation density is defined by the total length of the dislocation lines per unit volume. Therefore, it has the dimension $[m/m^3] = [m^{-2}]$. If all dislocations are straight and parallel, the dislocation density exactly corresponds to the number of points of intersection of dislocations per unit area of the crystal surface. In this case the dislocation density ρ is related to the dislocation spacing by $\rho = 1/d^2$ (Fig. 3.16). This definition of dislocation density and the corresponding assumptions are used to determine the dislocation density by etch pitting (see Section 3.3.2). The dislocation density, in general, is difficult to determine, especially after deformation. Therefore, slight geometrical deviations from the ideal arrangement of dislocations are of minor importance for the measurements of dislocation densities by the etch pitting method.

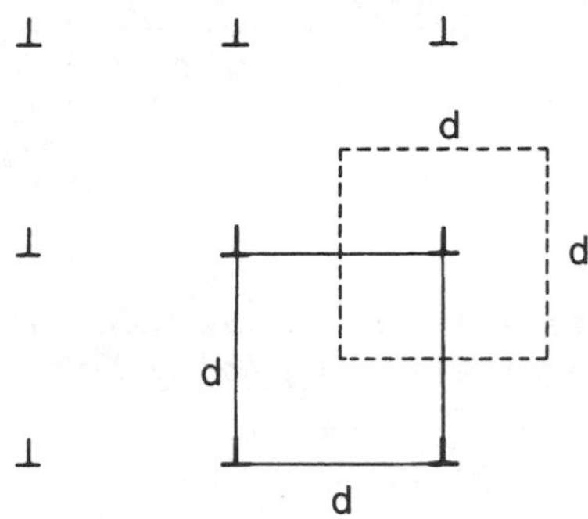

Fig. 3.16. If dislocations are arranged with mean spacing d, the dislocation density is $\rho = 1/d^2$.

From a thermodynamic point of view an annealed crystal ought to be free of dislocations, because the energy per atom of a dislocation line is comparable to the energy of formation of an interstitial atom (for instance about 5 eV for copper, see Chapter 5). Correspondingly, the concentration c_d^a of dislocation atoms ought to be negligibly small according to Eq. (3.16). If the spacing of atoms along the dislocation line is b we obtain the concentration of dislocated atoms for a dislocation density ρ

$$c_d^a = \frac{\rho/b}{1/b^3} = \rho b^2 \tag{3.24}$$

Because of the high enthalpy of formation, c_d^a and, therefore, also ρ ought to be negligibly small in thermal equilibrium. Invariably, however, even in very carefully annealed and in perfectly grown crystals dislocation densities of the order of 10^{10} m^{-2} are found. The existence of these dislocations is caused by the process of crystal growth and it is difficult to remove dislocations once they are generated, since they are in a state of mechanical equilibrium although in thermal non-equilibrium. A real crystal practically never attains thermal equilibrium.

3.3.2 Evidence of dislocations

By high resolution imaging techniques the atomic structure of crystals and surfaces can be revealed and, correspondingly, also crystal defects. Edge dislocations can be imaged by atomic resolution transmission electron microscopy (Fig. 3.17a) and screw dislocations are revealed at surfaces by scanning tunneling microscopy (3.17b). Such measurements require difficult specimen preparation and, thus, are only conducted if necessary. However, even with conventional transmission electron microscopy (TEM) dislocations can be imaged easily (Fig. 3.18a) because they appear as dark lines in a bright field image. The reason for these dark lines is the distortion of the lattice planes in the vicinity of the dislocation core, by which the transmitting electrons satisfy locally the Bragg condition for diffraction (Fig. 3.18b). The diffracted electron beam degrades the intensity of the transmitted beam so that material in the vicinity of the dislocation appears dark in the transmitted image.

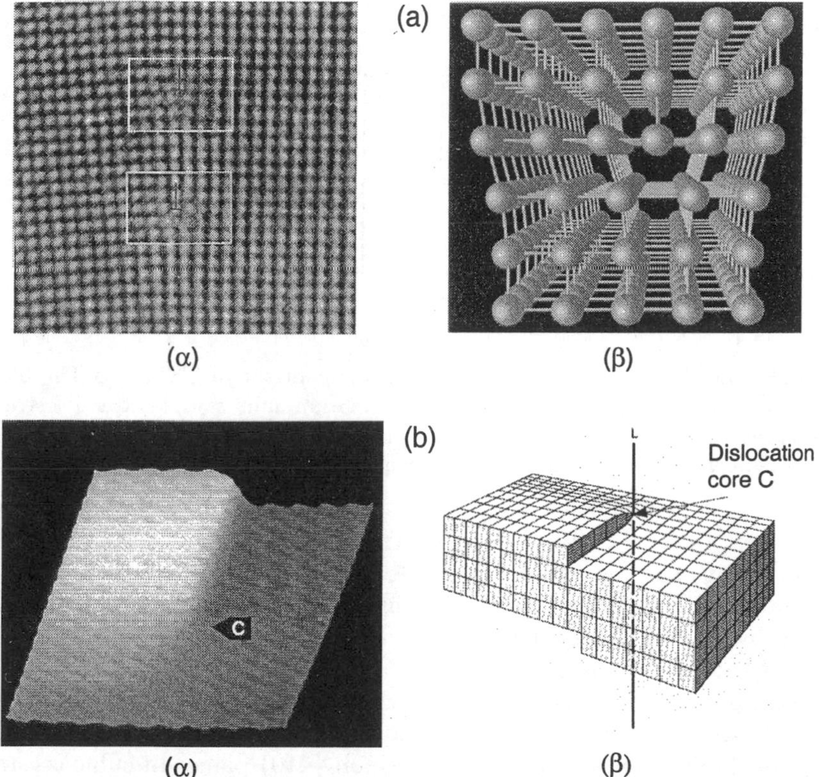

Fig. 3.17. a: (α) Atomic resolution TEM micrograph of $SrTiO_3$ crystal containing two $\langle 100 \rangle$ edge dislocations marked in the figure [3.3]. (β) Schematic atomic arrangement of an edge dislocation. b: (α) Atomic resolution micrograph obtained on the surface of a GaAs crystal by scanning tunneling microscopy revealing a screw dislocation at location c [3.4]. (β) Schematic atomic arrangement of a screw dislocation. (The arrangement in (β) is rotated when compared to the image in (α).)

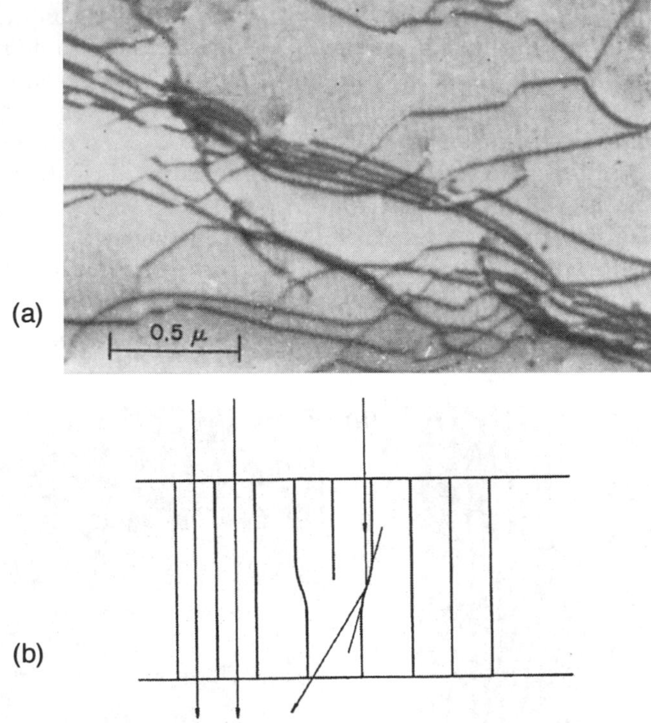

Fig. 3.18. Imaging of dislocations by amplitude contrast in TEM. (a) The black lines in the bright field are dislocation lines in an aluminum sample after 1% strain. (b) How the core of a dislocation causes diffraction contrast. The curvature of lattice planes near dislocations diffracts electrons, weakening the primary transmitted beam.

Dislocations can also be made visible by light optical microscopy. Since the points of intersection of dislocation lines with a surface are preferentially attacked by specific chemical solutions, so-called etch pits are formed on the surface, which are easily recognized in the optical microscope (Fig. 3.19). These etch pits have a shape that is characteristic of the crystallography of the etched surface. For instance, they form equilateral triangles on {111} planes, squares on {100} planes, and cube on edge on {110} planes in cubic crystals (Fig. 3.19). Deviations from the ideal orientation will distort the geometry of the etch pits in a characteristic way so that the orientation of a surface can be determined from the shape of the etch pits. For not too high dislocation densities, for instance in annealed or slightly deformed crystals, etch pits are the simplest method to determine the dislocation density.

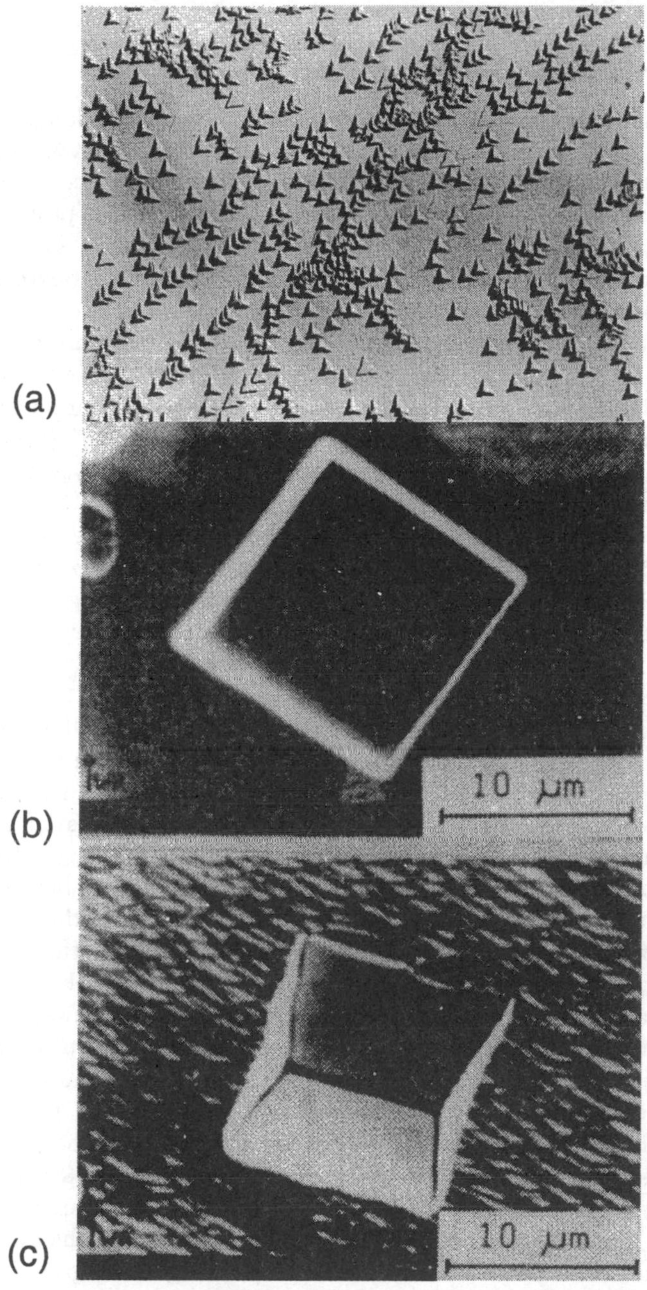

Fig. 3.19. Etch pits on a {111} surface of bent copper (a); on a {100} surface (b) and a {110} surface (c) in recrystallized Al-0.5%Mn.

3.4 Grain Boundaries

3.4.1 Terminology and Definitions

Grain boundaries are the lattice defects which have been longest known but which are least understood. A grain boundary separates two regions of the same crystal structure but of different orientation. In very coarse grained materials it can be discerned by the naked eye if the surface is properly prepared (Fig. 3.20a). Our deficiency of fundamental knowledge of grain boundaries is mainly due to their complex structure, which requires an extensive mathematical description for its macroscopic characterization. Even in 2-dimensions four parameters are needed to define mathematically a grain boundary (Fig. 3.20b), namely a rotation angle φ that describes the orientation difference between the adjacent crystals (orientation relationship), an angle Ψ that defines the spatial orientation of the grain boundary "plane" (grain boundary orientation) with respect to one crystal and the components t_1, t_2 of the translation vector $\mathbf{t}$ that characterizes the displacement of the two crystals with respect to each other. In 3-dimensions (the real case) one even needs eight parameters to unambiguously define a grain boundary, namely three terms for the orientation relationship, for instance the Euler angles $\varphi_1, \Phi, \varphi_2$, two parameters for the spatial orientation of the grain boundary plane by means of the normal to the grain boundary plane $\mathbf{n} = (n_1, n_2, n_3)$, with respect to one of the adjacent crystals (keeping in mind that $|\mathbf{n}| = \mathbf{1}$) and finally the three components of the translation vector $\mathbf{t} = (t_1, t_2, t_3)$. The properties, in particular energy and mobility of a grain boundary, are, in principle, a function of all eight parameters. Five of these eight can be influenced externally, i.e. orientation relationship and spatial orientation of the grain boundary. The translation vector will be constrained by the atomic nature of the crystals so as to minimize the total energy; however, the vector $\mathbf{t}$ need not be unambiguous as evident from computer simulations. To determine the dependency of grain boundary properties, for instance the mobility, on the 5 macroscopic parameters, it would be necessary to keep all parameters but one fixed and to systematically vary that free parameter. In real life, however, only a few of the external parameters are systematically varied, and usually this is the orientation relationship in terms of a fixed rotation axis and a variable angle of rotation, and, for a given rotation axis and rotation angle, the inclination of the grain boundary with respect to a reference position. The grain boundary energy as function of the orientation of the grain boundary plane is commonly represented in the "Wulff plot". The "Wulff plot" is constructed by drawing a line from the origin of the coordinate system in the direction of the grain boundary normal with a length corresponding to the grain boundary energy (Fig. 3.21). Hence, points closer to the origin represent low energy grain boundaries. If there are low energy boundary orientations such as coherent twin boundaries, it is likely that a grain boundary will tend to align itself parallel to such orientations, at least piecewise to minimize its total energy (facetting).

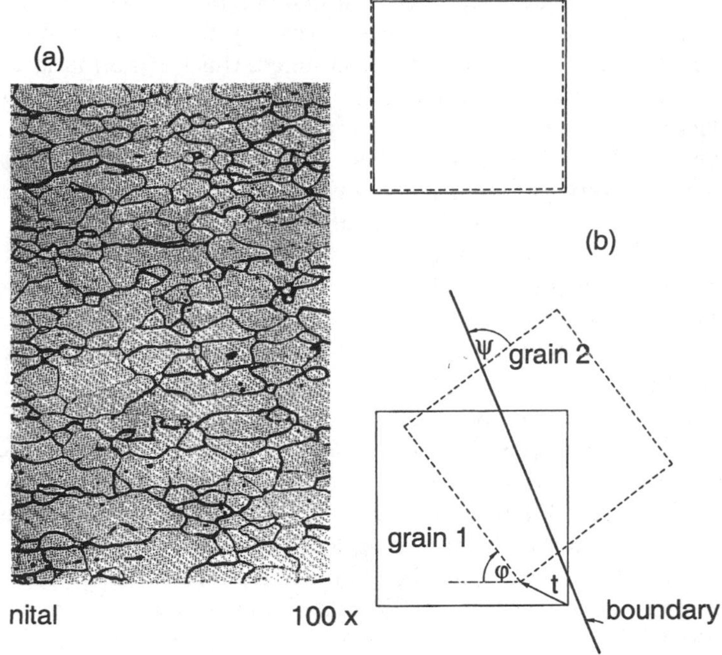

Fig. 3.20. (a) Partially etched grain boundaries in steel. (b) Four parameters are needed to describe mathematically a two-dimensional grain boundary.

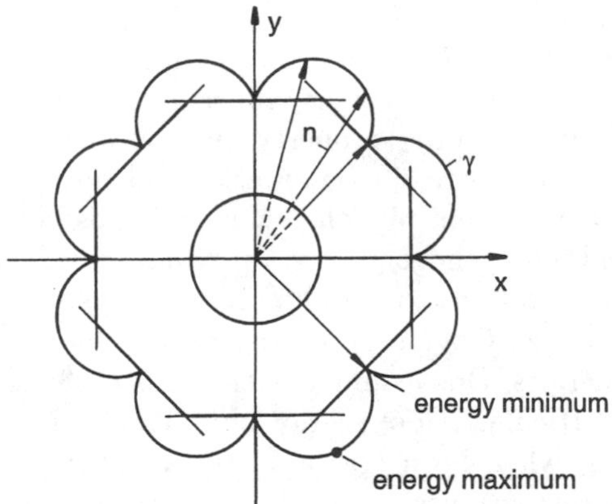

Fig. 3.21. A two-dimensional "Wulff-plot". The inner circle is the circle of orientation of the axes normal to grain boundaries. The outer curve indicates the corresponding grain boundary energy γ. The inner polygon represents the energetically most advantageous grain shape for a given grain boundary energy pattern (facetting).

The orientation relationship between two crystal lattices is a transformation, which has to be applied to one of the crystals to make both crystal lattices coincide. If a common origin is assumed, this transformation is a pure rotation, since the relative positions of the crystal axes in both crystals are the same. There are many ways to define a rotation. Frequently the three Euler angles are used, but it is easiest to picture a grain boundary when the rotation is represented in terms of an axis and an angle of rotation. In many instances it is very important to know the dependency of a property on the rotation angle for a given rotation axis. In this case it would be desirable to keep the crystallographic orientation of the grain boundary plane constant in order to obtain the dependency on the rotation angle only. If the grain boundary plane is perpendicular to the rotation axis, the boundary is referred to as a twist boundary (Fig. 3.22a). In this case the choice of the grain boundary plane is unambiguous, no matter what the rotation angle is. Grain boundaries are called tilt boundaries when the rotation axis is parallel to the grain boundary plane. Since there is an infinite number of possible planes parallel to a given direction, there is an infinite number of tilt boundaries for a given rotation. If the adjacent crystals are mirror images of each other (mirror plane = grain boundary plane), then the grain boundary is referred to as a symmetric tilt boundary (Fig. 3.22c). All other tilt boundaries are called asymmetric tilt boundaries (Fig. 3.22b). In a symmetric tilt boundary the grain boundary plane has equivalent Miller indices with respect to both adjacent crystals; for instance in Fig. 3.22c the boundary normal would be $(210)_1$ and $(\bar{2}10)_2$ for a $36.9°[001]$ symmetric tilt boundary. By definition, the normal to the grain boundary plane must be perpendicular to the rotation axis for tilt boundaries. It is impossible to keep the Miller indices of the grain boundary plane constant when changing the angle of rotation. At least with respect to one of the crystals they have to change in the asymmetric tilt boundary. Therefore, to investigate boundary structure - property relationships, it is sensible to confine oneself first to symmetric tilt boundaries and then to treat asymmetric tilt boundaries as a deviation from the symmetric position.

As an example the dependency of grain boundary energy on the angle of rotation for a $\langle 110 \rangle$ rotation axis in aluminum is given in Fig. 3.23 for symmetrical tilt boundaries. Obviously, there are orientation relationships with particularly low energies. One example is the $70.5°\langle 110 \rangle$ orientation relationship, which is a particularly low energy boundary, namely a coherent twin boundary. However, when both crystals have a $\{311\}$ plane parallel to the grain boundary, a very low grain boundary energy is also observed in many fcc metals.

3.4.2 Atomic Structure of Grain Boundaries

3.4.2.1 Low angle grain boundaries

If the misorientation between adjacent grains is small (low angle grain boundary), then the boundary is entirely comprised of a periodic crystal dislocation

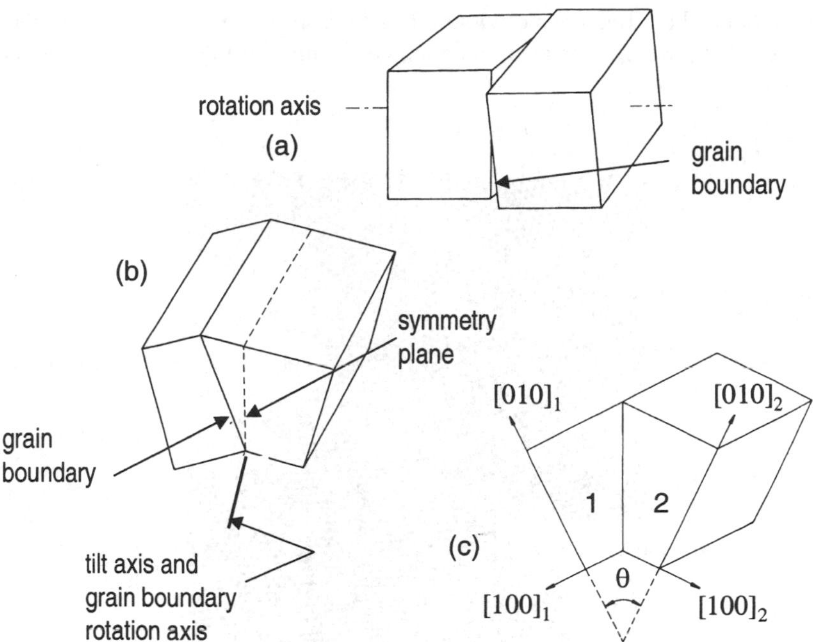

Fig. 3.22. Orientation of grain boundaries and rotation axes for different types of grain boundaries. (a) twist boundary; (b) asymmetrical tilt boundary; (c) symmetrical tilt boundary.

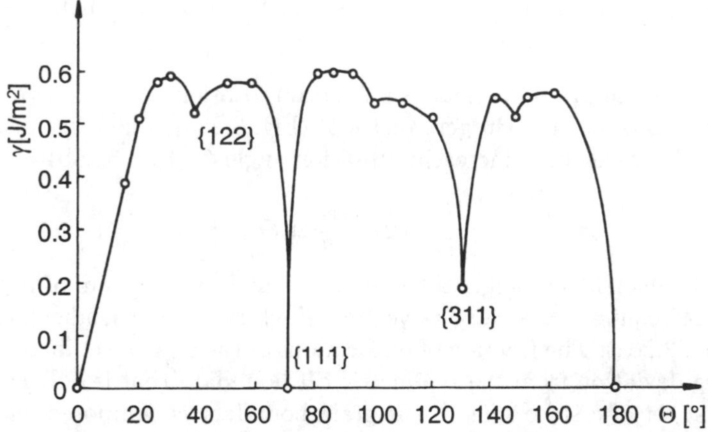

Fig. 3.23. Energy dependence of symmetrical ⟨110⟩ tilt boundaries in Al on the tilt angle Θ. The indices given in the figure are Miller indices of the corresponding grain boundary planes (see text) (after [3.5]).

arrangement. This becomes obvious already from a simple bubble model of a boundary (Fig. 3.24), but also has been confirmed by high resolution electron microscopy.

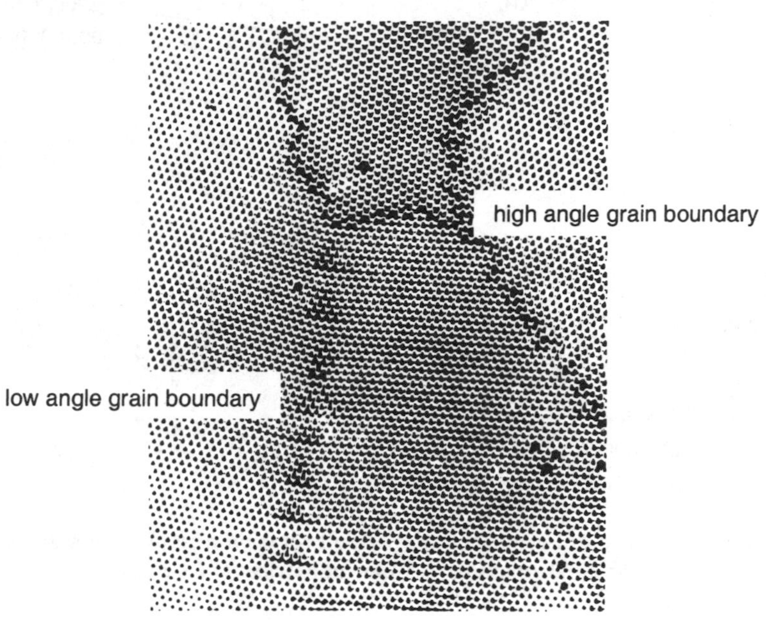

Fig. 3.24. High-angle and low-angle tilt boundaries in a soap bubble model.

Symmetric tilt grain boundaries for certain rotation axes consist of a single set of edge dislocations (Burgers vector b) (Fig. 3.25a), where the dislocation spacing d decreases with increasing rotation angle Θ (Fig. 3.25b).

$$\frac{b}{D} = 2\sin\frac{\Theta}{2} \cong \Theta \tag{3.25}$$

For asymmetric low angle tilt boundaries at least two sets of edge dislocations are required, the Burgers vectors of which are perpendicular to each other (Fig. 3.26a). The fraction of dislocations of the second set increases with increasing deviation from the symmetric tilt boundary that is only comprised of the first set of dislocations. If the grain boundary is composed only of the second set of dislocations, another symmetric tilt boundary is obtained (Fig. 3.26b) which is perpendicular to the symmetric tilt boundary comprised only of the first set of dislocations (Fig. 3.25a). Low angle twist boundaries require at least two sets of screw dislocations (Fig. 3.27a). Mixed boundaries are composed of dislocation networks of three Burgers vectors. This dislocation model of low angle grain boundaries, which is due to Read and Shockley, has been

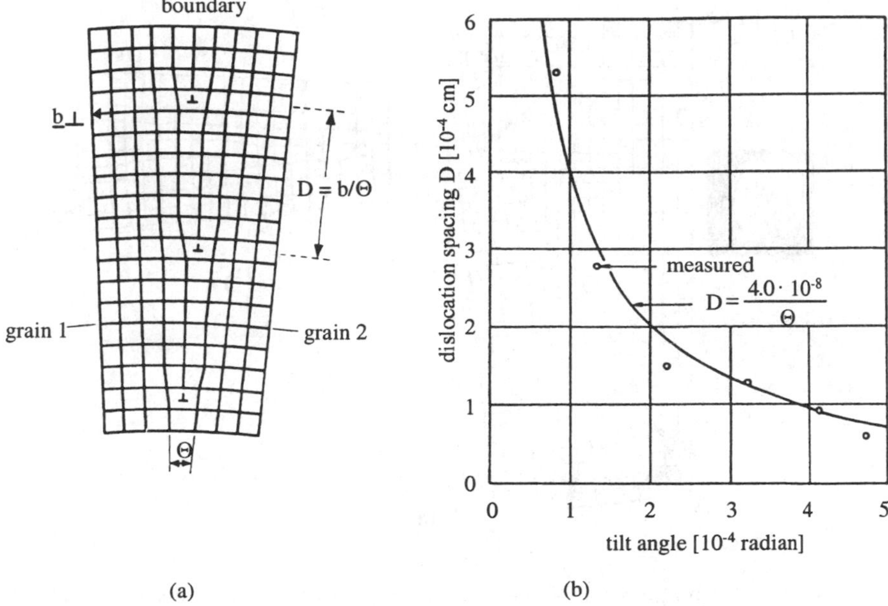

Fig. 3.25. (a) Configuration of the dislocations forming a symmetrical ⟨100⟩ low-angle grain boundary with tilt angle Θ in a simple cubic crystal. (b) Measured and calculated values of dislocation spacing in a symmetrical low-angle tilt boundary in Germanium.

frequently confirmed by transmission electron microscopy (Fig. 3.27b) and is supported by measurements of grain boundary energy (Fig. 3.28).

3.4.2.2 High angle grain boundaries

For small angles of rotation, the specific grain boundary energy (energy per unit area) increases exactly as predicted by the dislocation model. For angles of rotation in excess of 15°, however, measurements of grain boundary energy reveal no further change with increasing rotation angle, in contrast to the dislocation model which would predict an energy decrease for large angles of rotation (Fig. 3.28). For rotation angles larger than 15° the dislocation model fails, because the dislocation cores tend to overlap. Thus, the dislocations lose their identity as individual lattice defects. Therefore, grain boundaries with rotation angles in excess of 15° are distinguished from the low angle grain boundaries and are termed high angle grain boundaries. At first glance a high angle grain boundary appears to be a zone with a random atomic arrangement (Fig. 3.24). In fact, early models of grain boundary structure assumed that grain boundaries were an undercooled liquid. We will show below that this is a misconception and that grain boundaries do have a well defined atomic

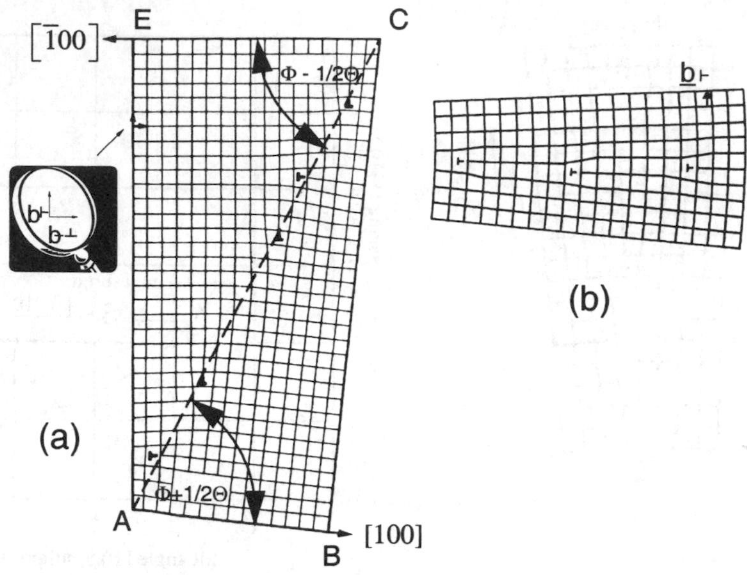

Fig. 3.26. (a) Configuration of the dislocations forming an asymmetrical low-angle tilt boundary with tilt angle Θ and inclination Φ. (b) Symmetrical low-angle grain boundary formed exclusively from 2nd family dislocations.

structure. Our current understanding of the structure of high angle grain boundaries has been derived from geometrical concepts, based on dislocation models of low angle grain boundaries. A fundamental reason for the failure of the lattice dislocation concept for larger angles of rotation is the requirement of a strictly periodic dislocation arrangement to minimize grain boundary energy. The spacing of dislocations, however, changes discretely, namely at least by one atomic distance. As a consequence the angle of rotation $\Theta = b/d$ also changes in steps rather than continually. For small angles $b \ll D$ so that Θ changes quasi-continuously. For larger rotations, however, the orientation difference between two consecutive periodic dislocation arrangements becomes substantial. If, for instance, there is an arrangement with a dislocation spacing $d = 4b$, then $\Theta = 14.3°$. Changing the dislocation spacing to 3 interatomic distances requires an angle of rotation of $\Theta = 19.2°$. Therefore, the problem arises of what is the grain boundary structure for $14.3° < \Theta < 19.2°$, or, in general, between rotations that represent a periodic arrangement of (primary) crystal dislocations?

In a perfect crystal the atoms have a defined (average) position, which is determined by the minimum of the free energy. Any deviation from this position necessarily increases the free energy. Therefore, it can be assumed that the crystal will try to keep the atoms as much as possible in their ideal position, which must also be the case in the grain boundary. There are orientation relationships where certain crystallographic planes continue through the

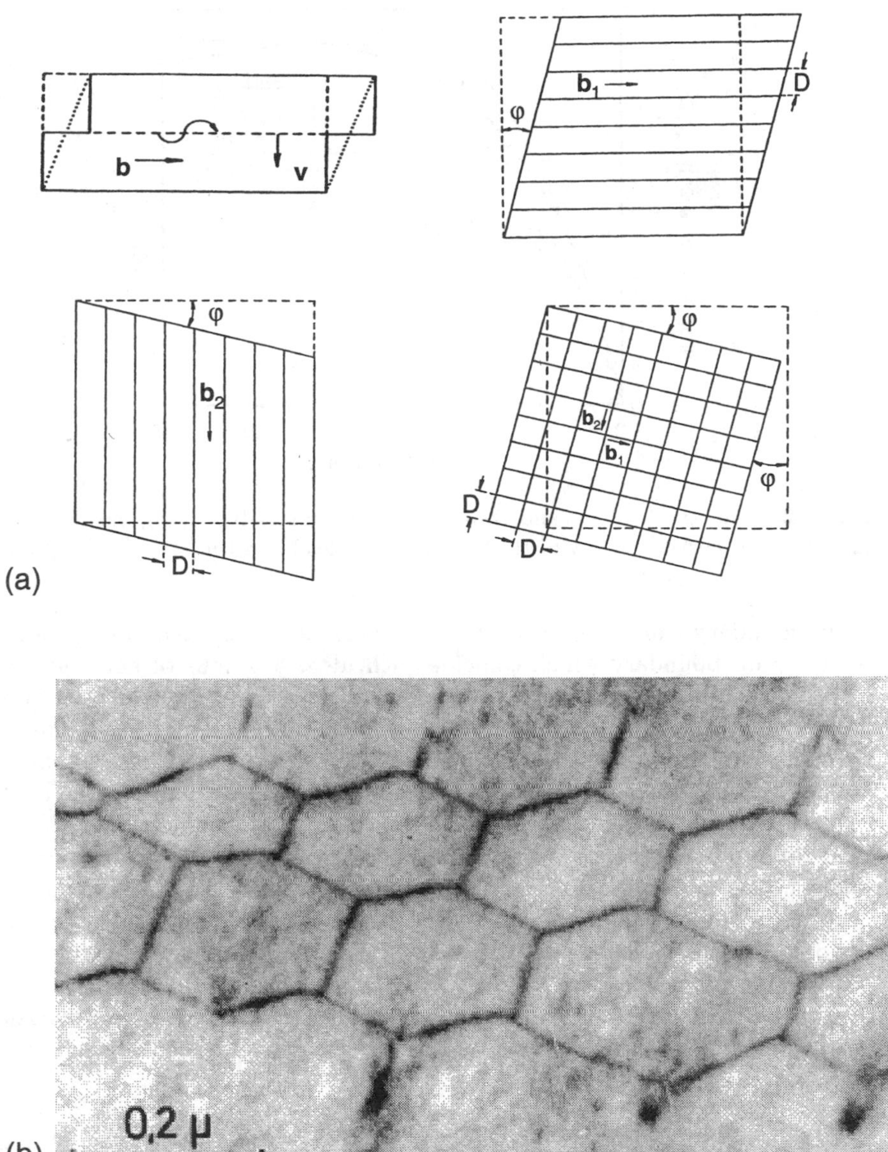

Fig. 3.27. (a) Explanation of the basic dislocation configuration of a low-angle twist boundary. A single family of parallel screw dislocations results in a shear deformation. But two perpendicular families of dislocations result in a rotation. (b) TEM image of a low-angle twist boundary in α-Fe. The hexagonal dislocation configuration is composed of screw dislocations with three different Burgers vectors [3.6].

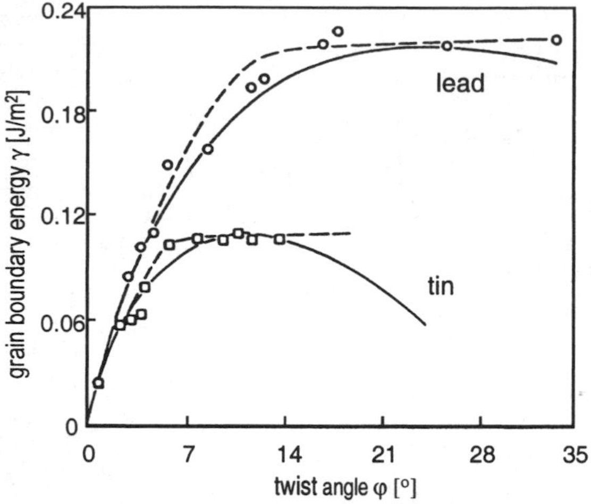

Fig. 3.28. Measured (points and dashed lines) and calculated (using the dislocation model) (solid curves) energy of tilt boundaries in lead and tin [3.7].

grain boundary from one crystal to the other, i.e. there are atomic positions in the grain boundary which coincide with ideal positions of both adjacent lattices. Such lattice points are called coincidence sites. Since the orientation relationship between the adjacent crystals is described by a rotation, one can determine under what conditions coincidence sites will occur. A simple example (Fig. 3.29) is a rotation of 36.87° about a $\langle 100 \rangle$ axis in a cubic lattice (equivalent to 53.13° because of the $90°\langle 100 \rangle$ crystal symmetry). If we consider the atomic positions of both adjacent lattices in a (100) grain boundary plane, i.e. perpendicular to the rotation axis (right part in Fig. 3.29) then the occurrence of many coincidence sites is evident. Since both crystal lattices are periodic, the coincidence sites also must be periodic, i.e. they also define a lattice, the coincidence site lattice (CSL). The elementary cell of the CSL is larger than the elementary cell of the crystal lattice, of course. As a measure of the density of coincidence sites or for the size of the elementary cell of the CSL, we define the quantity

$$\Sigma = \frac{\text{volume elementary cell of CSL}}{\text{volume elementary cell of crystal lattice}} \tag{3.26}$$

For the rotation $36.87°\langle 100 \rangle$ is $\Sigma = a(a\sqrt{5})^2/a^3 = 5$, i.e. every fifth lattice site is a coincidence site (Fig. 3.29).

Fig. 3.29 is only a very simple 2-dimensional case. In reality the coincidence site lattice is a 3-dimensional lattice, the generation of which can be imagined as follows. We take a crystal lattice, each lattice point of which carries two atoms, for instance one round and one triangular as in Fig. 3.29. Now we rotate the triangular atoms while the round atoms remain unchanged. Of course,

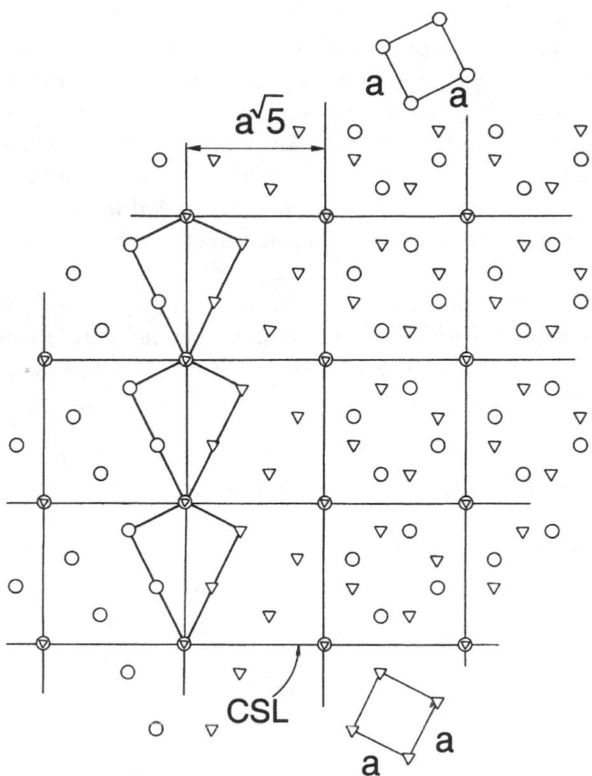

Fig. 3.29. Coincidence site lattice (CSL) and structure of a 36.9° ⟨100⟩ ($\Sigma = 5$) grain boundary in a cubic crystal lattice. Right side of figure: grain boundary plane ∥ plane of the paper (twist boundary); left side of figure: grain boundary plane ⊥ plane of the paper (tilt boundary).

the origin for the axis of rotation is a lattice point. After this rotation there are again lattice points where triangular and round atoms coincide. These are the coincidence sites and because of the periodicity of the crystal lattices they generate a 3-dimensional lattice, the CSL. To apply this crystallographic construct to grain boundaries, we have to define the spatial orientation of the grain boundary plane. Having defined this plane, we remove on one side of the plane the round atoms, on the other side of the plane the triangular atoms. This generates a bicrystal with a boundary, and the structure of the boundary is given by the atoms located in the boundary.

If good fit of the atoms - and coincidence sites represent atoms with ideal fit - is associated with a low energy, we can expect that the grain boundary strongly favors running through coincidence sites rather than non-coincidence sites. Grain boundaries between crystallites which have an orientation relationship corresponding to a high density of coincidence sites are called CSL boundaries or special boundaries. A smaller Σ value (which is always an odd integer) implies a more ordered grain boundary. Low angle grain boundaries can be characterized by $\Sigma = 1$, since almost all lattice points, except for the atoms of the dislocation cores, are coincidence sites. Grain boundaries between crystallites with a twin orientation relationship are defined by $\Sigma = 3$, and in the coherent twin boundary all lattice sites are coincidence sites. This is not a contradiction to $\Sigma = 3$, since the CSL is a 3-dimensional lattice which also extends perpendicular to the grain boundary, and only every third plane parallel to the coherent twin boundary is in perfect coincidence. This occurs for the fcc lattice when the coherent twin boundary is a $\{111\}$ plane. Because of the stacking sequence ABC of $\{111\}$ planes, a coincidence of the 3-D crystal lattices will produce coincidence sites only in every third plane parallel to the coherent twin $\{111\}$ boundary.

Table 3.2. Rotation angles Θ for lattice coincidence with $\Sigma<100$ in cubic lattices with $\langle 100\rangle$-rotation axis.

Θ	Σ	Θ	Σ
8.80	85	25.99	89
10.39	61	28.07	17
12.68	41	30.51	65
14.25	65	31.89	53
16.26	25	36.87	5
18.92	37	41.11	73
22.62	13	42.08	97
25.06	85	43.60	29

If we want to apply this crystallographic concept to real grain boundaries, there is the fundamental problem that the coincidence site lattice occurs only for very special rotations, and Σ does not change monotonically with the angle of rotation (Table 3.2). This problem is entirely complementary to the requirement of a discrete change of the angle of rotation in order to obtain a periodic crystal dislocation arrangement as described above. In essence, a strictly periodic arrangement of crystal dislocations is absolutely identical with the relaxed structure of a CSL grain boundary (Fig. 3.30). However, even for tiny deviations from the exact rotation relationship the long range coincidence is lost. Just as a crystal tries to compensate a small misorientation by a periodic arrangement of crystal dislocations, we expect that the bicrystal will try to maintain its ideal fit and to compensate deviations from this perfect fit by localized perturbations, i.e. dislocations. These dislocations must have a Burgers vector that conserves the CSL just as lattice dislocations conserve the crystal lattice when forming a low angle grain boundary. As the most trivial example, the CSL will not be changed, if dislocations are introduced for which the Burgers vectors are lattice vectors of the CSL. Equivalently, it is possible for the Burgers vector to be a vector of the crystal lattice. However, the elastic energy of dislocations increases with the square of the Burgers vector (see Ch. 6.4). Therefore, the energy of the grain boundary would increase dramatically if dislocations with a very large Burgers vector were incorporated in the grain boundary. However, the energy of a grain boundary is only determined by the density of coincidence sites, not their location. As a consequence we can relax the requirement that the location of the coincidence sites has to be conserved. There are very small vectors which conserve the size of the CSL if the locations of the coincidence sites are allowed to change. The displacement vectors which satisfy this condition define the so-called DSC-lattice[2]. The DSC lattice is the coarsest grid that contains all lattice points of both crystal lattices (Fig. 3.31). Of course, all translation vectors of the CSL and the crystal lattices are also vectors of the DSC lattice, but the elementary vectors of the DSC lattice are much smaller. Since the dislocation energy increases with the square of the Burgers vector, only base vectors of the DSC lattice qualify for Burgers vectors of the so-called secondary grain boundary dislocations (SGBDs). Dislocations with DSC Burgers vectors are referred to as SGBDs, in contrast to primary grain boundary dislocations, which are crystal lattice dislocations, the periodic arrangement of which generates the CSL. SGBDs are confined to grain boundaries, since their Burgers vectors are not translation vectors of the crystal lattice and their introduction into the crystal lattice would cause a local disruption of the crystal structure. With regard to their geometry and correspondingly, to their elastic properties, secondary grain boundary dislocations can be treated like primary dislocations. Much as primary dislocations can compensate a misorientation in a perfect crystal by a low angle grain boundary, secondary grain boundary dislocations can compensate an orientation difference to a CSL relationship while conserving the CSL. Since SGBDs also have an elastic strain field as does any dislocation, they can be imaged in a TEM (Fig. 3.32). The larger the orientation difference from the exact coincidence rotation, the smaller the spacing of the SGBDs according to Eq. (3.25).

A special property of the SGBDs is that at the location of the dislocation core, the grain boundary usually has a step (Fig. 3.33). This step is a consequence of the fact that the CSL is displaced when an SGBD is introduced. If an SGBD moves along the grain boundary, the step moves along with the dislocation and thus, the grain boundary is displaced perpendicular to its plane, i.e. the grain boundary will migrate by the distance of the step height. A fundamental property of all dislocations is that they cause a shear deformation upon their motion. Therefore, the motion of a grain boundary dislocation always will cause a combination of grain boundary migration and grain boundary sliding. In the case that the Burgers vector of the SGBD is

[2] DSC is the abbreviation for "displacement shift complete". This means that the CSL will displace as a whole, if one of the two adjacent crystal lattices is shifted by a translation vector of the DSC lattice.

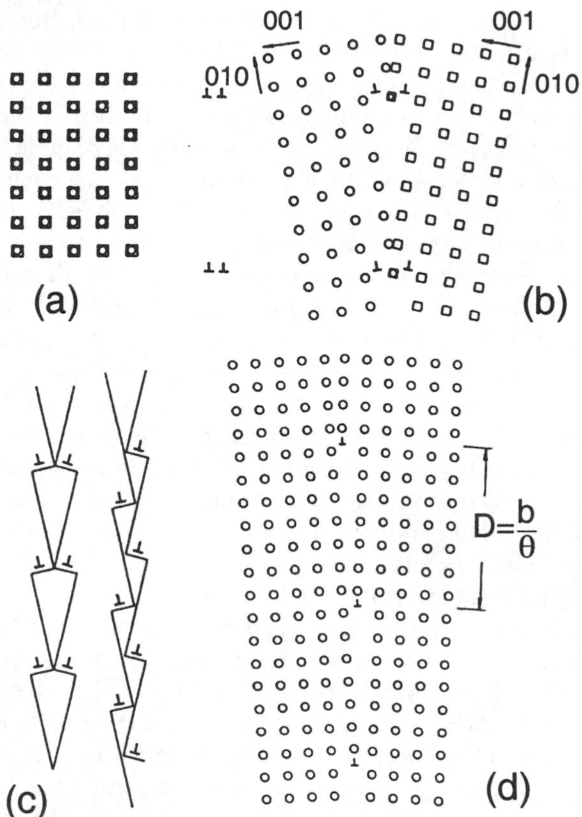

Fig. 3.30. Relationship between the coincidence site lattice and the primary dislocation structure at a grain boundary. If two identical, interlocking lattices (a) are turned symmetrically towards each other about an axis perpendicular to the plane of the page (b), a coincidence site lattice forms. The coincidence points are marked by overlapping circles and squares. The associated configuration of the resulting double dislocation is relaxed along the boundary (c), and the structure of a symmetrical low-angle tilt boundary forms (d).

parallel to the grain boundary plane, the dislocations need only to glide to cause the grain boundary to be displaced. This is the case, for instance, in symmetrical tilt grain boundaries (Fig. 3.33a). If the Burgers vector is inclined to the grain boundary plane, the dislocation can only move by a combination of glide and climb (Fig. 3.33b), which requires the diffusion of vacancies, and is, therefore, a thermally activated process.

The concept introduced is based on geometrical arguments only. Such a consideration cannot make predictions about the force equilibrium of the atomic arrangements considered. This problem can only be solved by computer simulations, which allow us to calculate the position of the atoms at an

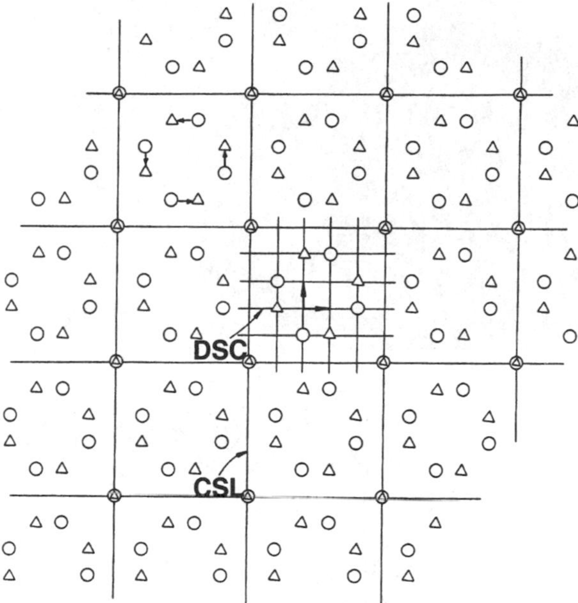

Fig. 3.31. Coincidence site lattice (CSL) and DSC lattice at 36.9° ⟨100⟩ rotation in a cubic lattice.

equilibrium of interatomic forces, i.e. by relaxation of the geometrical arrangement (Fig. 3.34). In contrast to the basis of the geometrical considerations, coincidence is almost always lost upon relaxation, but the periodicity remains and that means the conceptual frame work is still correct. A more detailed analysis reveals that the arrangement of atoms in the grain boundary can be described by polyhedra, and surprisingly enough, only seven different polyhedra are necessary to describe all possible arrangements of atoms in a grain boundary (Fig. 3.35). These polyhedra represent characteristic structures of the grain boundary and are therefore referred to as structural units. Computer simulations prove that grain boundaries of particularly low energy consist of only one type of polyhedron. If the orientation relationship is slightly changed, other structural units (polyhedra) will be introduced into the structure, and these new structural units are identical with the cores of the SGBDs (Fig. 3.36). With increasing misorientation the density of the new structural units increases until they are the majority type and therefore, on further rotation, the grain boundary structure will eventually consist of only this type of polyhedron. This structural unit concept of grain boundaries and its dependence on orientation relationship represents our current conception of grain boundary structure. The computer simulations of grain boundary structure have also been confirmed by high resolution electron microscopy (Fig. 3.37).

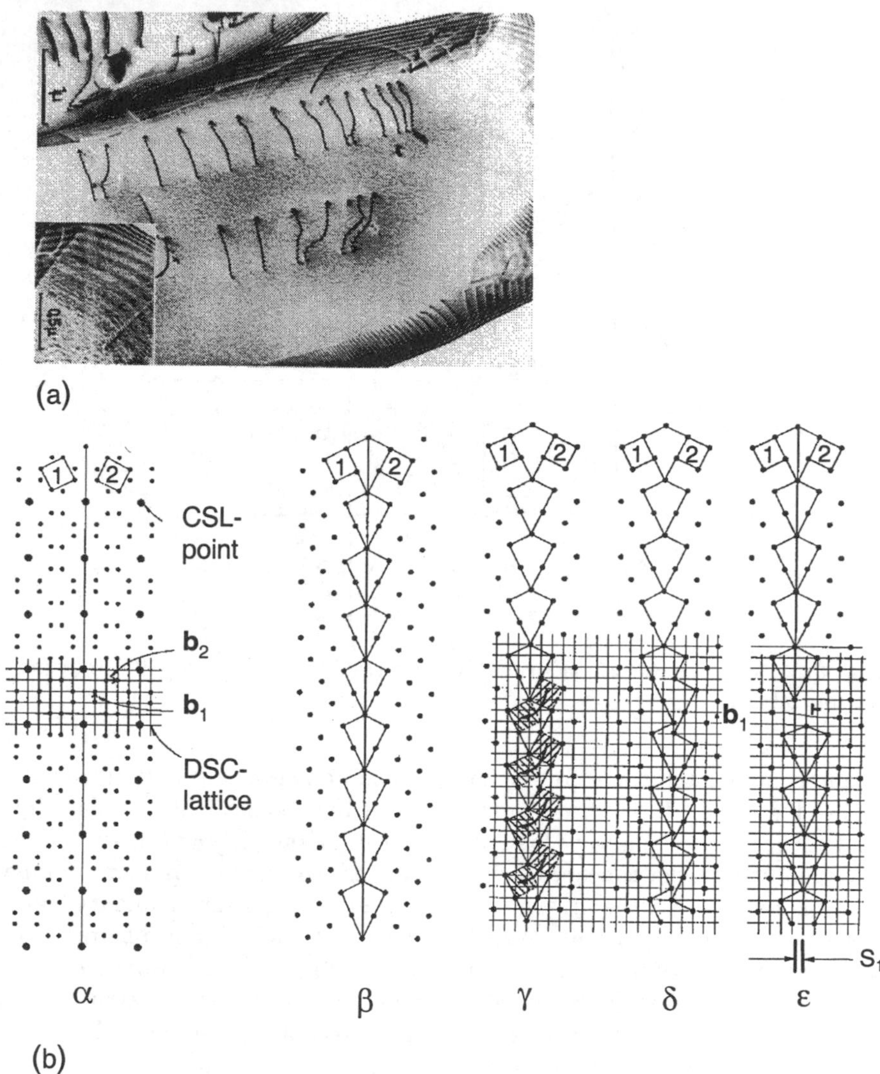

Fig. 3.32. (a) Grain boundary dislocations in a tilt boundary in stainless steel (after [3.8]). (b) The pattern of the generation of a grain boundary edge dislocation. (α) Positions of atoms (small dots), coincidence sites (big dots) and the DSC-lattice. (β) Orientation of grain boundary and positions of atoms at the grain boundary. (γ) Rearrangement of material makes the grain boundary shift in sectors. (δ) Partially shifted grain boundary. (ε) Generation of a grain boundary edge dislocation by moving atoms along the grain boundary.

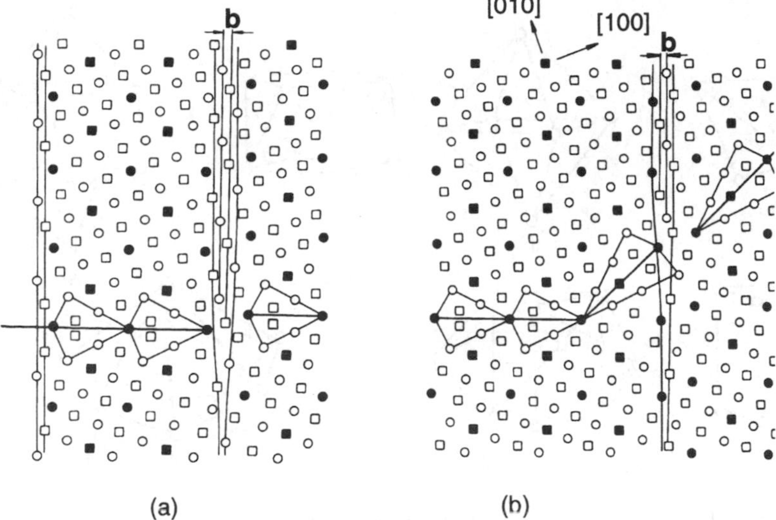

Fig. 3.33. Atomic configuration of a grain boundary edge dislocation at a $\Sigma = 5$ grain boundary in an fcc lattice. (a) Burgers vector parallel to the grain boundary; (b) Burgers vector inclined to the grain boundary.

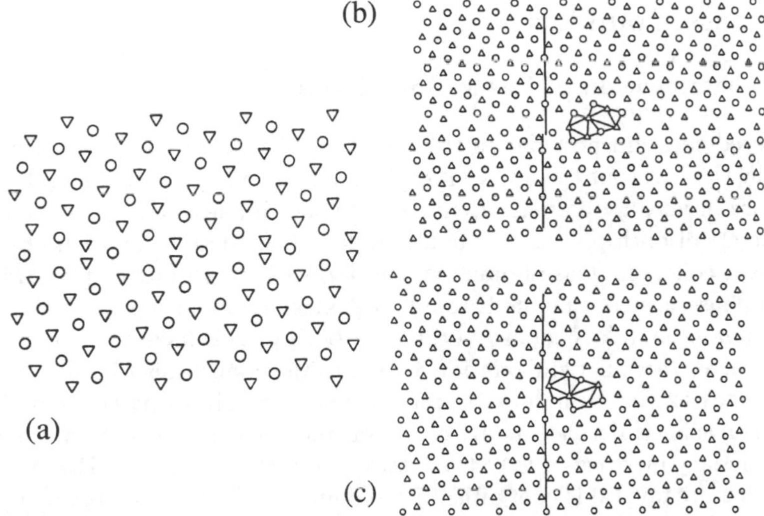

Fig. 3.34. Structure of a symmetrical 36.9°⟨100⟩ ($\Sigma = 5$) tilt boundary in aluminum, calculated by computer simulation. (a) Configuration after rigid rotation of the crystallites. (b) and (c) Structure of grain boundaries after relaxation. The staggered vertical lines at the grain boundary indicate the shift of the crystallites. Hence, for a given misorientation there can be more than one structure ([3.9]).

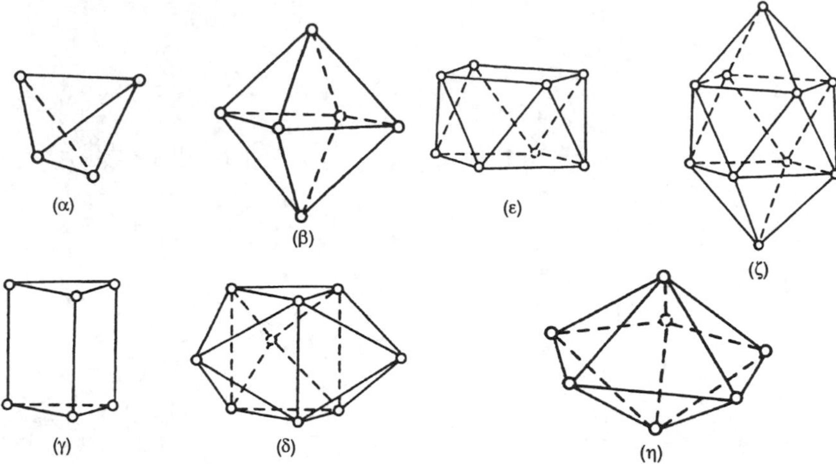

Fig. 3.35. The seven different Bernal structures that grain boundaries (in structures formed of hard spheres) can consist of: (α) Tetrahedron; (β) Octahedron; (γ) Trigonal prism; (δ) Truncated trigonal prism; (ε) Archimedean square antiprism; (ξ) Truncated Archimedean square antiprism; (η) Pentagonal double pyramid.

3.5 Phase Boundaries

3.5.1 Classification of Phase Boundaries

Compared to grain boundaries the structure of phase boundaries is complicated by the fact that the adjacent crystallites can also have a different crystal structure besides a different orientation. In the simplest case only the lattice parameters of both phases are slightly different. If there is no orientation difference, a coherent phase boundary will form where all lattice planes in the parent phase are continuous through the phase boundary (Fig. 3.38). A coherent phase boundary is also obtained if both phases have the same crystal structure but are in twin relation to each other, since also in this case all lattice sites in the boundary belong to both crystallites at the same time. With increasing difference of lattice parameter, the elastic energy of the interface boundary increases due to the lattice parameter mismatch. Eventually, it becomes more favorable energetically to compensate this mismatch by the formation of edge dislocations which will reduce the so-called coherency stresses (Fig. 3.39a,b). Since in this case not all lattice planes are continuous through the interface, this boundary type is called partially coherent.

If both phases have different crystal structures, the coherency of the interface is completely lost and an incoherent phase boundary is obtained (Fig. 3.40). As in the case of grain boundaries, we can also expect in this case that nature will prefer arrangements which are energetically most favorable. If the

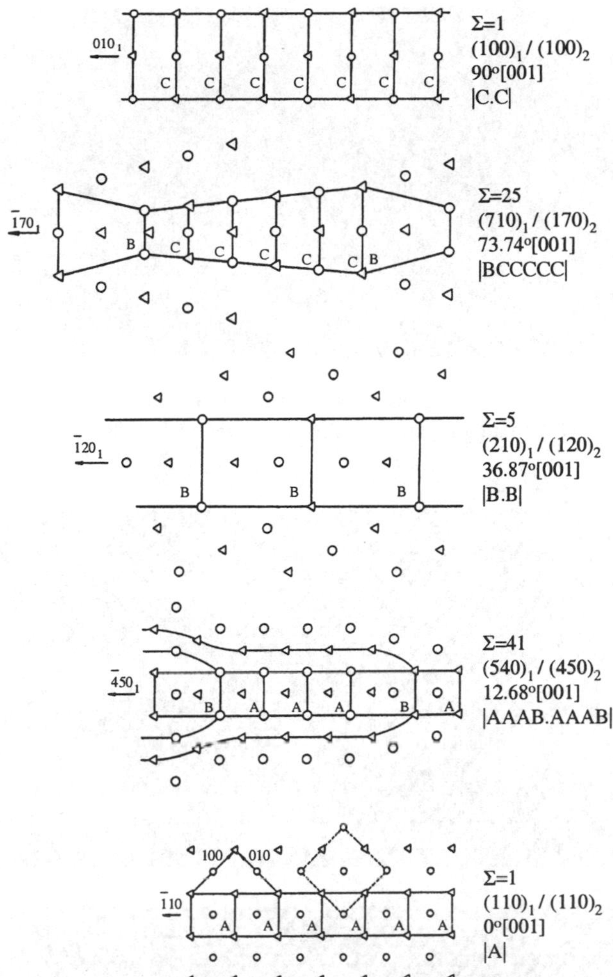

Fig. 3.36. Calculated dependence of grain boundary structure on the tilt angle of a symmetrical ⟨100⟩ tilt boundary in aluminum for various tilt angles. For every tilt angle there is a certain arrangement of structural units (A,B,C). If this arrangement is interrupted, this forms a grain boundary dislocation (illustrated for $\Sigma = 41$).

elastic energy plays an important role, then arrangements in the interface with good atomic fit will be preferred.

Real interfaces, in particular interfaces in man-made materials, for instance composites, usually are not in equilibrium and can develop very complicated interface structures. In particular, large perturbations around the interface and inhomogeneities in the adjacent lattices can be observed (Fig. 3.41).

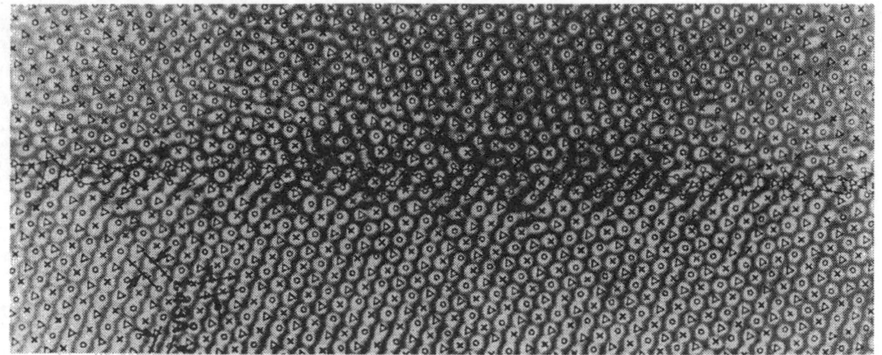

Fig. 3.37. There is good agreement of the structure calculated by computer simulation (symbols) with the TEM image of the structure of a 21.8° ⟨111⟩ grain boundary in gold [3.10].

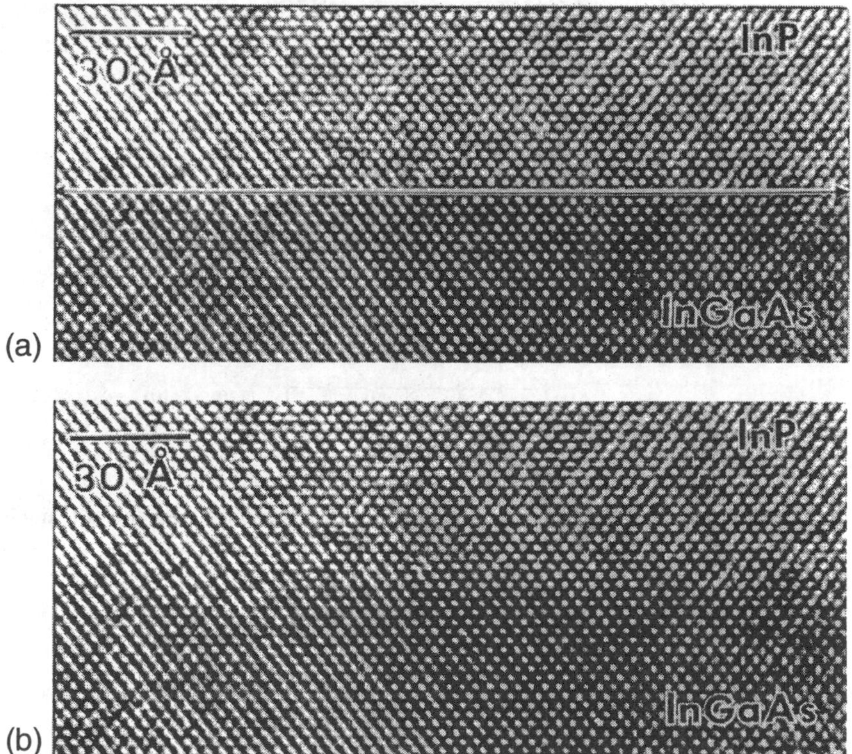

Fig. 3.38. Atomistic structure of a coherent interface of InP with InGaAs. In figure (a) the position of the interface is indicated, in (b) it is virtually not discernible [3.11].

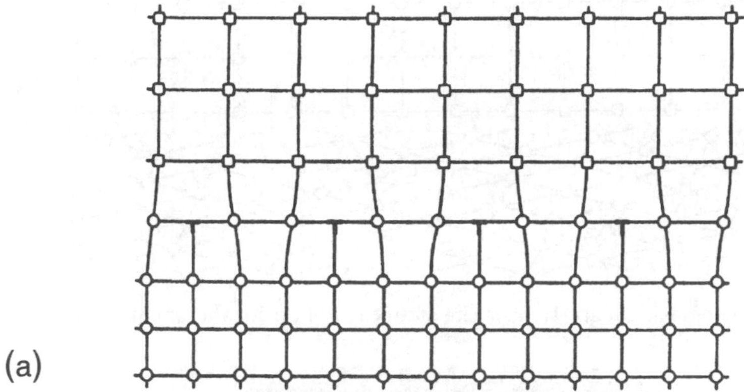

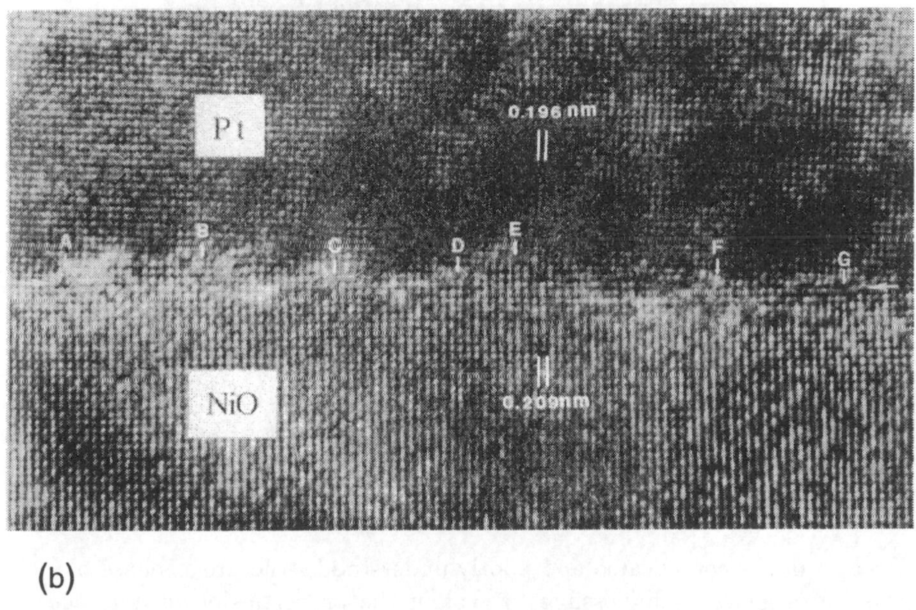

Fig. 3.39. A semi-coherent interface. (a) Schematic illustration. (b) TEM image of a semi-coherent interface joining Pt and NiO. A lattice plane ends at each position marked with a letter, i.e. there is an edge dislocation [3.12].

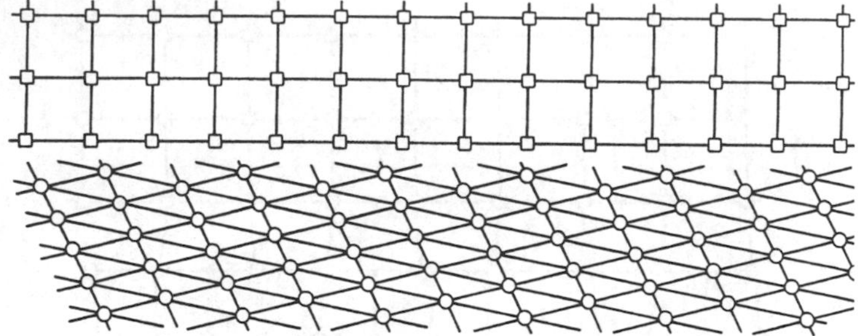

Fig. 3.40. Schematic illustration of the structure of an incoherent interface.

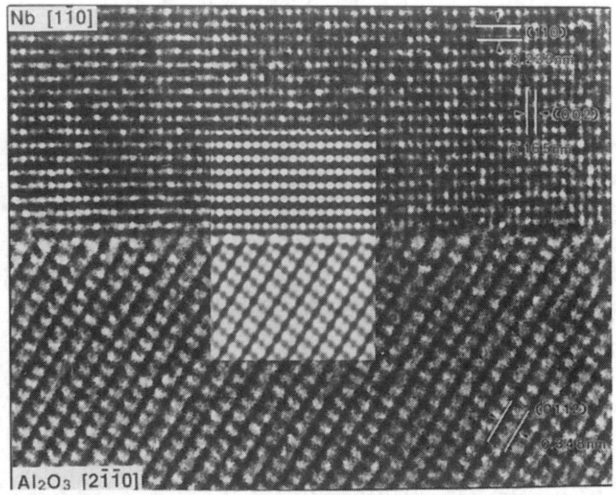

Fig. 3.41. Structure of an incoherent interface of Nb and Al_2O_3 (high resolution TEM image) [3.13].

3.5.2 Phenomenological Characterization of Phase Boundaries

Because of the complicated and poorly understood structure of phase boundaries, it is generally impossible to explain the properties of an interface on the basis of its atomic arrangement. In such case phenomenological models are a viable option, where the behavior of an interface is associated with a macroscopic property. One such property is the interfacial tension γ, which is essentially identical with the specific excess interface free energy. It has dimensions $[J/m^2] = [N/m]$, i.e. a force per unit of length. This interface tension becomes obvious if we consider an inflated balloon. Cutting the balloon along a line on its surface would lead to a rapid growth of the cut and the balloon would rupture. The force per unit length needed to keep the two surfaces created by the cut in contact is the surface tension or, in the context of internal interfaces, the interfacial tension.

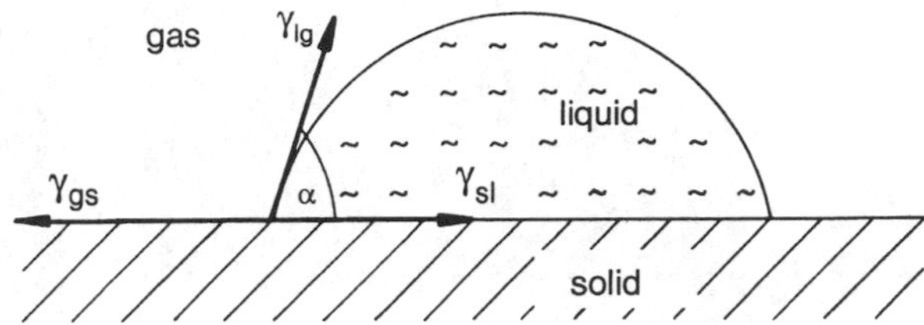

Fig. 3.42. Equilibrium shape and wetting angle α of a drop of liquid on a hard surface.

The interfacial tension determines the equilibrium shape of interfaces in phase mixtures. Let us consider the equilibrium shape of a droplet on a solid surface (Fig. 3.42). Along the lines of contact of the phases the surface tensions act as forces in order to establish the configuration with the smallest energy. Horizontal equilibrium of forces is obtained if

$$\gamma_{gs} = \gamma_{sl} + \gamma_{lg} \cdot \cos\alpha \tag{3.27}$$

from which we can calculate the contact angle α. For $\alpha = 0°$ the droplet will spread on the surface as a film, and we obtain complete wetting. For $\alpha = 180°$, the droplet will assume the shape of a sphere, which is the case of complete non-wetting. The real case, of course, is in between these two limits. Depending on application smaller or larger contact angles may be desirable. For example: "Rain-X" on windshields designed for $\alpha = 0°$, helps keep rain off! The magnitude of α can be affected by chemical composition. In immiscible systems α is usually large. If the phases, however, have a tendency to chemically react, α can be very small. Small values of α are desirable for composites, because this is associated with excellent adhesion between fiber and matrix. On the other hand chemical reactions and the formation of compounds degrade the mechanical properties of metallic and ceramic composites because they raise the tendency to brittleness. For complete non-wetting in composites poor adhesion is obtained and, therefore, unsatisfactory load transfer from the matrix to the fiber in fiber reinforced materials.

Small contact angles can also be undesirable in some cases, for instance in the case of inclusions with low melting temperatures. Examples are Bi in brass or FeS in steel. The bismuth segregates to the grain boundaries of brass. If the temperature is increased above the melting temperature of bismuth, liquid bismuth will spread along the grain boundaries because of its low surface tension which causes the well-known problem of hot shortness. Another commercially very important example is that of FeS inclusions in steel (Fig. 3.43). The hot shortness of α-brass can be avoided by addition of lead, since the interface tension between melt and crystal increases with increasing content of lead and, correspondingly, the contact angle increases (Fig. 3.44).

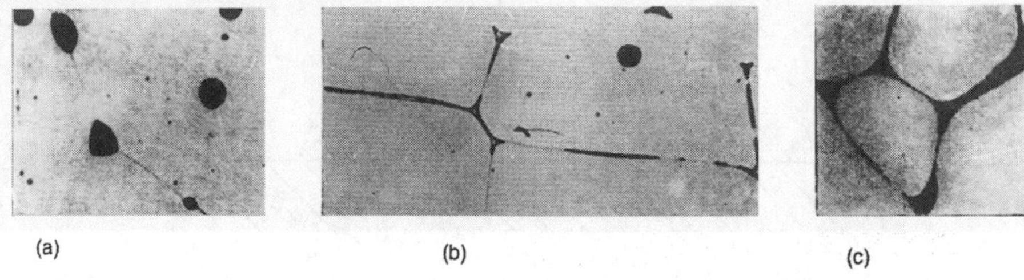

Fig. 3.43. Wetting of grain boundaries by the solid and liquid state. (a) Solid inclusions of lead in brass. (b) A film of liquid bismuth will totally wet the grain boundaries in copper. (c) FeS melt on grain boundaries in steel [3.14].

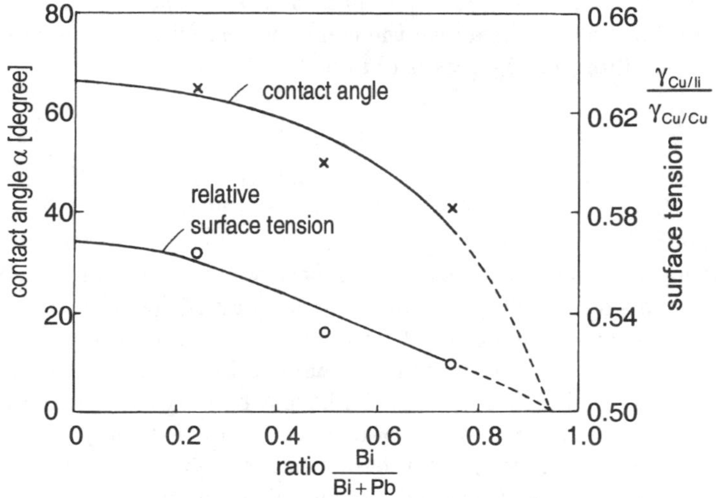

Fig. 3.44. Dependence of surface tension (interface energy) on composition in the CuPbBi system (after [3.15]).

The example of a liquid droplet on a solid surface (Eq. (3.27) is the special case of the equilibrium of three phases (Fig. 3.45). For the general case one can show (Herring equation)

$$
\begin{aligned}
&\frac{\gamma_{23}}{(1+\varepsilon_2 \cdot \varepsilon_3)\sin\alpha_1 + (\varepsilon_3 - \varepsilon_1)\cos\alpha_1} \\
&= \frac{\gamma_{13}}{(1+\varepsilon_1 \cdot \varepsilon_3)\sin\alpha_2 + (\varepsilon_1 - \varepsilon_2)\cos\alpha_2} \\
&= \frac{\gamma_{12}}{(1+\varepsilon_1 \cdot \varepsilon_2)\sin\alpha_3 + (\varepsilon_2 - \varepsilon_3)\cos\alpha_3}
\end{aligned}
\tag{3.28a}
$$

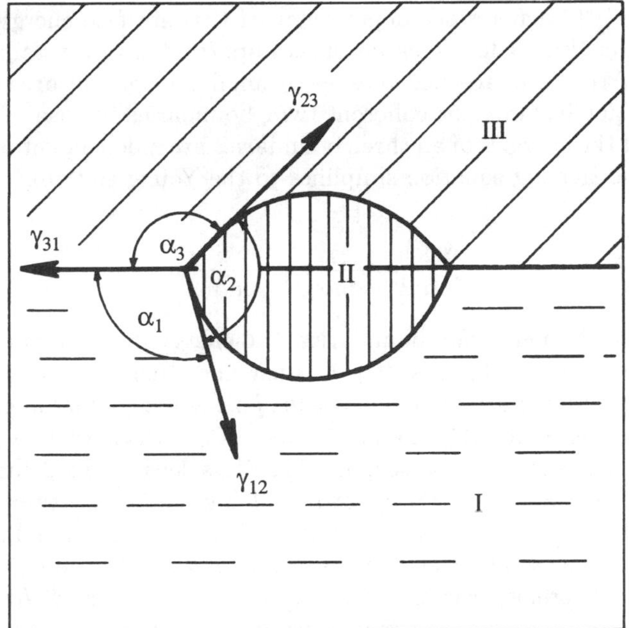

Fig. 3.45. Equibrium between surface tension γ_{ij} and the corresponding contact angle α_k in a three phase equilibrium.

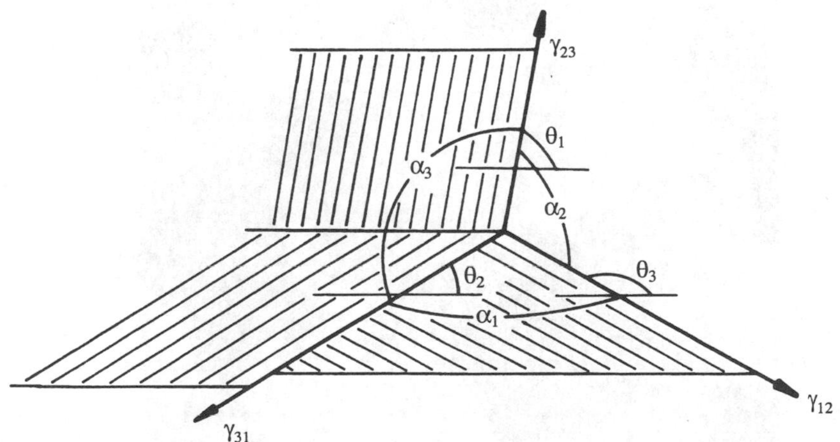

Fig. 3.46. Equilibrium of forces at a grain edge. The equilibrium angle α_i depends on both grain boundary energy γ_{ij} and grain boundary orientation Θ_k.

where $\varepsilon_i = \frac{\partial ln \gamma_{hkl}}{\partial \theta_i}$ denotes the dependency of the interface energy on interface orientation (Fig. 3.46). This becomes important if in crystalline phases the energy of the interface becomes very small for specific orientations of the interface, for instance for coherent twin boundaries in case of the grain boundaries. If the energies of all three boundaries are independent of interface orientation the Herring equation simplifies to the Young equation

$$\frac{\gamma_{12}}{\sin\alpha_3} = \frac{\gamma_{23}}{\sin\alpha_1} = \frac{\gamma_{31}}{\sin\alpha_2} \tag{3.28b}$$

For many high angle grain boundaries, the energy does not vary strongly with misorientation. In this case γ_{ij} = const and, therefore, $\alpha_k = 120°$. In equilibrium microstructures of homogeneous phases the contact angle of grain boundaries is, therefore, 120° in most cases (Fig. 3.47). In two-phase microstructures, the shape of the second phase is a lense shaped inclusion on boundaries corresponding to Eq. (3.28b) and Fig. 3.45. An example for the equilibrium of immiscible phases gives Fig. 3.43a for lead in brass. In the grain interior the lead has a spherical shape, since in this case the spherical surface is minimal and, therefore, the total surface energy is minimized. At the triple junctions, the shape of the lead inclusions is given by the force equilibrium according to Eq. (3.28b).

Fig. 3.47. Structure of annealed polycrystalline aluminum. Most contact angles are around 120°.

4

Alloys

4.1 Constitution of Alloys

Matter is known to occur in three different states as gas, liquid, or solid. Each element has a specific melting temperature T_m, which separates the solid and liquid temperature regimes and a boiling temperature T_b, for the transition from liquid to gas. At T_m and T_b two states of matter are in thermodynamic equilibrium. Melting and boiling temperatures also depend on pressure p, although this dependency may be very mild as in metals. Accordingly, the existence of a state (phase) is represented by a certain range in the $p-T$ diagram (Fig. 4.1). The lines in this diagram separate the ranges of existence of two phases and define the conditions where these two phases are in equilibrium. At the triple point all three phases are in equilibrium with each other. Between the triple point and the critical point (cr.P.), the transition from the liquid to the vapor phase is discontinuous. Beyond this critical point the phase transition from liquid to gas proceeds continuously. For a given pressure there is a defined melting temperature and a defined boiling temperature, namely the points of intersections of the respective isobar (line of constant pressure) and the phase boundaries of the phase diagram (Fig. 4.1). The range of existence of phases in thermal equilibrium follows Gibbs phase rule

$$f = n - P + 2 \tag{4.1a}$$

where n is the number of components, P the number of phases, and f the number of degrees of freedom. The components are the building blocks of the system, for instance the kinds of atoms in case of elements and their mixtures, or stable chemical compounds in more complex systems. For a pure element $n = 1$. A phase is a physically unique substance, not necessarily of constant chemical composition, for instance in a solid solution the composition maybe different. In a multicomponent system additional phases can occur by chemical reactions. The number of degrees of freedom denotes the number of variables in the system, that under the given conditions can be changed. In a one-component system ($n = 1$) the Gibbs phase rule means that, in the range of a single phase ($P = 1$), two parameters can be changed, namely pressure and temperature. At the triple point, however, $P = 3$ and $f = 0$, i.e. only at these specific values of pressure and temperature can all three phases be in equilibrium with each other.

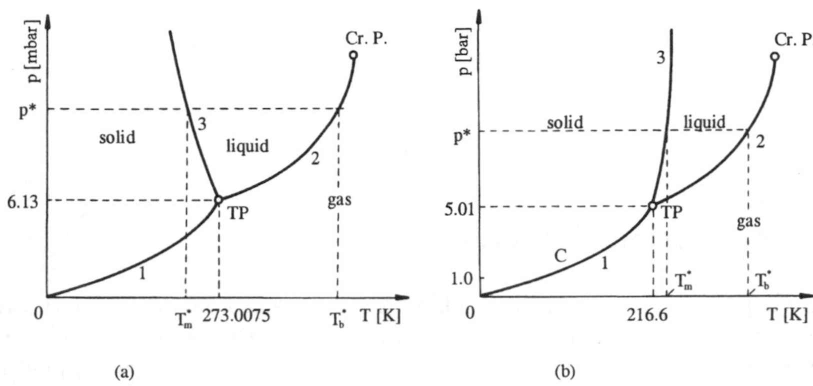

Fig. 4.1. Phase diagrams of water (a) and carbon dioxide (b). For a given pressure p^* there is a certain melting point T_m^* and boiling point T_b^*. The decrease of T_m with increasing pressure is a characteristic property of water. TP = triple junction point, Cr.P. = critical point.

Since the melting temperature and boiling temperature of metals depend only very little on pressure, and since in most commercial applications the pressure is the ambient pressure and not subject to change, the Gibbs phase rule is commonly used in the form

$$f = n - P + 1 \qquad (\text{p} = \text{const.}) \tag{4.1b}$$

corresponding to the isobaric section in Fig. 4.1. According to this modified rule, $f = 0$ at the melting temperature, i.e. only at the melting temperature can the liquid and solid phases be in mutual equilibrium. In the following we will disregard the pressure dependence and use Eq. (4.1b).

In binary alloys (two-component systems) $n = 2$. Besides the temperature another degree of freedom is introduced, namely the composition as expressed in terms of concentration which is defined differently depending on application. In engineering literature the weight concentration c_w (weight %) is most common. It is defined as the weight fraction of the element B of the total weight. In physics the atomic concentration is preferred, which is the fraction of B-atoms of all $(A + B)$-atoms (c_B [atom %] or c_B^a). If the atomic weight of A and B are Y_A and Y_B, respectively, the atomic concentration c_A^a of the atomic element A can be calculated from the weight concentration c_A^w as

$$c_A^a = \frac{c_A^w/Y_A}{c_A^w/Y_A + c_B^w/Y_B} = \frac{c_A^w}{c_A^w + c_B^w\left(\frac{Y_A}{Y_B}\right)}$$

and c_B^a correspondingly.

The range of existence of equilibrium phases in a two-component system is usually represented in a $T-c$-diagram, also referred to as a phase diagram. For $P = 1$ now two degrees of freedom exist, namely temperature and composition. A distinct difference from the one-component system is obtained for $P = 2$, i.e. for the transition from the liquid to the solid phase, namely $f = 1$. Hence, for a given concentration there is no defined melting temperature, rather a finite range where liquid and solid are in equilibrium (Fig. 4.2). Correspondingly, in this range, the liquid and solid phases do not necessarily have the same composition at a given temperature. The line which denotes the composition of the liquid phase with changing temperature in the $T-c$-diagram is referred to as the liquidus line. The corresponding line for the solid phase is called solidus line.

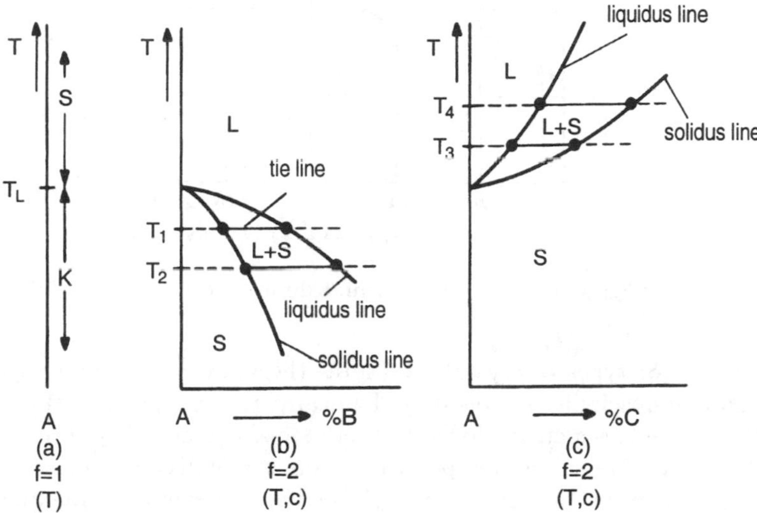

Fig. 4.2. At a given pressure a one component system (a) has a certain melting point, in a binary system, however, there is a melting range limited by liquidus and solidus lines. These can either fall (b) or rise (c) with concentration. The tie line connects equilibrium concentrations.

If an alloy with composition c is cooled down from the liquid state, solidification starts when the liquidus temperature is reached and does not finish until the solidus temperature is reached. Between the liquidus and solidus temperatures the system is comprised of a mixture of liquid and solid phases. The line that connects the concentrations of the liquid and solid phases in equilibrium at constant temperature is referred to as a tie-line (Fig. 4.2). The concentration in the liquid phase can be larger or smaller than in the solid phase. Correspondingly, the liquidus temperature (and the solidus temperature) increase or decrease with increasing concentration.

There are different specific types of phase diagrams depending on the behavior of the solid state. We will consider these different cases in the following. Let us consider the case where there is complete solubility in the liquid and the solid state, i.e. in the melt and in the crystal there is only one phase. In the solid state this is referred to as solid solution. In this case the two-phase regime of the partially solidified melt extends continuously between the two pure components. An example is the system Ag-Au (Fig. 4.3).

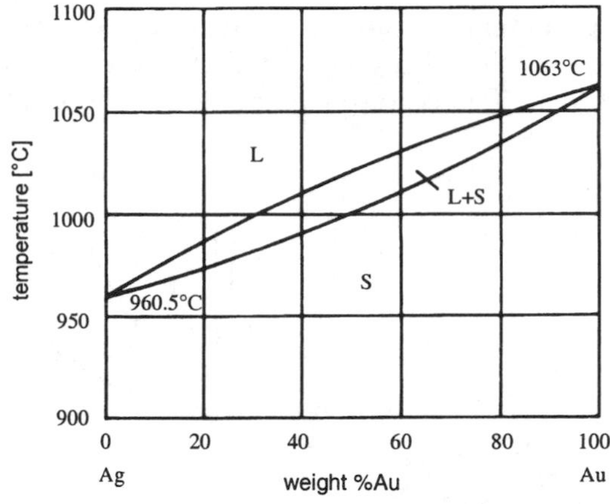

Fig. 4.3. Phase diagram of the AuAg system, exhibiting complete miscibility [4.1].

For all major types of crystal structures there are many binary systems with complete miscibility (Table 4.1). Typically, the two-phase regime has a shape that resembles a cigar. However, there are also cases where the liquidus temperature (and the solidus temperature) increases or decreases on both sides of the phase diagram. In such case the phase diagram exhibits a maximum or a minimum (Figs. 4.4 and 4.5). A maximum is observed usually in complex systems with intermetallic phases. At the maximum or minimum solidus and liquidus line have to intersect, i.e. there is a defined melting temperature.

Qualitatively an increase or decrease of the melting temperature reflects a strengthening or weakening of the atomic bonding forces, and thus, the tendency to decomposition or to the formation of intermetallic compounds in the solid state. If the alloying element promotes this tendency eventually a miscibility gap will be observed. A miscibility gap means that there is a composition range over which the components do not form a solid solution but decompose to form a mixture of two or more phases. Examples are the binary alloys of copper with gold (Fig. 4.5) or silver (Fig. 4.6). The change

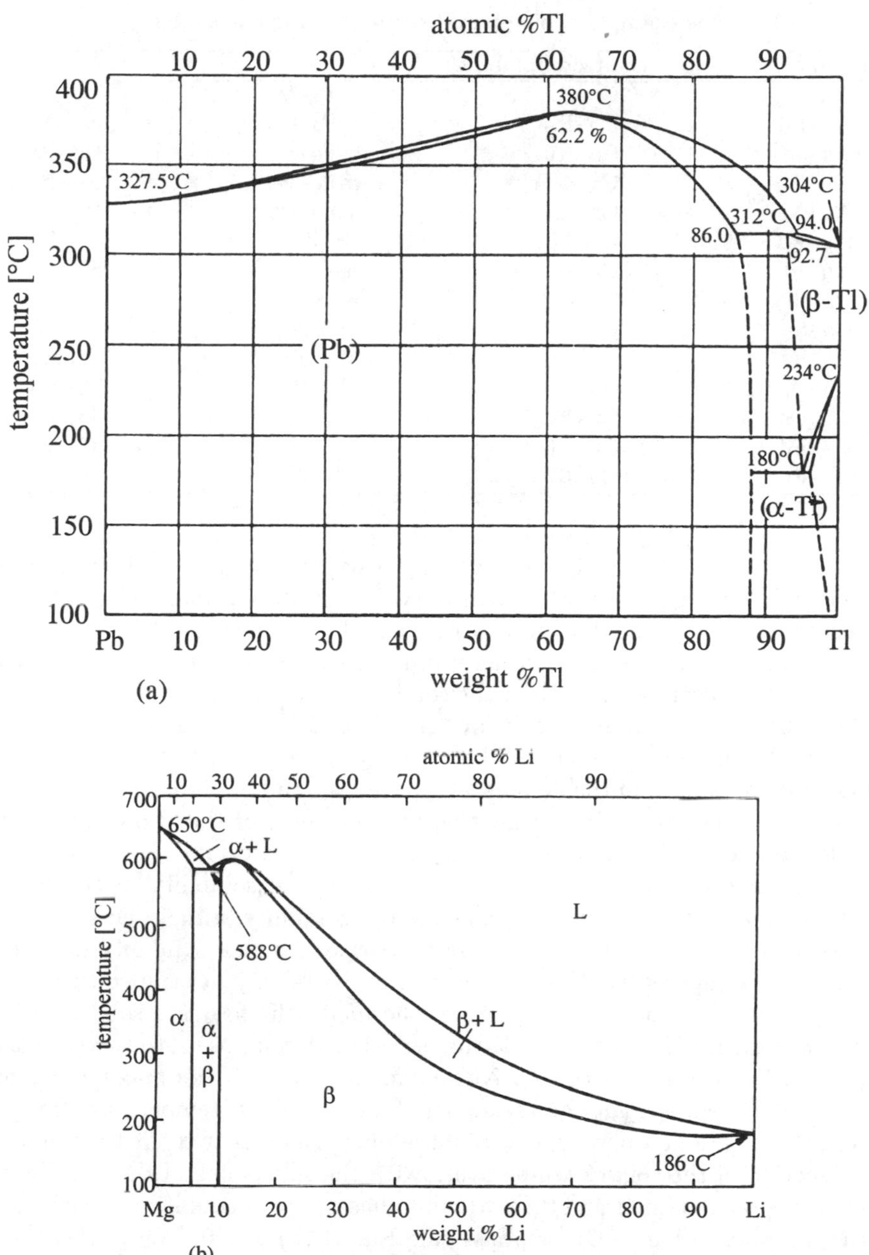

Fig. 4.4. Phase diagrams with a maximum: (a) PbTl; (b) MgLi. In both cases there is a miscibility gap in the solid state, typical behavior for phase diagrams with a maximum [4.1].

Table 4.1. Some example of binary systems with complete solubility.

Binary sistems with complete solubility			
Au-Ag	Co-Re	α-Fe-V	Ni-Pd
Ag-Pd	Co-Rh	γ-Fe-Co	Ni-Pt
As-Sb	Co-Ru	α-Fe-Ni	Pd-Rh
Au-Cu	Cr-α-Fe	α-Fe-Pd	Pd-Pt
Au-Ni	Cr-Mo	γ-Fe-Pt	Pt-Rh
Au-Pd	Cr-Ti	Hf-Zr	Se-Te
Au-Pt	Cr-W	Ir-Pt	Si-Ge
Bi-Sb	Cs-K	K-Rb	Ta-β-Ti
Ca-Sr	Cs-Rb	Mn-Ni	Ta-W
Co-Ir	Cu-Mn	Mo-Ta	Ti-Mo
Co-Ni	Cu-Ni	Mo-W	Ti-Nb
Co-Os	Cu-Pd	Nb-Ta	Ti-V
Co-Pd	Cu-Pt	Nb-Mo	Ti-Zr
Co-Pt	Cu-Rh	Nb-W	

from complete solubility in the system Cu-Au to a large miscibility gap in the system Cu-Ag is associated with the different atomic volumes of gold and silver, even though the difference is very small (see Section 4.3).

In binary systems with strongly limited solubility of the components a miscibility gap can occur both in the solid and in the liquid state. Such a system is referred to as a monotectic system. One example of virtually complete lack of solubility in liquid and solid state is the system Fe-Pb (Fig. 4.7). In the solid state as well as in the liquid state, pure lead and pure iron are separated from each other. Between the melting temperatures of the two components liquid lead and solid iron are in thermodynamic equilibrium.

Most metallic systems, however, show complete solubility in the liquid state. In contrast many systems show limited solubility in the solid state. Because of thermal activation, the tendency towards higher solubility increases with rising temperature. If a miscibility gap exists only at temperatures distinctly below the solidus temperature, the melt solidifies as a solid solution and only on further cooling will this solution decompose into a two-phase mixture, for instance in the Au-Ni system (Fig. 4.8). If the maximum temperature of the miscibility gap is higher than the solidus temperature for any given composition, a new type of a phase diagram is created. At the point of intersection of the solidus temperature with the solvus line of the miscibility gap, three phases (melt and the two solid phases) are in equilibrium with each other and according to Gibbs phase rule Eq. (4.1b) $f = 0$. Hence, there is a defined temperature at which the remaining melt completely solidifies. The range of concentration within which this three phase equilibrium is attained, is given by the points of intersection of the miscibility gap with the solidus temperature.

If the phase diagram of solidification has the shape of a cigar (monotonic change of solidus temperature with composition) a so-called peritectic phase

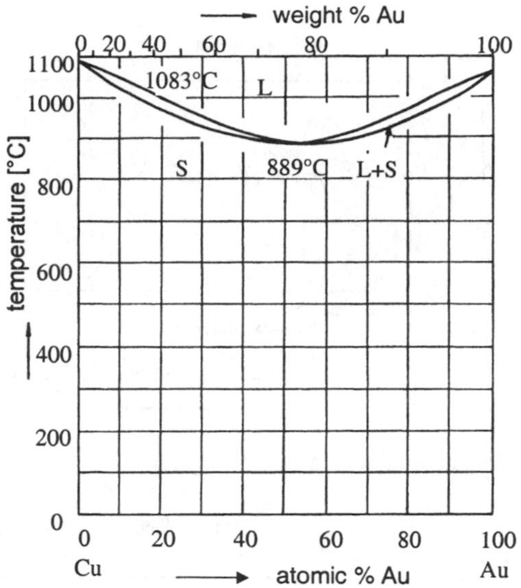

Fig. 4.5. Phase diagram with a minimum (example: CuAu). In the solid state there is continuous solid solution [4.1].

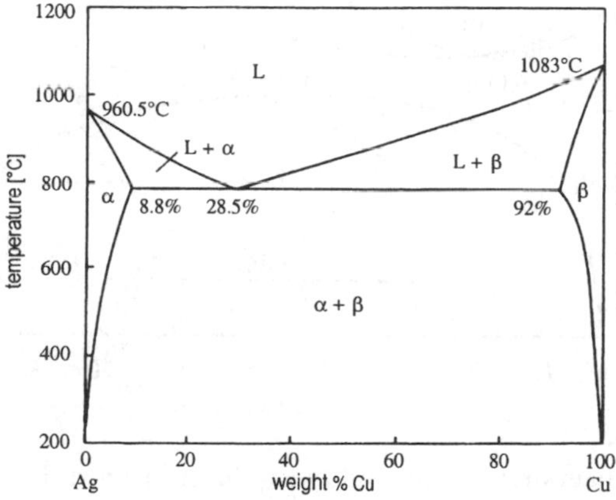

Fig. 4.6. Eutectic phase diagram (example: AgCu) [4.1].

diagram is obtained, as schematically illustrated in Fig. 4.9. A peritectic system is characterized by a reaction in which a solid phase α_2 with composition c_p decomposes at the peritectic temperature T_p during melting in liquid melt (L) and solid phase α_1 and vice versa. This is expressed by the peritectic reaction[1] equation

[1] In literature the different phases are also referred to as α and β.

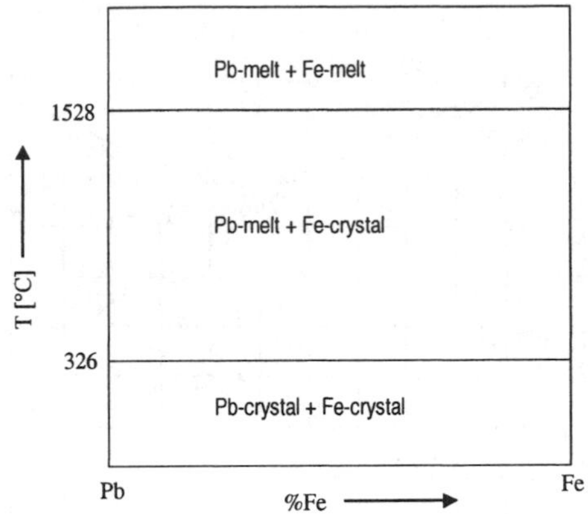

Fig. 4.7. Monotectic phase diagram of PbFe [4.1].

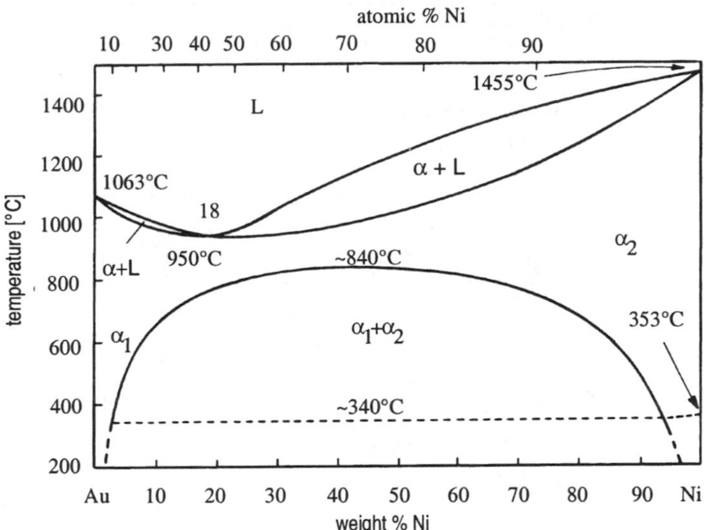

Fig. 4.8. Phase diagram with miscibility gap in the solid state. From 840 °C to 950 °C there is continuous solid solution, at lower temperature two solid phases coexist (α_1 and α_2) [4.1].

$$L + \alpha_1 \rightarrow \alpha_2$$

The peritectic temperature T_p is between the melting temperatures of both pure components. Peritectic systems usually occur if the melting temperatures of the components are very different. An example is the system Pt-Re (Fig. 4.10).

If the solidus line has a minimum, a eutectic phase diagram is obtained for the case of a miscibility gap for the the solid state. At the point of inter-

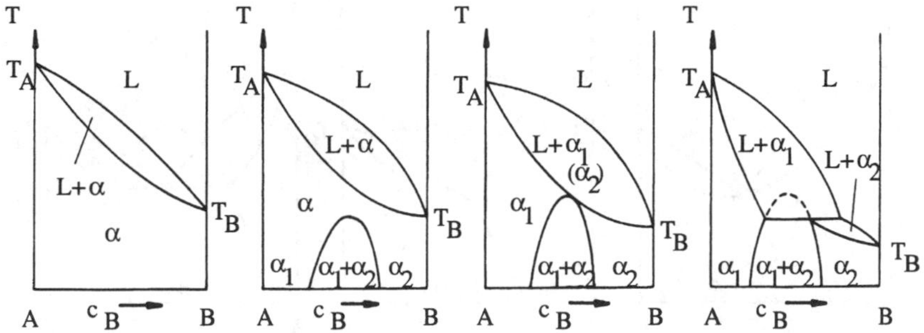

Fig. 4.9. Schematic illustration of the evolution of the peritectic phase diagram due to a widening miscibility gap in the solid state.

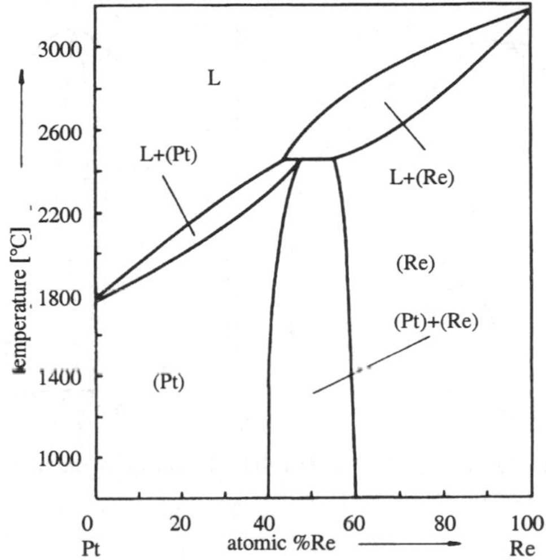

Fig. 4.10. Example of a peritectic phase diagram: PtRe [4.1].

section of miscibility gap (solvus) and solidus line there is again a three-phase equilibrium and, therefore, a unique temperature, the eutectic temperature T_E, where the remaining melt with the eutectic composition c_E completely decomposes into two solid phases α_1 and α_2 (Fig. 4.11). The eutectic reaction, therefore, reads

$$L \rightarrow \alpha_1 + \alpha_2$$

An example is the system Cu-Ag (Fig. 4.6).

If the solidus line has a maximum, there is a tendency for intermetallic compounds to form during solidification of the melt (Fig. 4.12). The intermetallic phase can either have a finite range of existence with variable composition as in the system Sb_2Te_3 (Fig. 4.13a) or occurs only in the perfectly stoichiometric composition, for instance $CaMg_2$ (Fig. 4.13b). Intermetallic

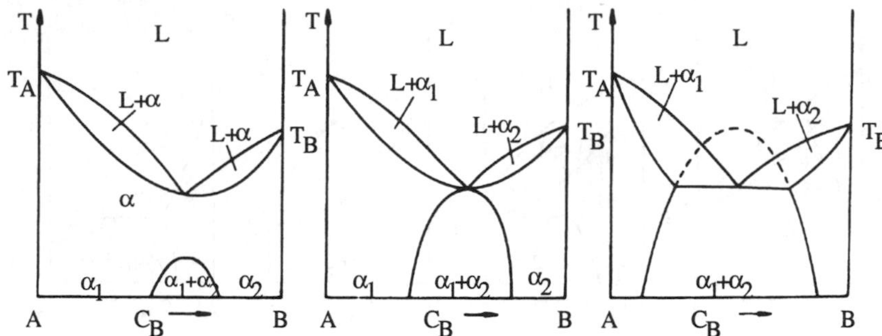

Fig. 4.11. Schematic illustration of the evolution of the eutectic phase diagram resulting from a widening miscibility gap in the solid state.

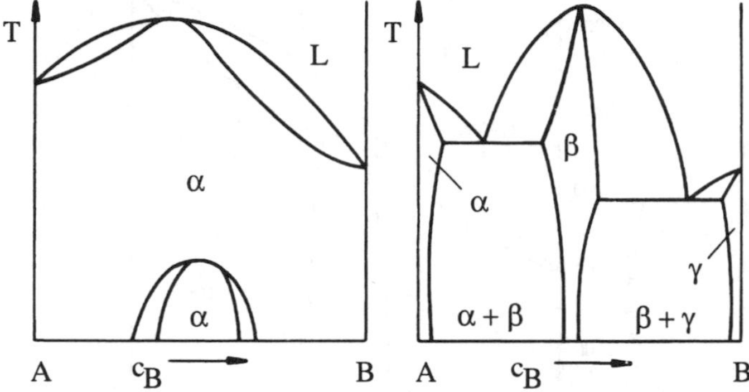

Fig. 4.12. Schematic diagram illustrating the formation of intermetallic phases due to a widening miscibility gap in the solid state.

phases can also form from the solid state, i.e. not directly from the melt. Also in this case the intermetallic phase can have a finite solubility range of existence or occur only in stoichiometric composition. Example are δ-brass (Fig. 4.14) or $NiBi_3$ (Fig. 4.15).

All other possible types of phase diagrams can be derived from these basic types of phase diagrams. They can be combined in a very complicated way as, for instance, in the system Cu-Zn (Fig. 4.14).

A phase diagram is useful for obtaining information on the state of an alloy in thermal equilibrium for a given temperature and composition. As an illustration let us consider the solidification of a binary alloy. The system may be comprised of the elements A and B and has the average concentration c_0 of B-atoms. We consider a phase diagram (Fig. 4.16) with complete solubility in the liquid and solid state. At very high temperatures only the liquid state is stable and the system exists as a melt. On cooling, a solid solution α with composition c_1 begins to crystallize when the liquidus temperature T_1 is reached. On further cooling, the volume fraction of α increases while that of the melt

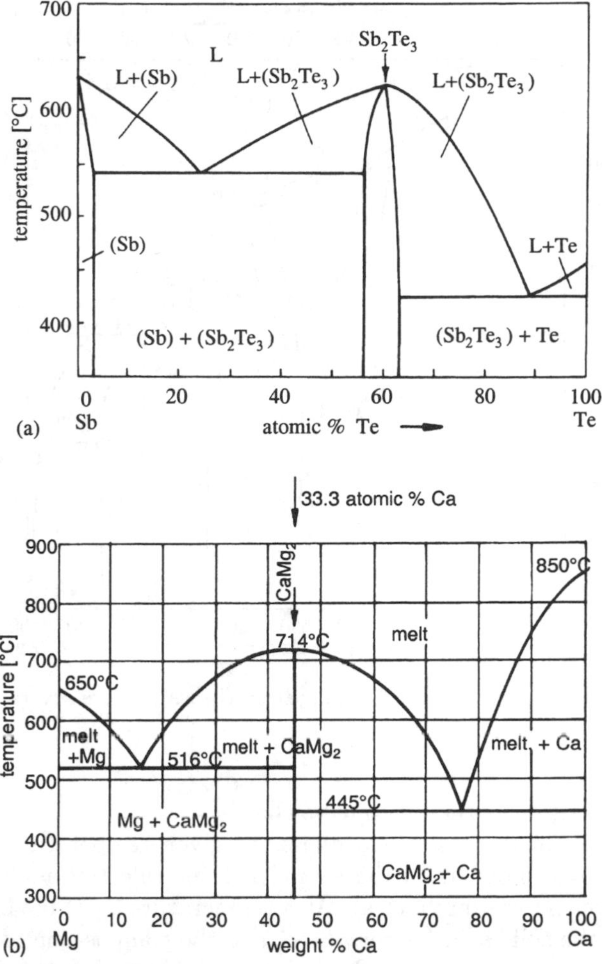

Fig. 4.13. Examples of phase diagrams with intermetallic phases, that may cover an extended range of compositions, e.g. SbTe (a), or only occur stoichiometrically (line phase), e.g. MgCa (b) [4.1].

decreases. At the same time the compositions of the solid solution and the melt change in such a way that the concentrations of B-atoms in the solid solution and in the melt decrease with decreasing temperature, corresponding to the concentration dependence of solidus and liquidus temperature. At an intermediate temperature T_1' the solid solution has the concentration c_1' and the melt c_2'. Eventually, the solidus temperature is reached and the terminal melt with concentration c_2'' is in equilibrium with a solid solution of composition c_0. The entire volume is now in the solid state and does not change composition on further cooling. In the two-phase regime both the fraction

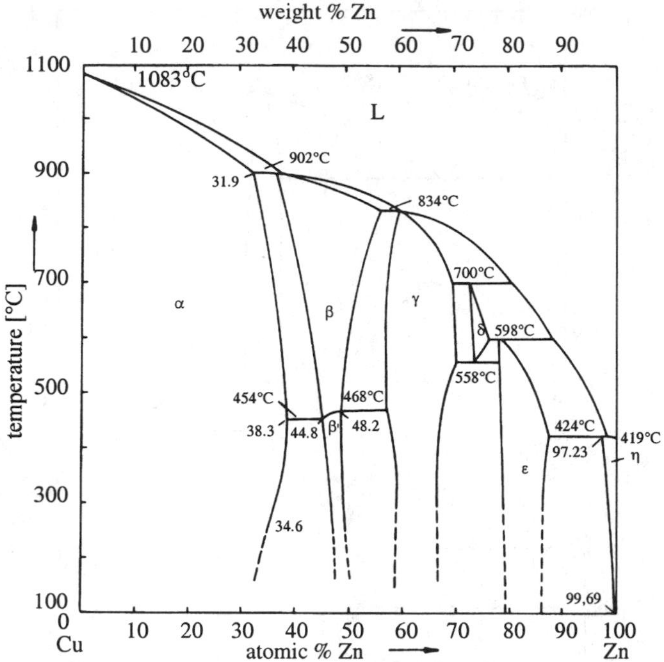

Fig. 4.14. Phase diagram of the CuZn system (brass), where several intermetallic phases occur [4.1].

and the composition of the two phases change.

The fraction of the respective phases at given temperature and composition can be calculated from the so-called lever rule (by analogy with the balance of moments in mechanics). At a temperature T_1' (Fig. 4.16) the concentration of the solid solution may be c_1' and the composition of the melt c_2''. If the average composition is c_0, the fraction liquid m_S and the fraction solid m_α are given by

$$m_\alpha = \frac{c_0 - c_2'}{c_1' - c_2'}$$

$$m_s = \frac{c_1' - c_0}{c_1' - c_2'}$$

$$\frac{m_\alpha}{m_s} = \frac{c_0 - c_2'}{c_1' - c_0}$$

This rule applies to all two-phase regimes, in particular for the fraction of solid phases in a miscibility gap, where c_1' and c_2' are the compositions of the two phases which are in mutual equilibrium at that temperature.

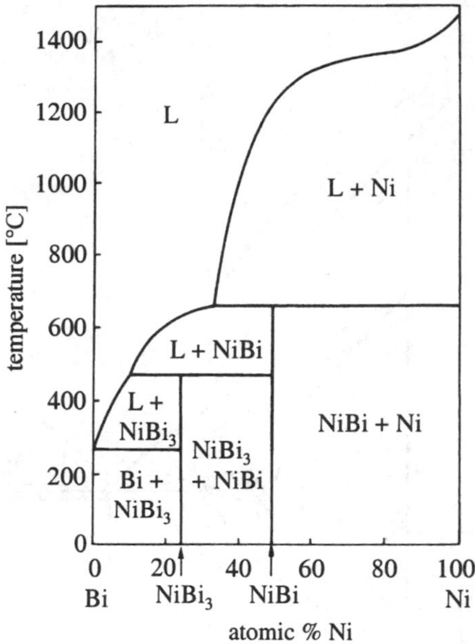

Fig. 4.15. Example of a phase diagram with hidden melting of intermetallic line phases [4.1].

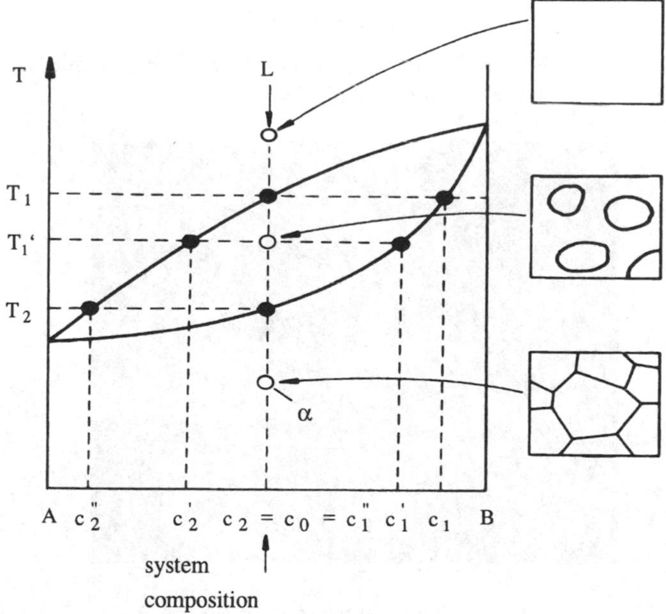

Fig. 4.16. Illustration of the solidification of binary alloys. The microstructures of the three states liquid, semi-solid and solid are schematically illustrated (c.f. text).

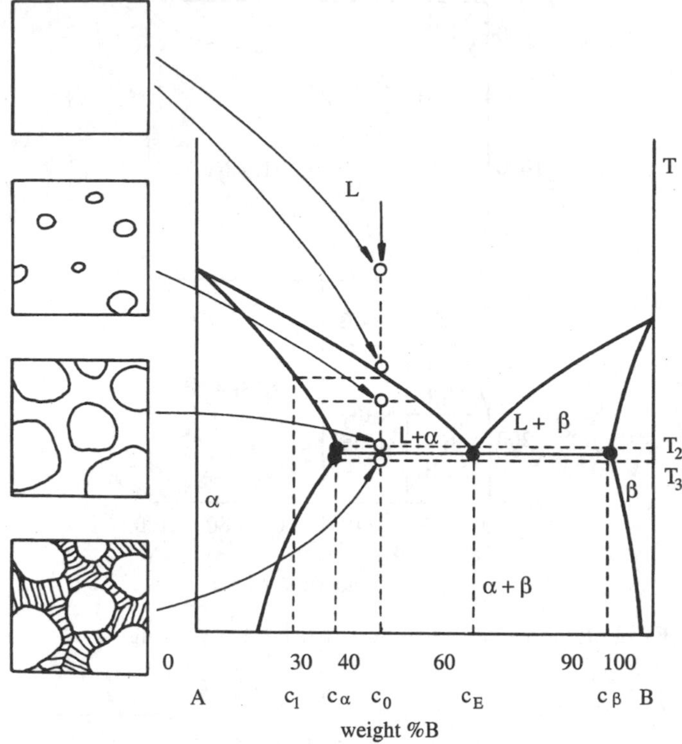

Fig. 4.17. Schematic diagram illustrating the development of the microstructure of a sub-eutectic alloy ($c_1 < c_E$) during solidification.

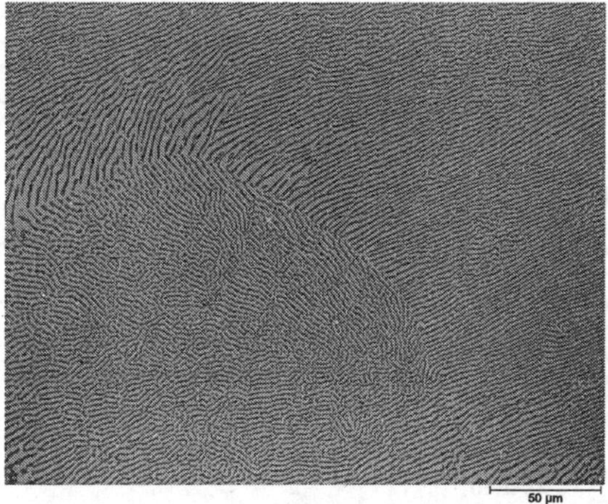

Fig. 4.18. Example of a eutectically solidified microstructure ($c = c_E$) in the AlZn system (95.16 weight% Zn, 4.84 weight % Al) [4.2].

In a eutectic system (Fig. 4.17) solidification initially proceeds completely analogously to the case of complete solubility if the composition is $c_\alpha < c_0 < c_E$. At the liquidus temperature the solid solution α solidifies with a composition c_1 from the melt with composition c_0. With decreasing temperature the fractions and compositions of the phases change according to the concentration dependence of the solidus and liquidus temperatures. When the eutectic temperature is reached, solid solution α and melt have attained the compositions c_α or c_E, respectively, and are in equilibrium with a solid solution crystal β of composition c_β. The terminal melt of composition c_E will then solidify concurrently into α and β with c_α or c_β, respectively, until the entire volume is solidified. Since both crystals solidify simultaneously but with different compositions, a lamellar solidification morphology is obtained, and the lamellar spacing depends on solidification rate (see Chapter 8). The completely solidified microstructure consists of primary solidified α solid solution crystals between which a lamellar structure of α and β crystals is spread out. If the average composition of the binary system is the eutectic composition c_E, one obtains a completely lamellar microstructure without any primary crystals (Fig. 4.18). In the solid state the composition of the two solid phases depends on temperature and, correspondingly, the volume fraction of both phases changes upon further cooling provided the physical mechanisms (diffusion, see Chapter 5) can operate.

4.2 Thermodynamics of Alloys

The phase diagrams can be derived and interpreted from fundamental thermodynamic considerations. At constant temperature T and constant pressure p, thermodynamic equilibrium is determined by the minimum of the Gibbs free energy G, where

$$G = H - TS \quad G = G_{min} \quad (\text{T, p = const.}) \tag{4.2}$$

H - enthalpy, S - entropy, T - temperature, p - pressure.

In Chapter 9 we will discuss in more detail the free energy of an alloy in the framework of the quasi-chemical model of a regular solution. In the following we will anticipate the most important results of such analysis in order to explain the types of phase diagrams observed.

For the temperature dependence of the free energy the entropy plays an important role, since with increasing temperature according to Eq. (4.2) the term $(-TS)$ becomes increasingly important. The larger TS the lower is G ($S > 0$, see below), and G attains a minimum in equilibrium. The increasing dominance of entropy with increasing temperature is also the reason for the melting of a solid as obvious from thermodynamic considerations of the temperature dependence of the free energy of pure metals (Fig. 4.19). The free energy of crystal G_c and melt G_m changes differently with temperature, hence the respective curves will intersect. The phase with the smallest free energy at a given temperature is the stable phase, i.e. the crystalline phase at low temperatures but the liquid phase at high temperatures. At the melting temperature $G_m = G_c$; both phases coexist in thermodynamic equilibrium. In pure elements the entropy is only caused by thermal motion of the atoms. These vibrations also contribute to the entropy of alloys, but there is another contribution to the entropy, the configurational entropy, which results from the multitude of different arrangements of atoms in a lattice. This was already introduced when the concentration of thermal vacancies was derived (see Section 3.2.2). In alloys the configurational entropy is usually referred to as the entropy of mixing S_m. For $N_A + N_B = N$ atoms and the atomic concentrations $c_A^a = N_A/N, c_B^a = N_B/N \equiv c$ we obtain for the entropy of mixing (see Chapter 9)

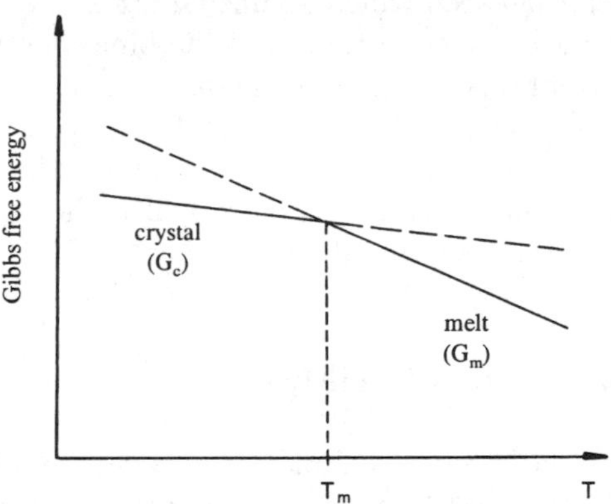

Fig. 4.19. Schematic representation of the temperature dependence of Gibbs free energy for the solid and liquid phases. $G_c = G_m$ at the melting point.

$$S_m = -Nk \{c \ln c + (1-c) \ln (1-c)\} \tag{4.3}$$

$S_m > 0$ because $c < 1$, and correspondingly, $-TS = -T(S_\nu + S_m) \approx -TS_m < 0$. The curve $S_m(c)$ (Fig. 4.20) is symmetrical with regard to $c = 0.5$, and approaches the pure components with infinite slope[2]

$$\lim_{c \to 0;1} \frac{\partial S}{\partial c} = \pm \infty \tag{4.4}$$

[2] Eq. (4.4) is also the reason why it is impossible to obtain pure elements by purification of alloys. Because of $\left.\frac{\partial H}{\partial c}\right|_{c=0} < \infty$ we obtain $\lim\limits_{c \to 0} \frac{\partial G}{\partial c} = -\infty$, i.e. with increasing purity the slope of the free energy curve becomes larger and for $c \to 0$ even infinitely large so that terminal purification can never be obtained.

The dependency $G(c)$ depends on $H(c)$. Under simplifying assumptions (quasi-chemical model of regular solution, see Chapter 9) $H(c)$ can be described as a parabola. Depending on the curvature of the parabola and the temperature, either H or S dominate $G(c)$. At very high temperatures (in particular in the melt) S always dominates and $G(c)$ follows a curve with a single minimum as function of composition. At low temperatures, i.e. in the solid phase, $G(c)$ has a similar shape as at high temperatures in the case of complete solubility. In contrast, for limited solubility $G(c)$ reveals a maximum in addition to two minima (Fig. 4.20).

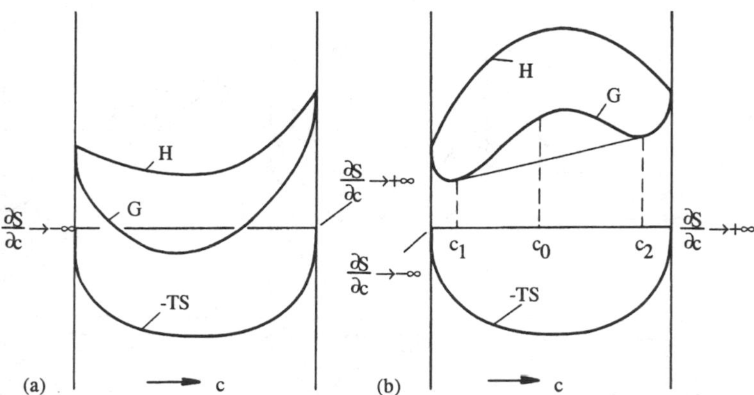

Fig. 4.20. Composition dependence of Gibbs free energy G and its component H and $(-TS)$ at constant temperature for (a) continuous miscibility; (b) miscibility gaps at $c_1 < c < c_2$.

The type and shape of a phase diagram can be qualitatively derived from the $G(c)$ curves of the respective phases at different temperatures. Since H does not change markedly with temperature, the curve $G(c)$ is only shifted vertically with temperature by the term $(-TS)$. Because only the relative position of the $G(c)$ curves is of importance, we only vary the $G(c)$ curve of one phase while the free energy of the other phase is kept unchanged. In the following, $G(c)$ of the melt will be kept constant as a reference while $G(c)$ of the crystal changes relative to $G(c)$ of the melt, i.e. with falling temperatures it will shift to smaller values.

Let us first consider the case of complete solubility (Fig. 4.21). At very high temperatures $G_L < G_c$ for all compositions, and the system exists as a liquid phase over the entire compositional range. With decreasing temperature eventually the case is obtained where $G_c = G_L$ for $c = 0$ or $c = 1$, i.e. at the melting temperature of one of the components. On further cooling one obtains separate ranges of composition, where either the melt or the crystal has the lower free energy. Between these compositional ranges the smallest free energy

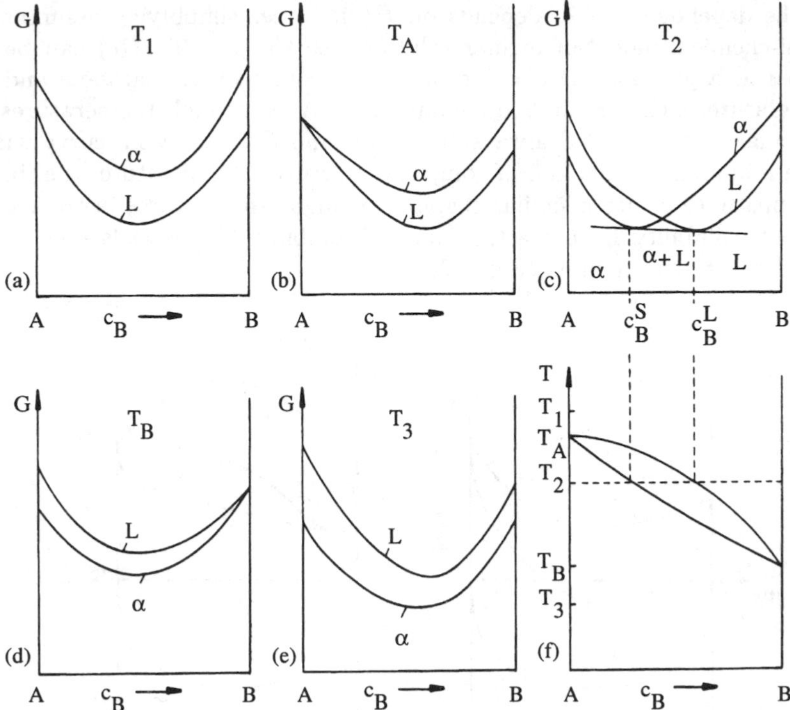

Fig. 4.21. Composition dependence of Gibbs free energy of crystal and melt of a binary alloy with continuous miscibility for different temperatures. (a) in a melt ($T_1 > T_A$); (b) at T_A (melting temperature of A); (c) for $T_B < T < T_A$; between c_B^S and c_B^L the system has two phases; (d) at T_B (melting temperature of B) ; (e) in the solid ($T_3 < T_B$). The composition ranges of the occurring phases correspond to an isothermal section through the phase diagram (f) (c.f. text).

is obtained for a two-phase mixture of crystal and melt. The free energy of a multi-phase mixture is given by the common tangent to the free energy curves of melt and crystal (see Ch. 9). According to the dependency $G(c)$, the range of existence of the constituent phases corresponds to an isothermal section through the two-phase regime of a cigar shaped phase diagram. On further cooling the points of intersection of the tangent, i.e. the concentration range of the two-phase regime, is shifted, until eventually at the melting temperature of the component with lower melting temperature $G_c < G_L$ for $0 \leq c \leq 1$ and $G_L = G_c$ for $c = 1$. Below this temperature $G_c < G_L$ holds for the entire range of composition and, therefore, only the solid phase exists. Plotting the ranges of existence of the constituent phases as function of temperature yields the phase diagram.

If a miscibility gap exists in the solid state, the $G(c)$ curve of the solid has two minima. By application of the same procedure as discussed for the case of complete solubility one obtains an eutectic or peritectic phase diagram (Figs. 4.22 and 4.23). If intermetallic phases exist a third free energy minimum appears in the solid state. Depending on the relative position of the minima with respect to each other the intermetallic phases are either obtained directly from the melt or by a peritectic reaction (Figs. 4.24, 4.25).

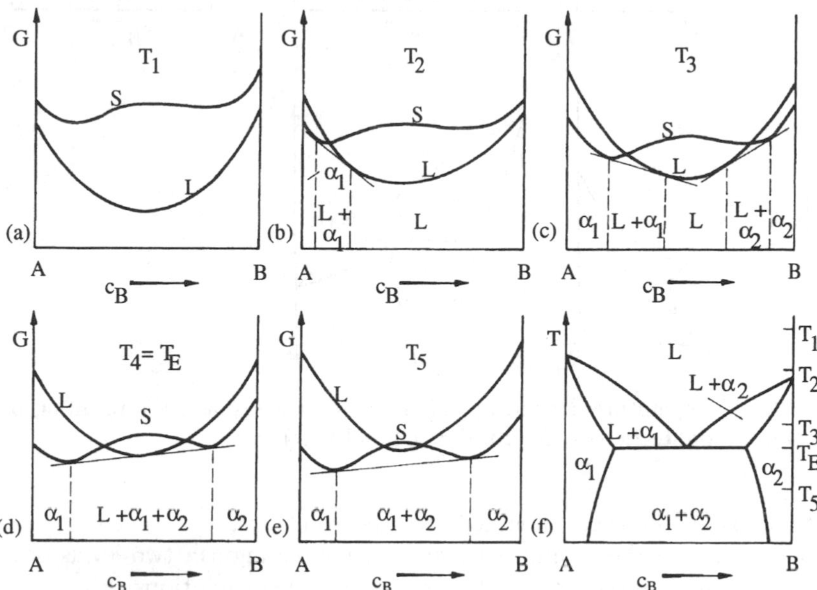

Fig. 4.22. Temperature dependence of Gibbs free energy and the resulting phase diagram in a eutectic system (c.f. Fig.4.6 and text).

4.3 Solid Solutions

If solute atoms are added to a pure metal there is always a range of concentration where solid solutions form because of the entropy of mixing as discussed in Sec. 4.2. The range of solubility can vary from virtually zero solubility to complete solubility of the alloying elements. Solution means in this context that the solute atoms are incorporated into the matrix crystal such that a mixture of alloying elements is obtained on an atomistic scale.

Since the solid phase in metallic materials is usually crystalline, the solid solution is also called a solid solution crystal. If the solubility range of a phase extends up to the pure component, i.e. to the limits of the phase diagram, this is called primary solid solution, or terminal solid solution. Intermetallic phases with a finite compositional range are discriminated from primary solid

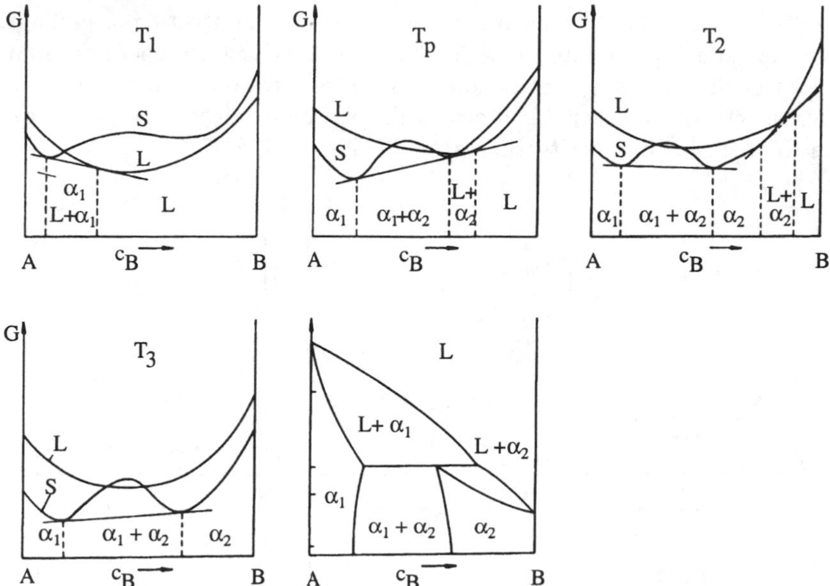

Fig. 4.23. Temperature dependence of Gibbs free energy and the resulting phase diagram in a peritectic system (c.f. Fig. 4.4 and text).

solutions by calling them intermediate solid solutions.

Depending on the atomic arrangement we distinguish two kinds of solid solutions, namely interstitial and substitutional solid solutions (Fig. 4.26). In interstitial solid solutions the solute atoms are located on the interstitial sites of the matrix lattice; substitutional atoms occupy regular lattice sites of the matrix.

Because of the small size of the interstitial sites, interstitial solid solutions occur only for alloying atoms of small atomic volume, in commercial alloys mainly the elements H, B, C, and N. The interstitial sites are usually smaller than the size of even the smallest alloying atoms, however, therefore interstitial atoms cause elastic distortions, the energy of which increases rapidly with increasing atomic size. The elastic distortion strongly decreases the solubility because the elastic energy increases the free energy of the solid solution and, therefore, destabilizes the solid solution in comparison to other phases. If second phases occur in the phase diagram the terminal solubility is determined by the relative position of the free energy curves $G(c)$, because the solubility limit is determined by the point of intersection of the common tangent to the $G(c)$ curves (Fig. 4.27). This influence becomes evident in the system Fe-C. The carbon atoms are located in the octahedral interstitial sites of the fcc lattice of γ-Fe and of the bcc lattice of α-Fe (Fig. 4.28). The size of an octahedral interstitial site in fcc lattices is $r^{\gamma}_{oct}/R_{Fe} = 0.41$ and, therefore, much larger than in the bcc lattice with $r^{\alpha}_{oct}/R_{Fe} = 0.16$. The ratio of the

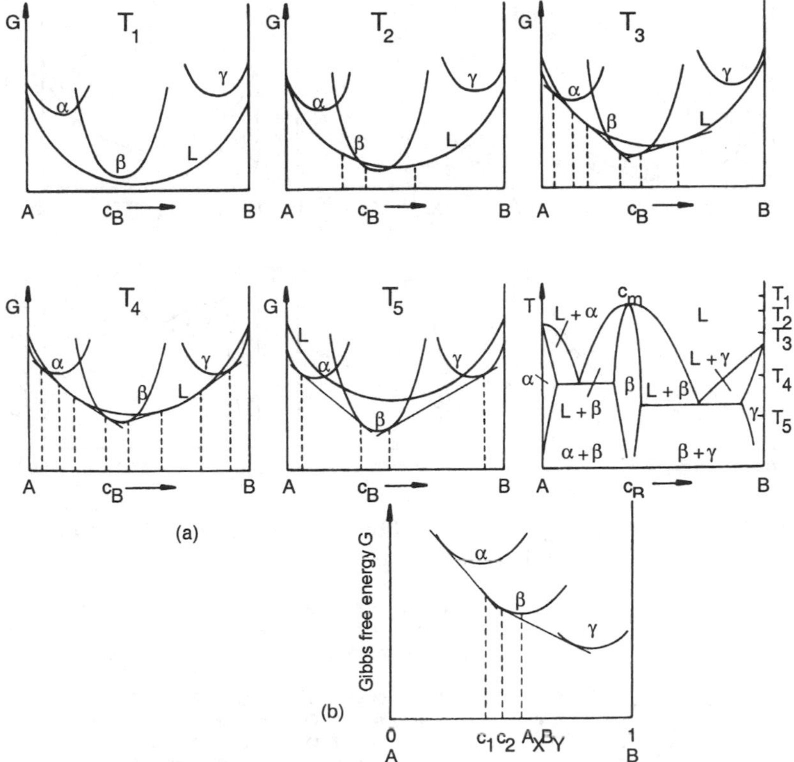

Fig. 4.24. (a) Temperature dependence of Gibbs free energy and the resulting phase diagram in a system with intermetallic phases that are formed in the melt (c.f. Fig. 4.21 and text). (b) Gibbs free energy of three phases α, β, γ. Phase β is stable between c_1 and c_2, but is not stable at its most stable composition A_xB_y.

atomic radii of carbon and iron atoms is $R_C/R_{Fe} = 0.61$. Obviously, the carbon atom is larger than the volume of the interstitial site, in particular much larger than the interstitial site in α-Fe. The phase diagram (Fig. 4.29) reveals the conspicuous effect of this difference on solubility. The solubility limit in fcc γ-Fe ($c^{\gamma}_{max} = 2.08$ weight %) is larger by two orders of magnitude than in bcc α-Fe ($c^{\alpha}_{max} = 0.02$ weight %).

The temperature dependence of the terminal solubility can be described by an Arrhenius law (Q - heat of solution)

$$c_{max} = c_0 \, e^{-\frac{Q}{kT}} \tag{4.5}$$

as demonstrated for C in Fe for interstitial alloys and copper in zinc for substitutional solid solutions in Fig. 4.30. This dependency can be easily explained in terms of the entropy of mixing as will be discussed in more detail in Chapter 9.

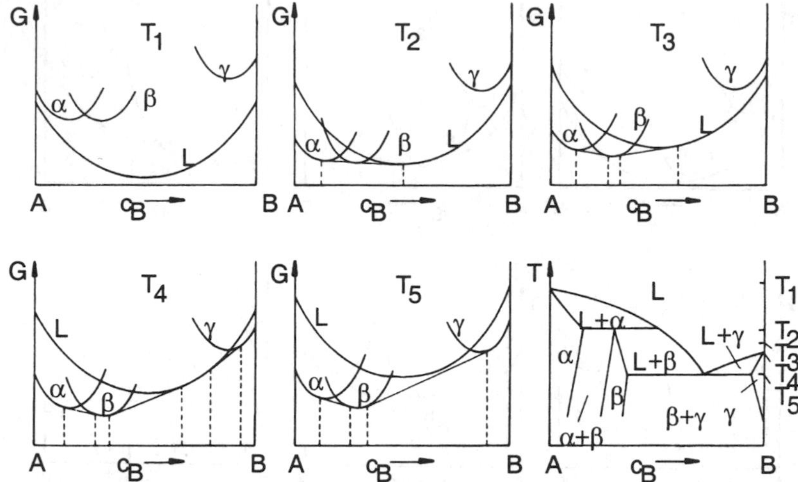

Fig. 4.25. Temperature dependence of Gibbs free energy and the resulting phase diagram in a system with an intermetallic phase.

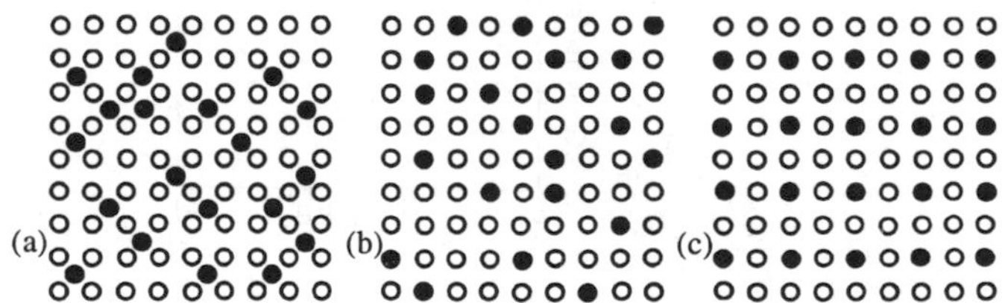

Fig. 4.26. Different types of solid solution crystals. (a) Interstitial solid solution, (b) substitutional solid solution with random distribution, (c) ordered substitutional solid solution.

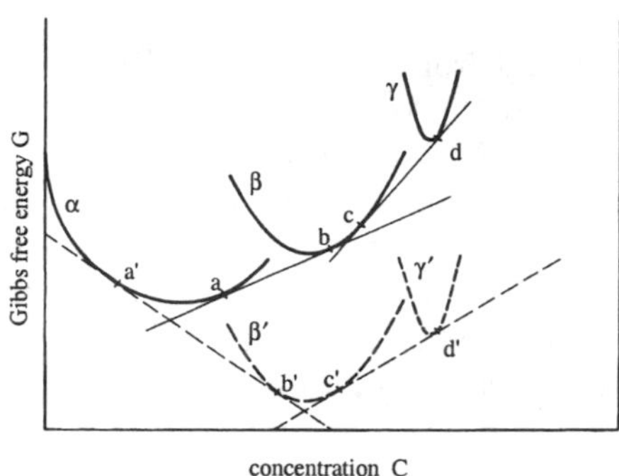

Fig. 4.27. The range of solubility of phases depends on the relative values of their Gibbs free energy. If, for example, the phase β' occurs, then the limit of solubility of the α phase is much lower than if β occurs, and α does not even occur in its most stable composition.

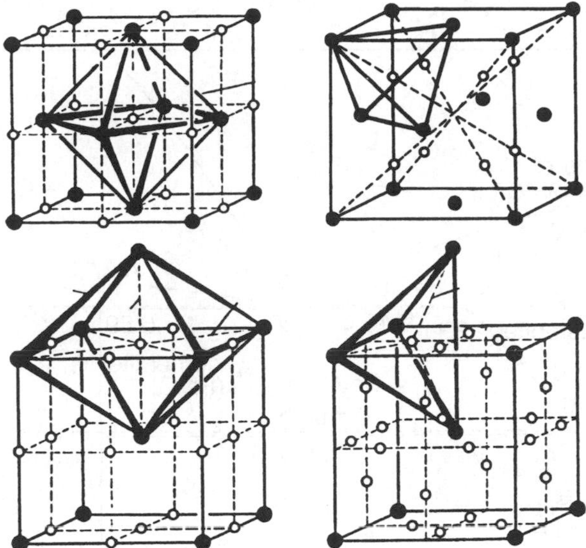

Fig. 4.28. Interstitials in the fcc and bcc lattice. (a) Octahedral interstice in fcc; (b) tetrahedral interstice in fcc, (c) octahedral interstice in bcc, (d) tetrahedral interstice in bcc. The open circles indicate the different but equivalent interstitial positions.

The majority of binary systems form substitutional solid solutions. Many of them show solubility over the entire compositional range, but in a large number of systems a miscibility gap is obtained in the solid state and, correspondingly, only a limited terminal solubility. It is self-evident that a miscibility gap will always appear if both alloying partners crystallize in different crystal structures. Irrespective of the crystal structure an incomplete solubility can have many physical causes. From a large number of studies on binary alloys Hume-Rothery formulated a set of rules to predict when substantial solubility can be expected:

1. The difference of atomic radii should not exceed 15%
2. The difference of electronegativity (chemical affinity) should be small
3. The number of valence electrons should not be very different

If any of these rules is violated one can expect only low solubility, and frequently even the existence of intermetallic compounds (Section 4.4).

The argument concerning the atomic size is easy to understand because of the associated elastic energy to force a solid atom into a matrix lattice site, quite analogous to the interstitial solid solutions discussed before. The limit between complete solubility and large miscibility gap is often sharply defined as evident for the systems Cu-Au and Cu-Ag in Section 4.1. All three metals crystallize in an fcc lattice. The system Ag-Au shows complete solubility (Fig. 4.3). Cu-Au reveals a minimum in the phase diagram (Fig. 4.6).

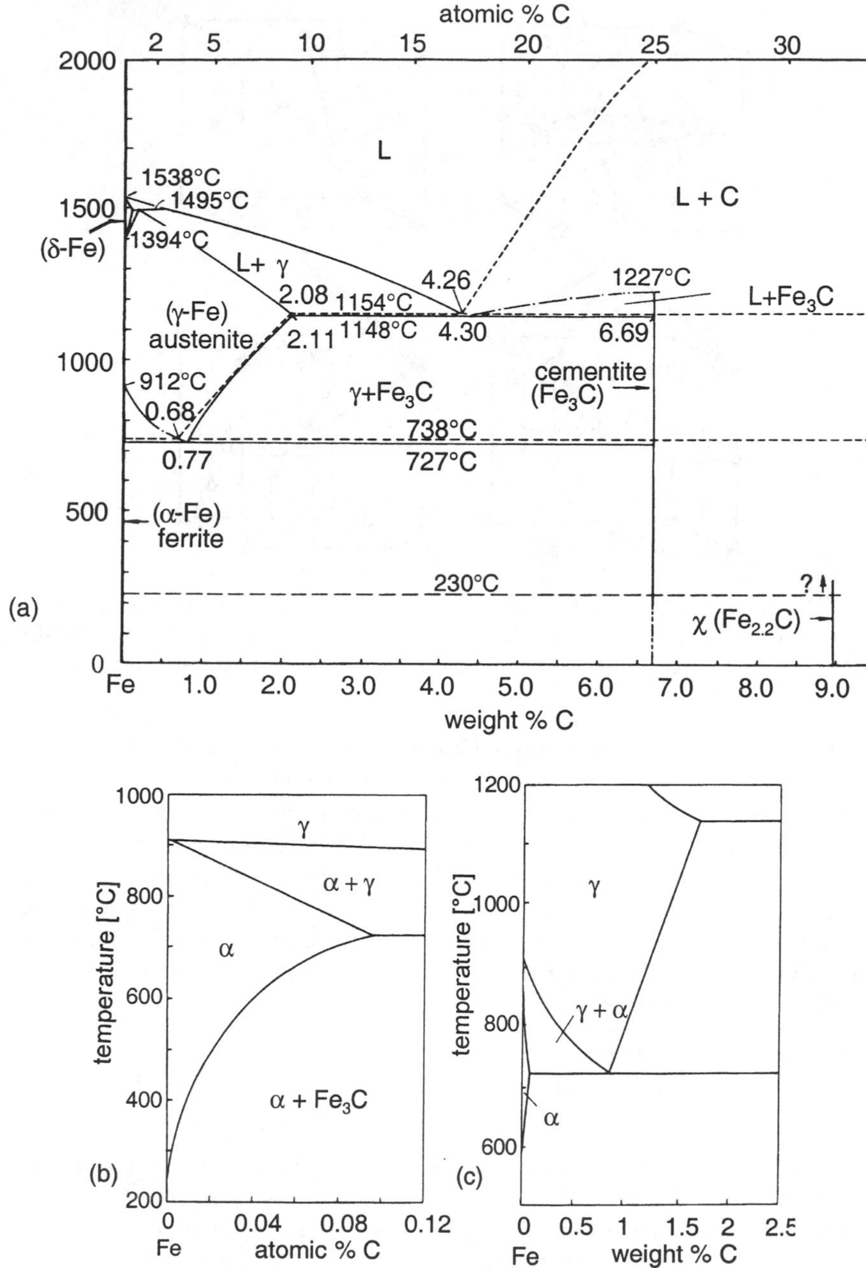

Fig. 4.29. Detail of the phase diagram of the FeC system. (a) The primary solubility of carbon increases with temperature in both α-Fe and γ-Fe. (b) The solubility of C in γ-Fe is considerably better than in α-Fe, due to the octahedral interstices being much smaller in the bcc lattice than in the fcc lattice (c) (after [4.1]).

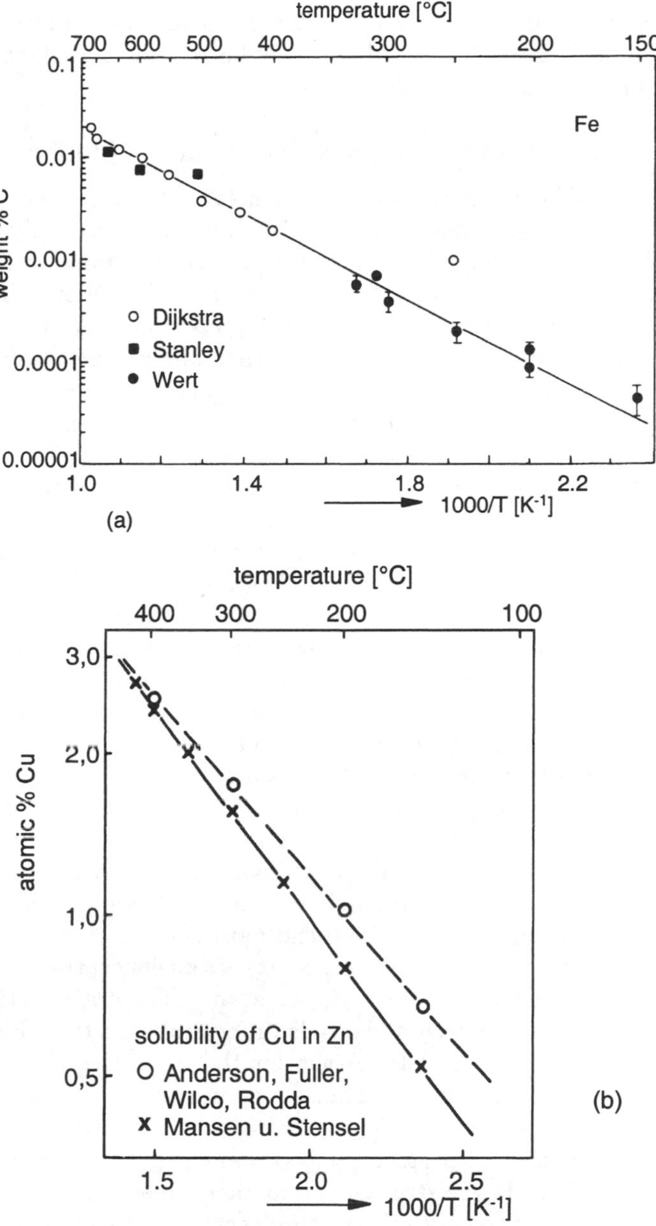

Fig. 4.30. Arrhenius plot of primary solubility of (a) C in Fe (interstitial solid solution) (after [4.3]) and (b) Cu in Zn (substitutional solid solution). In both cases temperature dependence of solubility is determined by a Boltzmann factor [$\exp(-Q/\mathrm{kT}]$, i.e. it increases rapidly with increasing temperature (after [4.4])

Although copper and silver are very similar chemically, their mutual solubility at room temperature is far less than 1%. The reason is a slightly different lattice parameter of gold and silver:

$$a_{Au} = 4.0786\text{Å}; \quad a_{Ag} = 4.0863\text{Å}; \quad a_{Cu} = 3.6148\text{Å}$$

The difference of lattice parameters between (a) silver and gold amounts to 0.19%, (b) between gold and copper 12.8%, and (c) between silver and copper 13%. Apparently, this minor difference in lattice parameter in the cases (b) and (c) leads to a complete change of solution behavior. The situation is already indicated by the minimum in the phase diagram of Cu-Au which signals a limit of complete solubility. A further diminishment of the atomic fit disfavors the solid solution with regard to a phase mixture.

A comparable atomic size and a correspondingly low elastic energy is a necessary but not a sufficient condition for extensive solubility as is evident from the other two rules of Hume-Rothery. The influence of the electronegativity is related to the tendency of formation of stoichiometric intermetallic compounds with increasing difference of electronegativity, because the heteropolar character of the bond increases. The appearance of such intermetallic compounds of course limits the solubility, and for very stable intermetallic compounds the terminal solubility can become very small (see Section 4.4).

The influence of the valence electrons is of a different nature. It is a common experience that the solubility of elements with a larger number of valence electrons, for instance when a bivalent element is alloyed to a monovalent matrix, is much larger than vice versa. The reason is the electronic band structure of solids according to the principles of quantum theory (see Chapter 10). Electrons have to comply with the Pauli principle which requires that every state of an electron can be occupied only by a single electron. Additional electrons have to assume higher energy states. In crystals there are adjacent electronic states with a large energy difference (band gap). Therefore, the energy of an additional electron can drastically increase at certain concentrations of valence electrons (number of valence electrons per atom). The critical electron density where this band gap occurs, depends on crystal structure. For instance, it is much larger for the bcc lattice than for the fcc lattice. Therefore, if an element with higher valency, for instance zinc in copper, is alloyed to an fcc solid solution and the critical valence electron concentration (VEC) for the fcc lattice is reached, the bcc structure becomes energetically more favorable, i.e. more stable than the fcc structure on further increase of electron concentration. The change of the crystal structure corresponds to the appearance of a new phase, and the primary solubility is determined by the stability of this secondary phase according to the common tangent in the free energy diagram. Intermetallic phases caused by supercritical VEC will be discussed in more detail in Section 4.4. The effect of the VEC can be demonstrated if the primary solubility of various alloys of a base metal and solute atoms of different valency are compared, for instance Cu with Zn, Ga, Ge, and As (Fig. 4.31).

Cu is monovalent. Zn has valency 2 and As has valency 5. With increasing valency the solubility decreases. If the solubility limit is plotted versus the VEC (instead versus atomic concentration), good agreement of the maximum solubility of the various alloys becomes obvious. The remaining minor differences are due to the different stability of the intermetallic compounds in the respective alloy system.

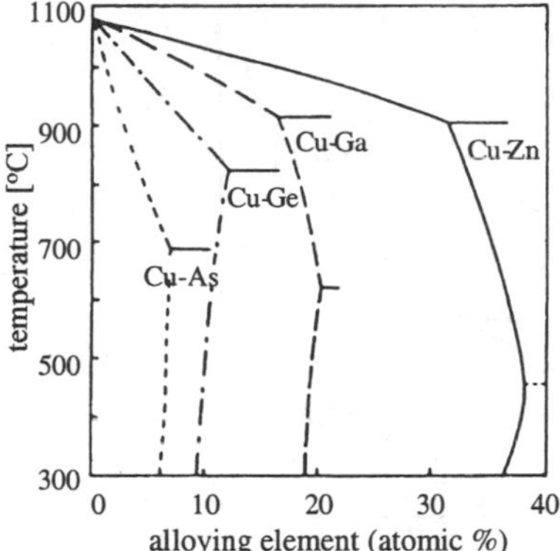

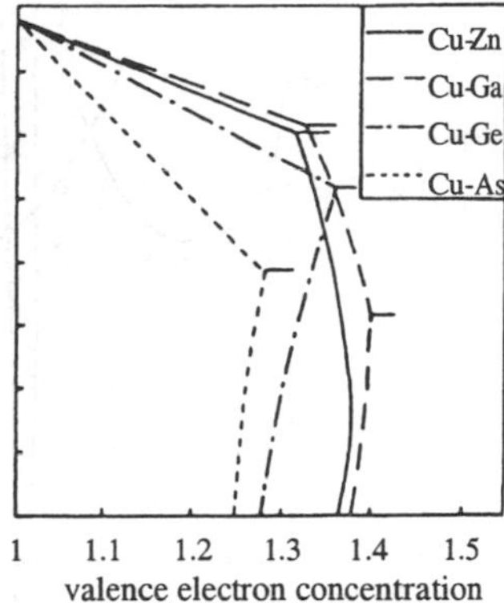

Fig. 4.31. Primary solubility limit of select copper alloys with different elements of higher valence. The greatly differing primary solubilities become quite comparable for all elements, if they are plotted over valence electron concentration rather than over atomic concentration.

4.4 Intermetallic Compounds

4.4.1 Overview

In many binary metallic systems new phases appear at intermediate concentrations and their range of existence does not extend to the pure components. Frequently, they are characterized by two special features, namely their non-stoichiometric composition and also their extended compositional range. The term intermetallic compound, therefore, is misleading, because it associates a stoichiometric composition depending on the valency of the partners, as in chemical compounds. It would be more appropriate to refer to these phases as intermetallic phases or intermediate solid solutions but this terminology has not been widely adapted in literature. The reasons for the apparently non-chemical nature of bonding in intermetallic compounds are not easily derived from physical principles. The compounds can form for many reasons and usually there are several causes and circumstances which make them appear. In the following we will try to use structural arguments to rationalize the existence and composition of frequently observed intermetallic compounds.

Quite generally, and also for intermetallic compounds, the existence and the compositional range of intermetallic compounds is determined by the relative position of their free energy curve (Fig. 4.32) with respect to other phases in the system. Because of the tangent rule at a given temperature some phases may not occur at all (δ) or at least not in their most stable (for instance stoichiometric) composition ($\alpha, \gamma, \varepsilon$).

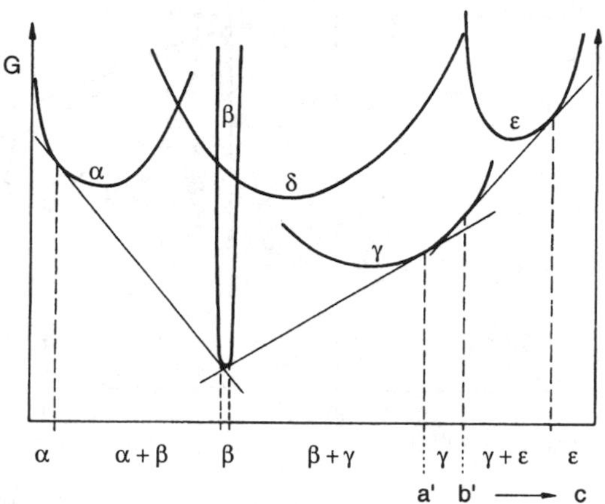

Fig. 4.32. Where intermetallic phases occur existence and solubility also depend on the relative positions of free energy curves. In this example α does not occur in its most stable phase and the δ phase is not formed.

4.4.2 Ordered Solid Solutions

If in a binary alloy the bonding between unlike partners is stronger than between like atoms, each atom will try to maximize the number of unlike atoms as its next neighbors. This corresponds to the case of a regular solution with a large negative exchange energy (see Chapter 9):

$$H_0 = H_{AB} - (H_{AA} + H_{BB})/2 \ll 0.$$

For specific stoichiometric compositions strictly periodic arrangements of atoms are attained this way, where each atom has an environment of a maximum number of alloying atoms. For instance, for an atomic concentration $c^a = 0.5$, i.e. a composition of type AB, this is the case in a bcc lattice, which is composed of two simple cubic sublattices. If each sublattice is occupied by only one sort of atoms, then each A-atom has only B-neighbors and vice versa (Fig. 4.33a). Such a perfectly ordered atomic distribution and arrangement is referred to as long range order, since this arrangement extends to macroscopic dimensions across many unit cells. By analogy the fcc lattice is composed of four sublattices so that three sublattices can be occupied with atoms of kind A and one sublattice with atoms of kind B, to represent a compound of type AB_3 for a composition 25:75 (Fig. 4.33b).

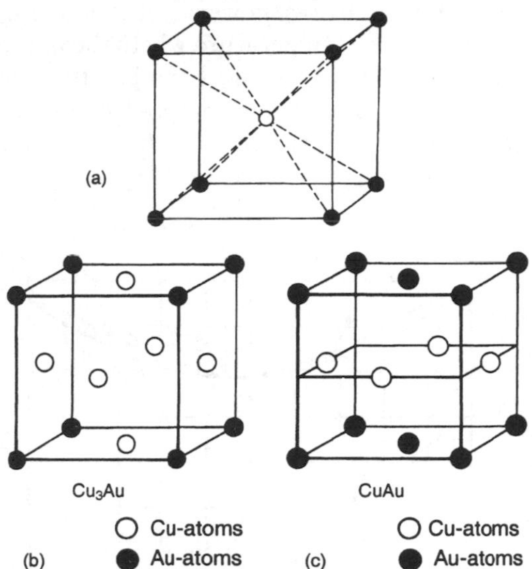

Fig. 4.33. Ordered atom distributions of the AB-type can be realized in the CsCl lattice (a) ($B2$ structure), those of the AB_3 in the Cu_3Au lattice (b) ($L1_2$ structure), that are consistent with a bcc or fcc solid solution crystal with random atom distribution. A face-centered cubic solid solution of the AB-type (c) cannot become ordered without losing its cubic structure. If the fcc lattice is subdivided into two sub-lattices for A and B, layers are formed. Due to differing atomic radii this results in a tetragonal crystal lattice.

On the other hand the fcc lattice cannot sustain a long range ordered structure of type AB without losing its cubic symmetry. In principle, two sublattices can be occupied with A-atoms and the remaining two sublattices with B-atoms, however, irrespective of how the sublattices are distributed, a layered structure is always obtained (Fig. 4.33c), that breaks the cubic symmetry and generates a tetragonal crystal structure ($a = b \neq c$). This structure is found in Cu-Au which indicates that long range ordering generates phases with other crystal structures[3].

Thermal motion of the atoms and, therefore, entropy tends to distroy ordered atomic distributions. Therefore, the degree of order decreases with

[3] In this context it is noted that the Cu_3Au ($L1_2$) and the CsCL (B2) structures are no longer fcc or bcc structures, in contrast to the disordered solid solution crystals of the same composition. Therefore, long range order is always associated with a change of the crystal structure. Although the lattice sites cannot be discriminated from the disordered structure, they are not equivalent anymore.

increasing temperature. If the exchange energy is very small, the tendency to order is not strongly pronounced, and already at temperatures far below the melting temperature long range order is lost. Examples are the various ordered phases in the system Cu-Au. For instance, Cu-Au is ordered at temperatures below 390°C. At higher temperatures, however, it exists as a disordered solid solution crystal (Fig. 4.34). The temperature where the ordered state changes to the disordered state is called the critical temperature. In other systems, for instance the important Ni-Al system, ordered phases like Ni_3Al and in particular NiAl are ordered up to the melting temperature, i.e. the critical temperature exceeds the melting temperature (Fig. 4.35).

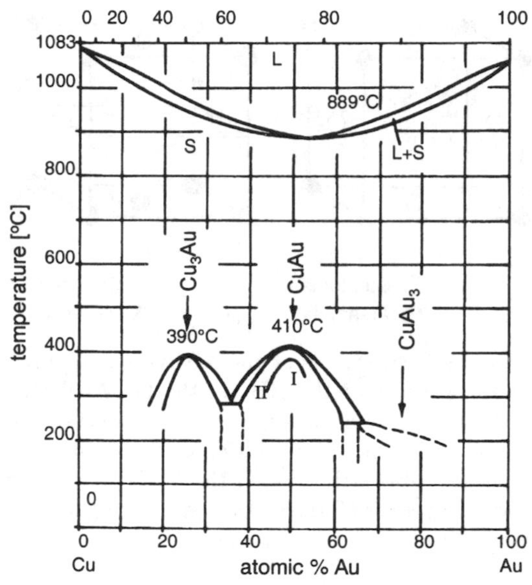

Fig. 4.34. Phase diagram of the system CuAu that forms several solid state ordered phases at low temperature. However, these turn to a disordered state well below the melting point [4.1].

The degree of long range order can be quantified according to a proposal of Bragg-Williams by the long range order parameter

$$s = \frac{p - x}{1 - x} \tag{4.6a}$$

where p is the fraction of A-atoms on A-sublattices, and x is the fraction of A-atoms in the alloy.

This concept is the simplest for an alloy of type AB if the solid solution has a bcc crystal structure. In this case one sublattice can be associated with each element and the long range order parameter reads

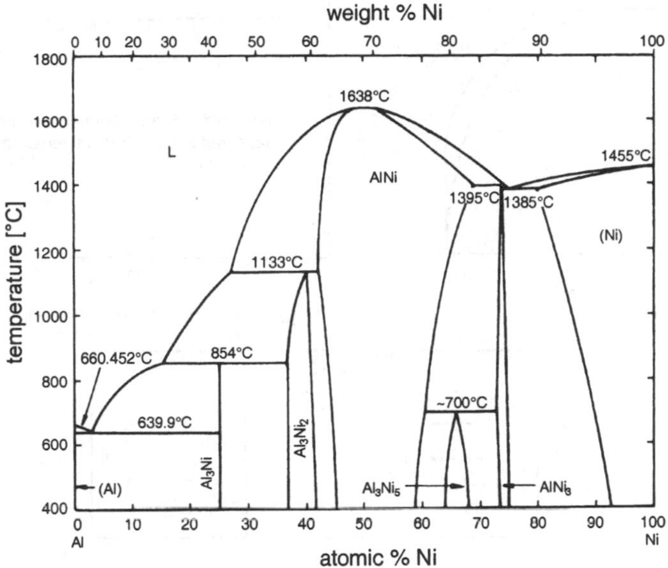

Fig. 4.35. Phase diagram of the system NiAl. The phases Ni_3Al ($L1_2$) and NiAl ($B2$) are ordered up to the melting point [4.1].

$$s = 2p - 1 \tag{4.6b}$$

where p again is the fraction of A-atoms on the A-sublattice. Correspondingly, s will change in the limits

$$-1 \leq s \leq 1,$$

because for perfect order $p = 1$ and, therefore, $s = 1$. The state of complete disorder is described by $s = 0$, i.e. $p = 0.5$ for a composition AB, this is a statistical atomic arrangement. The case $s = -1$ reflects also a perfectly ordered state except that the atoms are arranged on the wrong sublattices. Consequently, only the absolute value of s indicates the degree of order. Values of $s = -1$, however, lead to problems in some cases as will be discussed later. The magnitude of s can be expected to depend on temperature and on the exchange energy H_0. The temperature dependence $s(T)$ can be calculated, for instance, in the quasi-chemical model (see Chapter 9), if the thermodynamic functions H and S are expressed in terms of the order parameter s to yield $G(s)$, and equilibrium is attained for $dG/ds = 0$. This results in the dependency shown in Fig. 4.36.

There are several experimental methods to determine the long range order parameter, for instance the occurrence of so-called superlattice reflections in the Debye-Scherrer diagram. Because of the different scattering properties of

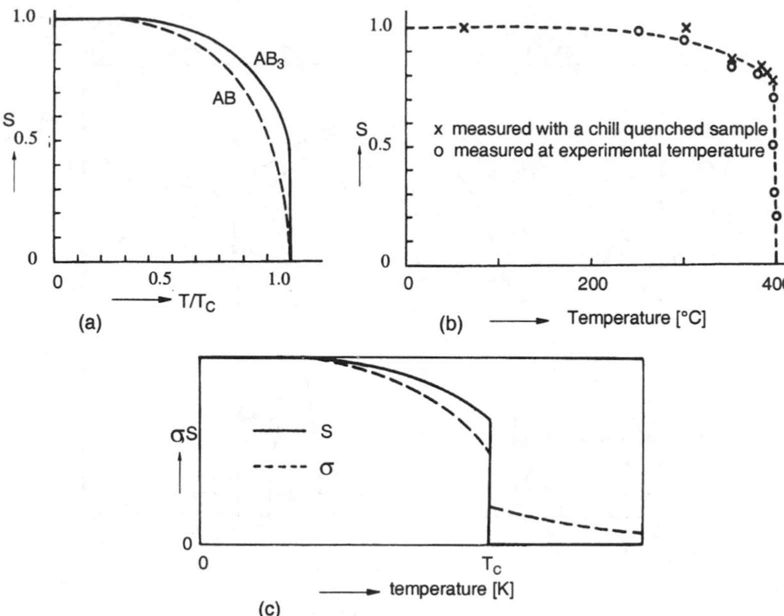

Fig. 4.36. Temperature dependence of the order parameters. (a) Long range order parameter s for ordered alloys of the AB- and AB_3-type. Above temperature T_c (critical temperature) the system is disordered. For AB_3 the temperature dependence of s at T_c is discontinuous (1st order phase transition). (b) Measured $s(T)$ curve for Cu_3Au. The measured values show good correspondence with the calculated curve (dashed line). (c) Comparison of the temperature dependence of the long range order parameter s and the short range order parameter σ. $\sigma > 0$ even above T_c (after [4.5]).

the two elements of a binary alloy, the extinction rules of X-ray diffraction (see Ch. 2.5.2) do not pertain to an ordered structure and X-ray reflections occur that are systematically absent in disordered solid solutions (Fig. 4.37). The occurrence of these superlattice reflections can also be explained from the unit cell of the long range ordered superstructure of the crystal (Fig. 4.38), which gives a larger lattice parameter and, according to Bragg's law, reflections at smaller angles will appear. The formation of superlattices also causes a drastic decrease of the electrical resistivity (Fig. 4.39) since the electrical resistivity is primarily caused by perturbations of the periodic arrangement, for instance by solute atoms in a solid solution. A superlattice, in contrast, is strongly periodic and despite having the same number of solute atoms, the electrical resistivity is substantially lower than in a random solid solution of the same composition.

The definition of the long range order parameter can yield contradictory results, if in different parts of a crystal A and B atoms occupy different sublat-

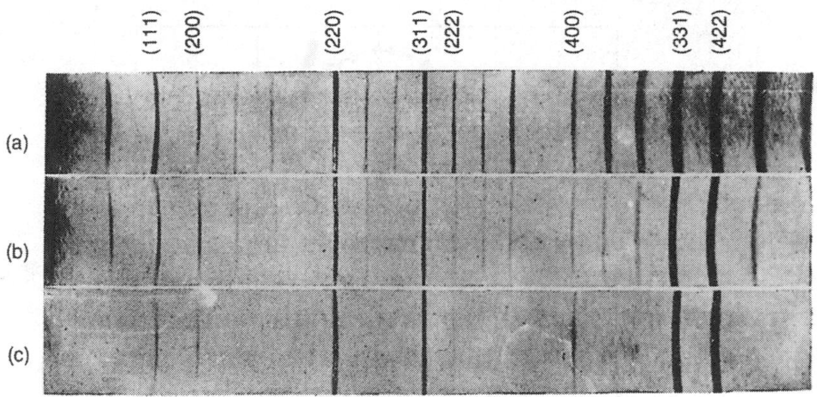

Fig. 4.37. Occurrence of long range ordered phases (in this case Cu_3Au) can be confirmed by superlattice lines in a Debye-Scherrer-image. (a) $T \ll T_c$; (b) $T < T_c$; (c) $T > T_c$.

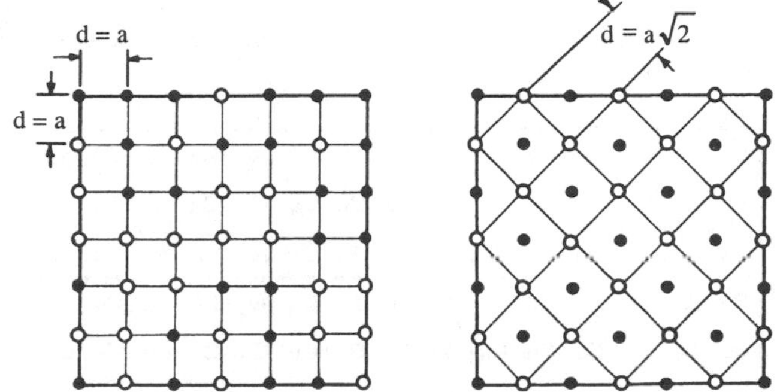

Fig. 4.38. Superlattice lines can be explained easiest for the simple case of a cubic lattice. With atoms arranged randomly the largest lattice plane spacing distance is $d = a$. An ordered arrangement (AB in this case) leads to a different unit cell, and the largest lattice plane spacing becomes $d = a\sqrt{2}$. This results in an additional reflection at a smaller angle, satisfying the Bragg condition.

tices (Fig. 4.40). This can occur if, during the transition from the disordered to the ordered state, the ordered state is generated concurrently at various nucleation sites, where the choice of the sublattice occurs at random and, therefore, different. If these differently arranged although perfectly ordered regions eventually impinge, the order parameter changes from $s = 1$ to $s = -1$ at the interface. These interfaces are called antiphase boundaries, and the perfectly ordered but differently occupied sublattices are called domains. They can be imaged by TEM (Fig. 4.41). On average the different domains occur with equal frequency so that the mean long range order parameter vanishes, which, however, is physically wrong because essentially the entire volume of the crystal is perfectly ordered except for the antiphase boundaries.

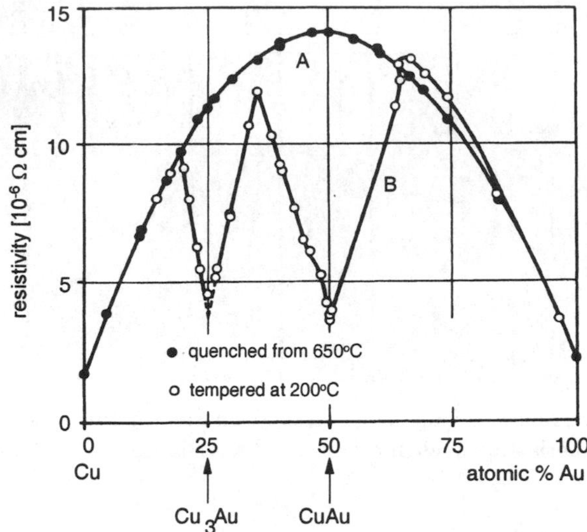

Fig. 4.39. When superlattices occur, electrical resistivity decreases sharply, e.g. Cu_3Au and CuAu (after [4.7]).

Cu	Au	Cu	Au	Cu	Au	Au	Cu	Au	Cu	Au	Cu	Au	Cu	Au
Cu	Cu	Cu	Cu	Cu	Cu	Cu	Cu	Cu	Cu	Cu	Cu	Cu	Cu	Cu
Cu	Au	Cu	Au	Cu	Au	Au	Cu	Au	Cu	Au	Cu	Au	Cu	Au
Cu	Au	Cu	Au	Cu	Au	Au	Cu	Au	Cu	Au	Cu	Au	Cu	Au
Cu	Cu	Cu	Ca	Cu	Cu	Cu	Cu	Cu	Cu	Cu	Cu	Cu	Cu	Cu
Cu	Au	Cu	Au	Cu	Au	Au	Cu	Au	Cu	Au	Cu	Au	Cu	Au
Cu	Au	Cu	Au	Cu	Cu	Cu	Cu	Cu	Au	Cu	Cu	Cu	Cu	Cu
Cu	Cu	Cu	Cu	Au	Cu	Au	Cu	Cu	Cu	Cu	Au	Cu	Au	Cu
Cu	Au	Cu	Au	Cu	Cu	Cu	Cu	Cu	Au	Cu	Cu	Cu	Cu	Cu
Cu	Cu	Cu	Cu	Au	Cu	Au	Cu	Cu	Cu	Cu	Au	Cu	Au	Cu
Cu	Au	Cu	Au	Cu	Au	Cu	Au	Cu	Au	Cu	Cu	Cu	Cu	Cu
Cu	Cu	Cu	Cu	Cu	Cu	Cu	Cu	Cu	Cu	Cu	Au	Cu	Au	Cu
Cu	Au	Cu	Au	Cu	Au	Cu	Au	Cu	Cu	Cu	Cu	Cu	Cu	Cu
Au	Cu	Au	Cu	Au	Cu	Cu	Cu	Au	Cu	Au	Cu	Au	Cu	Au
Cu	Cu	Cu	Cu	Cu	Cu	Cu	Au	Cu	Cu	Cu	Cu	Cu	Cu	Cu

Fig. 4.40. A long-range ordered crystal may be subdivided into regions (domains), that all exhibit complete long-range order, but that have differently occupied sublattices. The boundaries of these regions are called antiphase boundaries.

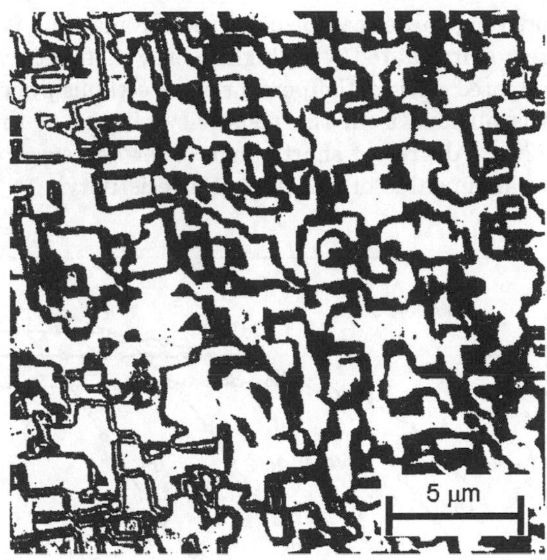

Fig. 4.41. Antiphase boundaries are visible in TEM, in this example $AuCu_3$.

This inherent contradiction in the concept of long range order can be avoided by defining order in terms of the short range order parameter

$$\sigma = \frac{q - q_d}{q_m - q_d} \tag{4.7}$$

where q is the fraction of B-atoms as next neighbors of A, q_d the fraction of B-atoms as neighbor of A in the disordered state, and q_m the fraction of B-atoms as next neighbors of A in the completely ordered state. For a given alloy q_m is constant, for instance for AB-alloys $q_m = 1$.

The short range parameter σ, therefore, is defined in terms of the next neighbor relationship of an arbitrary selected atom, and has the advantage over the long range order parameter that it does not depend on a long range correlation. Also it can be defined for any composition, even non-stoichiometric compositions and can be extended to the second next, third next, and so on neighbors (shells). In the case of long range order, i.e. $s = 1$ also $\sigma = 1$, however, σ varies only between $0 \leq \sigma \leq 1$. The temperature dependence $\sigma(T)$ can be calculated. It shows a similar dependency as $s(T)$ for $T \ll T_c$ (Fig. 4.36), but it does not go to zero for T_c and remains larger than zero even for $T \geq T_c$, i.e. $\sigma(T) \geq 0$ for $T \geq T_c$.

Short range order is more complicated to reveal experimentally in the absence of long range order. For this inelastic X-ray or neutron scattering is usually used. In contrast to long range ordered alloys, the electrical resistivity in many systems increases with increasing degree of short range order. The development of short range order can be observed if a material is quenched from high temperatures or irradiated with high energy elementary particles and, subsequently, tempered at low temperatures to establish short range order by diffusion. As is obvious from the resistivity change of the system gold-silver in Fig. 4.42, the degree of short range order decreases with increasing temperature (maximum value of the electrical resistivity).

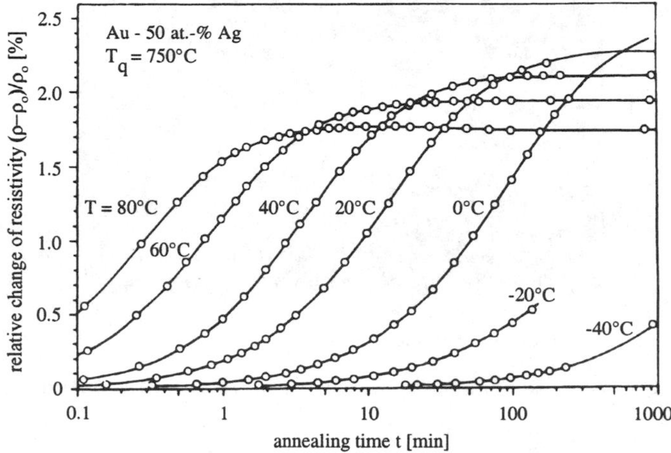

Fig. 4.42. In the AuAg system electric resistivity increases when short range order exists. If samples chill quenched from high temperature are tempered at low temperature, this results in short-range ordering, with the degree of order increasing with decreasing tempering temperature (after [4.7]).

4.4.3 Compound Phases

Pure metals are characterized by non-directional bonding and dense atomic packing. This is different for alloys, however. Only in very special cases do the components of an alloy form ideal solutions, i.e. $H_0 = 0$. Even if there is complete solubility in the solid state, there is always a tendency to long range order or decomposition, which will be considered in more detail in Chapter 9. This tendency is a consequence of the bonding character between the atoms and, therefore, depends on the specific alloy, because real bonds are always mixtures of bond types, such that one or the other type of bond will dominate depending on the constitutive components. This is also the reason underlying the Hume-Rothery rules of solubility and the violation of any of those rules increases the tendency for intermetallic compounds. If the polarity of the atoms, i.e. their relative location in the periodic system of elements, is very different, the ionic character of the bond becomes stronger.

Zintl-Limit

								Mg_4Al_3	Mg_2Si	Mg_3P_2	MgS	$MgCl_2$
								MgAl				
					Mg_2Co			Mg_3Al_4				
Mg_2Ca			MgCr	Mg_3Mn	Mg_2Ni	Mg_2Cu	Mg_7Zn_3	Mg_5Ga_2	Mg_2Ge	Mg_3As_2	MgSe	$MgBr_2$
			Mg_3Cr_2		$MgNi_2$	$MgCu_2$	MgZn	Mg_2Ga				
			Mg_3Cr				Mg_2Zn_3	MgGa				
			Mg_4Cr				$MgZn_2$	$MgGa_{1+x}$				
							Mg_2Zn_{11}					
Mg_9Sr		Mg_2Zr				Mg_3Ag	Mg_3Cd	Mg_5In_2	Mg_2Sn	Mg_3Sb_2	MgTe	MgI_2
Mg_4Sr						MgAg	MgCd	Mg_2In				
Mg_3Sr							$MgCd_3$	MgIn				
Mg_2Sr								$MgIn_2$				
Mg_9Ba	Mg_9La				Mg_2Pt	Mg_3Au	Mg_3Hg	Mg_5Tl_2	Mg_2Pb	Mg_3Bi_2		
Mg_4Ba	Mg_3La					Mg_5Au_2	Mg_2Hg	Mg_2Tl				
Mg_2Ba	Mg_2La					Mg_2Au	MgHg	MgTl				
	MgLa					MgAu	$MgHg_2$					

Structure typical for metallic compounds

Structure typical for salt-like compounds

(a)

IA	IIA	IIIB	IVB	VB	VIB	VIIB	VIIIB	VIIIB	VIIIB	IB	IIB	IIIA	IVA	VA	VIA	VIIA	VIII A
1 H																	2 He
3 Li	4 Be											5 B	6 C	7 N	8 O	9 F	10 Ne
11 Na	12 Mg											13 Al	14 Si	15 P	16 S	17 Cl	18 Ar
19 K	20 Ca	21 Sc	22 Ti	23 V	24 Cr	25 Mn	26 Fe	27 Co	28 Ni	29 Cu	30 Zn	31 Ga	32 Ge	33 As	34 Se	35 Br	36 Kr
37 Rb	38 Sr	39 Y	40 Zr	41 Nb	42 Mo	43 Tc	44 Ru	45 Rh	46 Pd	47 Ag	48 Cd	49 In	50 Sn	51 Sp	52 Te	53 I	54 Xe
55 Cs	56 Ba	57 La	72 Hf	73 Ta	74 W	75 Re	76 Os	77 Ir	78 Pt	79 Au	80 Hg	81 Tl	82 Pb	83 Bi	84 Po	85 At	86 Rn
87 Fr	88 Ra	89 Ac	104 Rf	105 Ha													

Zintl-Limit

(b)

Fig. 4.43. (a) Intermetallic phases of magnesium. Compounds formed with elements from groups IVA, VA, VIA, and VIIA are stoichiometric compounds, for the lower numbered groups of the periodic table (b) non-stoichiometric intermetallic phases occur. The Zintl boundary between columns IIIA and IVA divides these different types of intermetallic phases.

Surprisingly, there is a sharply defined limit, the Zintl limit, for the anionic partner, namely between the third and the fourth main group (IIIA- IVA) of the periodic system. Beyond this limit the ionic character of the bond dominates and stoichiometric salt-like compounds appear, the Zintl phases. They exist in specific lattice structures depending on stoichiometry and atomic size (see Chapter 2). The conspicuous discontinuity at the Zintl limit becomes obvious when we compare the known phases of a base-element with different elemental partners, for instance, Mg-alloys in Fig. 4.43. Mg alloys with partners of the groups IVA-VIIA form specific, stoichiometric compounds according to their valency. In contrast, the alloys of Mg with elements of other groups reveal a large variety of different phases where the valency is entirely irrelevant.

The strength of the polarity, i.e. the ionic bonding character decreases with decreasing electronegativity of the anion or a decreasing electropositivity of the cation, respectively. This becomes apparent from the stability of Zintl phases. The magnitude of the melting temperature serves as a rough measure of the stability (Fig. 4.44). The electronegativity usually decreases within a period with increasing group number and within a group with increasing period. Equivalently, the change of the electropositivity of the cation causes the same dependency (Fig. 4.44b). With decreasing group number and increasing period the electropositivity of the cation increases. The stability of a phase grows with increasing difference of electronegativity.

Depending on electron configuration also the covalent part of a bond can dominate. For pure elements a strictly covalent bond can be realized in the diamond lattice, since in such a crystal structure the directional bonds can be accomplished in a long range ordered arrangement. Alloys of type AB crystallize in related crystal structures, for instance the zinc blende and the wurtzite lattice (Fig. 4.45). The zinc blende lattice is a diamond lattice with a superstructure. The anions form an fcc lattice where every second tetrahedral interstitial site is occupied by one cation. Each cation, therefore, has four anions as next neighbors and vice versa. This arrangement can also be visualized as two interpenetrating fcc lattices, each of which contains only one kind of ions. The wurtzite lattice is very similar to the zinc blende lattice. It is also based on the diamond lattice with superstructure. In this case the anions form a hexagonal lattice. Table 4.2 gives examples of stoichiometric compounds corresponding to their respective crystal structures.

4.4.4 Phases with High Packing Density

If the bonds have a strong metallic character, the packing density of the atomic arrangement plays an important role. Crystal structures with high packing density, however, can only be obtained for specific compositions and atomic volumes of the alloy partner elements. In special cases a very high packing density of the alloy is obtained. An example are the Laves phases. They occur with composition AB_2 if the ratio of the atomic radii of the two

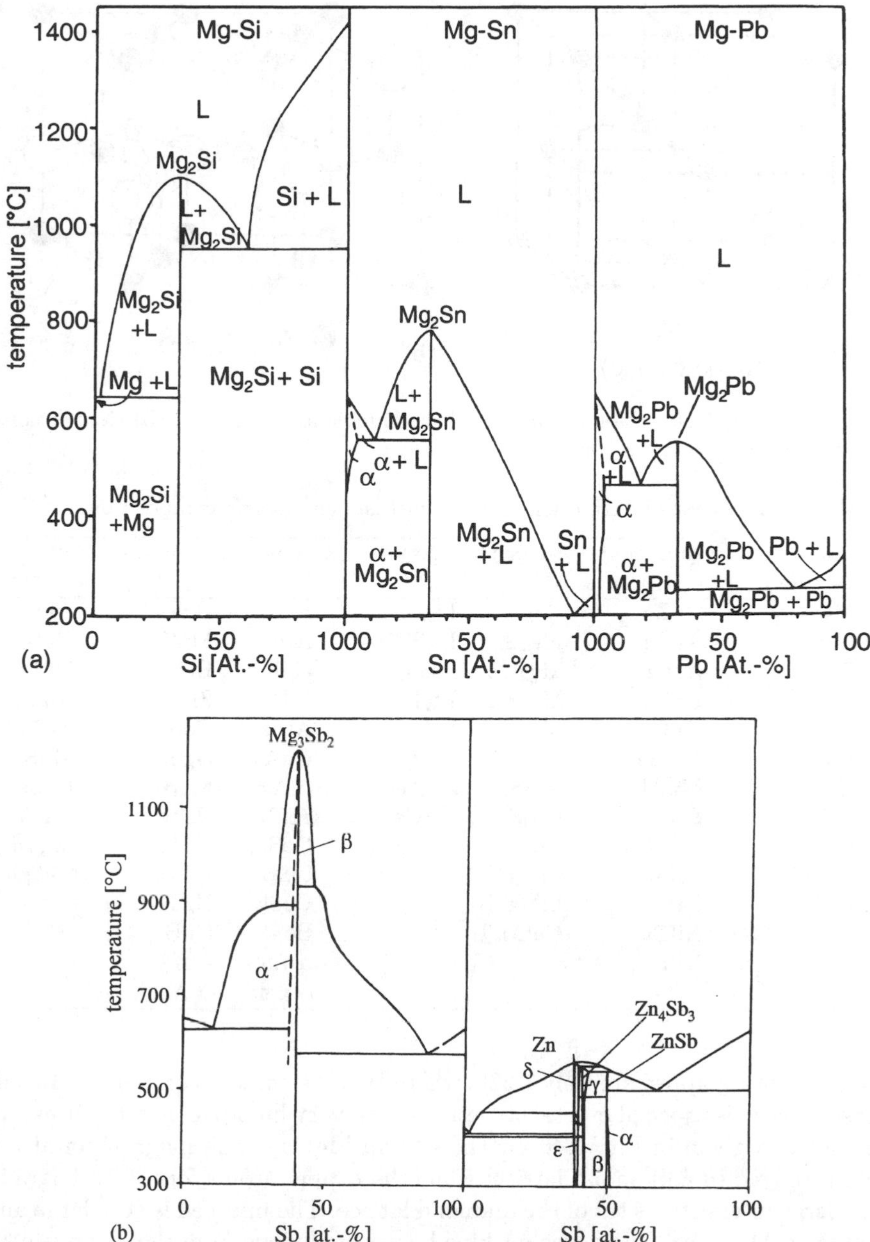

Fig. 4.44. Stability of intermetallic phases can be characterized (in a first order approximation) by their melting temperature. Evidently the stability of phases increases for increasingly electronegative anions (higher group number, lower period number) (a) and increasingly electropositive cations (lower group number, higher period number) (b) [4.1].

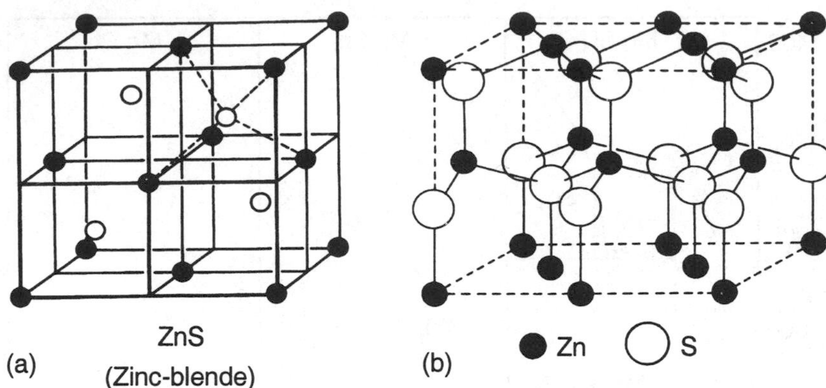

Fig. 4.45. (a) Conventional unit cell of the zincblende structure. (b) Conventional unit cell of the wurtzite lattice.

Table 4.2. Various intermetallic phases with stoichiometric composition.

Rock salt structure	Cesium chloride structure	Fluorite structure		Zincblende structure		Wurtzite structure
MgSe	CuZn	Be_2C	Li_3AlN_2	SiC	HgS	AlN
CaSe	AgZn	Mg_2Si	Li_5SiN_3	AlP	MnS	GaN
SrSe	AgCd	Mg_2Ge	$PtSn_2$	GaP	BeSe	InN
BaSe	AuZn	Mg_2Sn	Pt_2P	InP	ZnSe	β-ZnS
MnSe	AuCd	Mg_2Pb	$PtIn_2$	AlAs	CdSe	β-CdS
PbSe	MnHg	Li_2S	Ir_2P	GaAs	HgSe	MnS
CaTe	MnAl	Na_2S	$AuAl_2$	InAs	MnSe	CdSe
SrTe	FeAl	Cu_2S	Al_2Ca	AlSb	BeTe	MnSe
BaTe	CoAl	Cu_2Se		GaSb	ZnTe	MgTe
SnTe	NiZn	LiMgN		InSb	CdTe	α-AgI
PbTe	Nial	LiMgSb		GeSb	HgTe	
	NiGa	CuCdSb		BeS	CuBr	
	NiIn			α-ZnS	CuI	
				α-CdS	β-AgI	

components is approximately 1.225, where B is the smaller component. In this case there exist special crystal structures with very high packing densities. An example is given in Fig. 4.46 for the system $MgCu_2$. The magnesium atoms are arranged in a diamond lattice, while the copper atoms form tetrahedra in the large interstitial sites of the diamond lattice. The unit cell is very large and contains 24 atoms. The dense packing becomes obvious from the arrangement in the (110) plane (Fig. 4.46b). In this plane all the copper atoms are in contact if their spacing is $(a/4) \cdot \sqrt{2}$, and the magnesium atoms touch if their spacing is $(a/4) \cdot \sqrt{3}$. The highest packing density is obtained if both the copper atoms and the magnesium atoms are in contact, i.e. if the ratio of the radii $R_{Mg}/R_{Cu} = \sqrt{(3/2)} = 1.225$. In this case the packing density is 71%. Laves phases occur in many metallic systems. Besides the $MgCu_2$-lattice, shown in Fig. 4.46, also the hexagonal $MgZn_2$- and $MgNi_2$-lattices exist, which also have a high packing density for a ratio of 1.225 of the atomic radii. As can be seen from Table 4.3 usually the observed size ratio does not deviate from the ideal value of more than 10%.

Table 4.3. Various examples of Laves phases and the ratio of their component's atomic radii.

Laves phases					
$MgCu_2$-type	Ratio of radii	$MgNi_2$-type	Ratio of radii	$MgZn_2$-type	Ratio of radii
$CaAl_2$	1.38	$MgNi_2$	1.29	KNa_2	1.23
$MgCu_2$	1.25	Mg(CuAl)	1.18	$MgZn_2$	1.17
Mg(NiZn)	1.23	Mg(ZnCu)	1.21	Mg(CuAl)	1.18
$Mg(Co_{0.7}Zn_{1.3})$	1.21	$Mg(Ag_{0.4}Zn_{1.6})$	1.16	$Mg(Cu_{1.5}Si_{0.5})$	1.24
$Mg(Ni_{1.8}Si_{0.2})$	1.30	$Mg(Cu_{1.4}Si_{0.6})$	1.23	$Mg(Ag_{0.9}Al_{1.1})$	1.12
$Mg(Ag_{0.8}Zn_{1.2})$	1.14	β-$TiCo_2$	1.15	$CaMg_2$	1.23
$CeAl_2$	1.27	$Zr_{0.8}Fe_{2.2}$	1.26	Ca(AgAl)	1.37
$LaAl_2$	1.30	$Nb_{0.8}Co_{2.2}$	1.17	$CrBe_2$	1.13
$TiBe_2$	1.28	$Ta_{0.8}Co_{2.2}$	1.16	$MnBe_2$	1.16
$(FeBe)Be_4$	1.06			$FeBe_2$	1.12
$(PdBe)Be_4$	1.11			VBe_2	1.20
$CuBe_{2.35}$	1.13			$ReBe_2$	1.21
$AgBe_2$	1.27			$MoBe_2$	1.24
$(AuBe)Be_4$	1.14			WBe_2	1.25
Cd(CuZn)	1.15			WFe_2	1.11
α-$TiCo_2$	1.15			$TiFe_2$	1.14
$ZrFe_2$	1.26			$TiMn_2$	1.11
$ZrCo_2$	1.27			$ZrMn_2$	1.21
ZrW_2	1.13			$ZrCr_2$	1.25
$NbCo_2$	1.17			ZrV_2	1.18
$TaCo_2$	1.16			$ZrRe_2$	1.17
$BiAu_2$	1.26			$ZrOs_2$	1.20
$PbAu_2$	1.22			$ZrRu_2$	1.21
$NaAu_2$	1.33			$ZrIr_2$	1.19
KBi_2	1.30			$TaMn_2$	1.11
$CeNi_2$	1.47			$TaFe_2$	1.15
$CeCo_2$	1.44			$NbMn_2$	1.12
$CeMg_2$	1.14			$NbFe_2$	1.16
$GdMn_2$	1.37			$CaLi_2$	1.25
$GdFe_2$	1.41			$SrMg_2$	1.35
$LaMg_2$	1.16			$BaMg_2$	1.40
CuZnCd				$CaCd_2$	1.29
$Mg(Cu, Si)_2$	1.23			$CaAg_{1.9}Mg_{0.1}$	1.37
				$CaAg_{1.5}Mg_{0.5}$	
				$CaMg_{1.3}Ag_{0.7}$	

A high packing density is also obtained if one of the alloying elements is so small that it fits into the interstitial sites of the other component. This requires a ratio of atomic radii $r/R \leq 0.59$. For example, a compound of type AB can be formed if all octahedral interstitial sites of an fcc lattice are occupied (NaCl-lattice). It is typical for these interstitial phases that they have a very sharply defined upper solubility limit, because no more alloying atoms can be dissolved if all interstitial sites are occupied. Such phases are called Hägg-phases and, frequently, have a high stability. For instance, the Hägg-phase TaC (Fig. 4.47a) has the highest melting temperature among all solids, namely 3983°C. The matrix lattice is not necessarily the lattice of the pure component. For instance, tantalum is bcc, but in TaC the tantalum atoms occupy the lattice sites of an fcc lattice, while the C-atoms fill the octahedral interstitial sites of the fcc lattice (Fig. 4.47b). Hägg-phases not necessarily require that all sublattices are occupied. For instance, the Hägg-phase AB_4 is obtained if only the interstitial sites in the center of the fcc unit cell are occupied. Also the phase Ta_2C (Fig. 4.47) is a Hägg-phase but with a hexagonal structure.

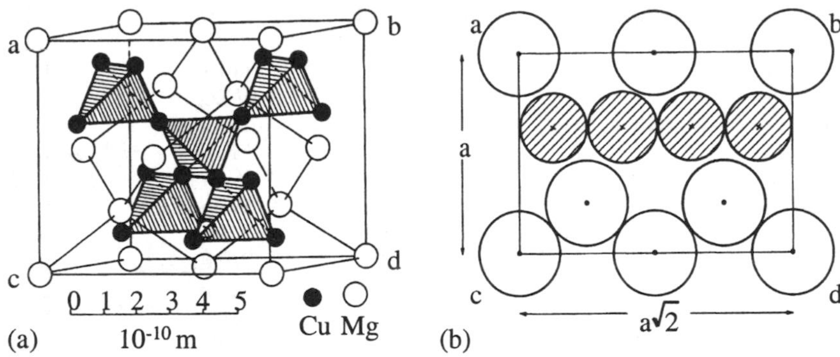

Fig. 4.46. (a) Unit cell of the $MgCu_2$ lattice (Laves phase). (b) Atomistic configuration in the (110) plane of the $MgCu_2$ lattice at closest packing.

4.4.5 Electronic Phases (Hume-Rothery-phases)

In his investigations Hume-Rothery found a large number of examples of binary alloys of elements of non-equal valency, where always the same sequence of crystal structures occurred with increasing concentration. He termed these phases $\alpha, \beta, \gamma, \delta, \varepsilon$ according to the Greek alphabet, where α is the component with the smallest valency. A classical example is the system Cu-Zn, i.e. brass shown in Fig. 4.14 (α = fcc). The β-phase has a CsCl-structure, the γ-phase is the γ-brass structure, and ε is a hexagonal close packed structure. This sequence of crystal structures with increasing concentration cannot be

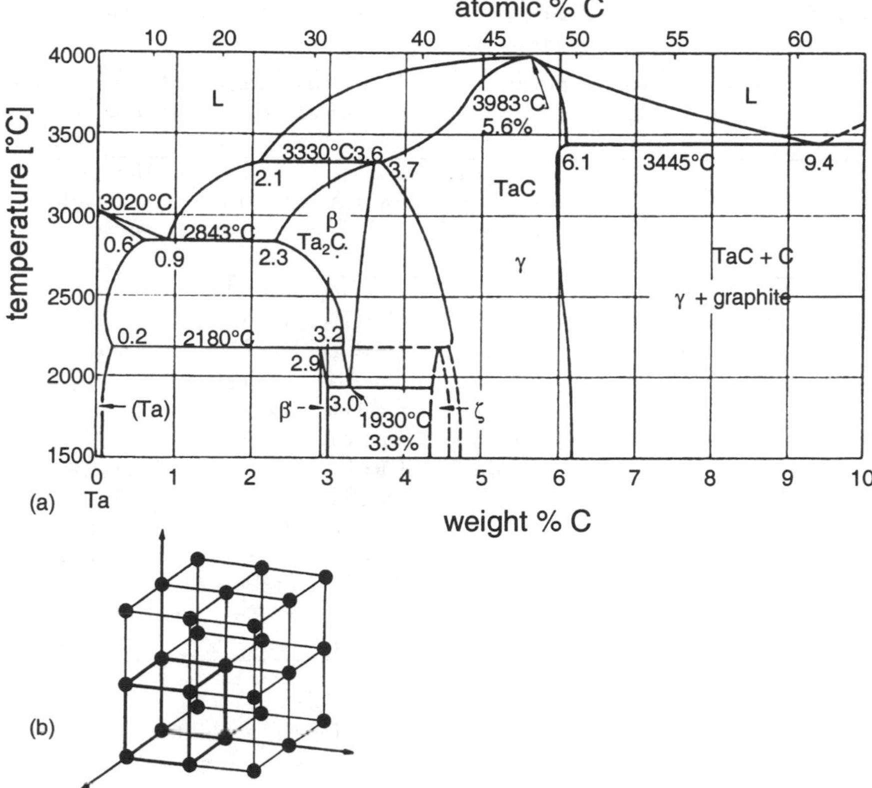

Fig. 4.47. (a) Phase diagram of the Ta-C system with the Hägg phases TaC (fcc) and Ta_2C (hexagonal). (b) Lattice of TaC; the C atoms occupy all octahedral interstitial lattice sites of the fcc lattice (NaCl structure) [4.1].

easily explained from simple structural arguments. Rather the stability of the phases depends on the valency via the valence electron concentration (VEC). The VEC in binary systems is defined by

$$\mathrm{VEC} = C_A \cdot N_{VA} + (1 - C_A) \cdot N_{VB} \tag{4.8}$$

where C_A is the fraction of A-atoms and N_{VA} and N_{VB} are the number of valence electrons of A and B, respectively. For instance, copper has valency 1, ($N_{VA} = 1$) and Zn has valency 2 ($N_{VB} = 2$). The composition CuZn, therefore, corresponds to a VEC $= 3/2$. The VEC obviously depends on composition. If by addition of an alloy partner with a higher valency, a certain value of the VEC is attained, the crystal structure becomes unstable. On further increase of concentration there is another critical value of the VEC, where yet another crystal structure becomes stable, and so on. Table 4.4 gives an overview of the Hume-Rothery phases in various binary systems and their

Table 4.4. Examples of Hume-Rothery phases, that have the same valence electron concentration for different compositions.

Valence electrons per atom $= 3/2$ $= 21/14$				Valence electrons per atom $= 21/13$		Valence electrons per atom $= 7/4$ $= 21/12$
CsCl structure (β)		β-manganese structure (ζ)	Hexagonal close-packed structure (μ)	γ-brass structure (γ)		Hexagonal close-packed structure (ϵ)
CuBe	AuMg	Cu_5Si	Cu_3Ga	Cu_5Zn_8	Au_5Cd_8	$CuZn_3$
CuZn	AuZn	AgHg	Cu_5Ge	Cu_5Cd_8	Au_9In_4	$CuCd_3$
Cu_3Al	AuCd	Ag_3Al	AgZn	Cu_5Hg_8	Mn_5Zn_{21}	Cu_3Sn
Cu_3Ga	FeAl	Au_3Al	AgCd	Cu_9Al_4	Fe_5Zn_{21}	Cu_3Ge
Cu_3In	CoAl	$CoZn_3$	Ag_3Al	Cu_9Ga_4	Co_5Zn_{21}	Cu_3Si
Cu_5Si	NiAl		Ag_3Ga	Cu_9In_4	Ni_5Be_{21}	$AgZn_3$
Cu_5Sn	NiIn		Ag_3In	$Cu_{31}Si_8$	Ni_5Zn_{21}	$AgCd_3$
AgMg	PdIn		Ag_5Sn	$Cu_{31}Sn_8$	Ni_5Cd_{21}	Ag_3Sn
AgZn			Ag_7Sb	Ag_5Zn_8	Rh_5Zn_{21}	Ag_5Al_3
AgCd			Au_3In	Ag_5Cd_8	Pd_5Zn_{21}	$AuZn_3$
Ag_3Al			Au_5Sn	Ag_5Hg_8	Pt_5Be_{21}	$AuCd_3$
Ag_3In				Ag_9In_4	Pt_5Zn_{21}	Au_3Sn
				Au_5Zn_8	$Na_{31}Pb_8$	Au_5Al_3

respective VECs. The significance of the VEC for the stability of the phases can be understood from a very simple model. As alluded to in Sect. 4.3 the valence electrons are added to the lattice gas on alloying. With increasing VEC the added electrons have to assume progressively higher energy states. For free electrons in a solid the energy is given by

$$E \sim \left(n_x^2 + n_y^2 + n_z^2\right) \tag{4.9}$$

where n_x, n_y, and n_z are the principle quantum numbers of the respective state (see Chapter 10). In a **n** space defined by the axes n_x, n_y, n_z, all states of same energy can be found on a sphere with radius $|\mathbf{r}| = (n_x^2 + n_y^2 + n_z^2)^{1/2}$, and all occupied states with a smaller energy are located inside this sphere. The highest energy is the so-called Fermi-energy and, therefore, the respective sphere is called Fermi-sphere. In crystals only specific energy ranges are allowed, and the respective **n** vectors define regions in **n**-space that are called Brillouin zones. At the boundaries of the Brillouin zones the electron energy changes discontinuously. Just for illustration let us assume a Brillouin zone has the shape of a cube. If the Fermi-sphere is smaller than the cube, additional electrons can easily be accepted and the Fermi-sphere grows. Eventually, the Fermi-sphere will touch the boundary of the Brillouin zone and the number of the remaining sites in the Brillouin zone decreases rapidly. Now additional electrons have to assume states of much higher energy in the next Brillouin zone. This is schematically depicted in Fig. 4.48 for the two-dimensional case

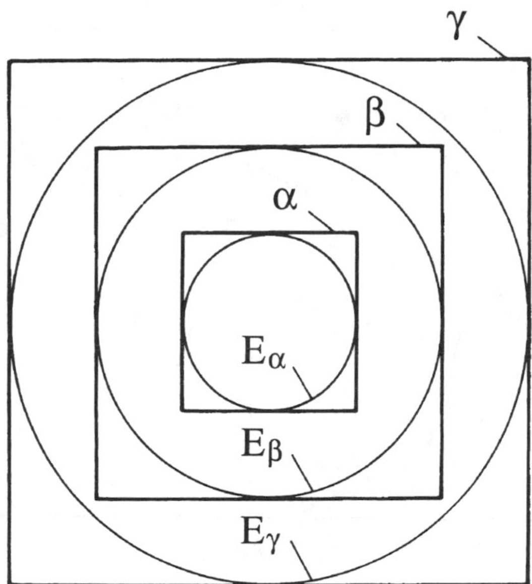

Fig. 4.48. Schematic diagram showing the differing sizes of the maximum Fermi-sphere (E_α, E_β, E_γ) and the first Brillouin zone for different crystal lattices (α, β, γ).

(circle and square rather than sphere and cube). Fig. 4.49 shows the density of states as function of their energy. At the boundaries of a Brillouin zone the density of available states in this zone decreases drastically.

The size of the Brillouin zone, however, depends on crystal structure. For instance, it is always larger for each of the successively observed Hume-Rothery phases. The more electrons can be accommodated in a lower Brillouin zone, the lower the total energy of the crystal and, correspondingly, the more stable the phase. The density of free electrons in a crystal is given by the VEC. This explains the occurrence of specific crystal structures at special values of the VEC as listed in Table 4.4. Of course, in reality the solution of the quantum mechanical problem of electron theory is much more complicated than presented, and frequently cannot be obtained in the form of a closed solution. However, the fundamental principle of electronic phases is captured well by this simple concept.

4.5 Multicomponent Systems

The general considerations for binary systems also hold for multicomponent systems. Ternary systems can be conveniently represented by choosing the binary alloys as terminal systems. The compositional range becomes two-dimensional and is best represented in form of an equilateral triangle, while the

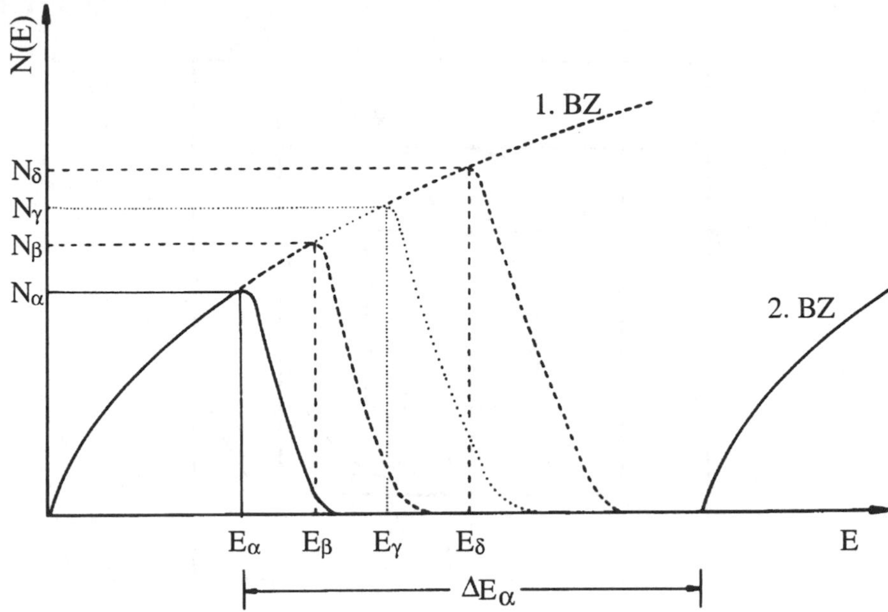

Fig. 4.49. When the Fermi-sphere reaches the surface E_α of the Brillouin zone of the crystal structure α, the density N of electrons that can still occupy this zone decreases sharply. The additional energy ΔE_α is necessary to incorporate electrons in the next Brillouin zone. For crystal structure β a higher density can be accommodated without occupying the next Brillouin zone. If certain valence electron concentrations are exceeded, this results in preference of lattices with larger Brillouin zones.

temperature defines the third dimension. In this 3D composition-temperature space the range of existence of phases can be plotted (Fig. 4.50a). For a two-dimensional representation the phase boundaries between the existent ranges of phases can be projected onto the concentration triangle, for instance the liquidus surface (Fig. 4.50b).

Quaternary systems are of special importance for ceramic materials. In this case already the concentration range is three-dimensional, i.e. an equilateral tetrahedron, and the representation of the regions of existence of phases with changing temperature cannot be plotted anymore in 3D. If the system contains chemical compounds, this space becomes conveniently subdivided. Then it is more reasonable to consider quasi-ternary or quasi-binary systems, where the terminal components consist of compounds (Fig. 4.51). Quasi-binary systems follow the same principles as true binary systems, as shown in Fig. 4.52 for some important ceramic systems.

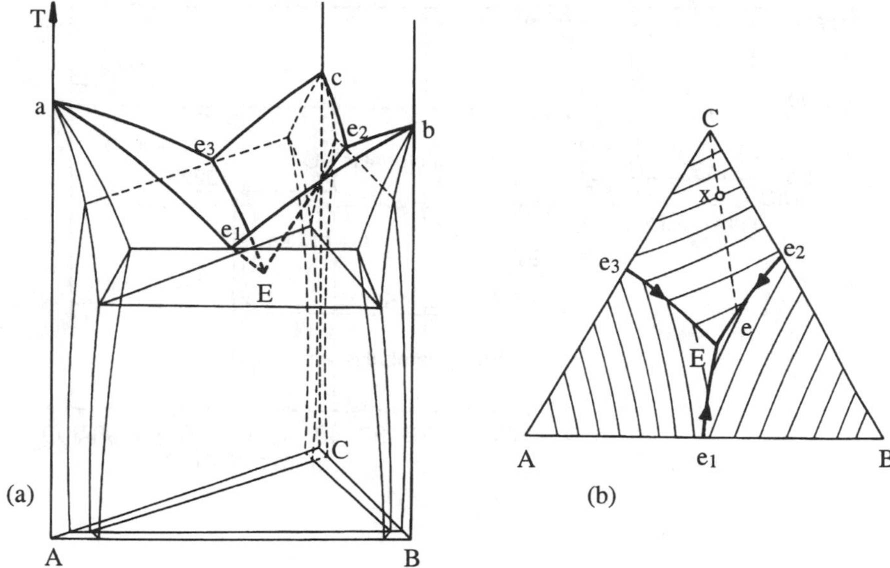

Fig. 4.50. (a) Schematic representation of a ternary system; (b) projection of the liquidus plane on the concentration triangle.

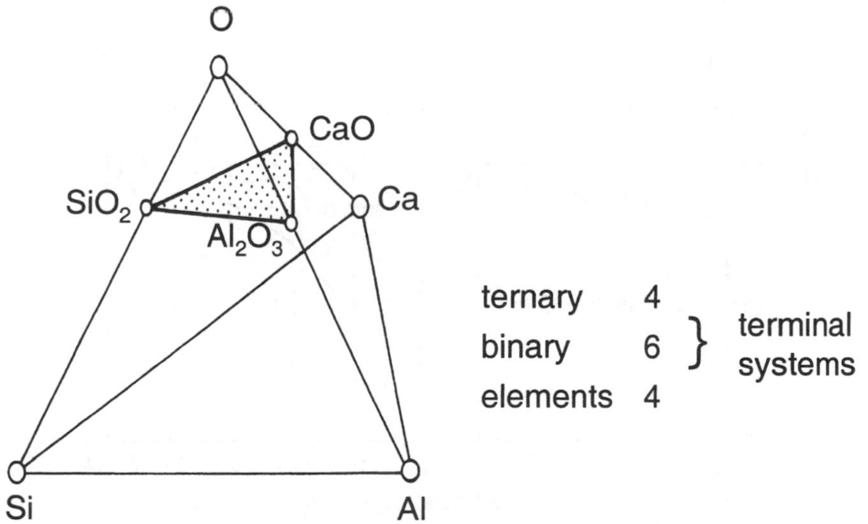

Fig. 4.51. Schematic representation of the quasi-ternary system of some important ceramics.

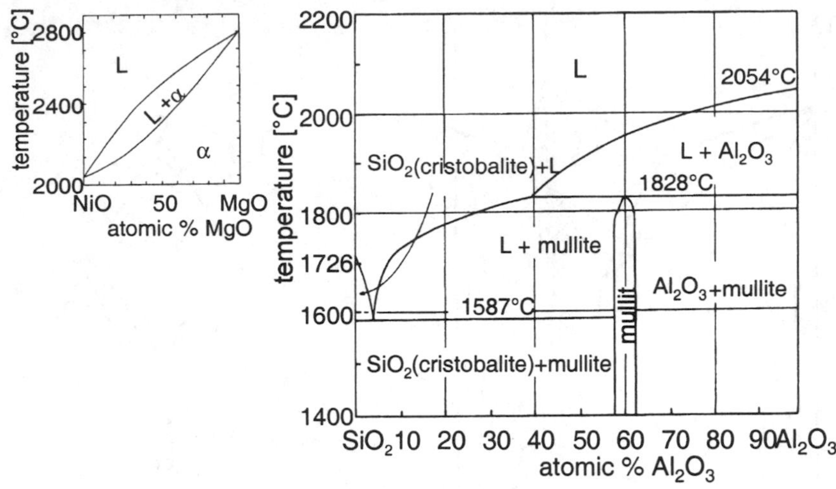

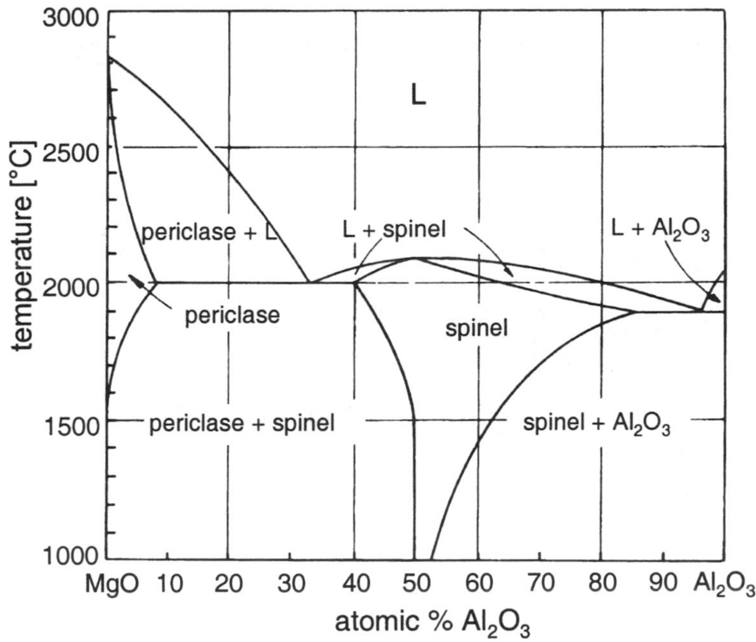

Fig. 4.52. Quasi-binary systems of several important ceramics [4,8].

5

Diffusion

5.1 Phenomenology and Basic Laws

Everybody is familiar with the experience that a drop of ink slowly dissolves in water or smoke spreads out in air. The reason for these phenomena is the motion of the molecules in a liquid or a gas. Although less obvious, atoms in a solid are also capable of leaving their lattice sites by thermal activation and to move through the crystal. This is referred to as solid state diffusion.

To begin with, we want to emphasize that diffusion is a process which is not due to the action of a specific force but rather that diffusion is a result of the random movement of atoms, i.e. a statistical problem. For a most general treatment of the problem we will first disregard the atomic nature of the diffusing particles and define their amount by the concentration c, i.e. the number of particles per unit volume [cm^{-3}].

According to the classical observations of Fick, a concentration gradient causes a flux of particles such that the concentration gradient becomes smaller and, finally, is eliminated. The diffusion flux density j_D [$\mathrm{cm}^{-2}\mathrm{s}^{-1}$], which is defined as the number of particles which pass through a unit area in a unit of time (Fig. 5.1a), is proportional to the concentration gradient (Fig. 5.1b). This relation is known as Fick's first law.

$$j_D = -D\frac{dc}{dx} \tag{5.1a}$$

or in 3 dimensions (3D) with $\nabla c \equiv \left(\frac{\partial c}{\partial x}, \frac{\partial c}{\partial y}, \frac{\partial c}{\partial z}\right)$

$$\mathbf{j}_D = -D \text{ grad } c \equiv -D\nabla c \tag{5.1b}$$

The proportionality constant D is referred to as the diffusion constant or diffusion coefficient and has dimensions [m^2/s]. The magnitude of the diffusion

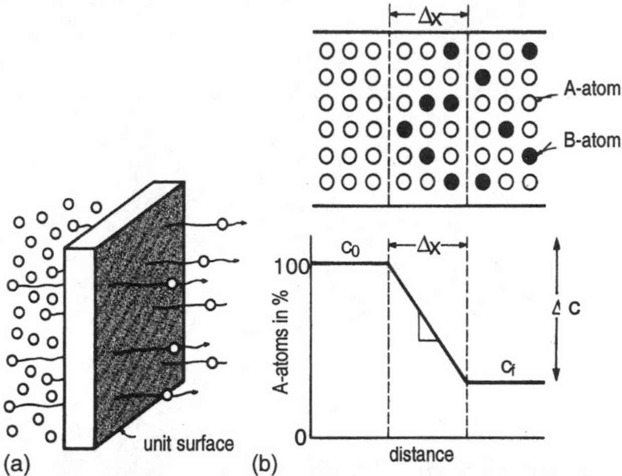

Fig. 5.1. Definition of the diffusion flux. (a) The flux density is the number of particles crossing a unit area per unit time. (b) The flux density of diffusion is proportional to the concentration gradient $\Delta c/\Delta x = (c_0 - c_f)/\Delta x$ (1st Fick's law).

constant determines the diffusion flux density for a given concentration gradient. The minus sign causes the flux to be directed from the higher to the lower concentration, i.e. opposite to the concentration gradient.

For most applications the flux density is of little interest because what one wants to know is the change of concentration in space with time. This information can be obtained by combining Fick's first law with the continuity equation which is in 1D and 3D, respectively,

$$\frac{\partial c}{\partial t} + \frac{\partial j}{\partial x} = 0 \quad or \quad \frac{\partial c}{\partial t} + \text{div}\mathbf{j} = 0 \tag{5.2}$$

with the scalar $\text{div}\mathbf{j} = \nabla \cdot \mathbf{j} \equiv \frac{\partial j_x}{\partial x} + \frac{\partial j_y}{\partial y} + \frac{\partial j_z}{\partial z}$, where $\mathbf{j}$ is a vector with the components (j_x, j_y, j_z).

Eq. (5.2) represents the fact that the total number of particles is conserved, i.e. that the difference of the fluxes into and out of a volume element corresponds to the change of concentration inside the volume element (Fig. 5.2). Combination of Eqs. (5.1a) and (5.2) yields Fick's second law in 1D and 3D, respectively,

$$\frac{\partial c}{\partial t} = \frac{\partial}{\partial x}\left(D\frac{\partial c}{\partial x}\right) \quad or \quad \frac{\partial c}{\partial t} = \nabla \cdot (D\nabla c) \tag{5.3}$$

Only in the case that the diffusion constant is independent of position $[D \neq D(x)]$ can we simplify Eq. (5.3) in 1D

$$\frac{\partial c}{\partial t} = D\frac{\partial^2 c}{\partial x^2} \tag{5.4a}$$

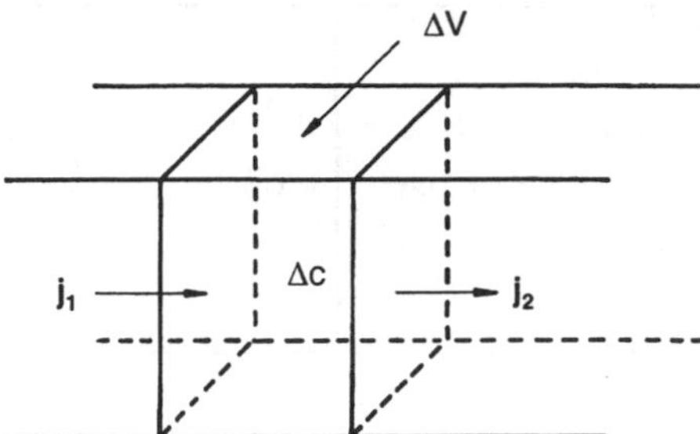

Fig. 5.2. The change of concentration Δc per unit time Δt in a unit volume ΔV is equal to the difference of the incoming current $\mathbf{j}_1$ and the outgoing $\mathbf{j}_2$ (equation of continuity or principle of conservation of mass).

or in 3D $[D \neq D(x, y, z)]$

$$\frac{\partial c}{\partial t} = D\nabla \cdot (\nabla c) = D\Delta c \equiv D\left(\frac{\partial^2 c}{\partial x^2} + \frac{\partial^2 c}{\partial y^2} + \frac{\partial^2 c}{\partial z^2}\right) \tag{5.4b}$$

with the (scalar) Delta operator

$$\Delta = \left(\frac{\partial^2}{\partial x^2} + \frac{\partial^2}{\partial y^2} + \frac{\partial^2}{\partial z^2}\right) \tag{5.4c}$$

With specific initial or boundary conditions this partial differential equation can be solved to give the concentration as function of spatial position and time $c(x, y, z, t)$. As an example let us consider two rods with different concentrations c_1 and c_2 which are joined at $x = 0$ and which are so long that mathematically they can be considered as infinitely long (Fig. 5.3). The concentration profile at $t = 0$ is discontinuous at $x = 0$. With the respective boundary conditions: $t = 0: \;\; c = c_1, \; x < 0; \; c = c_2, \; x > 0$ we obtain as a solution of Eq. (5.4a)

$$c(x,\ t) - c_1 = \frac{c_2 - c_1}{\sqrt{\pi}} \int_{-\infty}^{\frac{x}{2\sqrt{Dt}}} e^{-\xi^2}\, d\xi = \frac{c_2 - c_1}{2} \cdot \left(1 + \operatorname{erf}\left(\frac{x}{2\sqrt{Dt}}\right)\right) \tag{5.5}$$

where

$$\operatorname{erf}(z) = \frac{2}{\sqrt{\pi}} \int_0^z e^{-\xi^2} d\xi \tag{5.6}$$

is known as the error function (Fig. 5.4).

The concentration profile $c(x)$ is given in Fig. 5.3b for different times t. With increasing time the curves flatten out and after an infinitely long time the concentration becomes uniform at $c_\infty = 1/2\,(c_1 + c_2)$. In terms of atoms the

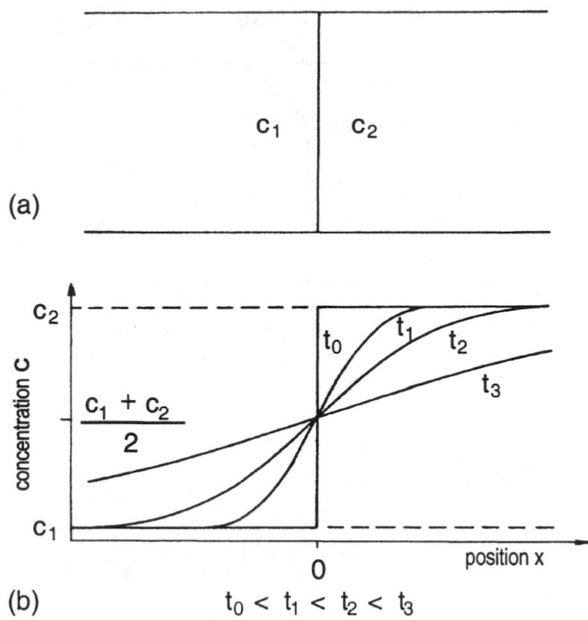

Fig. 5.3. Change in concentration in two semi-infinite rods for different diffusion times.

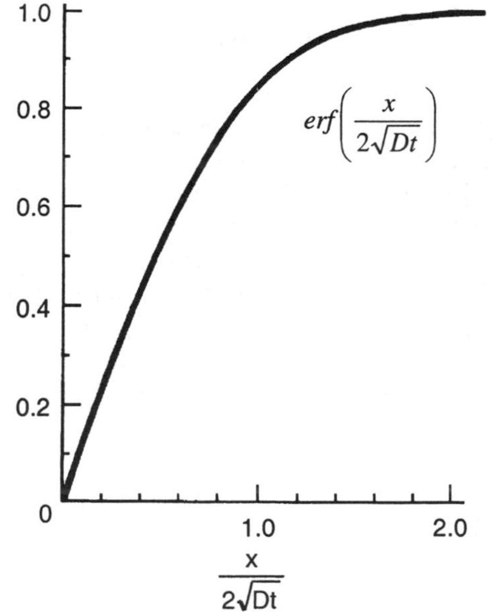

Fig. 5.4. Curve of the error function $erf(z)$ for $z = x/\left(2 \cdot \sqrt{Dt}\right)$.

example of a diffusion couple of Cu and Ni, Fig. 5.5, illustrates the evolution in atomistic arrangements.

The calculation of the concentration profile requires to solve the partial differential Eq. (5.4) for given boundary conditions [1]. Very often the boundary conditions are so complicated that a closed form solution such as the example given above cannot be obtained, so numerical approximations or computer simulations have to be used.

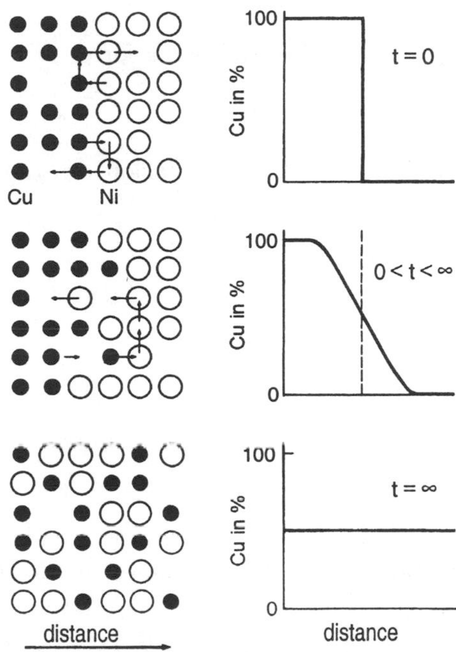

Fig. 5.5. Change in concentration in two semi-infinite rods of Cu and Ni caused by diffusion. The two types of atoms, that were originally separate, mix by changing places (left half of the figure). Concentration gradients are reduced and after a very long time an equilibrium distribution is reached (right half of the picture).

We mentioned before that diffusion is caused by the random movement of atoms due to thermal activation and that diffusion is not due to the action of a specific force. However, if diffusion occurs in a potential gradient, for instance in a locally varying elastic field or chemical potential, or in the case of an

[1] The diffusion equation (Eq. (5.4)) is mathematically equivalent to the equation of thermal conduction. Correspondingly, the solutions are identical for the same boundary conditions. Solutions for many problems and geometries can be found in the books of H.S. Carslav and J.C. Jaeger "Conduction of Heat in Solids" (1959) or J. Crank "Mathematics of Diffusion" (1956).

electrostatic interaction, the diffusion flux is superimposed on a convection flux due to the action of a force $\mathbf{F}$, which is related to the potential gradient $\boldsymbol{\nabla}\Phi$ by

$$\mathbf{F} = -\boldsymbol{\nabla}\Phi \tag{5.7}$$

The respective convection flux $\mathbf{j}_c$ is due to a homogeneous drift motion of atoms affected by the force. For a drift velocity v exactly vc atoms penetrate a unit area per unit of time, i.e.

$$\mathbf{j}_c = vc \tag{5.8}$$

The drift velocity is proportional to the acting force

$$v = B\mathbf{F} \tag{5.9}$$

where the mobility B is related to the diffusion constant by the so-called Nernst-Einstein-relation

$$B = \frac{D}{kT} \tag{5.10}$$

This yields

$$\mathbf{j}_c = \frac{D}{kT} \cdot (-\boldsymbol{\nabla}\Phi) \cdot c \tag{5.11}$$

and the total flux

$$\mathbf{j} = \mathbf{j}_D + \mathbf{j}_c = -D\left(\boldsymbol{\nabla}c + \frac{c\boldsymbol{\nabla}\Phi}{kT}\right) \tag{5.12}$$

Clearly Eq. (5.12) is an extension of Fick's first law. Combined with the continuity (Eq. (5.2)), we obtain for the diffusion in a potential gradient, assuming $D \neq D\,(x,y,z)$

$$\frac{\partial c}{\partial t} = D\boldsymbol{\nabla} \cdot \left(\boldsymbol{\nabla}c + \frac{c\boldsymbol{\nabla}\phi}{kT}\right) \tag{5.13a}$$

or in one dimension

$$\frac{\partial c}{\partial t} = D\frac{\partial^2 c}{\partial x^2} + \frac{D}{kT}\frac{\partial}{\partial x}\left(c\frac{\partial \Phi}{\partial x}\right) \tag{5.13b}$$

The partial differential equation of Fick's second law becomes substantially more complicated by the second term in Eq. (5.13b). A closed form solution can be obtained only in special cases.

The solution given in Eq. (5.5) shows a special feature which is common to all solutions of Eq. (5.4a). The concentration $c(x,t)$ always depends on $x/\sqrt{t}$. This is due to the fact that the partial differential equation of the variables x and t, Eq. (5.4a), can be written as a common or ordinary differential equation of the variable $\eta = \left(x/\sqrt{t}\right)$. Such a dependency allows us to conveniently interpret the mathematical solution of the diffusion equation as the displacement of a diffusion front. The diffusion front is defined by the location of a constant concentration $c(x,t) = k_R$, and this location or contour changes with time. The exact position is of course determined by the arbitrarily chosen value of the constant k_R. Once k_R has been set, the kinetics of diffusion can be easily described. The definition of such a diffusion front is also reasonable for physical reasons. Although the mathematical solution $c(x,t)$ will yield a finite value for any x, its magnitude may be much smaller than is accessible by measurement or even physically meaningless if it represents less than an atomic dimension in a crystal. The position of the diffusion front at time t defines the range X of diffusion. At a magnitude of $k_R = 1\%$, the vast majority of all diffusing particles are located behind the diffusion front. With $c(x,t) = k_R = 0.01$ we obtain

$$\frac{X}{2\sqrt{Dt}} \cong \sqrt{1.5}$$

or

$$X^2 = 6 \cdot Dt \tag{5.14}$$

This definition of the range of diffusion which yields the constant "6" in Eq. (5.14) is reasonable also from atomistic considerations (see Section 5.3).

The physical problem of diffusion is the analysis of the diffusion constant D. Fick's laws are concerned with the removal of concentration differences. The physical basis for the diffusion phenomena is the thermally activated motion of atoms in a crystal. This atomic hopping is not restricted to atoms in a concentration gradient, but also occurs in pure metals or homogeneous alloys. If we could color the individual atoms we could follow their movement as a function of time. The position of any particular atom would change with time, but the net flux through a given cross section would be zero, since the flux in opposite directions through that cross section would be equally large. Atomic transport in a pure metal or a homogeneous solid solution is referred to as self diffusion, which is characterized by the coefficient of self diffusion D^*. The quantity of D^* is difficult to measure, since moving and resting atoms cannot be chemically distinguished. The self diffusion coefficient is closely related, however, to the tracer diffusion coefficient D^T, which can be experimentally measured. In such experiments the movement of a radioactive isotope of the investigated element is measured. An isotope has the same electronic structure as the natural element and is, therefore, chemically identical and physically almost equivalent to a regular atom. A measurement of the radioactivity as function of space and time is sufficient to determine D^T and can be conducted with high accuracy. There is only a minor difference between D^T and D^* due to the slightly different atomic masses, which, however, can be corrected for. For all practical purposes the self diffusion coefficient is commonly determined by tracer experiments.

5.2 The Diffusion Constant

Exactly speaking, the diffusion constant is not a scalar, but a tensor of rank 2, i.e. the property depends on spatial direction. In high symmetry cubic lattices - note that 67 % of all metals crystallize in cubic lattices - the diffusion constant is isotropic, i.e. of the same magnitude in all directions. However, in crystal structures of lower symmetry, a directional dependence of the diffusion constant is quite common. In this case we have to write Fick's first law as a tensor equation

$$\mathbf{j}_D = -\mathbf{D}\nabla c \quad \text{with } \mathbf{D} = \begin{bmatrix} D_{11} & D_{12} & D_{13} \\ D_{21} & D_{22} & D_{23} \\ D_{31} & D_{32} & D_{33} \end{bmatrix} \tag{5.15}$$

or line by line

$$\begin{aligned} j_x &= -D_{11}\frac{\partial c}{\partial x} - D_{12}\frac{\partial c}{\partial y} - D_{13}\frac{\partial c}{\partial z} \\ j_y &= -D_{21}\frac{\partial c}{\partial x} - D_{22}\frac{\partial c}{\partial y} - D_{23}\frac{\partial c}{\partial z} \\ j_z &= -D_{31}\frac{\partial c}{\partial x} - D_{32}\frac{\partial c}{\partial y} - D_{33}\frac{\partial c}{\partial z} \end{aligned}$$

$D_{ij} \neq 0$ for $i \neq j$ represents a flux in direction i for a gradient in direction j. This is quite usual in structures with lower symmetry. In Table 5.1 the tensor elements for some important crystal systems are listed. Note that the major axes of the coordinate system have been chosen to be parallel to the crystal axes. A change of the coordinate system would cause non-zero off-diagonal elements to appear, of course. Table 5.2 gives some examples of anisotropic diffusion coefficients in non-cubic metals.

Table 5.1. Diffusion coefficient tensors of crystals with cubic, hexagonal and orthorhombic symmetry.

$$\mathbf{D}_{kub.} = \begin{bmatrix} D_1 & 0 & 0 \\ 0 & D_1 & 0 \\ 0 & 0 & D_1 \end{bmatrix} \qquad \mathbf{D}_{hex.} = \begin{bmatrix} D_{11} & 0 & 0 \\ 0 & D_{11} & 0 \\ 0 & 0 & D_{33} \end{bmatrix} \qquad \mathbf{D}_{ortho.} = \begin{bmatrix} D_{11} & 0 & 0 \\ 0 & D_{22} & 0 \\ 0 & 0 & D_{33} \end{bmatrix}$$

The diffusion constant is very sensitive to temperature, namely by a Boltzmann factor

$$D = D_0 \cdot \exp(-Q/kT) \tag{5.16}$$

where Q is the activation enthalpy and D_0 the pre-exponential factor. The relationship in Eq. (5.16) has been experimentally confirmed for many systems

Table 5.2. Diffusion coefficient of various non-cubic metals parallel (||) and perpendicular (⊥) to the basal plane.

Metal	Structure	$D_{0\parallel}$ $[cm^2/s]$	$D_{0\perp}$ $[cm^2/s]$	$Q_{\parallel}$ $[kJ/mol]$	$Q_{\perp}$ $[kJ/mol]$	$D_{\perp}/D_{\parallel}$ $T=0.8T_m$
Be	hcp	0.52	0.68	157	171	0.31
Cd	hcp	0.18	0.12	82.0	78.1	1.8
α-Hf	hcp	0.28	0.86	349	370	0.87
Mg	hcp	1.5	1.0	136	135	0.78
Tl	hcp	0.4	0.4	95.5	95.8	0.92
Sb	rhomb	0.1	56	149	201	0.098
Sn	diamond	10.7	7.7	105	107	0.4
Zn	hcp	0.18	0.13	96.4	91.6	2.05

over many orders of magnitude for D, for diffusion in both interstitial alloys like C in α-Fe (Fig. 5.6) and substitutional solutions, like gold in silver (Fig. 5.7). This temperature dependence appears obvious, since diffusion proceeds by thermally activated motion of atoms, and simple thermally activated processes always have a temperature dependent Boltzmann factor $\exp(-Q/kT)$. The activation energy Q depends on the elementary process of atomic motion and, in general, it will be different for different elements and crystal structures (Fig. 5.8). It can be obtained from the slope of an Arrhenius plot lnD versus $1/T$. From a comparison of Fig. 5.9 and Fig. 5.8 it is evident that small interstitial atoms diffuse substantially faster than atoms in substitutional alloys. The movement of ions in ceramic materials (for instance Mg^{++} in MgO or Ca^{++} in CaO (Fig. 5.8)) is more difficult, due to the electrostatic interaction of the ions, and, therefore, their diffusion coefficients are very small. The figures show that the diffusion coefficient of different substances at the same temperature can differ by many orders of magnitude.

The activation energy Q is found to be correlated with the melting temperature T_m of the diffusing element, i.e. Q increases with rising T_m (Fig. 5.10). Correspondingly, in solid solutions Q varies with the solidus temperature, and for non-monotonic variation of the solidus temperature, as in systems with a solidus minimum, the diffusion coefficient will assume the smallest activation energy and thus, a maximum value for the composition where the minimum occurs (Fig. 5.11).

The characteristic diffusion parameters of some metals and alloys are listed in Table 5.3. The activation energy of self diffusion ΔH_D for metals is related to the melting temperature such that $\Delta H_D/T_m \cong 1.5 \cdot 10^{-3} eV/K$ or with the heat of melting L_m we find $\Delta H_D/L_m \cong 15$. These empirical rules permit us to estimate the magnitude of the diffusion coefficient if no measurements are available. However, this has to be used with caution since there are exceptions, e.g. Germanium in Table 5.3. The pre-exponential factor D_0 for self diffusion is of the order of $1[\mathrm{cm}^2/\mathrm{s}]$. Table 5.3 shows that in solid solutions the diffusion of interstitial atoms proceeds faster than self diffusion by several

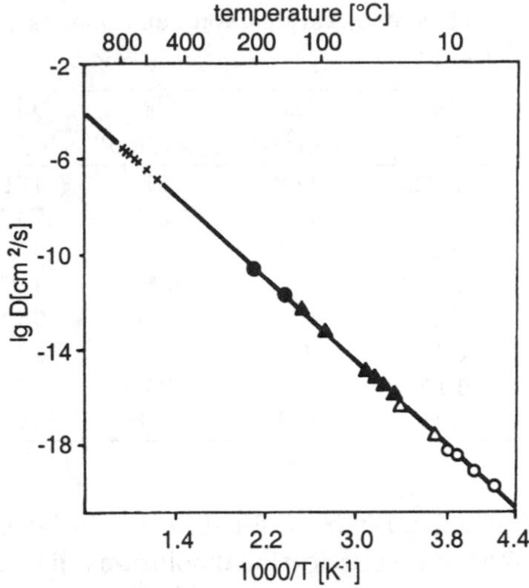

Fig. 5.6. Temperature dependence of the diffusion coefficient (Arrhenius plot) for carbon in α-Fe (after [5.1]).

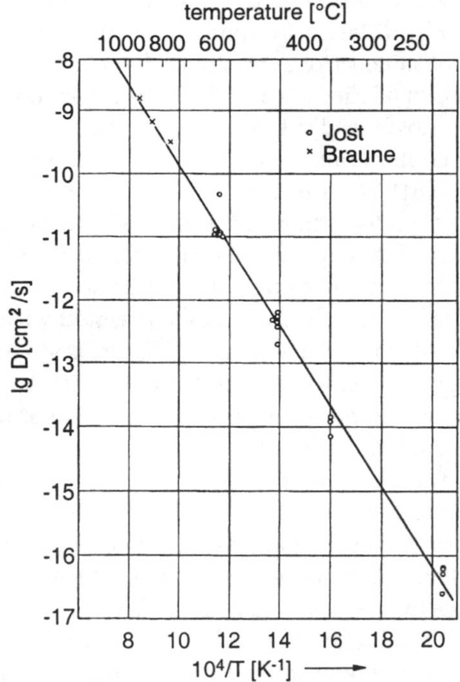

Fig. 5.7. Temperature dependence of the diffusion coefficient of gold in silver (after [5.2]).

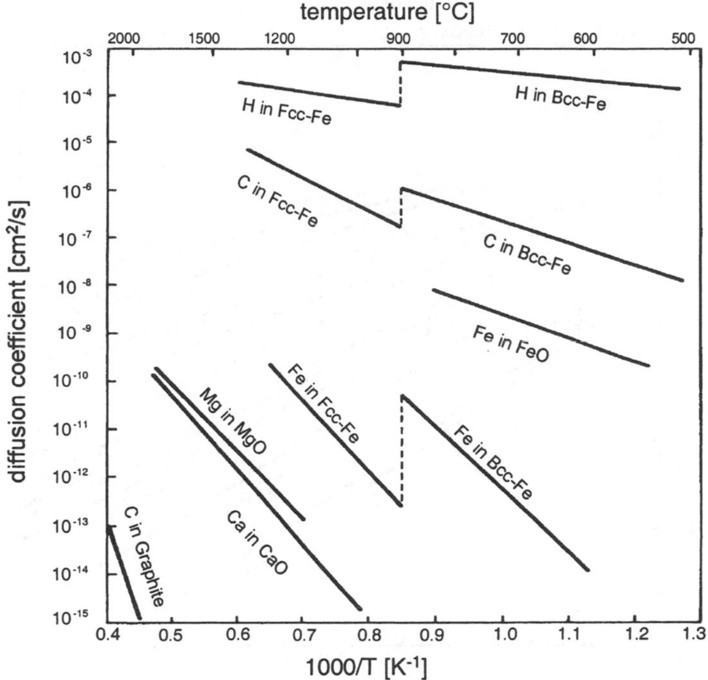

Fig. 5.8. Temperature dependence of self-diffusion and impurity diffusion in iron and several ionic crystals (after [5.3]).

orders of magnitude, while substitutional alloying elements diffuse at rates similar to atoms of the matrix material.

Frequently the diffusion constant is found to depend on concentration. Therefore, concentration gradients in alloys cause the diffusion constant to depend on location. In such cases the concentration profile becomes asymmetrical (Fig. 5.12) and Fick's 2nd law is represented by Eq. (5.3) rather than Eq. (5.4). Both D_0 and Q can depend on composition as evident from Fig. 5.13 for C in γ-Fe. The measurement of $D(c)$ is more difficult, but also very important for the diffusion process in concentrated alloys, i.e. in strong concentration gradients. This will be treated in more detail in Section 5.5.

5.3 Atomistics of Solid State Diffusion

The thermally activated jump motion of atoms constitutes the mechanism of diffusion in solids. For interstitial solid solutions it is easy to visualize that a foreign atom can change to a next neighbor interstitial site by sufficient thermal activation. The jump motion of substitutional atoms is complicated by the fact that the neighboring lattice site of a foreign atom is usually occu-

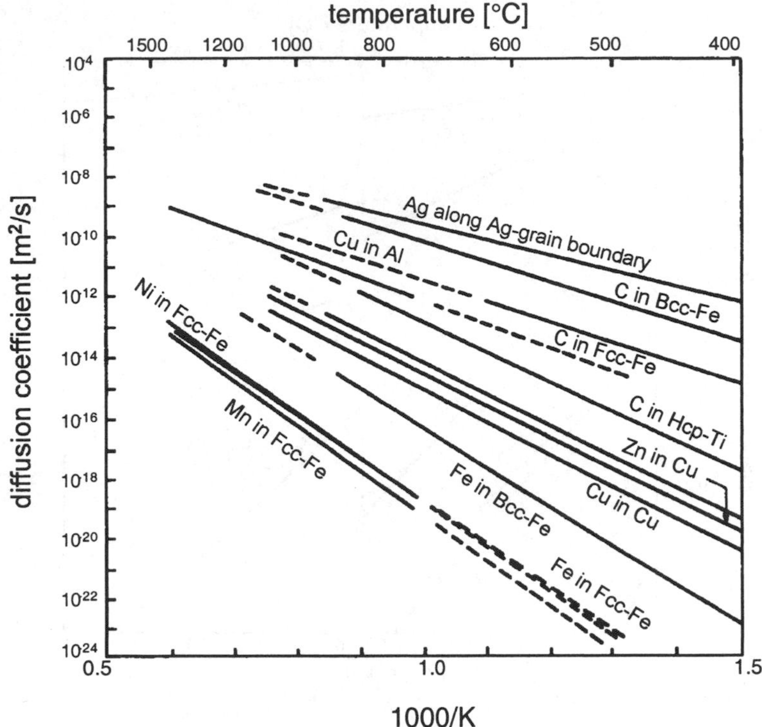

Fig. 5.9. Temperature dependence of diffusion in various metals (after [5.4]).

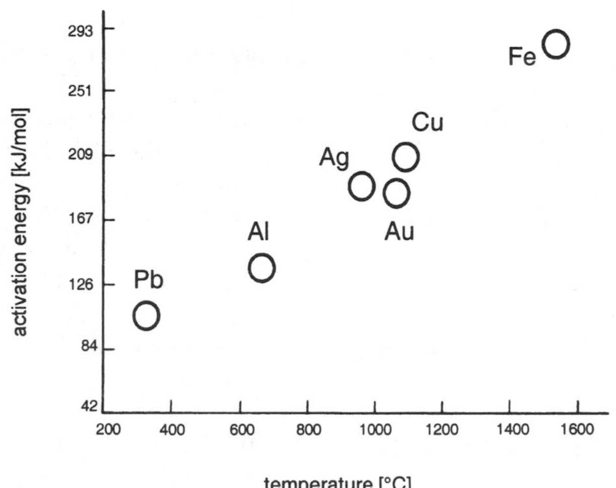

Fig. 5.10. Diffusion activation energy increases (nearly linearly) with rising melting temperature (after [5.5]).

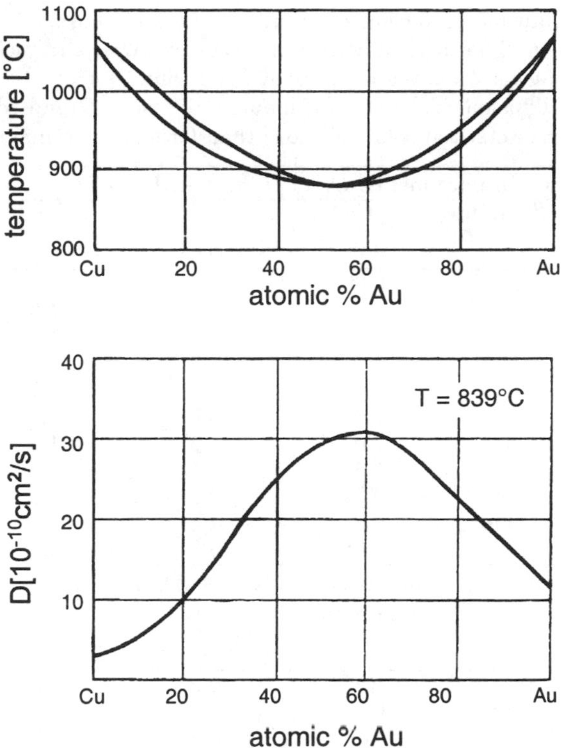

Fig. 5.11. Composition dependence of the diffusion coefficient in the CuAu system (after [5.6]).

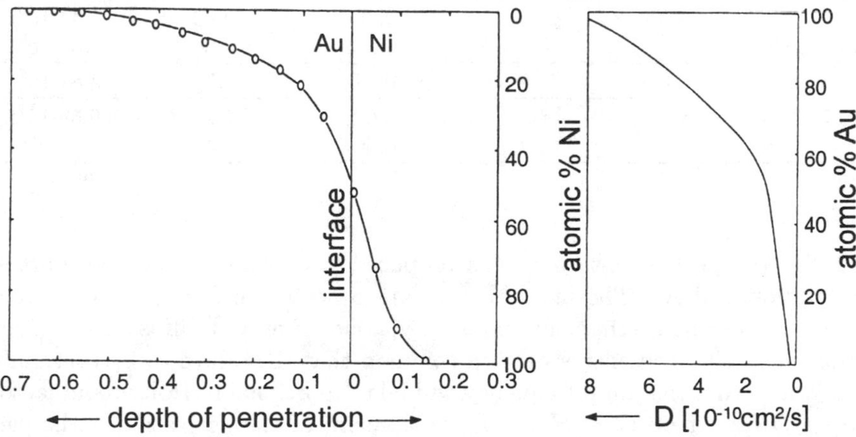

Fig. 5.12. (a) Concentration distribution due to interdiffusion of Au in Ni. (b) The diffusion coefficient is composition dependent (after [5.7]).

Table 5.3. Diffusion coefficient of several metals and alloys.
(a) Self-diffusion. The ratio of activation enthalpy ΔH_D and melting temperature T_m or heat of fusion L_m are approximately constant.
(b) Impurity diffusion. The ratio of impurity diffusion D and self-diffusion D^* is much larger for interstitial solid solutions than for substitutional solid solutions.

(a) Self-diffusion in pure metals $D^* = D_0 \exp(-\Delta H_D/kT)$

Metal	Structure	D_0 $[cm^2/s]$	ΔH_D $[eV]$	$\Delta H_D/T_m$ $[10^{-3}eV/K]$	$\Delta H_D/L_m$
Au	fcc	0.09	1.8	1.5	13.2
Ag	fcc	0.4	1.9	1.6	16.2
Cu	fcc	0.2	2.0	1.5	15.2
Ni	fcc	1.9	2.9	1.7	15.6
γ-Fe	fcc	0.4	2.8	1.6	17.4
W	bcc	1.9	5.6	1.6	16.9
α-Fe	bcc	2.0	2.5	1.4	15.5
Nb	bcc	1.3	4.1	1.5	14.8
Na	bcc	0.24	0.5	1.3	16.7
Mg	hep	1.3	1.4	1.5	18.5
Ge	dia	10.8	3.0	2.4	9.1

(b) Impurity diffusion in metals $D = D_0 \exp(-\Delta H/kT)$

Type of solution	Metal	Impurity atom	D_0 $[cm^2/s]$	ΔH $[eV]$	D/D^* $(1000K)$
Substitutional solid solutions	Ag	Au	0.26	2.0	0.25
		Cu	1.2	2.0	0.94
		Zn	0.54	1.8	4.3
	Cu	Au	0.69	2.2	0.49
		Ag	0.63	2.0	3.15
		Zn	0.34	2.0	1.7
	α-Fe	Co^{60}	0.2	2.4	0.35#)
Interstitial solid solutions	α-Fe	C	0.004	0.83	$1.44{\cdot}10^6$
		N	0.003	0.8	$1.55{\cdot}10^6$
	γ-Fe	C	0.67	1.6	$3.87{\cdot}10^6$
	Nb	O	0.021	1.2	$6.59{\cdot}10^{12}$
		C	0.004	1.4	$1.2{\cdot}10^{11}$

#) at 950 K

pied. Several mechanisms have been proposed to explain atomic transport in substitutional alloys. The most obvious are collected in Fig. 5.14, i.e. direct site exchange, ring mechanism, diffusion via vacancies and diffusion via interstitial sites. All mechanisms have in common that they have to be thermally activated so that the jump frequency will be proportional to Boltzmann factor $exp(-Q/kT)$, where $Q = H_f + H_m$ is the sum of the activation enthalpies for formation of the configuration (H_f) and atomic migration (H_m). The vacancy mechanism has the smallest sum Q and is, therefore, most likely to be activated. As a matter of fact experimental results overwhelmingly support this hypothesis.

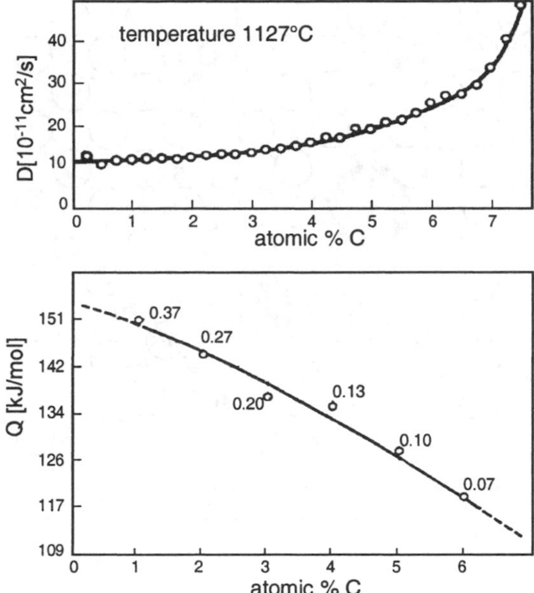

Fig. 5.13. Composition dependence of the diffusion coefficient of C in γ-Fe. Both D_0 and Q depend on concentration (after [5.8]).

The jump frequency $\Gamma\left[s^{-1}\right]$ of an atom is given by

$$\Gamma = \nu \exp\left(-G_m/kT\right)$$

This relation can be interpreted such that an atom attempts to overcome the energetically unfavorable barrier ν times per second, i.e. with an attack frequency ν (Fig. 5.15). Usually $\nu = \nu_D \approx 10^{13}s^{-1}$ is assumed where ν_D is the Debye frequency i.e. the vibration frequency of the atom. The probability for a successful jump is given by the Boltzmann factor $\exp\left(-G_m/kT\right)$, which decreases as the Gibbs free activation energy G_m increases. For instance, G_m is smaller for the interstitial mechanism than for the vacancy mechanism according to Fig. 5.14. There is a fundamental relationship between the jump frequency Γ and the diffusion coefficient D which holds independent of mechanism and crystal structure

$$D = \frac{\lambda^2}{6}\Gamma = \frac{\lambda^2}{6\tau} \tag{5.17}$$

In this equation λ is the jump distance of the diffusing atom and $\tau = 1/\Gamma$ is the time interval between two consecutive jumps. Eq. (5.17) can be readily derived from an atomistic consideration of the diffusion process as will be shown for the important case of the diffusion of C in α-iron. The carbon atoms are located on the octahedral interstitial sites of the body centered cubic lattice of α-Fe

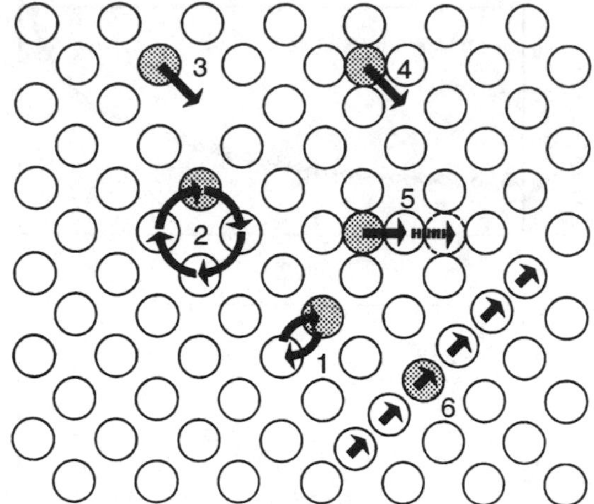

	Migration	Formation	Total
(1)	$8\,eV$	-	$8\,eV$
(2)		Small but less probable	
(3)	$1\,eV$	$1\,eV$	$2\,eV$
(4)	$0.6\,eV$	$3.4\,eV$	$4\,eV$
(6)	$0.2\,eV$	$3.4\,eV$	$3.6\,eV$

Fig. 5.14. Possible mechanisms of self-diffusion and their activation energy. (1) Neighboring atoms exchange sites; (2) ring mechanism; (3) vacancy mechanism; (4) direct interstitial mechanism; (5) indirect interstitial mechanism; (6) crowdion.

(see Chapter 2) (Fig. 5.16). For small carbon concentrations the carbon atoms do not affect each other, so their jumps can be considered to be independent. The net flux of carbon atoms between two next neighbor planes of the iron lattice say M and N in Fig. 5.16, reads

$$j = j_{MN} - j_{NM} \tag{5.18}$$

where j_{MN} denotes the flux from the plane M to the plane N and j_{NM} the flux in the opposite direction. With c_M^A and c_N^A, respectively, the number of atoms per unit area on the planes M and N, and Γ the jump frequency between next neighbor lattice sites, we obtain

$$j = c_M^A \Gamma \cdot \frac{1}{4} \cdot \frac{2}{3} - c_N^A \Gamma \cdot \frac{1}{4} \cdot \frac{2}{3}$$

The factor $1/4$ accounts for the fact that only $1/4$ of the possible jumps of a C atom lead to a flux in $+x$ (resp. $-x$) direction. Also, only $2/3$ of all C atoms can jump in x direction. This is because on the plane M there are three types of interstitial sites to be discriminated, i.e. the middle of the edges y

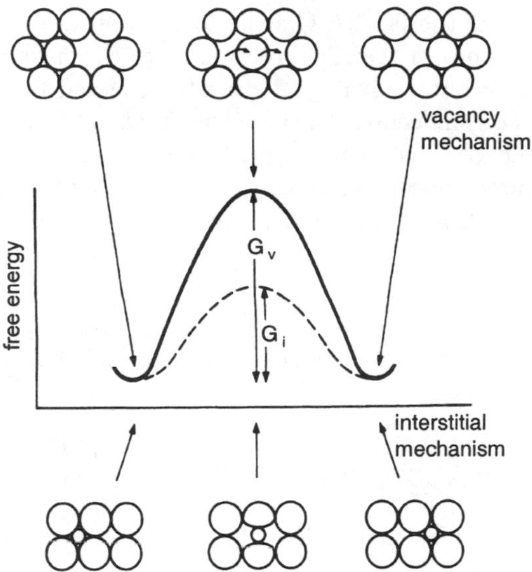

Fig. 5.15. Diagrammatic illustration of the thermally activated diffusion process. The diffusing atom must pass through an energetically less advantageous state in order to get to the neighboring lattice site. The enthalpy difference provides the activation energy. It is greater for the vacancy mechanism than for the interstitial mechanism.

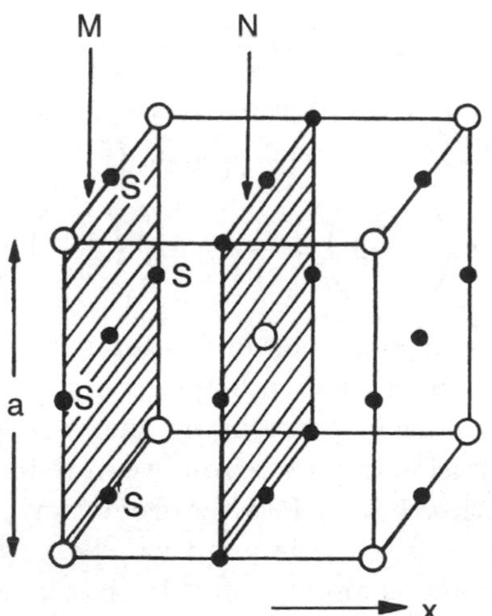

Fig. 5.16. Derivation of the diffusion coefficient of C in α-Fe. The filled circles indicate the interstitial lattice sites where C atoms may be located. The open circles indicate the Fe atoms. M and N are two neighboring planes along the direction of diffusion x.

and z and the center of the face. A C atom located on the center of the face on plane M cannot jump in the $\pm x$ direction since the next neighbor site is not an interstitial site but a regular lattice site occupied by an Fe atom. In contrast, the atoms on the centers of the edges, that is 2/3 of all C atoms on the plane M are capable of moving to next neighbor interstitial sites in the $\pm x$ direction. The area concentration c^A is the number of atoms in a volume of thickness $a/2$ and, therefore, related to the volume concentration c by

$$c^A = c \cdot \frac{a}{2} \tag{5.19}$$

where $a/2$ is the spacing between the next neighbor planes M and N. Finally, by Taylor series expansion

$$c_N = c_M + \frac{a}{2}\frac{dc}{dx} + \frac{a^2}{8}\frac{d^2c}{dx^2} + \ldots + \left(\frac{a}{2}\right)^n \cdot \frac{1}{n!}\frac{d^nc}{dx^n} + \ldots \tag{5.20}$$

where we can truncate the series after the linear term because of the low concentration difference of next neighbor planes in dilute alloys

$$j = \frac{1}{6}\Gamma\left(c_M - \left[c_M + \frac{a}{2}\frac{dc}{dx}\right]\right) \cdot \frac{a}{2} = -\Gamma\frac{a^2}{24}\frac{dc}{dx} \tag{5.21}$$

A comparison with Fick's first law (Eq. (5.1a)) yields

$$D = \frac{\Gamma a^2}{24} \tag{5.22}$$

Because $\Gamma = 1/\tau$ and the jump distance $\lambda = a/2$ we obtain for diffusion via interstitial sites the fundamental relation (Eq. (5.17)) or, in detail,

$$D = \frac{\lambda^2}{6\tau} = \frac{\lambda^2}{6}\nu_D \exp\left(-\frac{G_m^i}{kT}\right) = \frac{\lambda^2}{6}\nu_D \exp\left(\frac{S_m^i}{k}\right)\exp\left(-\frac{H_m^i}{kT}\right) \tag{5.23}$$

G_m^i - Gibbs free energy of migration of C atoms via interstitial sites, S_m^i - migration entropy, H_m^i - migration enthalpy.

In the case where diffusion proceeds by the vacancy mechanism, as for instance, during self diffusion in pure α iron, we have to modify the derivation given above. By definition $1/\tau = \Gamma$ is the frequency of successful jumps to a nearest neighbor lattice site. This jump can be successful only, however, if the nearest neighbor site is empty, i.e. if it is occupied by a vacancy. The probability that an arbitrarily picked lattice site is empty is given by the atomic vacancy concentration c_v^a since c_v^a is the fraction of lattice sites not occupied by atoms. If an atom has z next neighbors then the probability that any of the next neighbor sites is empty is given by $z \cdot c_v^a$. Therefore, the diffusion coefficient for diffusion via vacancies, for instance for self diffusion D^* reads

$$D^* = \frac{\lambda^2}{6\tau} = \frac{\lambda^2}{6} zc_v^a \, \nu_D \exp(-G_m/kT) = \frac{\lambda^2}{6} zc_v^a \, \nu_D \exp(S_m^v/k) \cdot \exp(-H_m^v/kT) \tag{5.24}$$

Here G_m is also Gibbs free energy of migration $G_m = G_m^v$ for a vacancy since the site exchange of an atom with a vacancy, which constitutes the atomic migration step, is the same process for both the atom and the vacancy. Since in thermal equilibrium (see Chapter 3, Eq. (3.16))

$$c_v^a = \exp\left(-\frac{G_f^v}{kT}\right)$$

with G_f^v - Gibbs free energy of formation of a vacancy we finally arrive at

$$\begin{aligned} D^* &= \frac{\lambda^2}{6} z\nu_D \exp\left(-\frac{G_f^v + G_m^v}{kT}\right) \\ &= \frac{\lambda^2}{6} z\nu_D \exp\left(\frac{S_f^v + S_m^v}{k}\right) \exp\left(-\frac{H_f^v + H_m^v}{kT}\right) \end{aligned} \tag{5.25}$$

or in comparison with Eq. (5.16) ($D = D_0 \exp(-Q/kT)$) the activation enthalpy for diffusion via vacancies

$$Q = H_f^v + H_m^v \tag{5.26}$$

and the pre-exponential factor

$$D_0 = \frac{\lambda^2}{6} \, z\nu_D \, \exp\left(\frac{S_f^v + S_m^v}{k}\right) \tag{5.27}$$

Eq. (5.26) provides a means to test the hypothesis that self diffusion proceeds via the vacancy mechanism, since H_f^v and H_m^v can be measured separately and the sum of both must be identical to the activation enthalpy for self diffusion. How to measure H_f^v was described in Chapter 3, for instance by measurement of the electrical resistivity of quenched specimens. The activation enthalpy H_m^v can be obtained from the annihilation of vacancies during annealing of quenched specimens (Fig. 5.17). During the anneal, the quenched-in surplus vacancies will anneal out until the equilibrium concentration corresponding to the annealing temperature is attained. Since the annihilation of surplus vacancies proceeds via vacancy diffusion and the change in the electrical resistivity ρ is proportional to the change in the vacancy concentration, H_m^v can be determined from the rate of change of resistivity

$$\frac{d\rho}{dt} \sim \frac{dc_v}{dt} = \frac{c_v}{\tau^-} \tag{5.28}$$

Here τ^- is the average time a vacancy needs to annihilate. According to Eq. (5.17) and Eq. (5.24) $\frac{1}{\tau^-} \sim D \sim \exp(-G_m^v/kT)$ and with a constant K

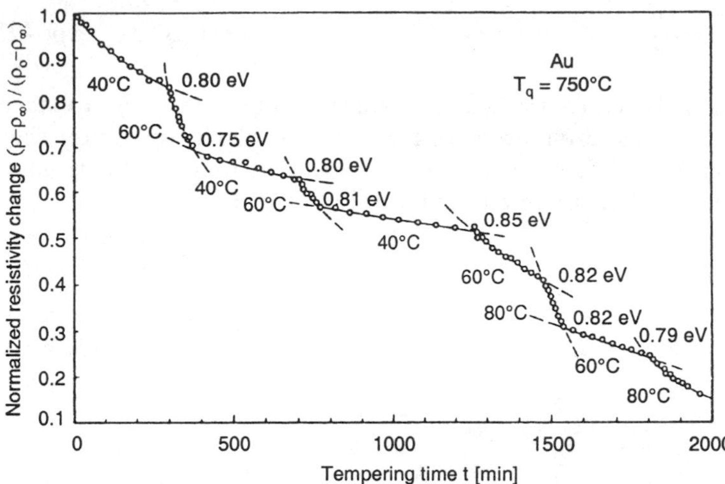

Fig. 5.17. Normalized resistivity change of gold wire after chill quenching from 750°C and tempering at the indicated temperatures. The migration energy of vacancies can be calculated from the drop in resistivity (see text).

$$\frac{dc_v}{dt} = Kc_v \exp\left(-\frac{G^v_m}{kT}\right) \tag{5.29}$$

For an instantaneous change of the annealing temperature, we obtain the ratio of the annealing rates at the time of the temperature change (c_v is the same)

$$\frac{\left.\frac{d\rho}{dt}\right|_{T_1}}{\left.\frac{d\rho}{dt}\right|_{T_2}} = \frac{\left.\frac{dc}{dt}\right|_{T_1}}{\left.\frac{dc}{dt}\right|_{T_2}} = \exp\left(\frac{H^v_m}{k}\left(\frac{1}{T_2} - \frac{1}{T_1}\right)\right) \tag{5.30}$$

All other constants and the entropy terms are eliminated. The correspondingly determined values of H^v_m for quenched gold are given in Fig. 5.17 for different temperature changes. Adding H^v_m determined from these experiments and H^v_f measured separately, we can compare the sum of both values with measured values of activation enthalpies of self diffusion (Table 5.4). Evidently, there is good agreement which supports the basic hypothesis that the mechanism of self diffusion or diffusion of foreign atoms in solid solutions proceeds via vacancies.

The site exchange of atom and vacancy can formally also be considered as diffusion of the vacancy, although the vacancy itself does not constitute a diffusing entity. In some cases, however, the history of the vacancy rather than that of the atoms is of interest, for instance, during the annealing of quenched-in or radiation-induced vacancies. Correspondingly, we can define a vacancy diffusion coefficient

Table 5.4. Formation enthalpy H_f^v and migration enthalpy H_m^v of vacancies in several metals. The sum is in good agreement with measured values of the activation energy Q for self-diffusion.

Metal	H_m^v [eV]	H_f^v [eV]	$H_m^v + H_f^v$ [eV]	Q [eV]
Au	0.83	0.95	1.78	1.76
Al	0.62	0.67	1.29	1.28
Pt	1.43	1.51	2.94	2.9
Cu	0.71	1.28	1.99	2.07
Ag	0.66	1.33	1.79	1.76
W	1.7	~ 3.6	~ 5.3	< 5.7
Mo	1.3	~ 3.2	~ 4.5	~ 4.5

$$D_v = \frac{\lambda^2}{6}\,\nu_D\,\exp\left(\frac{S_m^v}{k}\right)\exp\left(-\frac{H_m^v}{kT}\right) \tag{5.31}$$

Note that the activation energy for vacancy diffusion is different from the activation energy for diffusion via vacancies due to the fact that for the diffusion via vacancies, the vacancies have to be generated first and, therefore, the activation enthalpy for formation enters the total activation enthalpy for diffusion $Q^* = H_f^v + H_m^v$. In contrast, the activation energy for vacancy diffusion Q_v is given only by the activation enthalpy for migration $Q_v = H_m^v$. Therefore, $D_v \gg D^*$.

Eq. (5.17) can be utilized to link the continuum approach as manifested by Fick's laws, to the atomistic approach in terms of thermally activated jumps of atoms. For a motion of atoms by uncorrelated jumps (random walk), Eq. (5.17) can be associated with Eq. (5.14) which defines the macroscopic range of diffusion. For a jump distance λ and equal probability of a jump into all next neighbor sites, the distance R of an atom from its origin after n jumps will be on average

$$\overline{R_n^2} = n\lambda^2 \tag{5.32}$$

as can be seen from the following consideration. The vector $\mathbf{R}_n$ is a sum of the vectors of all individual jumps $\mathbf{r}_i$

$$\mathbf{R}_n = \mathbf{r}_1 + \mathbf{r}_2 + ... + \mathbf{r}_n = \sum_{i=1}^{n} \mathbf{r}_i \tag{5.33}$$

and

$$\begin{aligned} R_n^2 = \mathbf{R}_n \cdot \mathbf{R}_n &= \mathbf{r}_1 \cdot \mathbf{r}_1 + \mathbf{r}_1 \cdot \mathbf{r}_2 + \mathbf{r}_1 \cdot \mathbf{r}_3 + \ldots + \mathbf{r}_1 \cdot \mathbf{r}_n \\ &\quad + \mathbf{r}_2 \cdot \mathbf{r}_1 + \mathbf{r}_2 \cdot \mathbf{r}_2 + \ldots \quad \ldots + \mathbf{r}_2 \cdot \mathbf{r}_n \\ &\quad + \mathbf{r}_n \cdot \mathbf{r}_1 + \ldots \quad \ldots + \mathbf{r}_n \cdot \mathbf{r}_n \\ &= \sum_{i=1}^{n} \mathbf{r}_i \mathbf{r}_i + 2 \sum_{i=1}^{n-1} \mathbf{r}_i \mathbf{r}_{i+1} + 2 \sum_{i=1}^{n-2} \mathbf{r}_i \mathbf{r}_{i+2} + \ldots \qquad (5.34) \\ &= \sum_{i=1}^{n} \mathbf{r}_i^2 + 2 \sum_{j=1}^{n-1} \sum_{i=1}^{n-j} \mathbf{r}_i \mathbf{r}_{i+j} \\ &= \sum_{i=1}^{n} \mathbf{r}_i^2 + 2 \sum_{j=1}^{n-1} \sum_{i=1}^{n-j} |\mathbf{r}_i| \, |\mathbf{r}_{i+j}| \cos \Theta_{i,i+j} \end{aligned}$$

where $\Theta_{i,i+j}$ denotes the angle between the directions of the jumps i and $i+j$. Since the jump distance in a crystal is constant, i.e. $|\mathbf{r}_i| = \lambda$, we obtain the average

$$\overline{R_n^2} = n\lambda^2 \left(1 + \frac{2}{n} \overline{\sum_{j=1}^{n-1} \sum_{i=1}^{n-j} \cos \Theta_{i,i+j}} \right) \qquad (5.35)$$

Since the jumps are uncorrelated, the direction of any jump is independent of the direction of the previous jump. In such case there are on average as many jumps in a given direction as in the opposite direction, i.e. positive and negative $(\cos\Theta_{i,i+j})$ will occur equally often. As a result after many jumps the double sum in Eq. (5.35) will approach 0 and, correspondingly, we obtain Eq. (5.32). The average range $\sqrt{\overline{R_n^2}}$ increases in proportion to $\sqrt{n}$ and not in proportion to n. Since a macroscopic time interval t corresponds to $t = n\tau$ we obtain from the fundamental relation (Eq. (5.17)

$$D = \frac{\lambda^2}{6\tau} = \frac{\overline{R_n^2}/n}{6t/n} = \frac{\overline{R_n^2}}{6t} \qquad (5.36)$$

or

$$X^2 \equiv \overline{R_n^2} = 6\ Dt$$

in agreement with the Eq.(5.14) but derived from a totally different approach.

5.4 Correlation Effects

For interstitial diffusion or vacancy diffusion the assumption of an uncorrelated jump movement is certainly correct, i.e. each next neighbor site has the same probability to be jumped to. For the vacancy mechanism of diffusion, however, not all jumps are equally probable. Apparently, after a site exchange of an atom with a vacancy, the jump back into the vacancy is more probable than the jump in any other direction (Fig. 5.18). This imbalance of probability of jumps is accounted for by the correlation factor f which is defined as

$$f = \lim_{n\to\infty} \frac{\overline{R_n^2(Tr)}}{\overline{R_n^2(V)}} \tag{5.37}$$

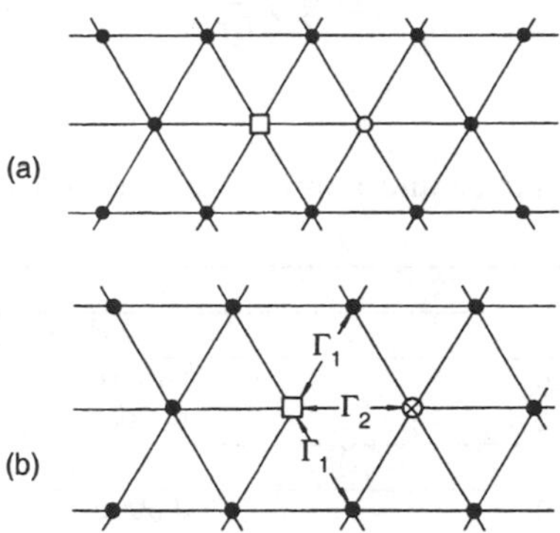

Fig. 5.18. Determining the correlation factor in a two-dimensional close-packed lattice. (a) selfdiffusion, (b) impurity diffusion with different jump frequencies Γ. Open square, open circle, and cirle with x indicate respectively the position of vacancy, tracer atom, and solute substitutional atom after the last jump.

where $\overline{R_n^2(Tr)}$ and $\overline{R_n^2(V)}$ denote the average square displacement after n jumps for tracer (Tr) or self diffusion, respectively vacancy diffusion (V). $\overline{R_n^2(Tr)}$ and $\overline{R_n^2(V)}$ are defined by Eq. (5.35). For the uncorrelated motion of vacancies Eq. (5.32) holds. The calculation of f is reduced to a calculation of the expression in the brackets of Eq. (5.35). For this all potential paths have to be considered by which the vacancy can circumvent the diffusing atom before the next site exchange with the atom takes place. A lengthy calculation finally yields

$$f = \lim_{n\to\infty} \left(1 + \frac{2}{n} \overline{\sum_{j=1}^{n-1} \sum_{i=1}^{n-j} \cos\Theta_{i,i+j}} \right) = \frac{1 + \overline{\cos\Theta_1}}{1 - \overline{\cos\Theta_1}} \tag{5.38}$$

where Θ_1 is the change of direction between the previous and the next jump. In first approximation we obtain

$$f = \frac{1 + \overline{\cos\Theta_1}}{1 - \overline{\cos\Theta_1}} = 1 + 2\overline{\cos\Theta_1} + 2\left(\overline{\cos\Theta_1}\right)^2 + \ldots \cong 1 - \frac{2}{z} \tag{5.39}$$

where z is the number of next neighbors, i.e. the coordination number. The final expression in equation (5.39) can be easily understood. If only a single jump, i.e. the jump back into the vacancy is considered, this jump happens with the probability $1/z$ and the reversal of direction corresponds to $\Theta_1 = 180°$, hence $\overline{\cos\Theta_1} = -1/z$. Truncation of the series expansion after the linear term in Eq. (5.39) yields the final expression for f. For pure metals, i.e. for self diffusion or tracer diffusion, Eq. (5.39) gives a good approximation as is evident from Table 5.5. The correlation factor modifies the fundamental equation (5.17) to become

$$D = f\frac{\lambda^2}{6\tau} \tag{5.40}$$

Table 5.5. Correlation factor and approximated values for several crystal structures.

Structure	z	f	1-2/z
		2-dim	
square	4	0.46705	0.5000
hexagonal	6	0.56006	0.667
		3-dim	
diamond	4	0.5000	0.5000
simple cubic	6	0.65549	0.667
bcc	8	0.72149	0.750
fcc	12	0.78145	0.833

For bcc, fcc, or hcp lattices the coordination number is 8, 12, and 12, respectively, and, therefore, f is only slightly different from 1 (Table 5.5). Therefore, correlation effects are of minor importance for self diffusion. In contrast, correlation effects can have a substantial influence on the diffusion coefficient for impurity diffusion. In particular, large effects can be expected if the foreign atom has a large binding energy to the vacancy, which causes the vacancy to change site with the foreign atom much more frequently than with a matrix atom.

Let us assume the exchange rate of the vacancy with the foreign atom to be Γ_2 and with the matrix atom Γ_1 and the hexagonal lattice geometry in Fig. 5.18b. With the constraint that the vacancy remains a nearest neighbor of the foreign atom and that only a single jump is considered we obtain

$$\overline{\cos\Theta_1} = -\frac{\Gamma_2}{\Gamma_2 + 2\Gamma_1} \tag{5.41}$$

or

$$f = \frac{1 + \overline{\cos\Theta_1}}{1 - \overline{\cos\Theta_1}} = \frac{\Gamma_1}{\Gamma_1 + \Gamma_2} \tag{5.42}$$

and

$$D = \frac{\lambda^2}{6}\,\frac{\Gamma_1\Gamma_2}{\Gamma_1 + \Gamma 2} \tag{5.43}$$

For a vacancy strongly bound to the foreign atom ($\Gamma_2 \gg \Gamma_1$)

$$D \cong \frac{\lambda^2}{6} \Gamma_1 < \frac{\lambda^2}{6} \Gamma_2 \tag{5.44}$$

In this case the diffusion coefficient for the foreign atom is not determined by the high jump frequency of the foreign atom but instead by the low jump frequency of the matrix atoms.
For the three dimensional case the geometry becomes more complicated but essentially Eq. (5.43) still holds. The ratio D^*/D in Table 5.3 gives a measure of the magnitude of the correlation factor. In special cases f can decrease the diffusion coefficient by several orders of magnitude.

5.5 Chemical Diffusion

The components of an alloy usually diffuse with different velocities as is evident for the example Au-Ni in Fig. 5.12 from the asymmetric concentration profile. This means that the diffusion coefficient in a concentration gradient according to Fick's first law

$$\tilde{D} = -\frac{j}{\left(\frac{\partial c}{\partial x}\right)} \tag{5.45}$$

is not constant but depends on composition: $\tilde{D} = \tilde{D}(c)$. $\tilde{D}$ is referred to as the chemical diffusion coefficient. It can be determined from the concentration profile if the boundary conditions of the diffusion problem can be expressed in the form $x/\sqrt{t}$. This is due to the fact that the partial differential equation of Fick's second law (Eq. (5.3)

$$\frac{\partial c}{\partial t} = \frac{\partial}{\partial x}\left(\tilde{D}\frac{\partial c}{\partial x}\right)$$

can be transformed to an ordinary differential equation by introduction of the variable $\eta = \frac{x}{\sqrt{t}}$

$$\frac{d}{d\eta}\left(\tilde{D}\frac{dc}{d\eta}\right) + \frac{\eta}{2}\frac{dc}{d\eta} = 0 \tag{5.46}$$

Let us consider, for example, the diffusion couple introduced in Section 5.1 consisting of two semi-infinite rods with the initial concentrations $c = 0$ $(x > 0)$ and $c = c_0$ $(x < 0)$: $t = 0$. These boundary conditions can be expressed as $c = c_0$ for $\eta = -\infty$ and $c = 0$ for $\eta = +\infty$. Now Eq. (5.46) can be integrated

$$-\frac{1}{2}\int_{c=0}^{c=c\prime} \eta dc = \left[\tilde{D}\frac{dc}{d\eta}\right]_{c=0}^{c=c\prime} \tag{5.47}$$

where c' is an arbitrary concentration with $0 < c' < c_0$. If the concentration profile $c(x)$ is known for a given time $t = t_1$, Eq. (5.47) can be utilized to determine $\tilde{D}(c)$. Since now $\eta = \frac{x}{\sqrt{t_1}}$ and

$$-\frac{1}{2}\int_0^{c'} xdc = \tilde{D}t_1 \left[\frac{dc}{dx}\right]_{c=0}^{c=c'} = \tilde{D}t_1 \left.\frac{dc}{dx}\right|_{c=c'} \tag{5.48}$$

because $dc/dx = 0$ for $c = 0$ (or for $x \to \infty$). Since also $dc/dx = 0$ for $c = c_0$ (or $x \to -\infty$) Eq. (5.48) yields

$$\int_0^{c_0} xdc = 0 \tag{5.49}$$

Eqs. (5.48) and (5.49) define the plane $x = 0$, the so-called Matano plane. To the right and left of this plane there are equal numbers of diffusing particles, i.e. the hatched areas in Fig. 5.19 are equal. With the coordinate system determined in this way

$$\tilde{D}(c') = -\frac{1}{2t_1} \left.\frac{dx}{dc}\right|_{c'} \int_0^{c'} xdc \tag{5.50}$$

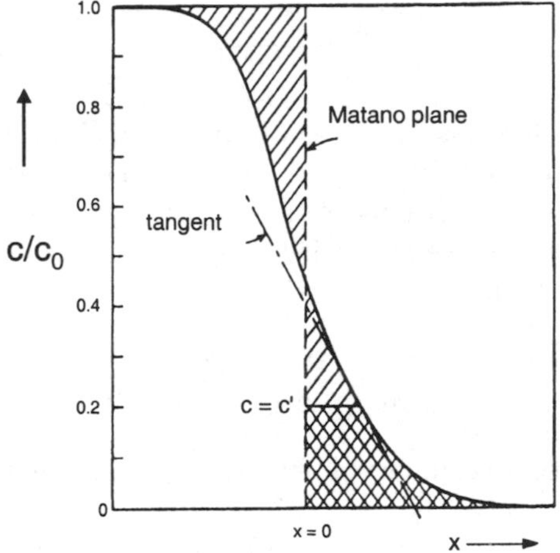

Fig. 5.19. Derivation of the chemical diffusion coefficient by the Matano-Boltzmann method (see text).

Eq. (5.50) allows to determine the chemical diffusion coefficient $\tilde{D}(c')$ for an arbitrary concentration c' if $c(x)$ at t_1 is known. For this one first defines the Matano plane, then determines the slope of $c(x)$ at $c = c'$ and, finally, evaluates graphically or numerically the integral $\int_0^{c'} xdc$, i.e. the cross hatched area in Fig. 5.19. The dependence $\tilde{D}(c)$ shown in Fig. 5.12 was determined this way.

The diffusion coefficient obtained by the procedure described above is the chemical diffusion coefficient, which accounts for the total diffusion flux comprised of all diffusing components. However, it does not yield information on the flux density of the individual components. The latter can be obtained by utilizing the Kirkendall effect. In the classical experiment of Kirkendall and Smigelskas, a core of 70/30 brass was covered by pure copper. Between core and cover molybdenum wires were inserted (Fig. 5.20). The refractory element Mo does not participate in the diffusion process at the annealing temperatures used, so that the Mo wires can be used as markers of the interface. Kirkendall found that during annealing the Mo wires of opposite interfaces had approached each other. This means, however, that the core had shrunk, i.e. substantially more zinc had diffused out of the brass than copper had diffused into the brass. This effect is again a proof for the vacancy mechanism of diffusion in substitutional alloys, since only a net vacancy flux can lead to such volume changes and, therefore, to a displacement of the Mo markers.

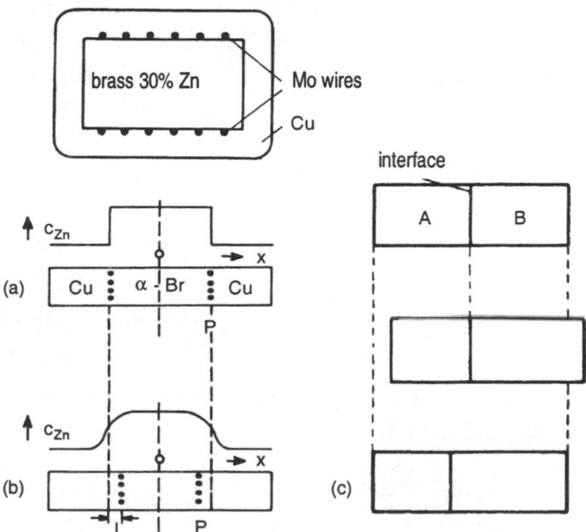

Fig. 5.20. Concentration distribution and marker shift due to the Kirkendall effect using the Cu-α-brass diffusion pair. (a) Before the experiment; (b) after diffusion annealing; (c) shift of the interface.

Darken used the experiments of Smigelskas and Kirkendall to formulate a phenomenological (thermodynamical) theory of chemical diffusion which allows one to obtain the chemical diffusion coefficients of the individual diffusing components. The major problem is to determine the coordinate system that defines the fluxes. The flux through any chosen cross section would be mainly determined by the motion of the atoms relative to this reference plane due to the change of the volume (Fig. 5.20). This flux, however, is not of interest, but rather we want to determine the diffusion flux of the individual components. An example that illustrates this problem is the diffusion of ink in running water, for instance in a river. Most of the flux that we would observe from a particular location on the bank of the river is the flux of the ink due to the motion of the water and not the diffusion flux of ink spreading out in the water. To determine the diffusion flux of the ink, therefore, we have to account for the flux of the ink molecules relative to the moving water molecules. This can be measured, for instance, by identifying the velocity of the water by floating markers, for instance, by little pieces of wood. The same problem applies to the determination of the diffusion flux density in a Kirkendall experiment.

In this example the ink flux j_{tot} measured from the bank of the river would be composed of the diffusion flux $j_D = -\tilde{D} \cdot \frac{\partial c}{\partial x}$ and the drift flux (convection flux) $j_c = vc$ or

$$j_{tot} = j_D + j_c = -\tilde{D}\frac{\partial c}{\partial x} + vc \tag{5.51}$$

where v is the drift velocity of the water as measured by a floating marker. Accordingly, Eq.(5.51) holds for the diffusion flux density of any component in a binary alloy in a concentration gradient. Let the number of atoms per unit volume in a binary alloy be c_1 and c_2. For each component the number of particles will not change and, therefore, the continuity equation applies (Eq. (5.2)):

$$\frac{\partial c_1}{\partial t} + \frac{\partial j_1}{\partial x} = \frac{\partial c_2}{\partial t} + \frac{\partial j_2}{\partial x} = 0 \tag{5.52}$$

The total number of particles per unit volume $c = c_1 + c_2$, therefore,

$$\frac{\partial c}{\partial t} = \frac{\partial c_1}{\partial t} + \frac{\partial c_2}{\partial t} = \frac{\partial}{\partial x}\left(\tilde{D}_1\frac{\partial c_1}{\partial x} + \tilde{D}_2\frac{\partial c_2}{\partial x} - v\left(c_1 + c_2\right)\right) \tag{5.53}$$

Since the total number of particles per unit volume does not change, $\partial c/\partial t = 0$. Integration of Eq. (5.53) yields

$$\tilde{D}_1\frac{\partial c_1}{\partial x} + \tilde{D}_2\frac{\partial c_2}{\partial x} - vc = 0 \tag{5.54a}$$

or

$$v = \frac{1}{c}\left(\tilde{D}_1\frac{\partial c_1}{\partial x} + \tilde{D}_2\frac{\partial c_2}{\partial x}\right) \tag{5.54b}$$

A relationship between $\tilde{D}$ and $\tilde{D}_1$ and $\tilde{D}_2$ can be obtained from a combination of Eq. (5.54b) with Eqs. (5.52) and (5.53).

$$\frac{\partial c_1}{\partial t} = \frac{\partial}{\partial x}\left(\tilde{D}_1\frac{\partial c_1}{\partial x} - \frac{c_1}{c}\tilde{D}_1\frac{\partial c_1}{\partial x} - \frac{c_1}{c}\tilde{D}_2\frac{\partial c_2}{\partial x}\right) \tag{5.55}$$

Because $c =$ constant, we can write

$$\frac{\partial c_1}{\partial x} = -\frac{\partial c_2}{\partial x}$$

$$\frac{\partial c_1}{\partial t} = \frac{\partial}{\partial x}\left(\frac{c_1\tilde{D}_2 + c_2\tilde{D}_1}{c} \cdot \frac{\partial c_1}{\partial x}\right) \tag{5.56}$$

By comparison with Fick's second law we obtain

$$\tilde{D} = \frac{c_1\tilde{D}_2 + c_2\tilde{D}_1}{c} = c_1^a\tilde{D}_2 + c_2^a\tilde{D}_1 \tag{5.57}$$

where c^a denotes the respective atomic concentration.

Equivalently Eq. (5.54b) and (5.55) yield

$$v = \left(\tilde{D}_1 - \tilde{D}_2\right)\frac{dc_1^a}{dx} \tag{5.58}$$

Eqs. (5.57) and (5.58) are known as the Darken equations which formulate the relationship between the macroscopic and atomic chemical diffusion coefficients. In other words, once $\tilde{D}$ and v have been determined experimentally, $\tilde{D}_1$ and $\tilde{D}_2$ can be calculated.

5.6 Thermodynamic factor

Fick's laws relate the observed diffusion flux to the concentration gradient with the understanding that nature strives to establish a homogeneous concentration distribution. From thermodynamics we know, however, that the thermodynamic equilibrium at constant temperature and pressure is determined by the minimum of Gibbs free energy G, and does not necessarily consist of a homogeneous concentration distribution. In a system with n components

$$G = U + pV - TS + \sum_{i=1}^{n}\mu_i N_i$$

where μ_i is the chemical potential and N_i is the number of particles of component i. As long as for a given p and T the chemical potential μ_i is not constant everywhere, there will be always a particle flux to establish thermodynamic equilibrium. To first order we can assume that the particle flux is proportional to the gradient of the thermodynamic variables as in Fick's first law, but with coefficients M_{ij} instead of the diffusion coefficient D. For instance, for the flux density of component 1 we obtain

$$j_1 = -M_{11}\frac{d\mu_1}{dx} - M_{12}\frac{d\mu_2}{dx} - \ldots - M_{1n}\frac{d\mu_n}{dx} - M_{1p}\frac{dp}{dx} - M_{1T}\frac{dT}{dx} \tag{5.59}$$

and similar equations for the other components, that is in total n equations. For a binary system at constant pressure and temperature

$$\begin{aligned} j_1 &= -M_{11}\frac{d\mu_1}{dx} - M_{12}\frac{d\mu_2}{dx} \\ j_2 &= -M_{21}\frac{d\mu_1}{dx} - M_{22}\frac{d\mu_2}{dx} \end{aligned} \tag{5.60}$$

The Darken equations follow from Eq. (5.60) and the condition that the off diagonal coefficients $M_{12} = M_{21} = 0$. With Eq.(5.45)

$$j_1 = -M_{11}\frac{d\mu_1}{dx} = -\tilde{D}_1\frac{dc_1}{dx}$$
$$j_2 = -M_{22}\frac{d\mu_2}{dx} = -\tilde{D}_2\frac{dc_2}{dx} \tag{5.61}$$

In Sec. 5.1 we showed that a potential gradient causes a convection flux. With the mobility $B = v/F$ and $F = -d\mu/dx$ we obtain according to Eqs. (5.8) and (5.9), for instance for component 1

$$j_1 = c_1 v = B_1 F_1 c_1 = -B_1 c_1 \frac{d\mu_1}{dx} = -M_{11}\frac{d\mu_1}{dx} = -\tilde{D}_1\frac{dc_1}{dx} \tag{5.62}$$

i.e. $M_{11} = B_1 c_1$ and

$$\tilde{D}_1 = B_1\frac{d\mu_1}{d\ln c_1} = B_1\frac{d\mu_1}{d\ln c_1^a} \tag{5.63}$$

The concentration dependence of the molar chemical potential is given by

$$\mu_1 = \mu_0(p,T) + \mathrm{RT}\ln(\gamma_1 \cdot c_1^a) \tag{5.64}$$

where R is the gas constant and μ_0 is that part of the potential which depends only on pressure and temperature. The quantity γ_1 is the so-called activity coefficient of the component 1, and depends on c_1. Therefore,

$$\frac{d\mu_1}{d\ln c_1^a} = \mathrm{RT}\left(1 + \frac{d\ln\gamma_1}{d\ln c_1^a}\right) \tag{5.65}$$

and

$$\tilde{D}_1 = B_1\mathrm{RT}\left(1 + \frac{d\ln\gamma_1}{d\ln c_1^a}\right) \tag{5.66}$$

An analogous relation holds for component 2.

The expression $1 + (d\ln\gamma_1/d\ln c_1^a)$ is referred to as the thermodynamic factor. In dilute alloys $\gamma_1 \cong$ constant (Raoult's and Henry's law) and, therefore,

$$D_1 = B_1\mathrm{RT} \tag{5.67}$$

But in concentrated non-ideal alloys $\gamma_1 = \gamma_1(c)$ and thus the thermodynamic factor is different from unity.

Diffusion in a concentration gradient is usually difficult to determine, but the dependency $\tilde{D}(c)$ is important for many materials science problems or predictions for materials processing. The tracer diffusion coefficient in homogeneous binary alloys is much easier to determine. In the latter case $dc/dx = 0$ but $dc_1^*/dx \neq 0$, where c_1^* is the concentration of the radioactive isotope. The tracer diffusion coefficient of the component of an alloy is, therefore, with Eq.(5.66)

$$\tilde{D}_1^* = B_1^* \mathrm{RT} \left(1 + \frac{d \ln \gamma_1^*}{d \ln c_1^{a*}}\right)_{c_1^a + c_1^{a*}} \tag{5.68}$$

Since the radioactive isotope is chemically identical with the natural element, γ_1 depends only on $c_1 + c_1^*$, which, however, is constant. Correspondingly the thermodynamic factor for tracer diffusion according to Eq.(5.68) is equal to unity or

$$\tilde{D}_1^* = B_1^* \mathrm{RT} = D_1^* \tag{5.69}$$

Since the isotope and the stable atom are chemically identical and physically almost identical their mobilities are practically identical. Therefore, for self diffusion

$$D_1^* = B_1 \mathrm{RT} \tag{5.70}$$

This yields finally

$$\tilde{D}_1 = D_1^* \left(1 + \frac{d \ln \gamma_1}{d \ln c_1^a}\right) \tag{5.71}$$

for the chemical diffusion coefficient in a concentration gradient, in other words for the chemical diffusion coefficient $\tilde{D}_1$ of a particular component. Evidently, this chemical diffusion coefficient in a concentration gradient is different from the self diffusion coefficient D_1^* of this component in a homogeneous alloy of equal composition if the thermodynamic factor is not equal to 1. The thermodynamic factor constitutes the fundamental relationship between the chemical and the self diffusion coefficient. Using the Gibbs-Duhem equation (for binary systems $c_1 d\mu_1 + c_2 d\mu_2 = 0$) Eq. (5.57) can be rewritten as

$$\tilde{D} = (D_1^* \, c_2^a + D_2^* \, c_1^a) \left(1 + \frac{d \ln \gamma_1}{d \ln c_1^a}\right) \tag{5.72}$$

$\tilde{D}$, D_1^* and D_2^* can be determined experimentally, and γ_1 can be obtained from thermodynamic measurements. Experimental measurements of $\tilde{D}$ according to Eq.(5.50) and calculated values of $\tilde{D}$ according to Eq. (5.72) are usually in good agreement.

Finally, we want to stress that the derivation yields for self diffusion

$$D^* = B \cdot \mathrm{RT}$$

as assumed in Sec. 5.1.

This equation is also referred to as the Einstein relation or the Nernst-Einstein relation.

The special property of the thermodynamic factor is that it can also become negative. This changes the sign of the diffusion constant and consequently the direction of the diffusion flux. For a negative thermodynamic factor the diffusion flux is aligned with the concentration gradient, the concentration difference is amplified, i.e. we observe uphill diffusion. This process is important for phase transformations in solid state, in particular, for spinodal decomposition (see Chapter 9.2.1.3).

As mentioned before, the observation of the Kirkendall effect strongly supports the hypothesis of the vacancy mechanism of diffusion in substitutional alloys. Correspondingly, with the flux of any component there is an equally large but opposite flux of vacancies. If the diffusion fluxes of all components in an alloy are the same, the net vacancy flux vanishes. For the case, however, where the diffusion fluxes of the components in an alloy are different there is a non-zero net vacancy flux that leads to an increased vacancy concentration at the origin of the faster diffusing material. Eventually, this can cause the formation of pores and substantial changes of volume (Fig. 5.21), which can drastically degrade the properties of the material. In particular, for the welding or hot dip coating of materials these phenomena can cause major problems and special measures have to be taken to avoid these effects.

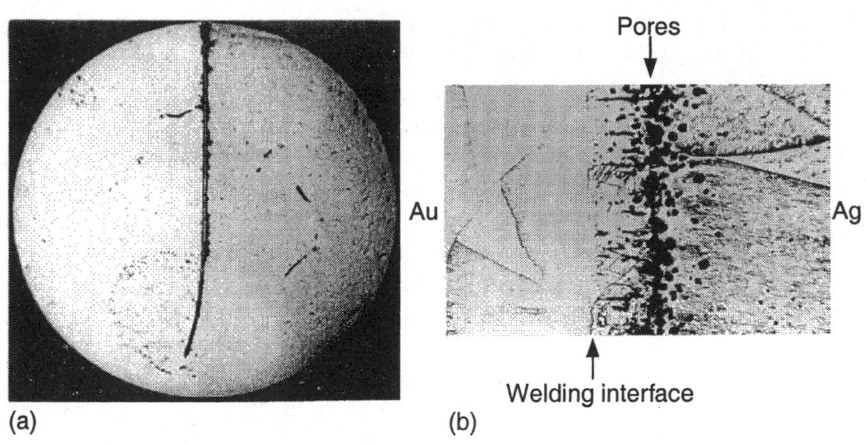

Fig. 5.21. Microstructure caused by the Kirkendall effect in the AgAu system. (a) A Mo foil cut to a point perpendicular to the plane of the picture is bent at the tip due to the different mass transport by diffusion [5.9]. (b) Due to the diffusion flux differential vacancies accumulate on one side of a welded joint, on the other side there is a large elastic stress (and recrystallization) [5.10].

5.7 Grain Boundary Diffusion

On a surface the motion of atoms is much less constrained than in the crystal interior and, therefore, can proceed much more rapidly than in the volume (Fig. 5.22). Rapid diffusion along surfaces can be easily demonstrated by the fast spread of radioactive material on a surface (Fig. 5.23). Surface diffusion is very important for the processing of advanced materials, for instance for coatings or deposition techniques in microelectronics. Internal surfaces, in particular grain boundaries and phase boundaries also constitute fast diffusion paths. This effect is felt, for instance, during high temperature creep and also plays an important role in superplasticity (see Chapter 6.8.1). At lower temperatures where volume diffusion is practically frozen, grain boundary diffusion still proceeds and can constitute the dominating atomic transport mechanism in a solid. This becomes immediately evident when comparing the self diffusion of single and polycrystals (Fig. 5.24). At low temperatures the polycrystal has a much higher diffusion coefficient than single crystals, since mass transport proceeds mainly via grain boundaries. At high temperatures when diffusion also occurs in the volume the measured diffusion coefficients in single and polycrystals are identical, since volume diffusion dominates due to the fact that the volume of grain boundaries is much smaller than the total bulk volume and, therefore, grain boundaries represent only a small fraction of the total cross section.

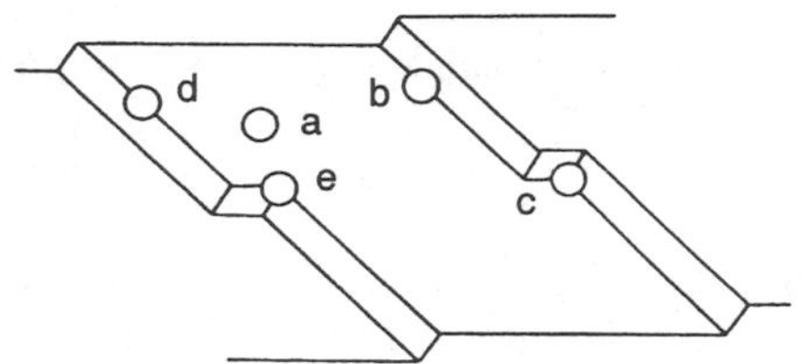

Fig. 5.22. Different layers of ad-atoms on the surface of a crystal. Motion of atoms is not significantly restricted.

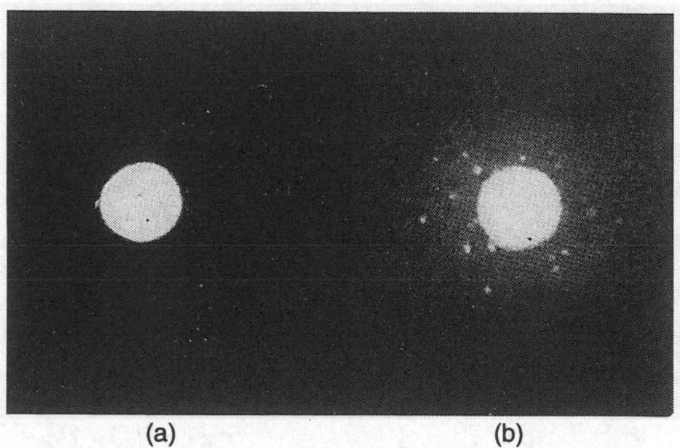

Fig. 5.23. Distribution of radioactive polonium on a silver surface. (a) At time $t = 0$; (b) after 4 days heat treatment at 480°C [5.11].

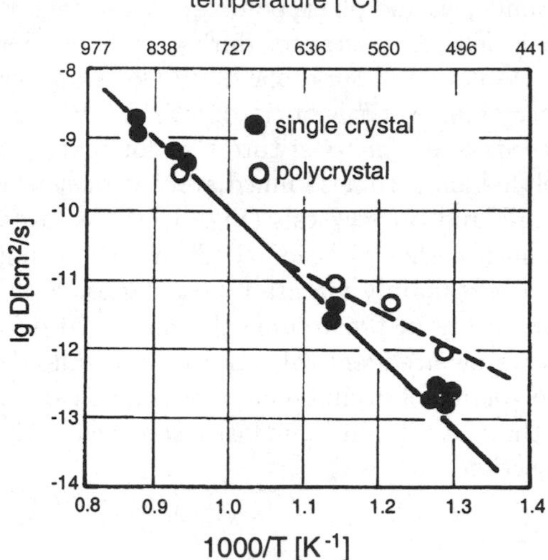

Fig. 5.24. Self-diffusion coefficient of single crystalline and polycrystalline Ag as a function of temperature (after [5.12]).

In each volume element of the grain boundary the diffusion flux is affected by both grain boundary diffusion and volume diffusion. For a grain boundary in the y direction (Fig. 5.25) the flux density change in a boundary volume element is in linear approximation

$$j_{y+dy} = j_y + dy \cdot \frac{dj_y}{dy} \tag{5.73}$$

and the continuity equation reads

$$\frac{\partial c}{\partial t} = \frac{1}{\delta \cdot dy \cdot z}\left\{\delta \cdot z \cdot \left[j_y - \left(j_y + dy\frac{dj_y}{dy}\right)\right] - 2 \cdot z \cdot dy\ j_x\right\} \tag{5.74}$$

where δ is the thickness of the grain boundary and z the depth of the (considered) volume element. Because

$$j_y = -D_{GB}\frac{dc}{dy} \quad \text{and} \quad j_x = -D_V\frac{dc}{dx} \tag{5.75}$$

with the coefficient of grain boundary diffusion D_{GB} and the volume diffusion coefficient D_v, we obtain Fick's second law for grain boundary diffusion

$$\frac{\partial c}{\partial t} = -\frac{\partial j_y}{dy} - 2 \cdot \frac{1}{\delta} j_x = D_{GB}\frac{\partial^2 c}{\partial y^2} + \frac{2D_V}{\delta}\frac{\partial c}{\partial x} \tag{5.76}$$

where we have assumed that the diffusion coefficient does not depend on position.

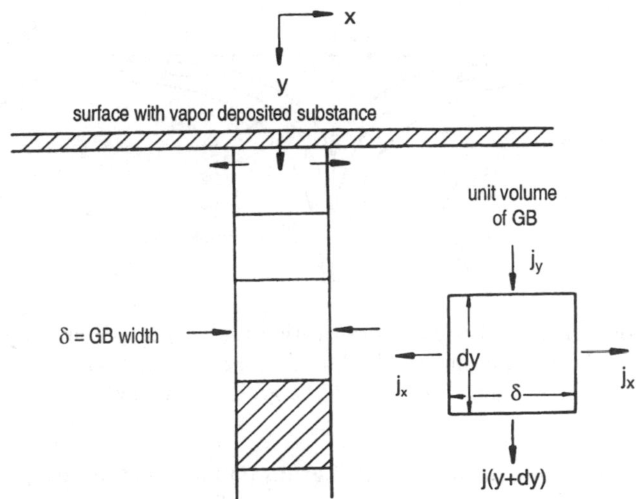

Fig. 5.25. Diffusion along a grain boundary of width δ in a $x - y$ section. The concentration change in a unit volume of the grain boundary is determined by the flux along the grain boundary (j_y) and into the volume (j_x).

Under strongly simplifying assumptions we obtain the so-called Whipple solution

$$c(x, y, t) = c_0 \exp\left[-y\frac{\sqrt{2}}{\sqrt[4]{\pi D_V t} \cdot \sqrt{\delta D_{GB}/D_V}}\right]\left\{1 - \operatorname{erf}\left(\frac{x}{2\sqrt{D_V t}}\right)\right\} \quad (5.77)$$

The concentration profile $c(x, y, t)$ can be experimentally determined by chemical analysis of serial sections, for instance by X-ray spectroscopy of thin layers removed parallel to the surface. If the results are plotted as contour lines of constant concentration they would follow a behavior as schematically drawn in Fig. 5.26, i.e. largest penetration depth in the boundary and decreasing penetration with increasing distance from the boundary.

We recognize from Eq. (5.77) that from the concentration distribution only the product $D_{GB}\delta$ can be determined, but never D_{GB} as a separate entity. From the temperature dependence of $D_{GB}\delta$, the activation enthalpy of grain boundary diffusion can be obtained. It is much smaller than the activation enthalpy for volume diffusion.

The grain boundary diffusion coefficient is not a material constant. It depends on the grain boundary character, i.e. on grain boundary structure (see Chapter 3). In low angle boundaries (misorientation $< 15\,°$), D_{GB} is of the same magnitude as D_V. For a given axis of rotation, D_{GB} grows with increasing angle of rotation (Fig. 5.27). Correspondingly, the penetration depth of a tracer atom becomes larger. For some special boundaries with strongly

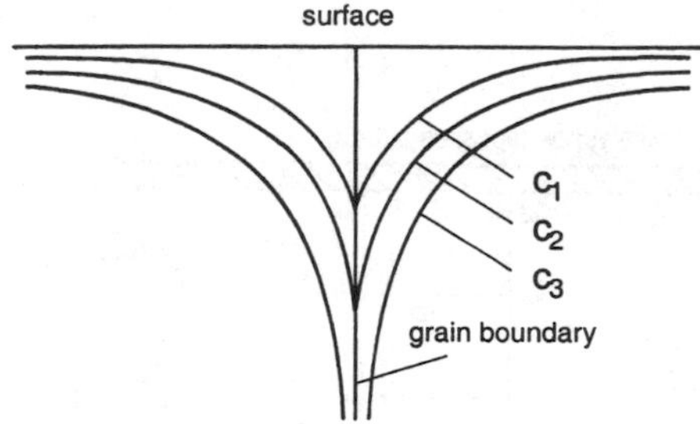

Fig. 5.26. Contour lines of equal concentration ($c_1 > c_2 > c_3$) due to diffusion of a substance from the surface into the bulk and along a grain boundary. The contour lines reach much further into the solid at the grain boundary.

ordered and, therefore, densely packed grain boundary structure, as described by a low value of Σ (see Chapter 3), very small values of D_{GB} are observed (Fig. 5.28). The anisotropy of grain boundary diffusion for low angle grain boundaries can be understood from its dislocation structure. Symmetrical low angle tilt boundaries, for instance, are comprised of an equidistant arrangement of parallel edge dislocations. Parallel to the dislocation lines, diffusion is increased owing to the dilatation of the lattice at the dislocation core (pipe diffusion). Therefore, we expect the diffusion parallel to the dislocation lines, i.e. parallel to the direction of the rotation axis to be faster than diffusion perpendicular to the dislocation lines where the diffusion flux has to proceed predominantly through the volume. In fact such an anisotropy has been experimentally confirmed (Fig. 5.29). The spacing of dislocations in a low angle boundary decreases with increasing angle of rotation. For a misorientation of $\approx$15 °, the dislocation cores are expected to overlap (see Chapters 3.4.2 and 6.4), and the grain boundaries ought to comprise a continuous slab of disordered material. Indeed, the anisotropy decreases with increasing angle of rotation, but surprisingly also for misorientations far larger than 15 °, there remains a substantial anisotropy of diffusivity.

If we compare diffusion through the volume, parallel to grain boundaries and on surfaces, the smallest activation energy is obtained for surface diffusion and the highest for volume diffusion. Under the assumption of the same value of the pre-exponential factor D_0, we obtain in an Arrhenius plot three straight lines with different slopes, as plotted in Fig. 5.30 for the example of diffusion of thorium in tungsten.

5.8 Diffusion in Nonmetals: Ionic Conductors

In ionic crystals the constraints of electrical charge neutrality disfavors the formation of individual vacancies, and vacancy pairs (Schottky defects) are formed, which consist of an anion vacancy and a cation vacancy (Fig. 5.31),

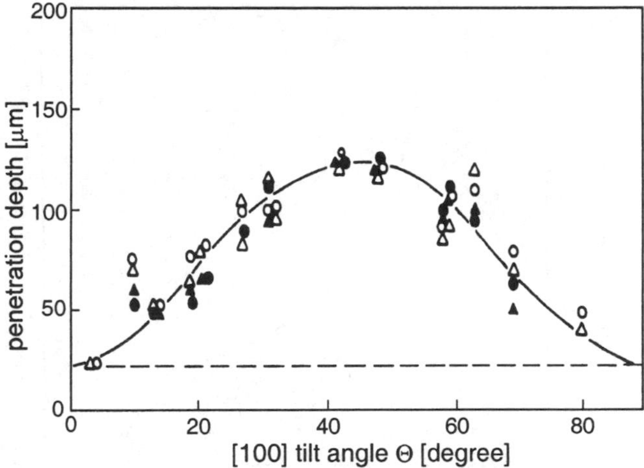

Fig. 5.27. Penetration depth of Ni tracer atoms along the grain boundary as a function of tilt angle of [100] tilt boundaries in Nickel at 1050°C (after [5.13]).

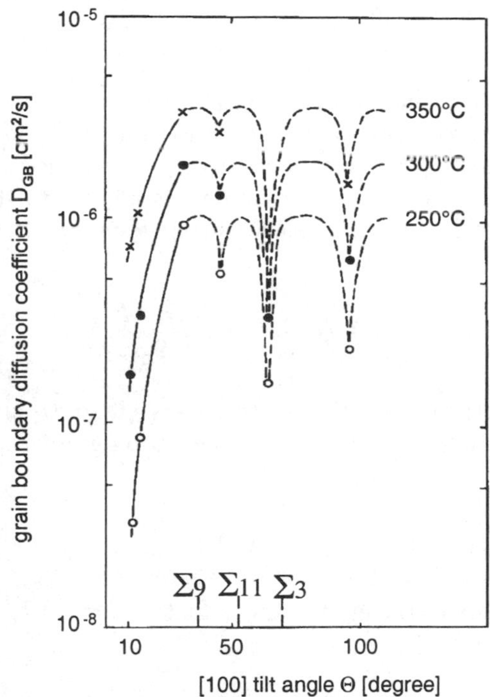

Fig. 5.28. Coefficient of grain boundary diffusion at [110] tilt boundaries in aluminium for different tilt angles. For certain grain boundaries (low Σ) the diffusion coefficient decreases sharply (after [5.13]).

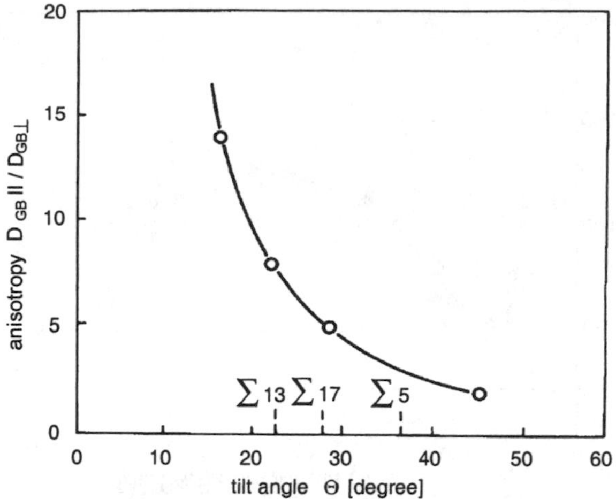

Fig. 5.29. Anisotropy of grain boundary diffusion at ⟨100⟩ tilt boundaries in silver at 450°C. There is anisotropy even for high-angle grain boundaries. $D_{||}$ and $D_{\perp}$ are the diffusion coefficients for diffusion along a grain boundary parallel or perpendicular to the tilt axis, respectively (after [5.14]).

or Frenkel defects (see Chapter 3). At finite temperature, ions in a crystal can move via the associated type of vacancy but as for self diffusion in metals, the net flux through a given cross section will vanish. However, if there is a concentration gradient or if an electrical field is applied to the crystal a flux of charge carriers will occur. The ionic current density of component i, for instance Na^+, in NaCl is given by

$$j_i = -D_i \frac{dc_i}{dx} - q_i c_i \frac{D_i}{\mathrm{kT}} \frac{d\Phi}{dx} \tag{5.78}$$

where Φ is the electrical potential, q_i the charge, and D_i the diffusion constant of the ion of type i. Without a concentration gradient, $dc/dx = 0$, the generated electrical current will be

$$I = q_i j_i = \frac{D_i q_i^2 c_i}{\mathrm{kT}} \left(-\frac{d\Phi}{dx} \right) \tag{5.79}$$

or using the definition of the electrical conductivity σ

$$\sigma = \frac{I}{-\left(\frac{d\Phi}{dx}\right)} \tag{5.80}$$

$$\frac{\sigma}{D_i} = \frac{c_i q_i^2}{\mathrm{kT}} \tag{5.81}$$

From measurements of the ionic conductivity the tracer diffusion coefficient Eq.(5.81) can be validated. At high temperatures there is usually good agree-

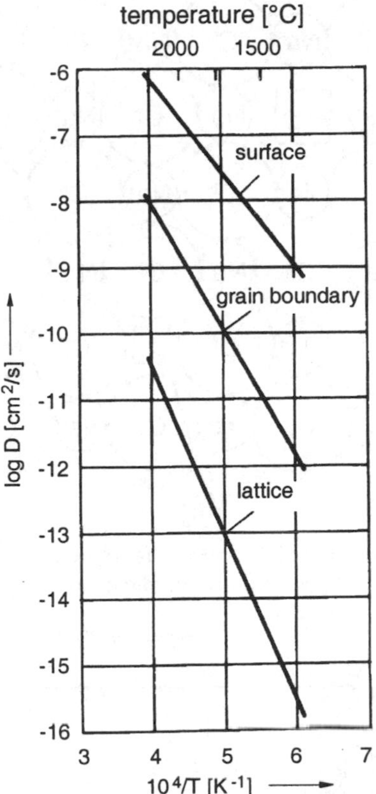

Fig. 5.30. Comparison of surface, grain boundary and lattice diffusion of thorium in tungsten for equal D_0.

ment (Fig. 5.32). For lower temperatures, however, deviations from the high temperature behavior are observed, and in particular a discontinuous slope in an Arrhenius plot is evident.

This deviation is due to impurities with different valency (for instance by divalent impurities in NaCl) as will be considered in the following. The addition of impurities with different valency causes the formation of structural vacancies in order to conserve charge neutrality. For instance, if divalent Ca^{++} is added to Na^+Cl^-, charge neutrality requires that for each added Ca^{++} ion, two Na^+ ions have to be removed, i.e. for each Ca^{++} ion a cation vacancy (cv) has to be formed so that the total number of cation vacancies will be

$$c_{++} + c_{av} = c_{cv} \tag{5.82}$$

where c_{++} is the concentration of divalent ions and c_{av} or c_{cv} are the concentrations of the anion or cation vacancies, respectively. In thermal equilibrium the concentration of vacancy pairs will be

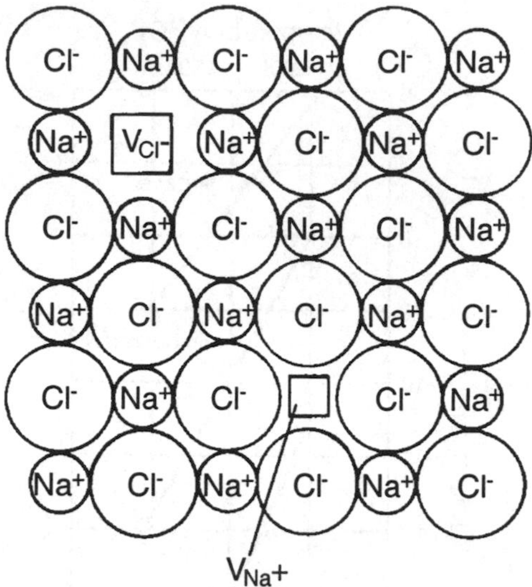

Fig. 5.31. Schottky defect in NaCl. Due to charge neutrality there must be an equal number of anion vacancies V_{Cl-} and cation vacancies V_{Na^+}.

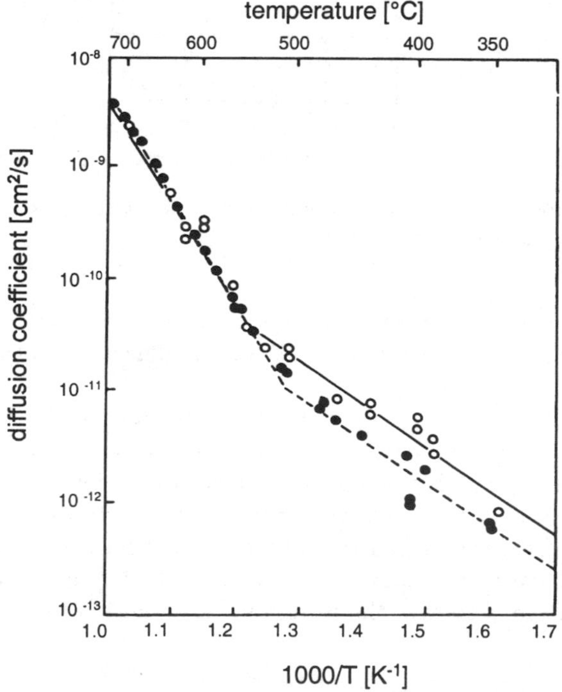

Fig. 5.32. Diffusion coefficient of Na^+ in NaCl. The solid circles were calculated from conductivity measurements, the open circles are derived from measuring tracer diffusion (after [5.15]).

$$c_{av}\, c_{cv} = \exp\left(-\frac{\Delta G_f^s}{\mathrm{kT}}\right) \tag{5.83}$$

where ΔG_f^s is the Gibbs free energy of formation of a vacancy pair (Schottky defect).

Eq. (5.83) is a thermodynamic requirement and, therefore, independent of the mechanism of vacancy formation. It holds for vacancy formation by thermal activation in pure compounds (concentrations c_{av}^0 and c_{cv}^0) as well as for structural defects in doped crystals (concentrations c_{av} and c_{cv}). With Eq. (5.82) we can rewrite Eq. (5.83) for ionic crystals with divalent impurities

$$c_{cv}\,(c_{cv} - c_{++}) = \exp\left(-\frac{\Delta G_f^s}{\mathrm{kT}}\right) = \left(c_{cv}^0\right)^2 = \left(c_{av}^0\right)^2 \tag{5.84}$$

The solution of the quadratic equation for $c_{cv} > 0$ is

$$c_{cv} = \frac{c_{++}}{2}\left\{1 + \sqrt{1 + \frac{4\left(c_{cv}^0\right)^2}{\left(c_{++}\right)^2}}\right\} \tag{5.85}$$

For very pure compounds or very high temperatures $c_{cv}^0 \gg c_{++}$ and, therefore, $c_{cv} \approx c_{cv}^0$. For heavily doped compounds or very low temperatures $c_{cv}^0 \ll c_{++}$ and thus $c_{cv} \approx c_{++}$. Since c_{cv}^0 decreases exponentially with falling temperature, in commercially pure materials there is always a temperature below which $c_{cv}^0 \ll c_{++}$.

The tracer diffusion coefficient of the cations can be written as

$$D_T = f \cdot \frac{\lambda^2}{6} z\, c_{cv} \exp\left(-\frac{G_m^c}{\mathrm{kT}}\right) \tag{5.86}$$

If $c_{cv} = c_{cv}^0 = exp\left(-\Delta G_f^s/2kT\right)$ (intrinsic range), the activation energy is given by

$$H_i = \frac{H_f^s}{2} + H_m^c \tag{5.87a}$$

For $c_{cv} \cong c_{++}$ (extrinsic range) c_{cv} is independent of temperature and

$$H_e = H_m^c \tag{5.87b}$$

Correspondingly there are always two ranges of the $\sigma(T)$ curve, and the temperature of the transition between these two ranges is higher, the higher the concentration of the impurity (Fig. 5.33). The difference of the temperature dependence of σ and D_T in the extrinsic range is caused by the fact that

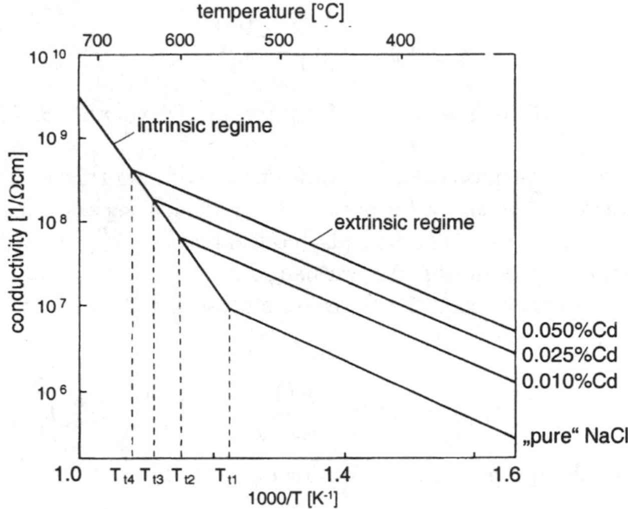

Fig. 5.33. Ionic conductivity of (commercially) pure and $CdCl_2$-doped NaCl (after [5.16]).

the structural vacancies are bound to the impurities and, therefore, do not contribute to σ, but do contribute to D_T since they provide potential lattice sites for diffusing tracer atoms.

The phenomena of diffusion and ionic conduction in ionic crystals are complex and have many aspects. For instance, depending on the type of compound, diffusion can also proceed by an interstitial mechanism, for instance Ag in AgBr. In principle, however, all these cases can be treated on the basis of the concept introduced above, although modified if complications such as special interactions occur or constraints are imposed.

6

Mechanical Properties

6.1 Basic Elements of Elasticity

If a solid is exposed to external forces that try to change its shape, it remains coherent but resists by the generation of an internal stress (Fig. 6.1). For instance, if we pull on a solid it would separate into two parts, if it were cut through in the center. The forces that need to be exerted to keep the two parts in contact correspond to the internal forces in the loaded solid. Let us use a simple model in which we visualize the solid as being composed of hard spheres (atoms) that are connected by springs (interatomic potential, Fig. 6.2). External tensile forces imposed on the solid will stretch the springs until their reaction force, which is proportional to the elongation of the springs, balances the external forces. The state of the stressed springs represents the internal forces.

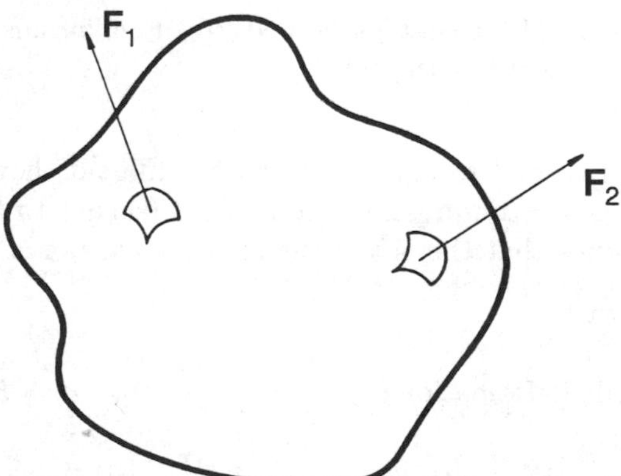

Fig. 6.1. The definition of stress. Forces **F** acting on the surface produce internal stress.

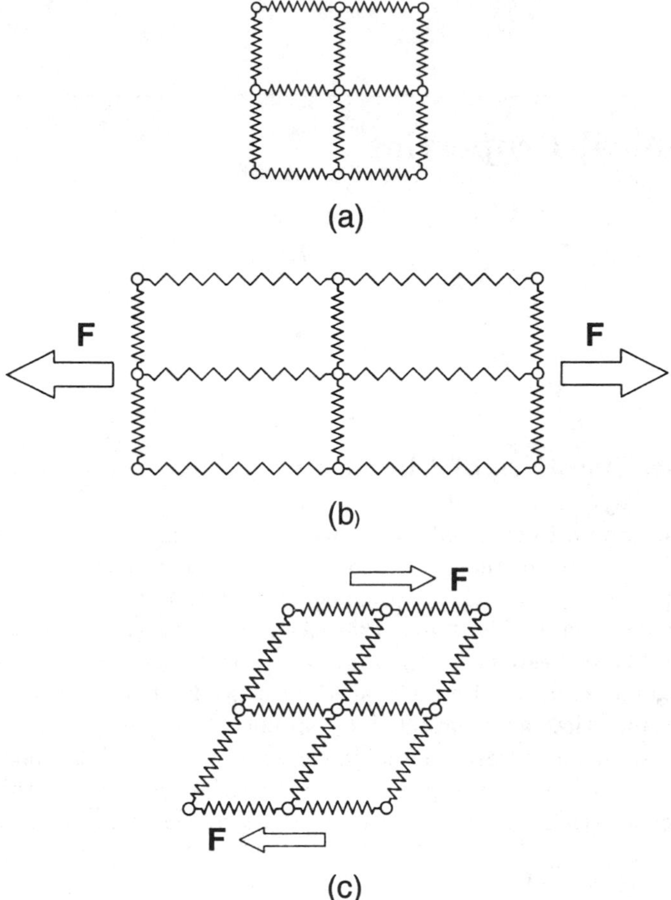

Fig. 6.2. Spring model of the elastic solid body. (a) Equilibrium; (b) strain under tensile stress; (c) shear under shear stress.

The reaction of a solid to external forces is different, however, for forces perpendicular (tension, compression) or parallel (shear) to the surface. We know from experience that the shape change for small forces is proportional to the force (Fig. 6.3)

$$\text{tensile deformation}: \quad \frac{F_\perp}{q} = E\frac{\Delta\ell}{\ell_0} \quad \text{or} \quad \sigma = E\varepsilon \tag{6.1a}$$

$$\text{shear deformation}: \quad \frac{F_\parallel}{q} = G\frac{\Delta x}{d} \quad \text{or} \quad \tau = G\gamma \tag{6.1b}$$

Here $F_\perp$ and $F_\parallel$ are the forces perpendicular and parallel to the surface, respectively, ℓ_0 and d are the length and thickness of the crystal, q the area

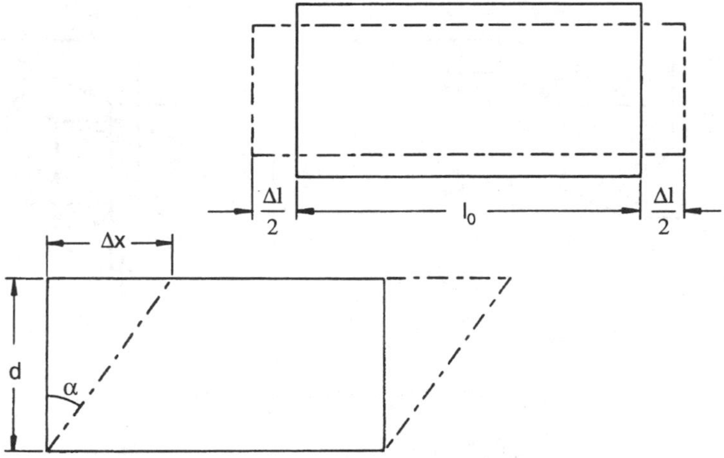

Fig. 6.3. The definition of strain. $\varepsilon = \Delta l/l_0$ and shear $\gamma = \Delta x/d = tan\alpha$.

on which the force acts, and $\Delta\ell$ or Δx are the length change or horizontal shift of the crystal (Fig. 6.3). Eqs. (6.1a,b) formulate Hooke's law: the normal stress σ is proportional to the strain ϵ. The proportionality constant is Young's modulus E. The shear stress τ is proportional to the simple shear γ. The proportionality constant in this case is the shear modulus G. Commonly $G < E$, which means that a material has a higher resistance against deformation by tension or compression than in shear. This is also obvious from the simple spring-sphere model of the solid. If the force is parallel to the spring, the elongation of the spring is smaller than if the force is perpendicular to the spring axes.

Consider a small cubic volume element in a solid. The stress state in the solid can be decomposed into forces per unit area perpendicular and parallel to the cube faces. A force parallel to a face comprises two components that are perpendicular to each other. A traction on any of the three cube faces is, therefore, composed of three stresses, namely one normal stress and two shear stresses (Fig. 6.4). Thus the stress state of a volume element is correspondingly described by nine components of stress. These nine components define the stress tensor $\boldsymbol{\sigma}$

$$\boldsymbol{\sigma} = \begin{bmatrix} \sigma_{xx} & \sigma_{xy} & \sigma_{xz} \\ \sigma_{yx} & \sigma_{yy} & \sigma_{yz} \\ \sigma_{zx} & \sigma_{zy} & \sigma_{zz} \end{bmatrix} \tag{6.2}$$

The magnitude of the individual components depends, of course, on the spatial orientation of the volume element. If a different orientation of the volume element in Fig. 6.4 were to be chosen, the stress components parallel and perpendicular to its faces would be different. This, however, does not change the

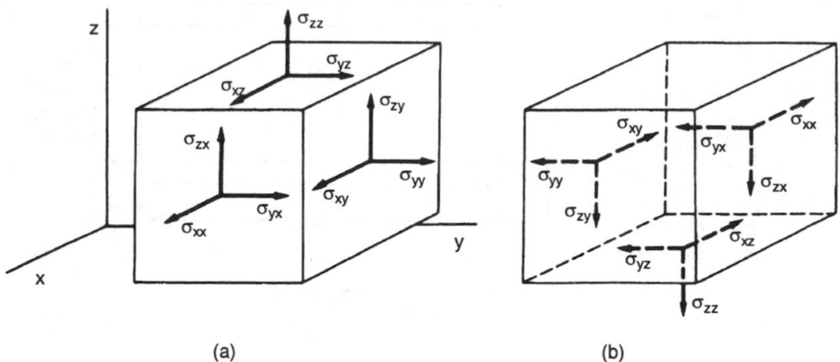

Fig. 6.4. The components of the three-dimensional stress tensor.

physical stress state because the stress state is a physical state that does not depend on the choice of the coordinate system. A different choice of the volume element, i.e. the different choice of the coordinate system, only changes the decomposition of the stress state parallel to different spatial directions. The change of a tensor upon change of the one coordinate system to another can be described mathematically by a tensor transformation. If the relationship between the coordinate system K_2 and the original coordinate system K_1 is given by a rotation, i.e. by a rotation matrix $\mathbf{A}$, the stress tensor $\boldsymbol{\sigma}_2$ in the rotated system can be calculated from the original stress tensor $\boldsymbol{\sigma}_1$ in the coordinate system K_1 thus

$$\boldsymbol{\sigma}_2 = \mathbf{A}^t \, \boldsymbol{\sigma}_1 \mathbf{A} \tag{6.3}$$

(A^t - transposed matrix: $A^t_{ij} = A_{ji}$).

There is always a choice of coordinate system in which only the normal stresses are non-zero, which means that the stress tensor has the following appearance:

$$\boldsymbol{\sigma} = \begin{bmatrix} \sigma_1 & 0 & 0 \\ 0 & \sigma_2 & 0 \\ 0 & 0 & \sigma_3 \end{bmatrix} \tag{6.4}$$

The stresses σ_1, σ_2, and σ_3 are referred to as principal stresses. The maximum shear stresses occur at 45° to the directions of the principal stresses. If $\sigma_1 > \sigma_2 > \sigma_3$, the maximum shear stress is given by

$$\tau_{max} = \frac{1}{2}\left(\sigma_1 - \sigma_3\right) \tag{6.5}$$

The stress tensor is always symmetrical, i.e., $\sigma_{ij} = \sigma_{ji}$ because of the equilibrium of the balance of moments in any volume element. For application to crystal plasticity it is often convenient to decompose the stress tensor into a hydrostatic part $\boldsymbol{\sigma}_H$ and a deviatoric part $\boldsymbol{\sigma}_D$, because only the stress deviator is relevant for the plastic deformation of crystals.

The hydrostatic stress is the average normal stress

$$p = \frac{1}{3}(\sigma_1 + \sigma_2 + \sigma_3) \tag{6.6}$$

and, in most metals, changing the mean stress makes no difference to their plastic deformation.

One can show that for any choice of the coordinate system $\sigma_{xx}+\sigma_{yy}+\sigma_{zz} = \sigma_1 + \sigma_2 + \sigma_3 =$ const., i.e.,

$$\boldsymbol{\sigma} = \boldsymbol{\sigma}_H + \boldsymbol{\sigma}_D = \begin{bmatrix} p & 0 & 0 \\ 0 & p & 0 \\ 0 & 0 & p \end{bmatrix} + \begin{bmatrix} \sigma_{xx} - p & \sigma_{xy} & \sigma_{xz} \\ \sigma_{yx} & \sigma_{yy} - p & \sigma_{yz} \\ \sigma_{zx} & \sigma_{zy} & \sigma_{zz} - p \end{bmatrix} \tag{6.7}$$

Hooke's law, Eq. (6.1), describes the reaction of a solid to a force acting on it in terms of a shape change, namely strain and shear. In analogy to the stress state we can also express the elastic shape change by a strain tensor $\boldsymbol{\varepsilon}$

$$\boldsymbol{\varepsilon} = \begin{bmatrix} \varepsilon_{xx} & \varepsilon_{xy} & \varepsilon_{xz} \\ \varepsilon_{yx} & \varepsilon_{yy} & \varepsilon_{yz} \\ \varepsilon_{zx} & \varepsilon_{zy} & \varepsilon_{zz} \end{bmatrix} \tag{6.8}$$

Here $\varepsilon_{xx}, \varepsilon_{yy}$, and ε_{zz} are (normal) strains and $\varepsilon_{xy}, \varepsilon_{xz}$, and ε_{yx} are shear strains [1]. We note that in comparison to the definition in Eq. (6.1) $\varepsilon_{xy} = 1/2\gamma_{xy}$. The strain tensor is also symmetrical, so a coordinate system can always be found in which the strain state can be described in terms of principal strains only

$$\boldsymbol{\varepsilon} = \begin{bmatrix} \varepsilon_1 & 0 & 0 \\ 0 & \varepsilon_2 & 0 \\ 0 & 0 & \varepsilon_3 \end{bmatrix} \tag{6.9}$$

Again the sum $\varepsilon_1 + \varepsilon_2 + \varepsilon_3$ or "trace" of the matrix is independent of the choice of the coordinate system and

[1] The shape change of a volume element is given by the so-called displacement gradient tensor $\mathbf{e}$, which can be decomposed into the strain tensor $\boldsymbol{\varepsilon}$ and the rotation tensor $\boldsymbol{\omega}$, i.e. $\mathbf{e} = \boldsymbol{\varepsilon} + \boldsymbol{\omega}$. For the shear deformation considered in Fig. 6.3

$$\mathbf{e} = \begin{pmatrix} 0 & \gamma & 0 \\ 0 & 0 & 0 \\ 0 & 0 & 0 \end{pmatrix} \boldsymbol{\varepsilon} = \begin{pmatrix} 0 & 1/2\gamma & 0 \\ 1/2\gamma & 0 & 0 \\ 0 & 0 & 0 \end{pmatrix} \boldsymbol{\omega} = \begin{pmatrix} 0 & 1/2\gamma & 0 \\ -1/2\gamma & 0 & 0 \\ 0 & 0 & 0 \end{pmatrix}$$

The strain tensor $\boldsymbol{\varepsilon}$ is the symmetrical part (i.e. $\varepsilon_{ij} = (e_{ij} + e_{ji})/2 = \varepsilon_{ji}$) of $\mathbf{e}$ and describes the pure deformation of the volume element under consideration. The rotation (pseudo)tensor $\boldsymbol{\omega}$ corresponds to the antisymmetric part (i.e. $\omega_{ij} = (e_{ij} + e_{ji})/2 = -\omega_{ji}$) of $\mathbf{e}$ and reflects the so-called rigid body rotation of the volume element. For a pure rigid body rotation no deformation is necessary; it can be entirely described by a rotation of the reference coordinate system. Therefore, the actual deformation is described in terms of the strain tensor $\boldsymbol{\varepsilon}$ only.

$$\varepsilon_1 + \varepsilon_2 + \varepsilon_3 = \varepsilon_{xx} + \varepsilon_{yy} + \varepsilon_{zz} = \frac{\Delta V}{V} \tag{6.10}$$

where $\Delta V/V$ is the relative volume change caused by the elastic deformation. In mathematical terms the hydrostatic strain (or stress) is an invariant of the tensor. With the stress and strain tensors defined above, Hooke's law can be reformulated as:

$$\boldsymbol{\sigma} = \mathbf{C}\boldsymbol{\varepsilon} \tag{6.11a}$$

$$\sigma_{ij} = \sum_{k,l=1}^{3} C_{ijkl}\, \varepsilon_{kl} \tag{6.11b}$$

Here $\mathbf{C}$ is the tensor of elastic constants, a tensor of rank 4 with $3^4 = 81$ elements, C_{ijkl}. However, because of crystal symmetry there are at most 21 different tensor elements (in case of triclinic symmetry). Therefore, the tensor of rank 4 can be conveniently rewritten as symmetrical matrix C_{ij} with elements C_{11} through C_{66}. This simplified representation is common in the materials and physics literature. In elastically isotropic materials the number of independent elastic constants is even reduced to two, for instance Young's modulus E and Poisson ratio ν. Poisson's ratio expresses the experience that a solid changes (diminishes) its cross section if it is elongated (Fig. 6.3a). Hence, a tensile stress in the x direction also causes a strain in y and z directions

$$\varepsilon_{yy} = \varepsilon_{zz} = -\nu\, \varepsilon_{xx} \tag{6.12}$$

The shear modulus G in an isotropic solid is related to Young's modulus and Poisson ratio by

$$G = \frac{E}{2(1+\nu)} \tag{6.13}$$

The elastic behavior of a material is defined by two independent elastic constants only in the case of exact elastic isotropy of the solid, that is, when the elastic properties do not depend on spatial direction. In polycrystals that have a random orientation distribution this is usually the case. In cubic single crystals or polycrystals with texture, the elastic behavior does depend on orientation. In this case Eq. (6.13) does not hold any more, and there are three independent elastic constants. For hexagonal crystal symmetry, the number of independent elastic constants increases to five.

6.2 The Flow Curve

Strictly speaking Hooke's law is valid only for very small strains ($\varepsilon < 10^{-4}$). At larger deformations there will be (initially small) deviations from the proportionality between stress and strain, which increase with increasing strain.

Macroscopically, this non-linearity can be hardly recognized and for technical purposes it is usually neglected. It is characteristic of elastic deformation, however, that a solid spontaneously reverts to its original undeformed shape when the forces are removed.

However, metallic materials can be deformed to strains much larger than the elastic limit before they break, because they can deform plastically. This means that a shape change remains after unloading. In a rough specification of materials with regard to plastic deformation we discriminate three types of materials based on fracture strain, ε_f,

- (a) brittle materials: $\varepsilon_f \leq 0.1\%$ (for instance ceramic materials, hard materials)
- (b) ductile materials: $\varepsilon_f \approx 10\%$ (metals and commercial alloys)
- (c) superplastic materials: $\varepsilon_f \approx 1000\%$ (special fine grained alloys).

The stress-strain behavior of materials is usually measured by uniaxial deformation, that is in a tensile test. For this a specimen is mounted into a mechanical testing machine where it is elongated with constant crosshead speed (Fig. 6.5), and both the elongation $\Delta\ell$ and the applied force F are continuously recorded. From the force F and the initial cross section q_0 we obtain the engineering stress $\sigma = F/q_0$. The elongation $\Delta\ell$ normalized by the initial length ℓ_0 defines the engineering strain $\varepsilon = \Delta\ell/\ell_0$.

If we plot σ versus ε we obtain the engineering stress-strain diagram. Fig. 6.6 gives some examples of engineering stress-strain diagrams of commercial materials. Although they vary greatly in detail, the essential features of the diagrams are constant (Fig. 6.7a). After exceeding a yield stress σ_y, i.e. the end of the elastic regime, the stress first increases with strain (strain hardening), but passes through a maximum UTS (ultimate tensile strength) at a strain ε_u (uniform strain) and decreases until the strain to fracture ε_f is attained. The yield stress is not uniquely defined because of the continuous transition from the elastic to the plastic range. Most frequently the yield stress is defined as the $\sigma_{0.2}$, which is the stress where the remaining strain would be 0.2%. To obtain the proof stress, a line parallel to the elastic regime is drawn such that it intersects the strain axis at $\varepsilon = 0.2\%$. $\sigma_{0.2}$ is determined by the intersection of this line with the flow curve. Alternatively σ_y can be determined by extrapolation of the terminal elastic regime and back-extrapolation of the incipient plastic regime. In this case σ_y is given by the point of intersection of the extrapolation lines (Fig. 6.8). Some materials in particular plain carbon steels show a different behavior at transition from the elastic to the plastic regime, namely a discontinuous stress behavior (Fig. 6.7b). After attainment of an upper yield stress σ_y^H the stress drops to the lower yield stress σ_y^L and remains approximately constant during the strain interval ε_L (Lüders strain), before regular strain hardening commences.

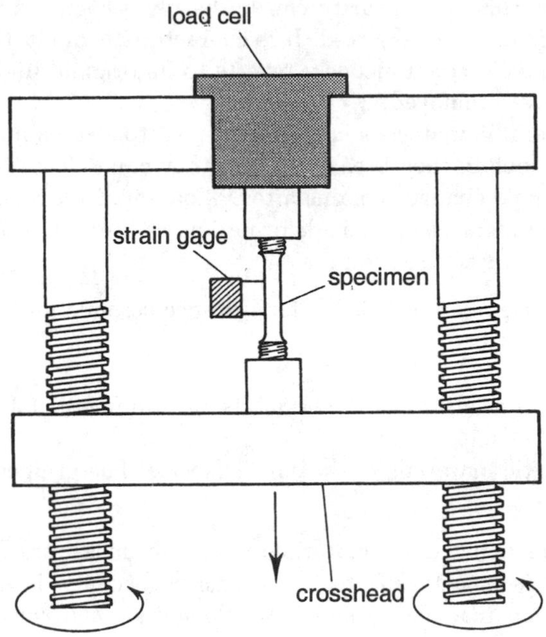

Fig. 6.5. The dynamic tensile test.

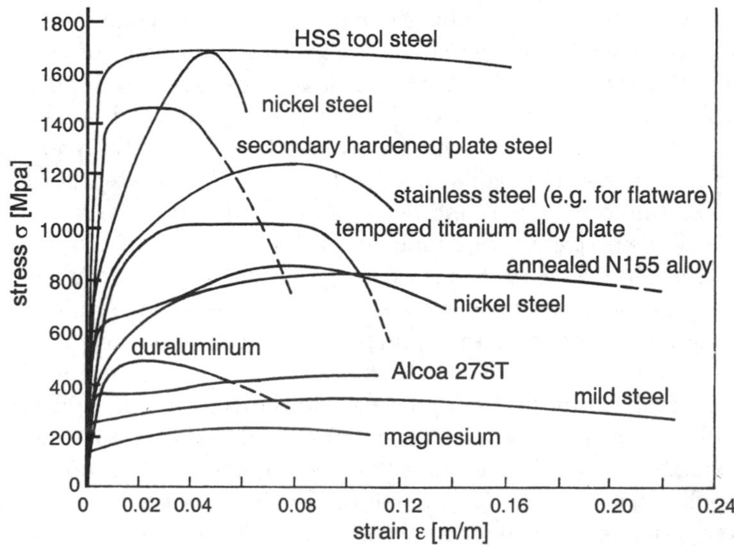

Fig. 6.6. Stress-strain diagrams of several engineering materials (after [6.1]).

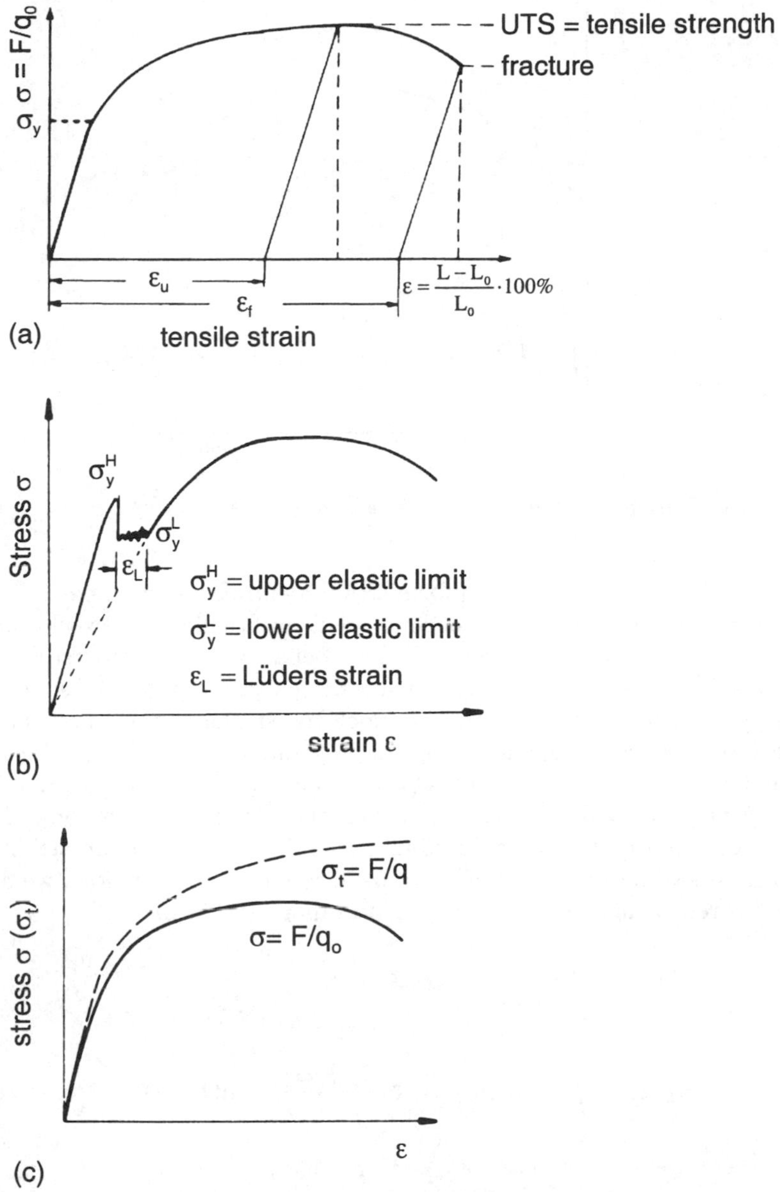

Fig. 6.7. Schematic stress-strain diagrams. (a) Engineering diagram showing the important characteristic quantities of materials testing. (b) Diagram showing a pronounced elastic limit and the appearance of flow lines (Lüders bands). (c) Engineering and true stress.

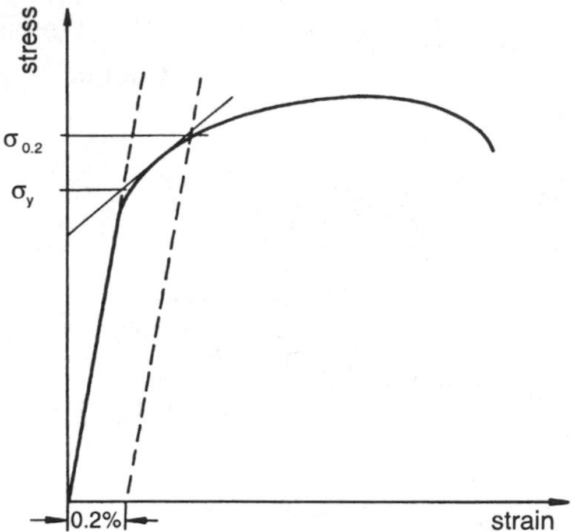

Fig. 6.8. Definition of the yield stress: σ_y and $\sigma_{0.2}$.

For strains larger than the uniform strain, deformation becomes unstable. The cross section of a tensile specimen will eventually decrease locally or "neck" and cause the sample to break. This behavior can be understood from the true stress-strain behavior. The engineering stress and strain were defined in analogy to elastic deformation, for which the strains are so small that the length and cross section change only negligibly and, therefore, stress and strain can be defined in terms of the initial dimensions without introducing a large error. During plastic deformation, however, the shape change becomes large, so that for a more physically meaningful definition of stress and strain, the actual cross section and length have to be accounted for. Therefore, we define the true stress σ_t and the true strain ε_t in differential terms

$$d\varepsilon_t = \frac{d\ell}{\ell} \tag{6.14}$$

$$\varepsilon_t = \int_{\ell_0}^{\ell} \frac{d\ell}{\ell} = \ln\frac{\ell}{\ell_0} = \ln\frac{\ell_0 + \Delta\ell}{\ell_0} = \ln(1+\varepsilon) \tag{6.15}$$

$$\sigma_t = \frac{F}{q} = \frac{F}{q_0} \cdot \frac{q_0}{q} \tag{6.16}$$

During plastic deformation the volume remains constant. Therefore,

$$\ell_0 \cdot q_0 = \ell \cdot q \tag{6.17}$$

or

$$\frac{q_0}{q} = \frac{\ell}{\ell_0} = 1 + \varepsilon \tag{6.18}$$

hence

$$\sigma_t = \sigma \cdot (1 + \varepsilon) \tag{6.19}$$

Correspondingly, the true stress-strain curve can be calculated from the engineering stress-strain curve. It is also referred to as a flow curve. The increase of force dF which is necessary to deform a tensile specimen by $d\varepsilon_t$ is

$$F = \sigma_t \cdot q \tag{6.20}$$

$$\frac{dF}{d\varepsilon_t} = q \cdot \frac{d\sigma_t}{d\varepsilon_t} + \sigma_t \frac{dq}{d\varepsilon_t} \tag{6.21}$$

Thus, the stability of plastic deformation is determined by physical hardening (slope of the flow curve) $d\sigma_t/d\varepsilon_t > 0$ and geometrical softening (reduction of cross section) $dq/d\varepsilon_t < 0$. If the hardening coefficient $d\sigma_t/d\varepsilon_t$ is large, deformation proceeds in a stable mode. A local reduction of the cross section during stable deformation will cause that cross section to harden more than the neighboring material and to suppress further deformation in this section until a uniform cross section is re-established. However, the hardening coefficient typically decreases with increasing strain, and so there is a critical strain, the uniform strain, where physical hardening and geometrical softening balance. At even larger strains geometrical softening dominates. If under these conditions the cross section is locally reduced, the physical hardening cannot compensate the geometrical softening anymore, and necking of the cross section will proceed at this location. The external load will decrease but because of the diminishing cross section the true stress in the constricted volume will further increase until the material breaks.

Correspondingly, deformation becomes unstable at the maximum of the engineering stress-strain diagram. If the maximum is not very pronounced the uniform strain is difficult to read from the engineering stress-strain curve. It is easy to find, however, if the true stress is plotted versus the engineering strain. The maximum of the engineering stress-strain curve is defined by

$$dF = 0 = q d\sigma_t + \sigma_t dq \tag{6.22}$$

Because of constant volume Eq. (6.17) $\ell_0 \; q_0 = \ell \cdot q = \text{const.}$. Therefore,

$$\ell \cdot dq + q \cdot d\ell = 0 \tag{6.23}$$

$$\frac{d\ell}{\ell} = -\frac{dq}{q} \tag{6.24}$$

$$\frac{d\ell}{\ell} = \frac{d\ell}{\ell_0} \cdot \frac{\ell_0}{\ell} = \frac{d\varepsilon}{(1 + \varepsilon)} \tag{6.25}$$

$$\frac{d\sigma_t}{d\varepsilon} = \frac{\sigma_t}{1+\varepsilon} \tag{6.26}$$

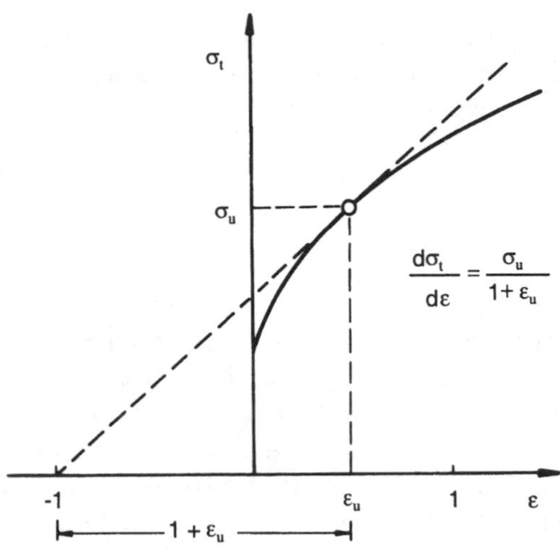

Fig. 6.9. Determining the uniform elongation ε_u by the Considère-criterion.

The tangent of the $\sigma_t - \varepsilon$ diagram which intersects at $\varepsilon = -1$ defines the true stress σ_t and the uniform strain ε_u at maximum load (UTS), i.e. the point of instability (Fig. 6.9). Eq. (6.26) is also referred to as the Considère-criterion of instability.

We have to note that after UTS/maximum load, deformation is, in principle, unstable but strain rate sensitivity, to be explained later, can delay the onset of actual instability. In particular during deformation at elevated temperatures, deformation can proceed to large strains after exceeding the Considère criterion (superplasticity, see Chapter 6.8.1).

The tension test has the disadvantage that it becomes unstable because of geometrical softening at relatively low strains. This problem can be avoided in a uniaxial compression test (Fig. 6.10a), since the cross section increases in this case, resulting in geometrical hardening. In fact, large strains can be reached in compression tests. A problem is caused by friction between the compression platens and the specimen surface which opposes expansion of the cross section at the contact surface. At large strains this causes "barrelling" since the cross section in the specimen center can expand more than at its ends. Consequently, a nonuniform cross section is obtained resulting in a loss of uniaxial deformation geometry.

Geometrical softening can also be avoided by using a torsion test (Fig. 6.10b), since the cross section does not change during deformation. Again,

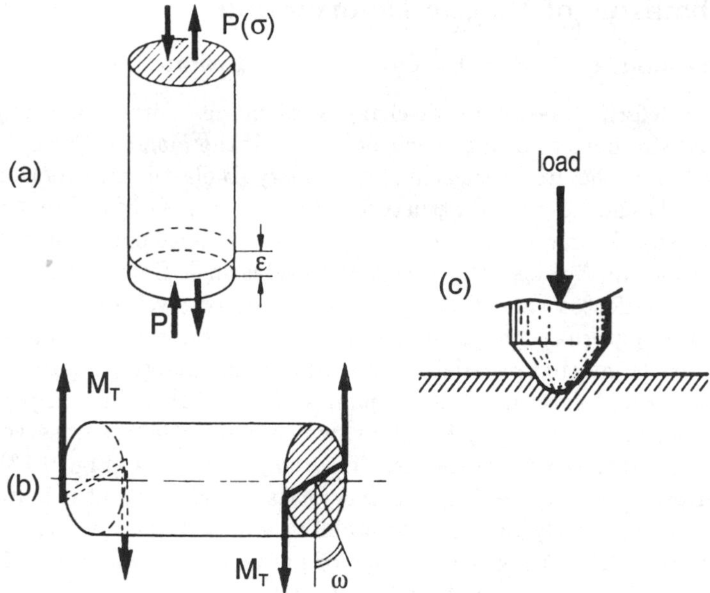

Fig. 6.10. Several methods of materials testing. (a) Tensile and compression testing; (b) torsion testing; (c) hardness testing.

large degrees of deformation can be obtained. The deformation during torsion is not constant in the cross section, though, and the (shear) strain varies from $\gamma = 0$ in the axial center of the cylinder to $\gamma = \gamma_{max}$ at the surface. This problem can be circumvented by using thin wall hollow cylinders, which are susceptible to buckling, however.

A very easy and widely used method of materials testing is the hardness test (Fig. 6.10c). An indenter is pressed into the surface of a sample under a defined load and the size of the remaining indent is measured after unloading. There are different methods which are distinguished by the geometry of the indenter, e.g. semispherical (Brinell-hardness), cone shaped (Rockwell-hardness) or pyramidal (Vickers-hardness) etc. The disadvantage of the hardness test is the physically undefined state of the material during the test, since it is deformed both elastically and plastically under a multiaxial stress state. For certain conditions the hardness can be correlated with the yield stress. However, even if this is not the case, hardness tests provide fast qualitative information which can be easily obtained without specific requirements of specimen geometry. It is particularly beneficial for investigation of changes of strength (e.g. hardness change during precipitation or recrystallization) or comparison of strength.

6.3 Mechanisms of Plastic Deformation

6.3.1 Crystallographic Slip by Dislocation Motion

If a material deforms plastically, it changes its shape. Correspondingly, the atoms of a crystal have to change their positions. If the macroscopic deformation is reflected at the atomistic scale, by analogy to elastic deformation, the crystal must change its crystal structure (Fig. 6.11a). X-ray diffraction investigations show, however, that the crystal structure does not change during plastic deformation, because the Bragg reflections in a Debye-Scherrer diagram remain unaffected (Fig. 6.12). A conservation of crystal structure concurrent with an external change of shape is possible only if complete blocks of a crystal are translated parallel to crystallographic planes by integer multiples of the atomic spacing in those planes (Fig. 6.11b). Such slip causes steps on the surface, so-called glide steps, which can indeed be observed, for instance on the surfaces of tensile deformed single crystals (Fig. 6.13). Microscopic investigations of deformed polycrystals reveal different orientations of slip lines in differently oriented crystallites, but within a crystallite the slip lines are parallel or consist of groups of parallel slip lines (Fig. 6.14). A crystallographic analysis proves that the slip lines extend parallel to specific low index crystallographic planes, for instance parallel to {111} planes in fcc crystals.

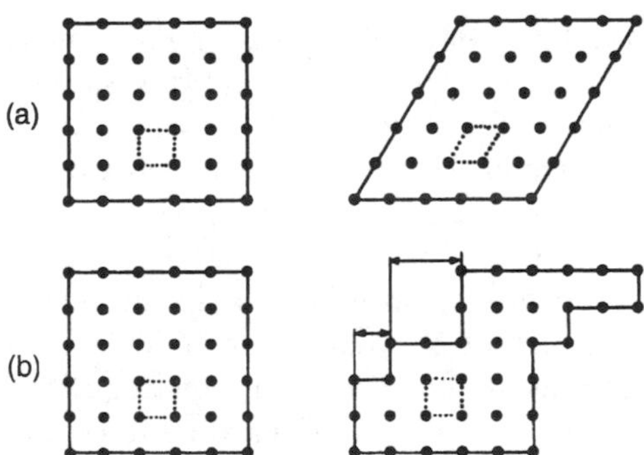

Fig. 6.11. Plastic deformation of crystals (dashed lines = unit cell). (a) By changing the crystal structure; (b) without changing the crystal structure.

The shear stress τ_{max}, that is required to displace two crystal blocks along a crystallographic plane by one atomic spacing b can be calculated (Fig. 6.15). To displace two atomic planes by a distance x, the shear stress τ has to be increased to a maximum, τ_{max}, where the two adjacent crystal planes are displaced by about 1/4 of the atomic spacing. At $x = b$ the stress again drops to zero, since another equilibrium configuration of the crystal lattice is reached. Such a dependency can be approximated by

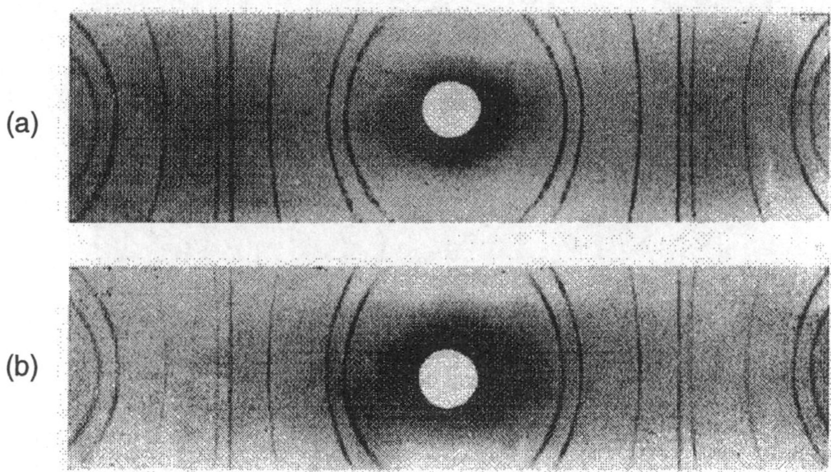

Fig. 6.12. Debye-Scherrer diagrams of (a) undeformed and (b) deformed copper [6.2].

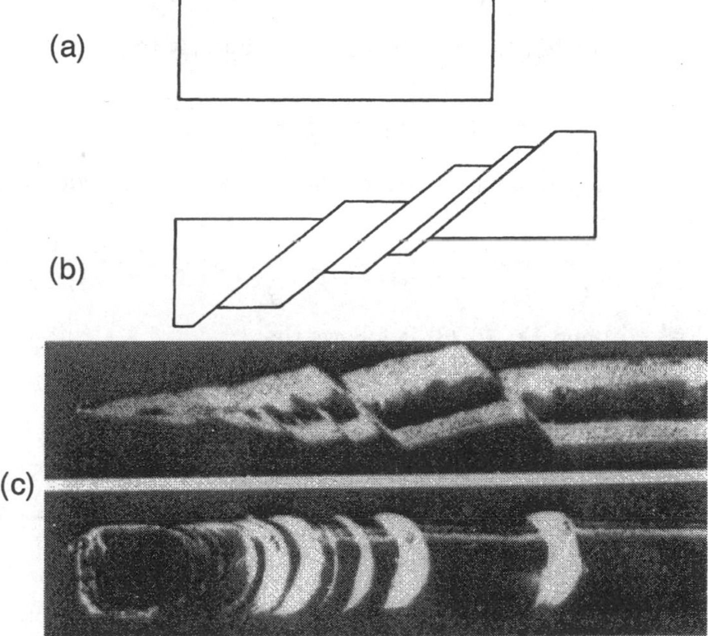

Fig. 6.13. Plastic deformation of a specimen under tensile stress by slip. (a) Before deformation; (b) after deformation; (c) stretched tin single crystal.

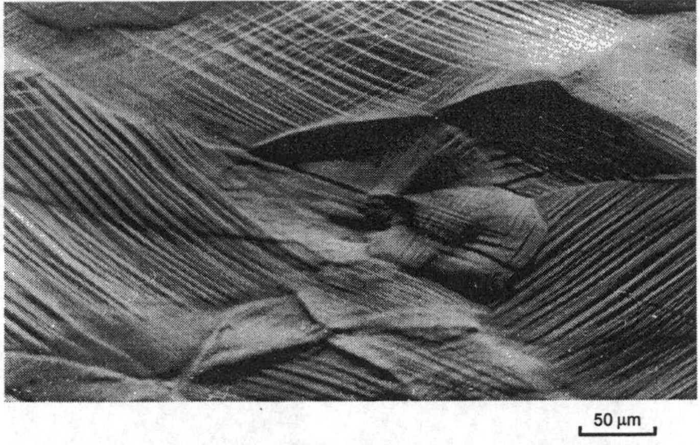

Fig. 6.14. Slip lines on the surface of rolled polycrystalline Fe_3Al.

$$\tau = \tau_{max} \sin\left(\frac{2\pi x}{b}\right) \tag{6.27a}$$

For small x the sine function can be expanded linearly to yield

$$\tau = \tau_{max} \sin\left(\frac{2\pi x}{b}\right) \cong \tau_{max}\frac{2\pi x}{b} \tag{6.27b}$$

For very small displacements x, Hooke's law applies, i.e., with the lattice spacing d

$$\tau = G\gamma = G \cdot \frac{x}{d} \tag{6.28}$$

From Eq. (6.27b) and Eq. (6.28) it follows that

$$\tau_{max}\frac{2\pi x}{b} = G\frac{x}{d} \tag{6.29}$$

and, therefore

$$\tau_{max} = \frac{G}{2\pi} \cdot \frac{b}{d} \tag{6.30}$$

If realistic interatomic potentials are used, a slightly smaller value for τ_{max} is obtained. This maximum $\tau_{max} = \tau_{th}$ is the theoretical shear strength and ought to correspond to the yield stress (critical resolved shear stress), in a shear test. The critical resolved shear stresses observed in metals and alloys are smaller by orders of magnitude, however, than the theoretical shear stress. For instance in copper, $\tau_{th} \approx 1.4$GPa, but the measured yield stress in copper

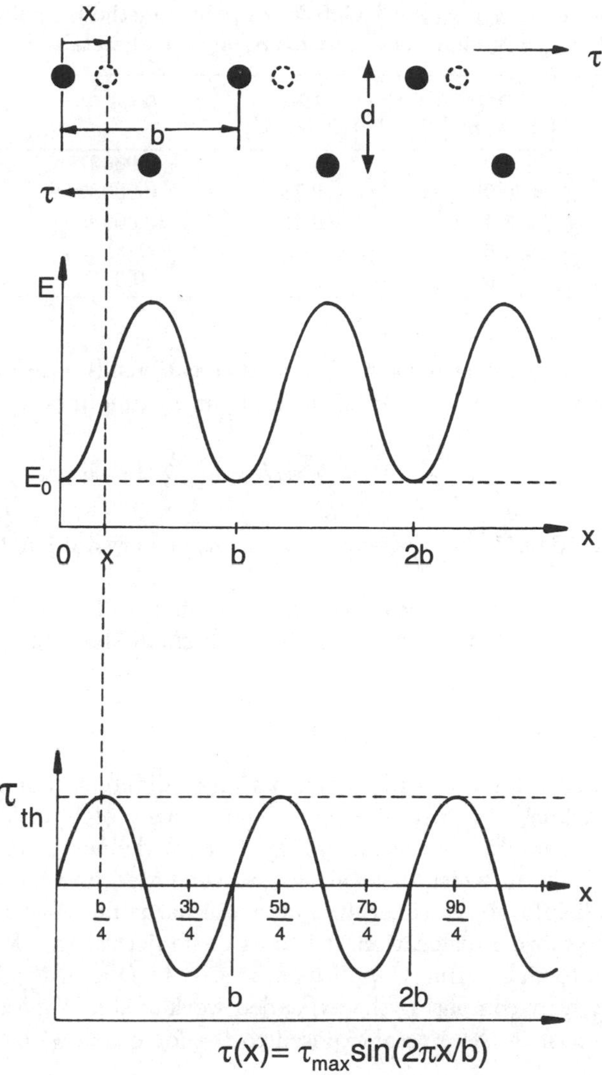

Fig. 6.15. Energy and stress curves for rigid shear yielding.

single crystals is only $\tau_0 \approx 0.5$MPa, i.e., about 4 orders of magnitude smaller than τ_{max} (Table 6.1). This contradiction is caused by the fact that we tacitly assumed the constraint that the motion of the atoms has to occur simultaneously. However, the atoms can move in stages. This kind of motion also occurs in nature in other cases, for instance the motion of a caterpillar (Fig. 6.16). In a crystal this mechanism corresponds to the motion of a dislocation (Figs. 6.16, 6.17). If a dislocation with Burgers vector b moves through a crystal of thickness ℓ_2, the crystal will be sheared by $\gamma = b/\ell_2$. This can be generalized

Table 6.1. Shear strength τ_{th} (calculated by a refined method), critical shear stress (experimental) $\tau_{exp} = \tau_0$ and fracture stress σ_f of several metals.

Material	τ_{th} $[10^9 N/m^2]$	τ_{exp} $[10^6 N/m^2]$	τ_{exp}/τ_{th}	σ_f $[10^6 N/m^2]$
Ag	1.0	0.37	0.00037	20
Al	0.9	0.78	0.00087	30
Cu	1.4	0.49	0.00035	51
Ni	2.6	3.2	0.0070	121
α-Fe	2.6	27.5	0.011	150

(Fig. 6.17) to an arbitrary motion of many dislocations. If n dislocations move by a distance dL the corresponding shear strain by amounts to

$$d\gamma = n \cdot \frac{dL}{\ell_1} \cdot \frac{b}{\ell_2} = \rho\; b\; dL \quad \text{or} \quad \dot{\gamma} = \rho \cdot b \cdot v \tag{6.31a}$$

(ρ - dislocation density, v - dislocation velocity, ℓ_1 - crystal length, see Chapter 3).

Equivalently, if the dislocations are immobilized after a slip distance L then $d\rho$ dislocations have to be generated in a small shear increment $d\gamma$ expressed as

$$d\gamma = b \cdot L \cdot d\rho \tag{6.31b}$$

Eqs. (6.31a) and (6.31b) describe deformation at different time scales. While Eq. (6.31a) defines the instantaneous displacement of a dislocation, with respect to its instantaneous velocity, Eq. (6.31b) defines deformation on a coarser time scale, in which dislocations are generated, moved, and immobilized. This is helpful for consideration of strain hardening (see section 6.5.2).

In order to move a dislocation, it has to be subjected to a force F, which can be related to an externally applied shear stress τ (Fig. 6.17). This relation can be obtained by comparing the expended work of slip W necessary to displace the top part of the crystal exposed to the force $(\tau\ell_1\ell_3)$ by the distance b

$$W = \tau\ell_1\ell_3 \cdot b \tag{6.32a}$$

Alternatively, the exact same deformation can be accomplished by the motion of a dislocation through the distance ℓ_1. If a force per unit length F acts on a dislocation of length ℓ_3

$$W = F \cdot \ell_3 \cdot \ell_1 \tag{6.32b}$$

Comparison of Eqs. (6.32a) and (6.32b) yields

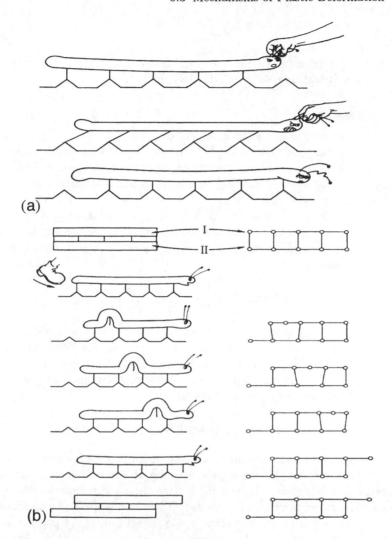

Fig. 6.16. Analogy of the motion of a caterpillar and the motion of a dislocation.

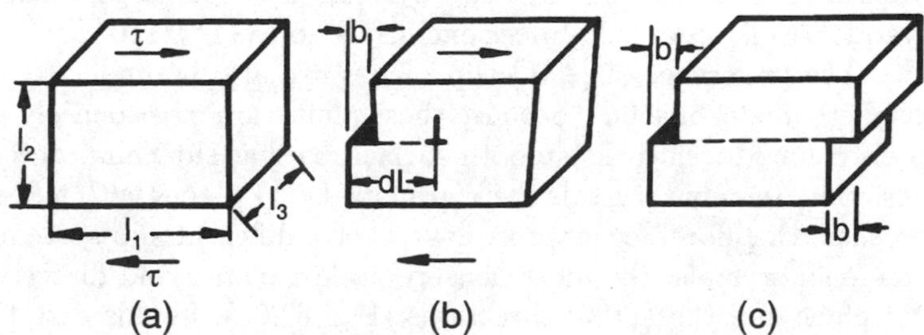

Fig. 6.17. Motion of a dislocation changes the shape of a crystal.

$$F = \tau b \tag{6.33}$$

For a general stress state $\boldsymbol{\sigma}$, the force per unit length on a dislocation with Burgers vector $\mathbf{b}$ and line element $\mathbf{s}$ is given by

$$\mathbf{F} = (\boldsymbol{\sigma} \cdot \mathbf{b}) \times \mathbf{s} \tag{6.34}$$

Eq. (6.34) is known as the Peach-Koehler equation.

Eqs. (6.31) and (6.34) provide the fundamental relations between the macroscopic stress and strain and the microscopic mechanisms of deformation. In simple terms the stress-strain diagram corresponds to a force-displacement diagram of the dislocations.

To move a dislocation on its slip plane, it has to pass through a higher energy configuration (Fig. 6.18). This requires a force corresponding to a shear stress on its slip plane Eq. (6.33). This Peierls stress τ_p can be derived with some approximations

$$\tau_p = \frac{2G}{1-\nu} \exp\left(-\frac{2\pi}{(1-\nu)}\frac{d}{b}\right) \tag{6.35}$$

The Peierls stress is proportional to the shear modulus but depends exponentially on the ratio of lattice spacing d to the Burgers vector b (slip direction). With increasing d and decreasing b the value of τ_p decreases. The spacing between next neighbor crystallographic planes with Miller indices {hkl} reads for cubic crystal symmetry (lattice parameter a)

$$d = \frac{a}{\sqrt{h^2 + k^2 + \ell^2}} \tag{6.36}$$

Therefore, d is largest for low index planes. In the case of fcc crystals, d is largest for {111} planes. Moreover, the spacing of atoms b is smallest along the close-packed directions. In the fcc lattice this is the case for the $\langle 110\rangle$ directions. The Peierls stress in fcc crystals, therefore, ought to be smallest for slip of dislocations with Burgers vector $\mathbf{b} = a/2\,\langle 110\rangle$ on {111} planes. In fact slip in fcc crystals is observed to proceed on {111} planes in $\langle 110\rangle$ direction, i.e. on $\{111\}\langle 110\rangle$ slip systems. Because of the exponential dependency of τ_p on (d/b) the Peierls stress in other slip systems is much larger. Therefore, in fcc crystals slip is observed almost exclusively on $\{111\}\langle 110\rangle$ slip systems (Fig. 6.19). The preference of {111} slip planes can also be understood from the hard sphere model of solids, because these planes are most densely packed and, therefore, have a relatively smooth surface, so that slip can proceed with the least friction. In cubic crystals there are four {111} planes with three $\langle 110\rangle$ directions each, therefore, fcc crystals have twelve different slip systems.

In hexagonal crystals the most densely packed planes and directions are the (0001) plane and the $\langle 11\bar{2}0\rangle$ directions (Fig. 6.20a). In this case there is only one slip plane, the basal plane with its three slip directions, i.e. three slip systems. This is only true, however, for hexagonal close-packed crystals or

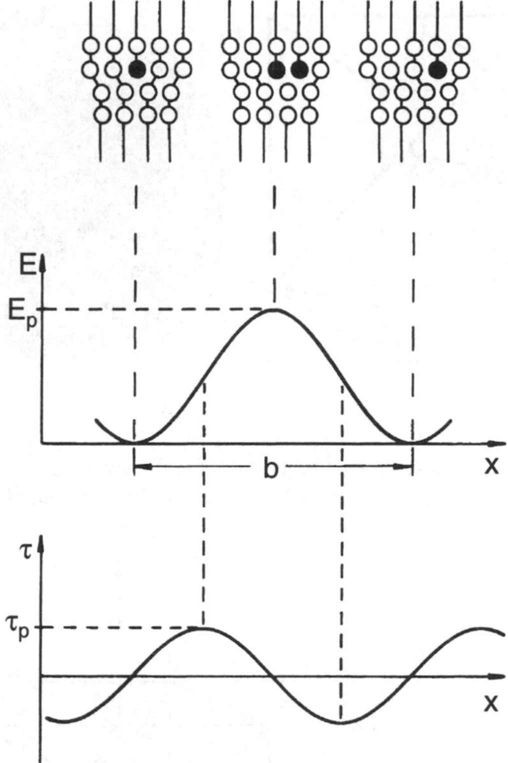

Fig. 6.18. Illustration of the definition of Peierls-Nabarro energy E_p and Peierls-Nabarro stress τ_p.

hexagonal crystals with an axis ratio $c/a \geq 1.63$. For crystals with $c/a < 1.63$ the basal plane is not the most densely packed plane any more, but comparable to the prism and pyramidal planes. In this case slip can also occur on the prism and pyramidal planes, and, correspondingly, many more slip systems are activated (Fig. 6.20b). An important example is the metal Ti and its alloys. Ti has a $c/a = 1.58$ and deforms by prism and pyramidal slip, which, therefore, is also referred to as the Ti mechanism.

In bcc crystal structures there is a maximum densely-packed direction, namely the $\langle 111 \rangle$ direction, along which the atoms are in contact, but there is no maximum densely-packed plane comparable to the $\{111\}$ planes in the fcc crystal or the basal plane in the hexagonal lattice. The most densely-packed planes are the $\{110\}$ planes although the packing density is only slightly different from the $\{112\}$ and $\{123\}$ planes (Fig. 6.21). Therefore, in bcc crystals besides $\{110\}\langle 111 \rangle$ frequently also $\{112\}$ $\langle 111 \rangle$ and $\{123\}\langle 111 \rangle$ slip systems are observed. In some cases it happens that there is no defined slip plane but

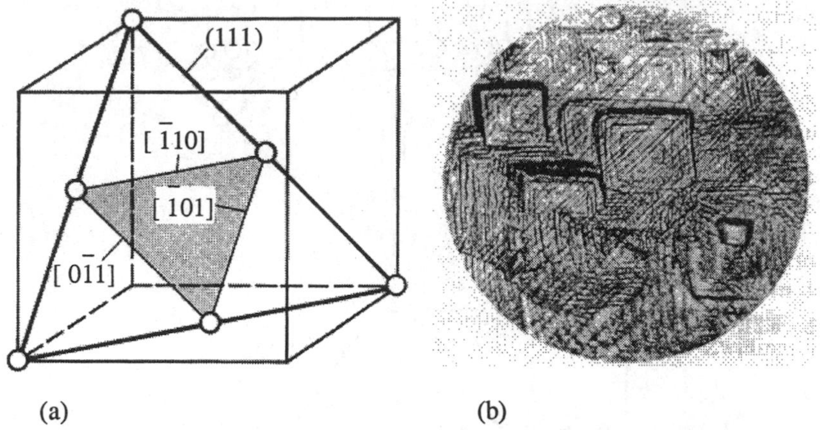

Fig. 6.19. Set of slip planes in the fcc lattice. (a) Every {111} plane contains three ⟨110⟩ directions. (b) Slip lines on a {100} plane in Cu [direction: ⟨110⟩] [6.5].

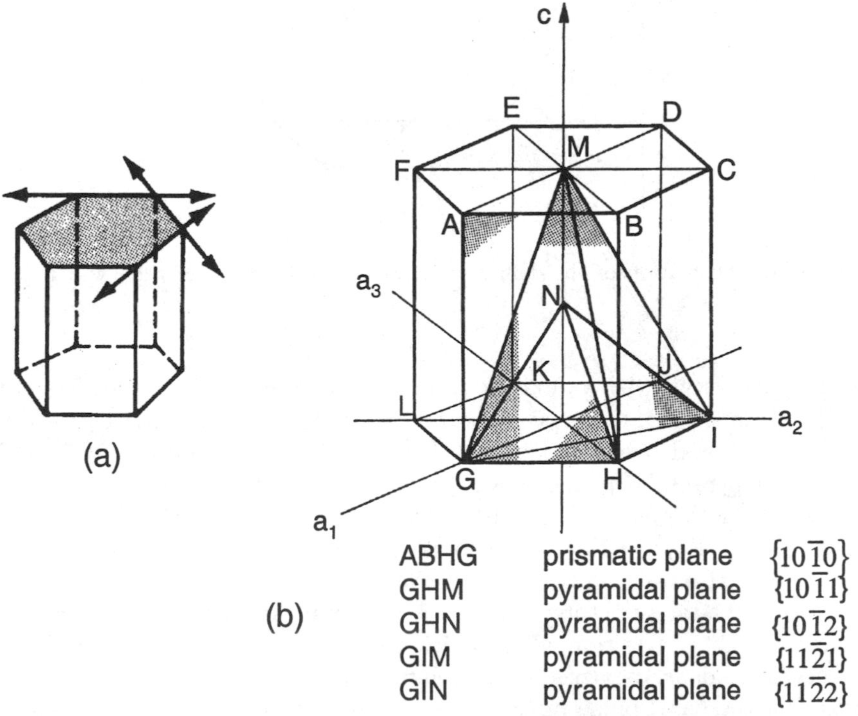

Fig. 6.20. Set of slip planes in hexagonal crystals. (a) Basal slip; (b) prismatic and pyramidal slip planes.

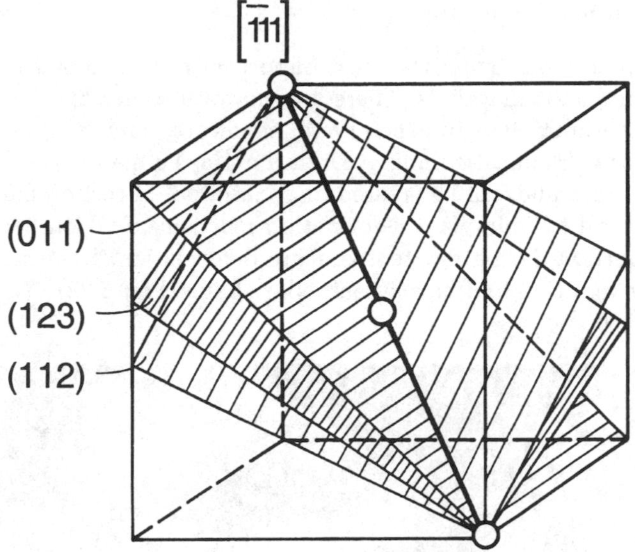

Fig. 6.21. Set of slip planes in the bcc lattice. Several slip planes share a ⟨111⟩ direction.

only a defined slip direction. In this case one can visualize deformation like an axial displacement of a stack of pencils and, therefore, this is referred to as "pencil glide" or {hkl}⟨111⟩.

The Peierls stress provides an explanation for why crystal structures with low crystal symmetry like many ceramics or intermetallic phases are brittle. In such cases slip planes and directions are not densely packed so the Peierls stress is high and can exceed the fracture stress with the consequence that no appreciable dislocation motion occurs prior to initiation of fracture.

An overview of slip systems in different crystal structures is given in Table 6.2.

Table 6.2. Slip systems of the most important lattice types.

Crystal structure	Slip plane	Slip direction	Number of non-parallel planes	Slip directions per plane	Number of slip systems
fcc	{111}	⟨1$\bar{1}$0⟩	4	3	12 = (4 × 3)
bcc	{110}	⟨$\bar{1}$11⟩	6	2	12 = (6 × 2)
	{112}	⟨11$\bar{1}$⟩	12	1	12 = (12 × 1)
	{123}	⟨11$\bar{1}$⟩	24	1	24 = (24 × 1)
hcp	{0001}	⟨11$\bar{2}$0⟩	1	3	3 = (1 × 3)
	{10$\bar{1}$0}	⟨11$\bar{2}$0⟩	3	1	3 = (3 × 1)
	{10$\bar{1}$1}	⟨11$\bar{2}$0⟩	6	1	6 = (6 × 1)

6.3.2 Mechanical Twinning

Crystallographic slip is by far the most important and dominant deformation mechanism in ductile materials. There are, however, other options for plastic deformation without changing the crystal structure, namely a shape change by diffusion or mechanical twinning. Diffusion plays a major role during high temperature creep and will be treated more in detail in section 6.8.2. In contrast, mechanical twinning is a deformation mechanism which becomes particularly important during low temperature deformation. Deformation twins are generally very thin and have a lenticular shape (Fig. 6.22).

Fig. 6.22. Deformation twins in zirconium.

Twinning is a shear deformation in which a crystalline volume is transformed into an orientation with mirror symmetry relative to the parent material (Fig. 6.23). The mirror plane is common to both twin and matrix and is referred to as the (coherent) twinning plane. All other boundaries between twin and matrix are termed incoherent twin boundaries. Because of its mirror symmetry to the matrix the twin has the same crystal structure. Crystallographically, twin and matrix are related by a 180° rotation about the normal to the twinning plane.

The geometry of mechanical twinning is characterized by the twinning plane {hkl} and the direction of shear (displacement) $\langle uvw \rangle$, i.e. by the twinning system $\{hkl\}\langle uvw \rangle$. The twinning systems of the most important crystal structures are listed in Table 6.3. The direction of displacement is the direction of the line of intersection of the twinning plane and the displacement plane

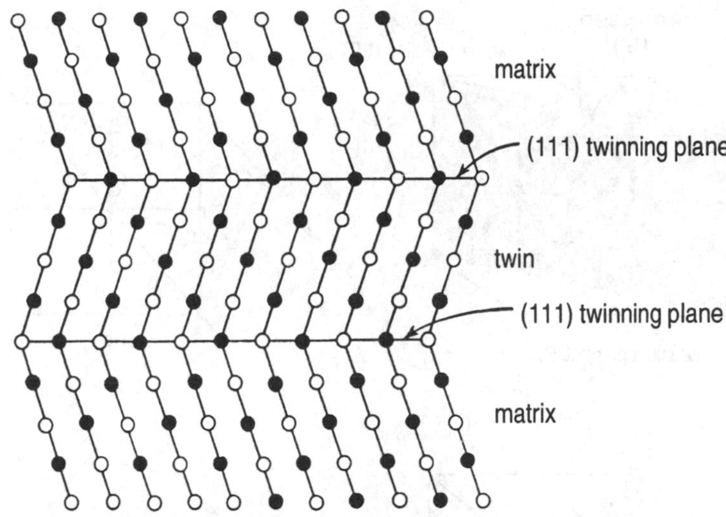

Fig. 6.23. Atomic configuration in matrix and twin of a fcc lattice.

perpendicular to it. In the displacement plane the motion of the atoms and the corresponding shear can be displayed. Fig. 6.24 shows the crystallographic location of the twin systems in cubic lattices and the motion of atoms during twinning.

Table 6.3. **(a)** Twinning system of the major lattice types; **(b)** c/a ratio of several hexagonal metals.

(a) Twinning elements of metallic crystals

Lattice type	Twinning plane	Shear direction	Displacement plane	Prototype
fcc	$\{111\}$	$\langle 112 \rangle$	$\{110\}$	Ag, Cu
bcc	$\{112\}$	$\langle 111 \rangle$	$\{110\}$	α-Fe
hcp	$\{10\bar{1}2\}$	$\langle 10\bar{1}1 \rangle$	$\{1\bar{2}10\}$	Cd, Zn

(b)	Cd	Zn	Mg	Co	Zr	Ti	Be
c/a	1.88	1.86	1.62	1.62	1.59	1.58	1.57

In contrast to crystallographic slip, the amount of shear during twinning is not variable but fixed. In cubic crystals the shear is $\gamma_t = \sqrt{2}/2$. The total shape change of a specimen, however, is determined not only by γ_t, but also by the volume of the twin. If the entire specimen were twinned, the shear would be $\gamma = \gamma_t$. If only a small volume fraction undergoes twinning, the total deformation is correspondingly reduced. Also, in contrast to crystallographic slip, the shear in a twin system has a defined sense, i.e. twinning requires the shear deformation in a specific direction and cannot occur in opposite

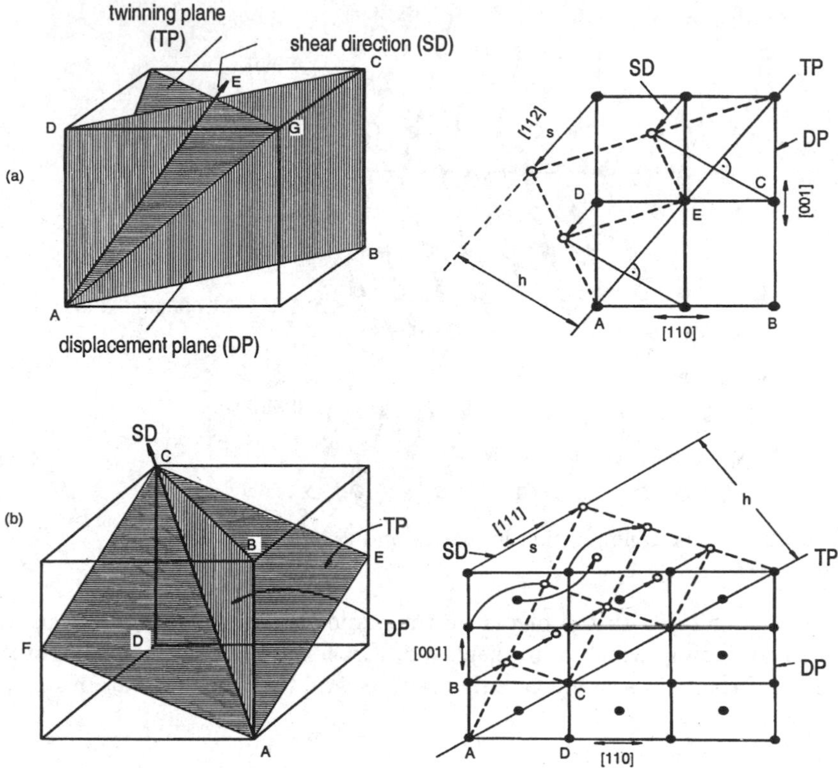

Fig. 6.24. Twinning in cubic crystals. (a) fcc; (b) bcc.

direction. This limits the operation of twinning systems in some cases, or the capacity to support significant plastic deformation (and be ductile).

During twinning the specimen changes its shape such that it becomes larger in one direction while it shortens in another direction (Fig. 6.25). The twin system will only be activated if the twinning shear is favored by the imposed shape change, i.e., during tensile deformation twinning has to elongate the specimen in the tensile direction. In cubic crystals there are twelve twinning systems. Hence, there is always one twinning system which will support the required shape change. In less symmetrical crystal systems this is not the case if the three major axes of the lattice are not equivalent. For example, consider the hexagonal lattice (Fig. 6.26). Depending on the c/a ratio the crystal becomes longer ($c/a < 1.73$) or shorter ($c/a > 1.73$) by twinning. If a hexagonal material is compressed perpendicular to the basal plane it can only deform by twinning if $c/a > 1.73$. This constraint has severe consequences for the formability of hexagonal materials. In anticipation of polycrystal deformation (section 6.5.2) we note that the deformation of polycrystals requires five independent slip systems and that crystals usually change their orientation during slip. Hexagonal crystals with $c/a > 1.63$ can only deform by slip on the three slip systems of the basal plane. Consequently, an arbitrary shape change requires twinning in addition to crystallographic slip. During the commercially most important forming process, rolling, a material becomes thinner and longer when passing through the rolls. As a result of crystallographic slip the grains rotate in such a way that the basal plane becomes approximately parallel to the rolling plane. Then the specimen can only become thinner by twinning during rolling, if $c/a > 1.73$. In case of $c/a < 1.63$, however, a sufficient number of slip systems is provided by pyramidal and prism slip (Fig. 6.27). For hexagonal materials with $1.63 < c/a < 1.73$, neither slip nor twinning can accomplish the necessary shape change during rolling. Therefore, these materials are difficult to deform plastically, i.e., they behave in a brittle manner. One example is magnesium which has a ratio $c/a = 1.624$ which is almost hexagonal close-packed and can only deform by basal slip. Single crystals of magnesium are ductile whereas polycrystals are brittle (see section 6.6).

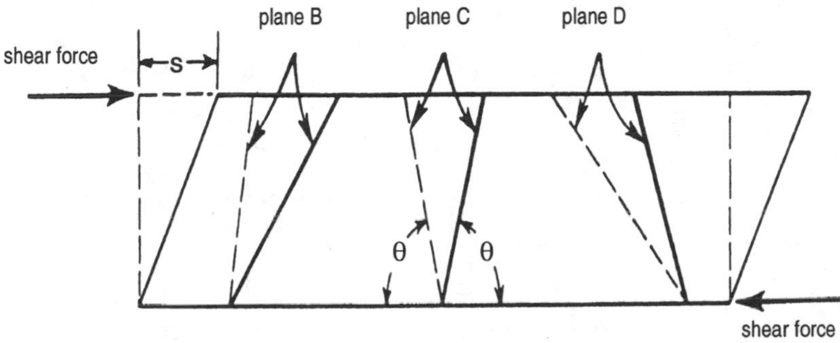

Fig. 6.25. Twinning stretches some directions (plane B), others are shortened (plane D).

Mechanical twinning is a coordinated process, similar to martensitic transformations (see Chapter 9) and proceeds with the speed of sound. This sets off a sound wave in the material which can be detected externally as a noise, which is known as "tin cry" in tin. In the stress-strain test, twinning is obvious as a serration of the flow stress curve (Fig. 6.28). After a sudden decrease the flow stress increases until the next twin is formed. The flow stress drop is due to the fact that twinning rapidly provides a deformation which is larger than the length change $\Delta\varepsilon = \dot{\varepsilon} \cdot \Delta t$ imposed by the mechanical testing machine. Consequently, the specimen is partially unloaded for a short time and afterwards the flow stress rises in the elastic regime.

The larger the distance between an atom and the twinning plane, the larger is the distance it has to move during twinning and thus, requires a higher energy. Therefore, twins are usually very thin to keep the elastic distortion small

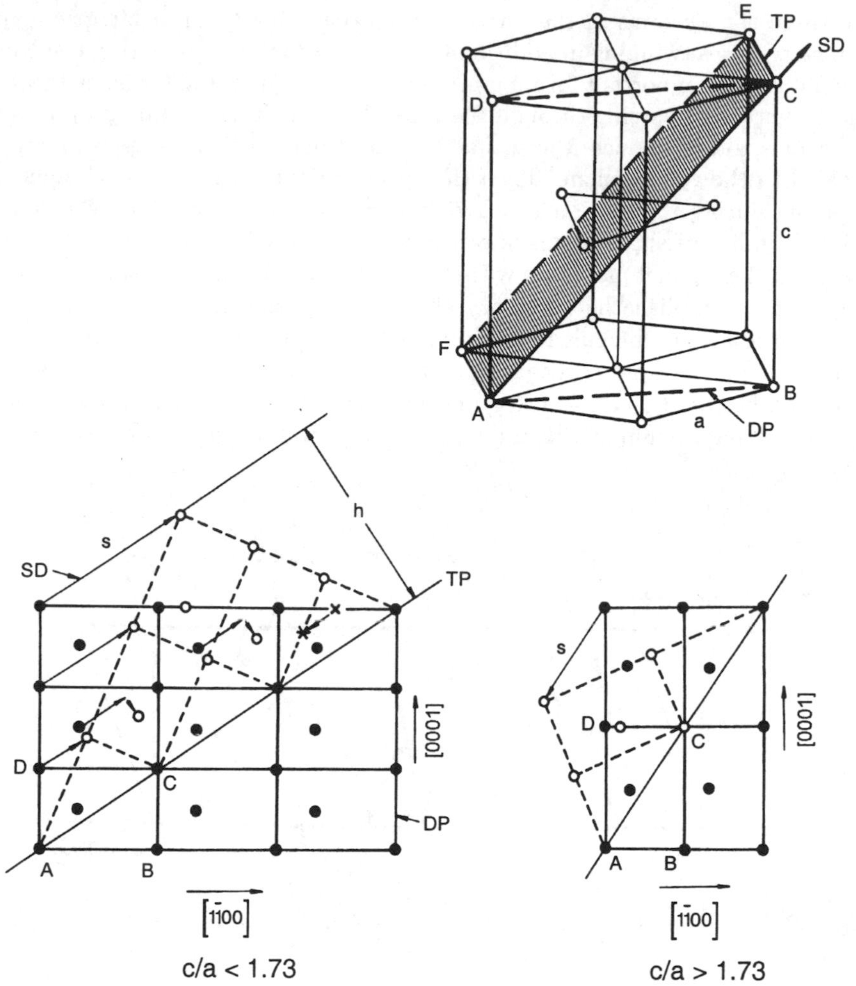

Fig. 6.26. Twinning in hexagonal crystals.

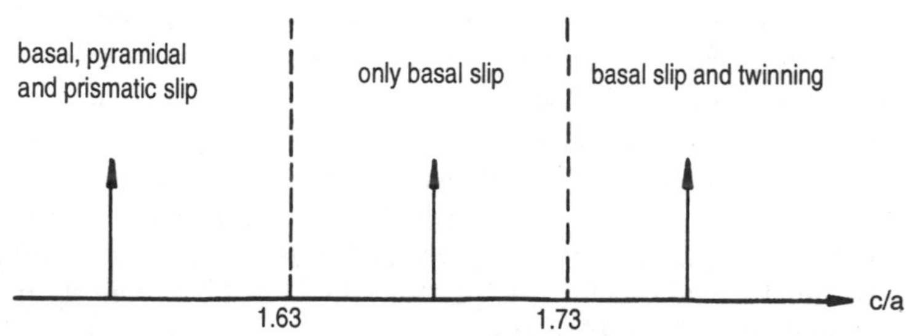

Fig. 6.27. Influence of the c/a ratio on the deformation mode in hexagonal crystals.

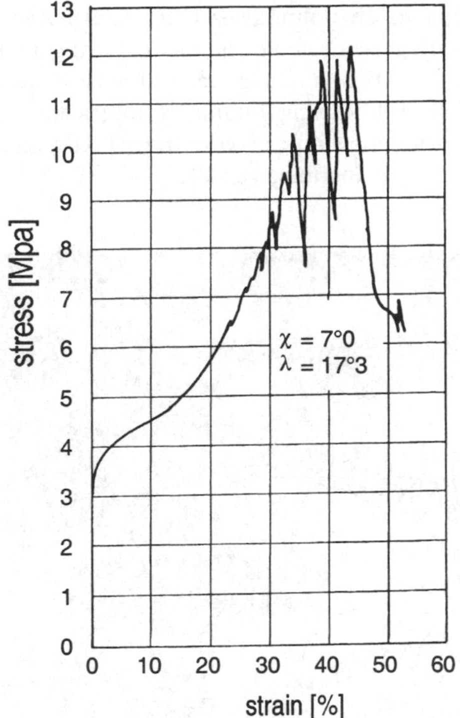

Fig. 6.28. Hardening curve of a Zn single crystal. At larger strain twinning sets in.

and thus minimize the elastic energy increase. For the same reason the critical shear stress for twinning is much larger than for crystallographic slip by dislocation motion. Since the flow stress increases during work hardening, the critical stress for twinning can be reached at higher degrees of deformation. For instance copper deforms at room temperature by slip only. At low temperatures (80K), where higher strength is obtained, mechanical twinning occurs. Of course, the occurrence of twinning depends on the energetics. For instance, the energy of the twin boundary must be provided , which varies from material to material. If the energy of the twin boundary is small (for instance silver), twinning is likely to occur. Materials that undergo twinning exhibit a higher strain hardening rate than materials with a high twin boundary energy (for instance aluminum). The mechanism of twinning can principally be described by dislocation motion. However, the required dislocations (Shockley partial dislocations) have a Burgers vector which is not a translation vector of the crystal but is a vector that transforms a crystal into its mirror image. These dislocations (partial dislocations) will be considered in more detail in Section 6.5.3).

Deformation twins are distinguished from annealing twins, which are generated during recrystallization or grain growth and are characterized by their conspicuously straight grain boundaries (Fig. 6.29). Atomistically both deformation and annealing twins have the same arrangement, but the formation of recrystallization twins is not associated with a shear deformation but rather with a growth fault during boundary migration. However, materials with strong tendency for mechanical twinning usually also show a high density of annealing twins (for instance brass).

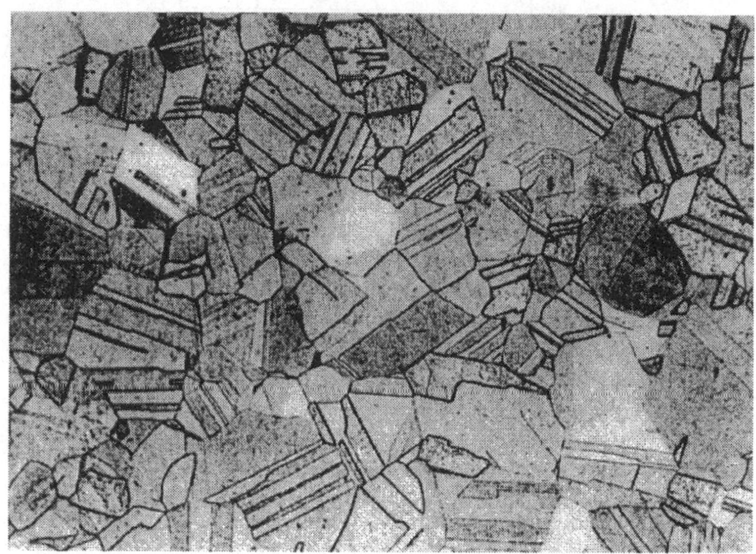

Fig. 6.29. Recrystallization twins in a CuZn alloy.

Twinning causes a characteristic discrete orientation change, namely a 180° rotation about the twinning plane normal, which is distinctly different from the gradual orientation change observed during crystallographic slip. Consequently the deformation texture (for instance rolling texture) in materials that undergo twinning during deformation (brass texture) is characteristically different from materials which only deform by crystallographic slip (copper texture, see Chapter 2).

6.4 The Critical Resolved Shear Stress

6.4.1 Schmid's Law

The yield stress σ_y characterizes the start of plastic deformation. It varies with material but even for the same material it can be different from specimen to specimen if single crystals of different orientations are deformed. In addition, the magnitude of the yield stress is, of course, influenced by the deformation temperature and higher temperatures generally result in lower yield stresses. The orientation dependence of the yield stress can be explained in context with the Peach-Koehler equation. The start of plastic flow is equivalent to the onset of large scale dislocation motion. A dislocation moves when it becomes subject to a force which has a component parallel to the slip plane in the slip direction. Hence, it is not the applied tensile stress, rather the resolved shear stress in the slip system that causes dislocation motion. This resolved shear stress τ is related to the tensile stress σ by

$$\tau = \sigma \cos\kappa \cdot \cos\lambda = m\sigma \tag{6.37}$$

where κ is the angle between tensile direction and slip plane normal and λ denotes the angle between tensile direction and slip direction (Fig. 6.30a). Eq. (6.37) can be easily derived. The stress $\sigma' = \sigma \cos\kappa$ in the slip plane ($\sigma' = \sigma \cdot A/A'$ with the cross section of the slip plane $A' = A/\cos\kappa$) and $\sigma' \cos\lambda$ is the component of σ' in slip direction. The factor $m = \cos\kappa \cos\lambda$ is called Schmid-factor and assumes values $0 \leq |m| \leq 0.5$ for tensile deformation. The acting force on a dislocation depends on the orientation of the slip system relative to the tensile axis. If there is more than one slip system in a crystal structure, the different slip systems usually have different Schmid-factors. For a given tensile stress the slip system with the largest Schmid-factor experiences the highest shear stress. Dislocation motion sets in if the force on the dislocation and, correspondingly, the resolved shear stress exceeds a critical value τ_0, which we shall calculate in section 6.4.2. This critical shear stress should be the same for all slip systems. This is the statement of Schmid's law. Experiments do confirm this hypothesis. If the measured yield stress is plotted versus $1/(\cos\kappa \cos\lambda)$, one obtains a straight line through the origin according to Eq. (6.37) in agreement with Schmid's law, i.e., $\tau_0 = \text{const.}$ (Fig. 6.30b).

With Schmid's law the active slip systems of a single crystal can be determined. The slip system with the highest Schmid-factor will reach the critical resolved shear stress first and, therefore, carry the plastic deformation. The crystallography of fcc single crystals shows (Fig. 6.31) that only one slip system (single slip) is activated if the tensile axis is parallel to an orientation in the interior of the standard crystallographic stereographic triangle. On the symmetry lines (001)-($\bar{1}$11), (001)-(011), (011)-($\bar{1}$11), however, two slip systems always have the same Schmid-factor (double slip). For the corner orientations as many slip systems are activated as there are stereographic triangles touching that corner, i.e., four slip systems for $\langle 110 \rangle$, six slip systems for $\langle 111 \rangle$, and eight slip systems for $\langle 100 \rangle$ (multiple slip).

The magnitude of the yield stress is, therefore, given by the critical resolved shear stress τ_0 required to activated a slip system, i.e. the force to move dislocations in the slip system. The minimum stress that is required to move a dislocation is the Peierls stress τ_p. However, in ductile materials the critical resolved shear stress is found to be either larger ($\tau_0 > \tau_p$) or smaller ($\tau_0 <$

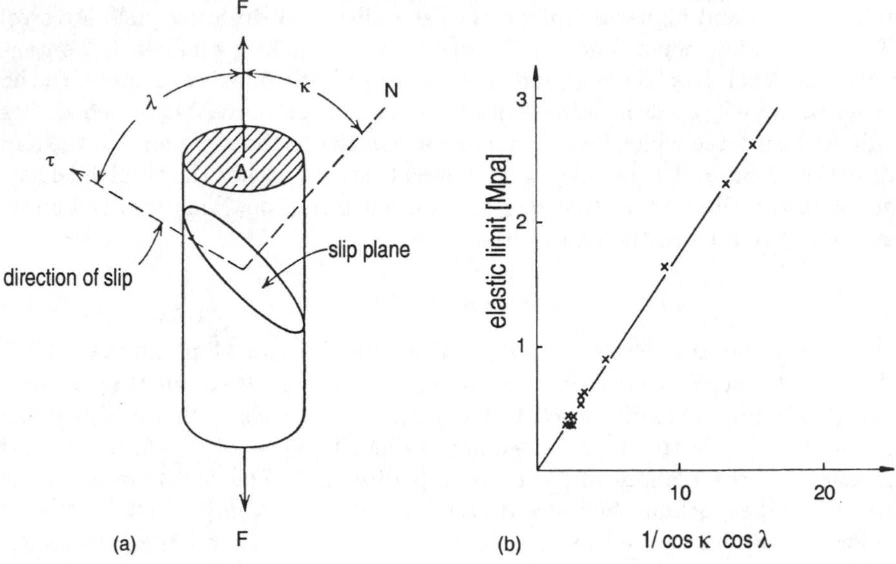

Fig. 6.30. (a) Determining the Schmid factor. (b) The elastic limit is proportional to the reciprocal Schmid factor (after [6.7]).

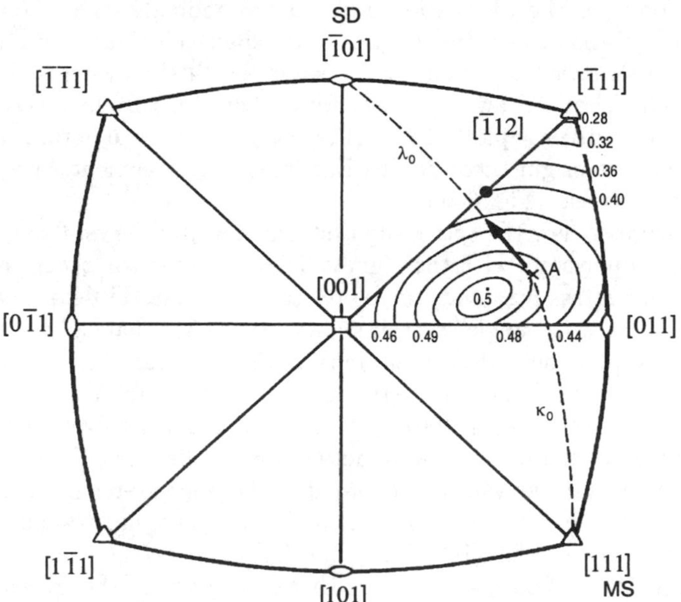

Fig. 6.31. Stereographic projection with contour lines of equal Schmid factor.(SD — slip direction, MS — main slip plane).

τ_p) than the Peierls stress. The observation $\tau_0 < \tau_p$ can be explained from the temperature dependence of the Peierls stress owing to thermal activation (section 6.4.3). Correspondingly, the Peierls stress determines the critical shear stress in such materials. Body-centered cubic metals and alloys are important examples. It was mentioned already that a material is brittle if its yield stress is higher than the fracture stress. Since the Peierls stress can be overcome by thermal activation, the critical resolved shear stress decreases with increasing temperature. Therefore, there is a specific transition temperature, namely the brittle-to-ductile transition temperature T_t (also, DBTT). Above T_t the yield stress σ_y is lower than the fracture stress σ_f ($\sigma_y < \sigma_f$) and, therefore, ductile plastic deformation occurs. With increasing Peierls stress the transition temperature T_t increases, of course. For α-iron it is about -100°C, for Bi which crystallizes in the less dense rhomboedric lattice it is +20°C. For the same reason many ceramic materials develop considerable plastic formability at high temperatures.

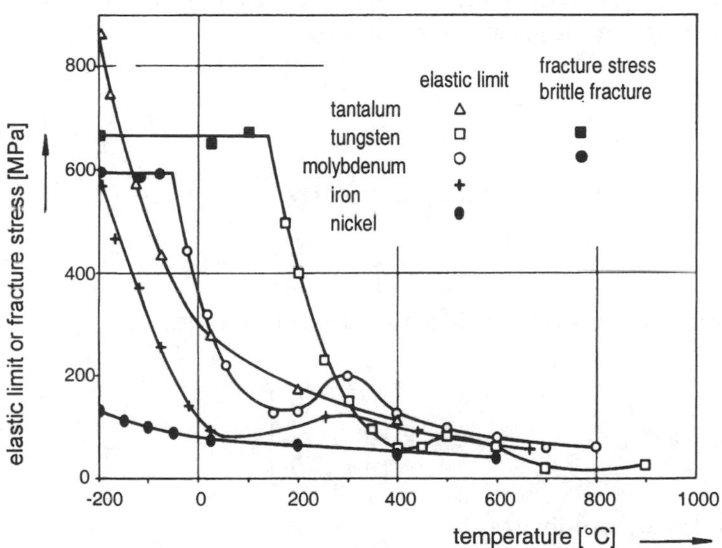

Fig. 6.32. Elastic limit of ductile (or fracture limit of brittle) bcc metals as a function of deformation temperature. Fcc Nickel is included for comparison (after [6.8]).

Fig. 6.32 also shows the yield stress of Ni plotted. Ni has an fcc crystal structure. So both the absolute value of τ_0 and its temperature dependence are much smaller than for the bcc metals. In fact, for Ni, as for all fcc and hexagonal materials $\tau_0 > \tau_p$. Apparently, there are other mechanisms which also affect dislocation motion. In pure metals this can only be the mutual interaction of dislocations via their elastic stress fields.

6.4.2 Dislocation Model of the Critical Resolved Shear Stress

6.4.2.1 Elastic Properties of Dislocations

A dislocation is a crystal defect, and its structure, therefore, is defined on an atomistic level. The introduction of a dislocation in a perfect crystal is associated with a shape change of the crystal which causes an elastic distortion of the lattice. Apart from the dislocation core, the elastic strain field of a dislocation can be associated with the elastic deformation of a hollow cylinder (Fig. 6.33). In chapter 3 we have described the creation of a dislocation in such a way that a crystal is partly cut along a crystallographic plane and both adjacent crystal parts are displaced in the plane of cut in radial (edge dislocation) or axial direction (screw dislocation). The stress field at a distance r from the dislocation core corresponds to the stress state in a thin cylindrical shell of radius r. For a screw dislocation in a cylindrical coordinate system there is only one stress component, namely the shear stress in a radial plane (θ = const.) in the axial diretion z, $\tau_{\theta z}$ (Figs. 6.33, 6.34). On unfolding the cylindrical shell we recognize the shear strain

$$\gamma_{\theta z} = \frac{b}{2\pi r} \tag{6.38a}$$

and according to Hooke's law the shear stress

$$\tau_{\theta z} = G\gamma_{\theta z} = \frac{Gb}{2\pi r} \tag{6.38b}$$

The stress tensor in cylindrical coordinates is as follows

$$\boldsymbol{\sigma}^{(s)}_{r\theta z} = \begin{bmatrix} 0 & 0 & 0 \\ 0 & 0 & \tau_{\theta z} \\ 0 & \tau_{\theta z} & 0 \end{bmatrix} \tag{6.39a}$$

and in cartesian coordinates

$$\boldsymbol{\sigma}^{(s)}_{xyz} = \begin{bmatrix} 0 & 0 & \tau_{xz} \\ 0 & 0 & \tau_{yz} \\ \tau_{xz} & \tau_{yz} & 0 \end{bmatrix} \tag{6.39b}$$

where

$$\tau_{xz} = G\gamma_{xz} = -\frac{Gb}{2\pi}\frac{y}{x^2+y^2} \tag{6.40a}$$

$$\tau_{yz} = G\gamma_{yz} = \frac{Gb}{2\pi}\frac{x}{x^2+y^2} \tag{6.40b}$$

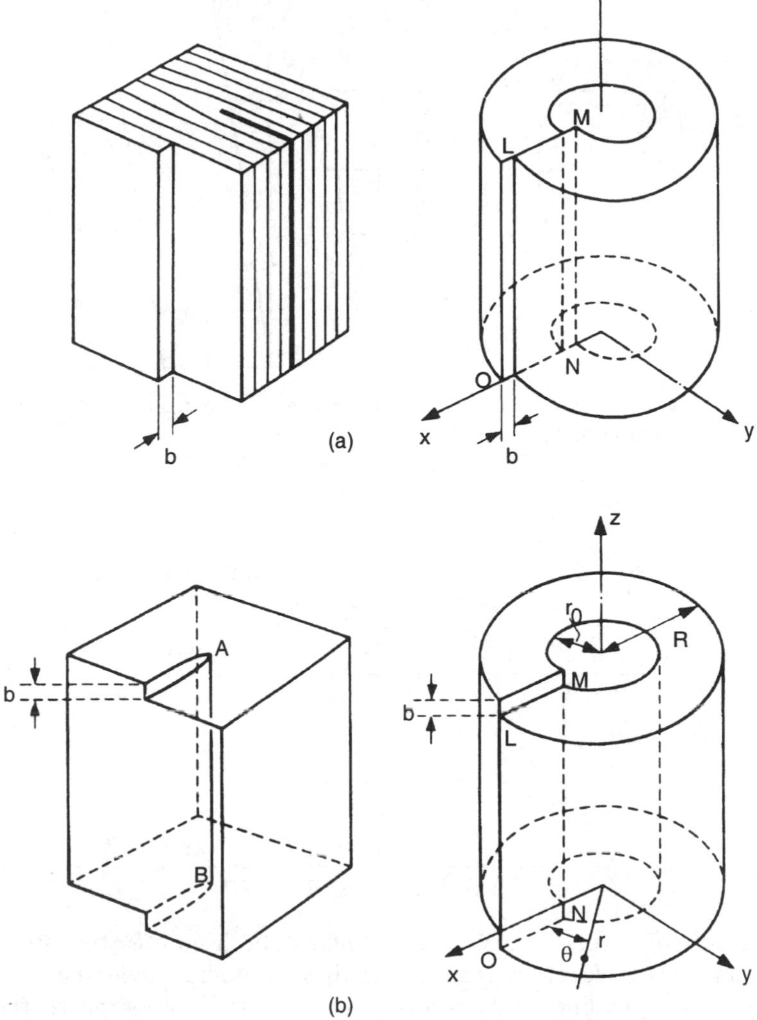

Fig. 6.33. Dislocations and the resulting state of internal stress. (a) Edge dislocation; (b) Screw dislocation.

For edge dislocations the stress state is more complicated (Fig. 6.35a). In contrast to the screw dislocation there are also normal stresses in addition to the shear stress component.

$$\boldsymbol{\sigma}_{xyz}^{(e)} = \begin{bmatrix} \sigma_{xx} & \tau_{xy} & 0 \\ \tau_{xy} & \sigma_{yy} & 0 \\ 0 & 0 & \sigma_{zz} \end{bmatrix} \tag{6.41}$$

where

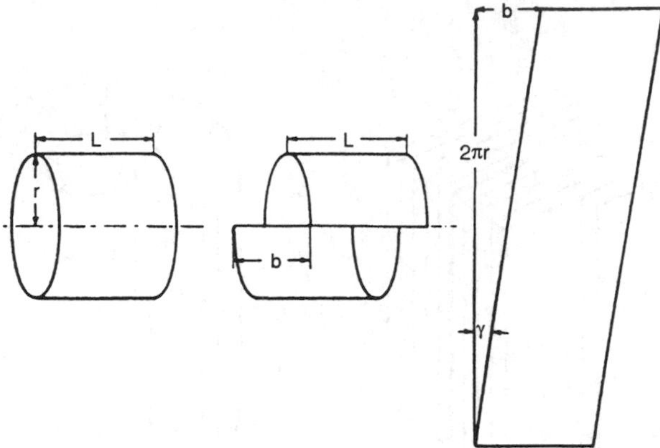

Fig. 6.34. How to calculate shear and the shear stress field of a screw dislocation by unwrapping a thin hollow cylinder.

$$\sigma_{xx} = -\frac{Gb}{2\pi(1-\nu)} \frac{y\,(3x^2+y^2)}{(x^2+y^2)^2} = -\frac{Gb}{2\pi(1-\nu)} \frac{\sin\theta\,(2+\cos(2\theta))}{r} \tag{6.42a}$$

$$\sigma_{yy} = \frac{Gb}{2\pi(1-\nu)} \frac{y\,(x^2-y^2)}{(x^2+y^2)^2} = \frac{Gb}{2\pi(1-\nu)} \frac{\sin\theta\cdot\cos(2\theta)}{r} \tag{6.42b}$$

$$\sigma_{zz} = \nu\,(\sigma_{xx}+\sigma_{yy}) \tag{6.42c}$$

$$\sigma_{xy} \equiv \tau_{xy} = \frac{Gb}{2\pi(1-\nu)} \frac{x\,(x^2-y^2)}{(x^2+y^2)^2} = \frac{Gb}{2\pi(1-\nu)} \frac{\cos\theta\cos(2\theta)}{r} \tag{6.43}$$

The existence of normal stresses can be understood from the atomistic structure of an edge dislocation (Fig. 3.7). Above the slip plane the lattice is squeezed, corresponding to compressive stresses, while underneath the dislocation core the lattice is dilated leading to tensile stresses.

In this continuum mechanics approach we have neglected the dislocation core, where Hooke's law does not hold anymore because of the large distortions. For an assessment of the radial size of the dislocation core r_0, one can utilize the fact that the elastic stresses can never exceed the theoretical shear stress (Eq. 6.30). In the case of a screw dislocation

$$\tau\,(r_0) = \frac{Gb}{2\pi r_0} \cong \tau_{th} \cong \frac{G}{2\pi} \tag{6.44}$$

hence

$$r_0 \cong b \tag{6.45}$$

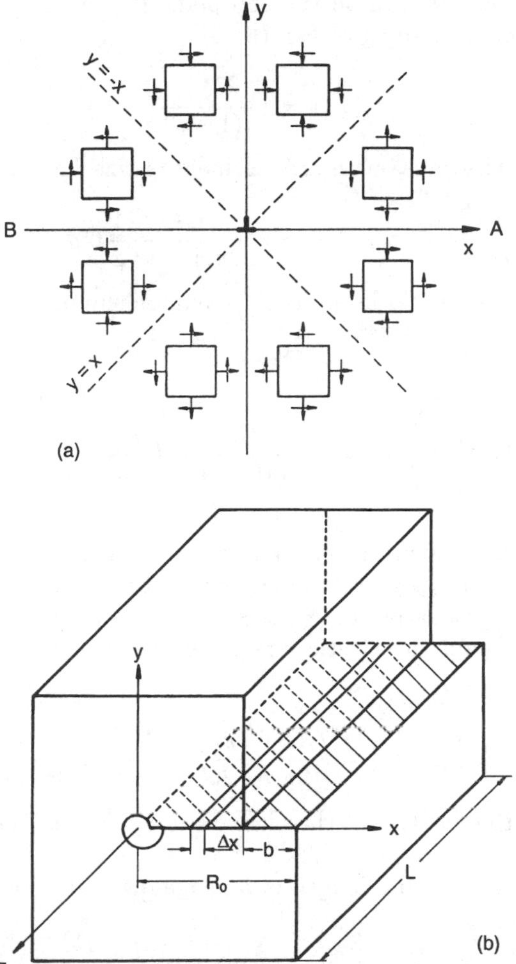

Fig. 6.35. (a) The stress field of an edge dislocation; (b) how to calculate the energy of an edge dislocation.

Surprisingly, the (macroscopic) continuum mechanical representation of a dislocation stress field extends to practically a single atomic distance from the dislocation core. Similar values are obtained for edge dislocations.

The elastic stress state is associated with an elastic energy. Let us consider a cylindrical shell of size R_0 about the core of an edge dislocation (Fig. 6.35b). To generate the edge dislocation, a displacement by an amount of b along the slip plane is necessary. The elastic energy of the dislocation is the work expended to generate the dislocation. A displacement by the amount ξ

induces an elastic stress field on the slip plane ($\theta = 0$) at a distance x from the dislocation core according to Eq. (6.43)

$$\tau_{xy}(x) = \frac{G\xi}{2\pi(1-\nu)}\frac{1}{x} \tag{6.46}$$

Correspondingly, the force on an area element of size Ldx at location x reads

$$F_x = \tau_{xy} \cdot L \cdot dx = \frac{G\xi}{2\pi(1-\nu)}\frac{1}{x}L\,dx \tag{6.47}$$

To displace both crystal blocks along the common slip plane by $\xi = b$ requires the work

$$E_{el} = \int_{r_0}^{R_0}\left(\int_0^b F_x(\xi)d\xi\right)dx = L \cdot \frac{Gb^2}{4\pi(1-\nu)}\int_{r_0}^{R_0}\frac{dx}{x} = L \cdot \frac{Gb^2}{4\pi(1-\nu)}\ln\frac{R_0}{r_0} \tag{6.48}$$

where r_0 is the size of the dislocation core and R_0 is the crystal size.

Again, this analysis neglects the energy of the dislocation core. If we assume that in the dislocation core the stress corresponds to the theoretical shear stress we obtain the elastic energy of the dislocation core (per unit length)

$$\frac{E_{\text{core}}}{L} \cong \frac{\tau_{th}^2}{2G} \cdot \pi r_0^2 = \left(\frac{G}{2\pi}\right)^2 \frac{\pi r_0^2}{2G} \cong \frac{Gb^2}{8\pi} \cong \frac{Gb^2}{4\pi(1-\nu)} \tag{6.49}$$

where we utilize the result that the elastic energy density (energy / volume) for a shear stress τ is given by $\tau^2/2G$.

The total energy per unit length of an edge dislocation is

$$E^{(e)} = \frac{E_{el}}{L} + \frac{E_{core}}{L} = \frac{Gb^2}{4\pi(1-\nu)}\left(\ln\frac{R_0}{r_0} + 1\right) \tag{6.50}$$

By analogy we obtain for a screw dislocation

$$E^{(s)} = \frac{Gb^2}{4\pi}\left(\ln\frac{R_0}{r_0} + 1\right) \tag{6.51}$$

The dislocation energy depends on the crystal size (R_0), although in a logarithmic fashion. Since the ratio R_0/r_0 is very large, $E^{(s)}$ or $E^{(e)}$ changes very slowly with R_0 . For instance, for copper with a grain size of 30 μm we obtain $r_0 \cong b \cong 3\text{Å}$, $R_0/r_0 = 10^5$, and $lnR_0/r_0 \cong 11$. Doubling R_0 results in $lnR_0/r_0 \cong 13$ which is only a 20% change. Therefore, a good approximation for the energy of a dislocation per unit length $E^{(e)} \approx E^{(s)} = E^{(d)}$

$$E^{(d)} \cong \frac{1}{2}Gb^2 \tag{6.52}$$

6.4.2.2 Interaction of dislocations

Dislocations interact via their elastic strain fields. The force of interaction is given by the Peach-Koehler equation (Eq. (6.34)). Let us consider the interaction between two parallel edge dislocations (Fig. 6.36). The force $\mathbf{F}_{12}$, which dislocation 1 with its stress field $\boldsymbol{\sigma}_1$ exerts on dislocation 2 with the Burgers vector $\mathbf{b}_2$ and the line element $\mathbf{s}_2$ reads

$$\mathbf{F}_{12} = (\boldsymbol{\sigma}_1\ \mathbf{b}_2) \times \mathbf{s}_2 \tag{6.53}$$

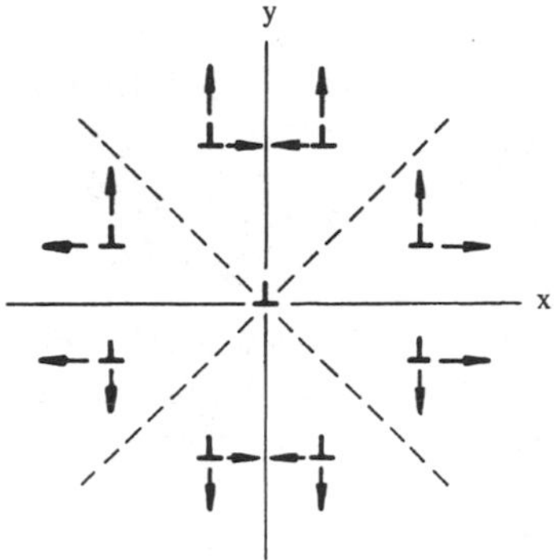

Fig. 6.36. Force between two parallel edge dislocations. The arrows indicate the direction of the force exerted by the resting dislocation.

If dislocation 1 intersects the origin of the coordinate system and extends parallel to the z direction, then $\mathbf{s} = [001]$ and $\mathbf{b}_1 = \mathbf{b}_2 = b[100]$ we obtain

$$\mathbf{F}_{12} = \begin{bmatrix} F_x \\ F_y \\ F_z \end{bmatrix} = \left(\begin{bmatrix} \sigma_{xx} & \sigma_{xy} & 0 \\ \sigma_{xy} & \sigma_{yy} & 0 \\ 0 & 0 & \sigma_{zz} \end{bmatrix} \cdot \begin{bmatrix} b \\ 0 \\ 0 \end{bmatrix} \times \begin{bmatrix} 0 \\ 0 \\ 1 \end{bmatrix} \right) = \begin{bmatrix} \sigma_{xx} b \\ \sigma_{xy} b \\ 0 \end{bmatrix} \times \begin{bmatrix} 0 \\ 0 \\ 1 \end{bmatrix} = \begin{bmatrix} \sigma_{xy} b \\ -\sigma_{xx} b \\ 0 \end{bmatrix} \tag{6.54}$$

In the slip plane parallel to the slip direction the force is $F_x = \sigma_{xy} b$. Perpendicular to the slip plane there is a force $F_y = -\sigma_{xx} b$. The force F_y does not cause a slip deformation but is important for the mechanism to leave the glide plane (climb process; see section 6.8.2). No force acts parallel to the dislocation line. With Eqs. (6.43) and (6.54) we obtain

$$F_x = \frac{Gb^2}{2\pi(1-\nu)} \frac{\cos\theta\ \cos(2\theta)}{r} \tag{6.55}$$

The direction of the force F_x depends on the relative position of the dislocations. In the following we assume that only dislocation 2 can move under the action of the force exerted by dislocation 1. At locations with $\theta = \pm 45°$ and $\theta = \pm 90°$ the force $F_x = 0$. The dislocation arrangement at $\theta = \pm 45°$ is metastable, however, because a small displacement from the ideal position causes the dislocation to move away from this position. In contrast, at the location $\theta = \pm 90°$ the arrangement is stable. As a result, if parallel edge dislocations are free to move on parallel slip planes they either repel each other or arrange themselves on top of each other. A periodic arrangement of many dislocations on top of each other corresponds to the structure of a symmetrical low angle tilt grain boundary (see Chapter 3), which is known to form during recovery processes by rearrangement of dislocations (see Chapter 7).

F_x is also the force which a dislocation experiences during its motion due to dislocations on parallel slip planes. For a large scale displacement of a dislocation, this force F_x has to be overcome, more precisely F_x^{max}, since F_x depends on the positions of the mobile dislocation. Since $\sigma_{xy} = F_x/b \sim Gb/r$ a shear stress of magnitude

$$\tau_{pass} = \alpha_1 \frac{Gb}{d} = \alpha_1 Gb\sqrt{\rho_p} \tag{6.56}$$

has to be applied, where α_1 is the geometrical factor and $d = 1/\sqrt{\rho_p}$ is the average spacing of the parallel (primary) dislocations (Fig. 6.37). Since τ_{pass} is the shear stress to move a dislocation past other parallel dislocations it is also referred to as the passing stress.

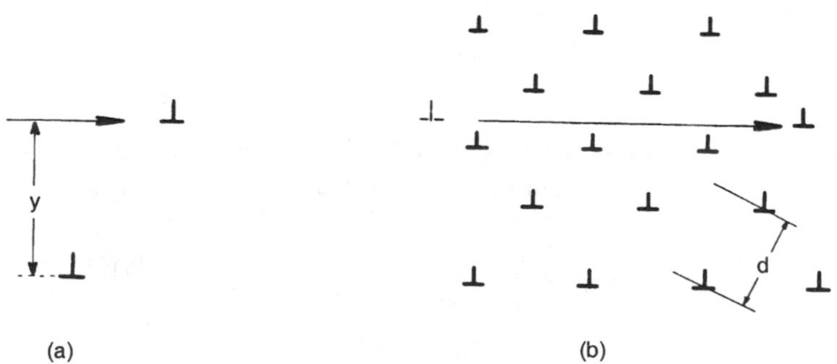

Fig. 6.37. Two dislocations passing each other (a) at a distance y that is proportional to the mean spacing d of parallel (primary) dislocations (b).

The passing stress is not the only obstacle which a mobile dislocation has to overcome. Dislocations of non-parallel (secondary) slip systems intersect the primary slip plane and have to be cut through by primary dislocations during their motion. During the cutting process in both the cutting and the cut dislocations, steps are generated of size and orientation of the Burgers vector of the reaction partner (Fig. 6.38). We discriminate two types of steps, namely kinks, which are parallel to the glide plane, and jogs that are inclined to the glide plane. Kinks, therefore, can be removed by dislocation motion while jogs are immobile and cause the formation of dislocation dipoles on dislocation motion (Fig. 6.38aγ). Only in the special case that the Burgers vector of the one dislocation is parallel to the dislocation line of the other dislocation can the formation of a step be avoided. Hence, the cutting process always creates at least one step which is associated with an increase of the energy of the affected dislocation, namely by the energy of the newly generated section of the dislocation line $1/2Gb^2 \cdot b$. This energy has to be provided from the work expended by the external stress field and more specifically during motion through the distance b where the step is produced. If the average free dislocation line length is ℓ_f the force acting at the point of dislocation intersection amounts to $F = \tau_c b \ell_f$ and consequently the energy balance is

$$\tau_c \, b\ell_f \cdot b = \frac{1}{2} Gb^2 \cdot b \tag{6.57a}$$

Hence we obtain for the cutting stress τ_c

$$\tau_c = \frac{1}{2} \frac{Gb}{\ell_f} \tag{6.57b}$$

The average free dislocation spacing is the average spacing of the dislocations intersecting the primary slip plane. These dislocations do not belong to the primary slip plane and, therefore, are called secondary dislocations or forest dislocations (the visualization being that they appear to the primary dislocations like trees in a forest). If the density of the forest dislocations is ρ_f then $\rho_f = 1/\ell_f^2$ or

$$\tau_c = \frac{1}{2} Gb\sqrt{\rho_f} \tag{6.58}$$

We obtain a similar relation as for the passing stress except that the relevant dislocation densities are different. Prior to plastic flow the dislocation densities on all slip systems should be approximately the same, and during deformation both ρ_f and ρ_p remain (law of similitude) a constant fraction of the total dislocation density $\rho = \rho_f + \rho_p$ (i.e., $\rho_f \sim \rho$ and $\rho_p \sim \rho$). We obtain for the critical resolved shear stress

$$\tau_0 = \alpha_1 \, Gb\sqrt{\rho_p} + \frac{1}{2} Gb\sqrt{\rho_f} = \alpha \, Gb\sqrt{\rho} \tag{6.59}$$

where α is a geometrical constant of the order of 0.5.

6.4.3 Thermally Activated Dislocation Motion

The measured critical resolved shear stresses are usually smaller than those calculated according to Eqs. (6.35) or (6.59). The reason is that we neglected

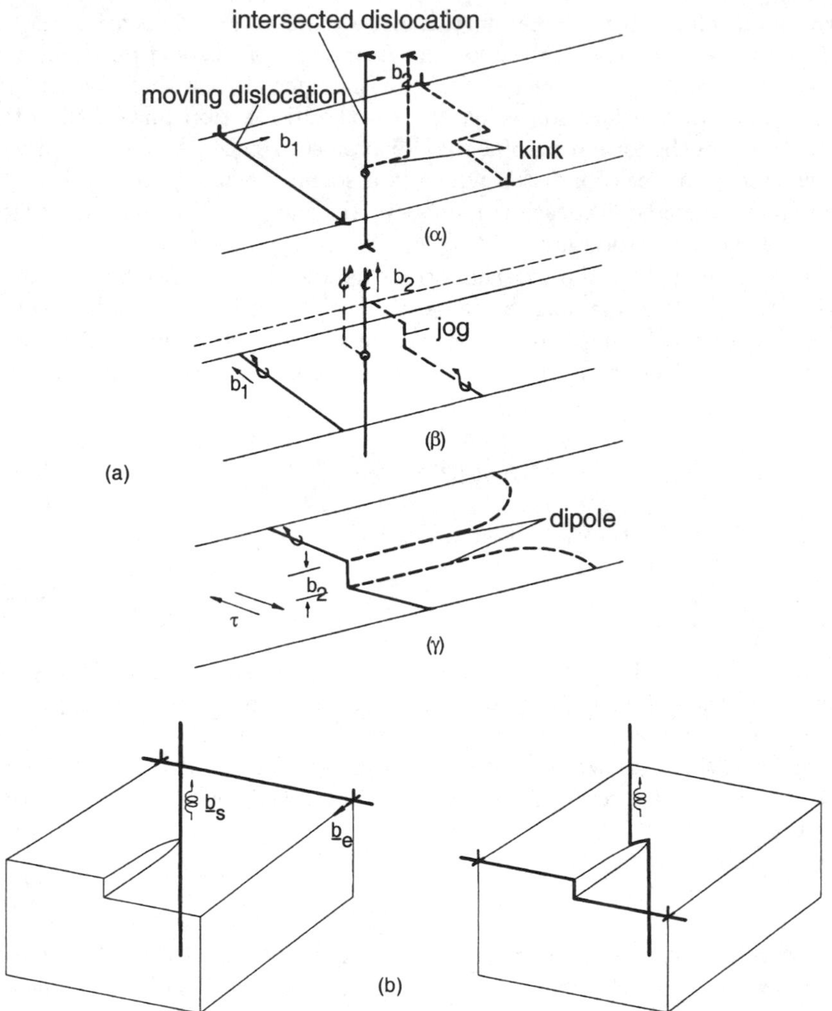

Fig. 6.38. (a) Steps formed by the intersection of dislocations (α) two kinks (β) a jog (γ) a dipole formed at a jog. (b) Illustration of the formation of a jog by intersection of an edge and a screw dislocation.

thermal activation although the measurements were taken at ambient temperature, i.e. $T \gg 0K$. The dyamics of deformation are described by Eq. (6.31a) (Orowan equation: $\dot{\gamma} = \rho_m b v$), where ρ_m is the mobile dislocation density and v is the dislocation speed. Please note that v is only an average value because the dislocation motion is jerky owing to the interaction with obstacles. If the stresses calculated by Eqs. (6.35) and (6.59) are applied, the dislocations can surmount the obstacles directly. This mechanical stress is, however, reduced because of thermal activation. A dislocation moves unconstrained between obstacles (time t_m) but has to wait in front of an obstacle (time t_w) until thermal activation can impart the energy necessary to overcome the obstacle at the applied shear stress. At an average obstacle spacing ℓ, the average dislocation speed is

$$v = \frac{\ell}{t_m + t_w} \approx \frac{\ell}{t_w} \tag{6.60}$$

because $t_m \ll t_w$. As outlined in Chapter 5

$$\frac{1}{t_w} = \nu_D \; exp\left(-\frac{\Delta G(\tau)}{\mathrm{kT}}\right) \tag{6.61}$$

and because of Eq. (6.31a)

$$\dot{\gamma} = \dot{\gamma}_0 \; exp\left(-\frac{\Delta G(\tau)}{\mathrm{kT}}\right) \tag{6.62}$$

where ν_D is the vibration frequency of the dislocation and $\Delta G(\tau)$ is the activation free energy to overcome the obstacle under the action of an applied stress τ.

If the obstacle to dislocation motion is the Peierls stress, dislocation motion proceeds in such a way that first a segment of the dislocation proceeds beyond the Peierls hill (Fig. 6.39). This generates two kinks, which move in opposite directions under the action of the applied stress. Eventually the whole dislocation line has moved by one Burgers vector. The activation energy for this process is the average value of line energy $E_L \cong 1/2Gb^2$ and the Peierls energy E_p to yield

$$\Delta G(0) = \frac{4b}{\pi}\sqrt{E_p \, E_L} \tag{6.63}$$

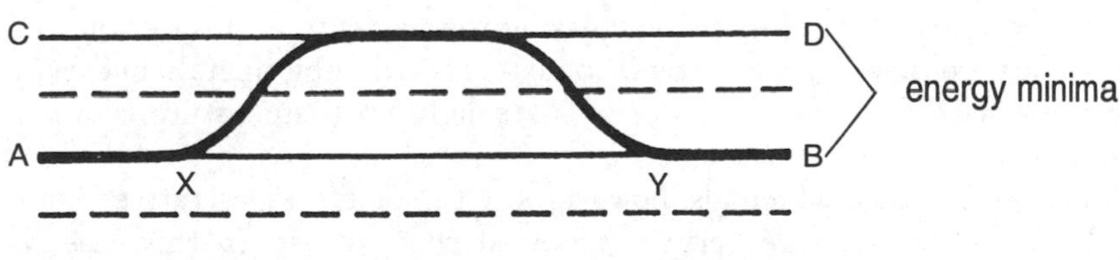

Fig. 6.39. Schematic illustration of the formation of a double kink.

The Peierl stress τ_p and Peierls energy E_p are related by

$$\tau_p = \frac{2\pi}{b^2} E_p \tag{6.64}$$

Without the external stress the activation energy

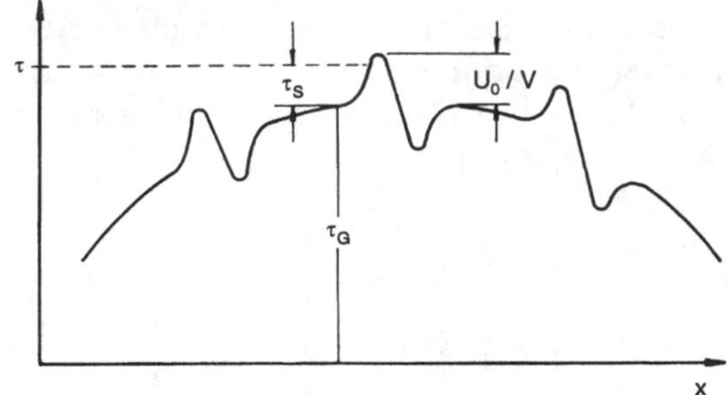

Fig. 6.40. Stress distribution along the slip plane of a dislocation.

$$\Delta G(0) = \frac{4b}{\pi} \cdot \frac{b}{\sqrt{2\pi}} \cdot \sqrt{\tau_p} \cdot \sqrt{E_L} \tag{6.65a}$$

has to be overcome. If a stress $\tau < \tau_p$ is applied, the thermal activation barrier is reduced to the stress differential $\tau_p - \tau$, hence only

$$\Delta G(\tau) = \frac{4b}{\pi} \cdot \frac{b}{\sqrt{2\pi}} \cdot \sqrt{\tau_p - \tau} \cdot \sqrt{E_L} \tag{6.65b}$$

has to be provided thermally . Combining Eqs. (6.62) and (6.65b) we eventually obtain for the critical resolved shear stress at temperature T and a strain rate $\dot{\gamma}$

$$\tau = \tau_p - AT^2 \left(\ln \frac{\dot{\gamma}}{\dot{\gamma}_0} \right)^2 \tag{6.66}$$

where A is a constant.

The critical resolved shear stress at $T=$ 0K corresponds to the Peierls stress $\tau_p = \tau$ and decreases with increasing temperature in proportion to T^2. The critical resolved shear stress in body-centered cubic metals and crystal structures with low symmetry depends strongly on temperature, since it is determined by the Peierls stress.

In fcc and hexagonal metals, however, the calculated Peierls stress is much smaller than the measured critical resolved shear stress. In this case τ_0 is determined by the passing and cutting stress due to the other dislocations in the crystal. Note that even very carefully grown single crystals or polycrystals after long annealing times still retain a dislocation density of about $10^{10}/\text{m}^2$ which is caused by the mechanisms of crystal growth during solidification or recrystallization. The passing stress is a long range stress field because of the dependency $\tau_{xy} \sim 1/r$ of the stress field of the dislocations. The corresponding activation energy is so large that no noticeable thermal activation could take place. The passing stress, therefore, depends only slightly on temperature, namely, via the temperature dependence of the shear modulus G (see Eq.

(6.56)), and is referred to as the athermal flow stress or τ_G. In addition to τ_G, also the cutting stress τ_c has to be overcome, which is a short range stress, since it acts only during the cutting process (i.e., during dislocation motion of an atomic spacing). In the worst case τ_G and τ_c add to give a yield stress

$$\tau_0 = \tau_G + \tau_c \tag{6.67}$$

At an applied stress τ, only the stress differential $\tau - \tau_G$ supports the cutting process. The activation energy is given by Eq. (6.57a) $\Delta G^0 = 1/2Gb^3$ which is reduced by the expended external work $W = \tau_c b^2 \ell_f$. Defining $b^2 \ell_f$ as the (so-called) activation volume, we obtain with Eq. (6.62)

$$\dot{\gamma} = \dot{\gamma}_0 \exp\left[-\frac{\Delta G^0 - (\tau_0 - \tau_G)\,V}{\mathrm{kT}}\right]$$

or

$$\tau_0 = \tau_G + \frac{1}{V}\left[\Delta G^0 - \mathrm{kT}\left(\ln\frac{\dot{\gamma}_0}{\dot{\gamma}}\right)\right] \tag{6.68}$$

with $\dot{\gamma} \ll \dot{\gamma}_0$.

Eq. (6.68) demonstrates that τ_0 depends mildly on temperature (Fig. 6.41), since only the fraction τ_c has to be thermally activated. For sufficiently high temperatures $\tau_0 \cong \tau_G$ is sufficient to move a dislocation. In this case the cutting process is completely thermally activated ($\Delta G(\tau) = \Delta G^0$), and $\tau_0 = \tau_G$ is independent of temperature (Fig. 6.41). At very high temperatures, i.e. above half the melting temperature, additional diffusion controlled processes become active, which also cause plastic deformation at a stress level $\tau < \tau_G$ (see Section 6.8).

6.5 Strain Hardening of fcc Single Crystals

6.5.1 Geometry of Deformation

Crystallographic slip is a shear deformation. In a pure shear test parallel to the slip system the orientation of a single crystal does not change (Fig. 6.11). In a uniaxial tensile or compression test, however, a crystal rotation is superimposed on the shear, to keep the specimen aligned in the tensile direction or parallel to the compression platens. Therefore, deformation is associated with an orientation change of the crystal with respect to the tensile or compression axis. During tensile deformation, the crystal rotates in such a way that the slip direction approaches the tensile axis. Conversely, in a compression test, the slip plane normal rotates towards the compression axis. For a representation of this rotation it is more convenient to describe the rotation of the tensile axis (or compression axis) relative to the crystal lattice in the standard stereographic projection. In this representation, the tensile (compression) axis

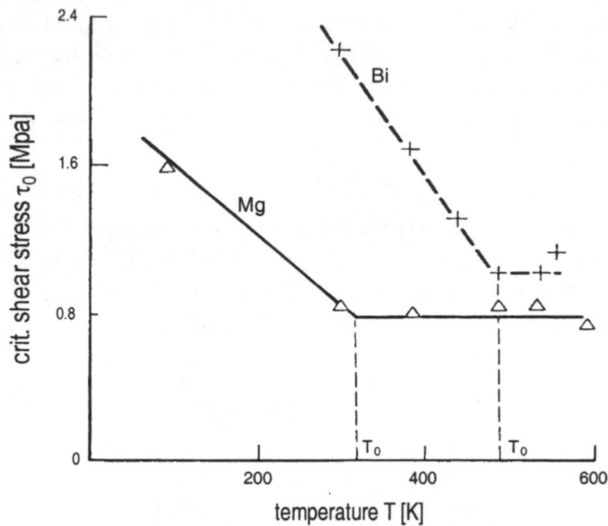

Fig. 6.41. Temperature dependence of critical shear stress in hexagonal Mg and Bi (after [6.9]).

moves on a great circle from its initial orientation into the slip direction (slip plane normal) (Fig. 6.42). In the following we will confine our consideration to tensile deformation. Compression is completely analogous. If more than one slip system is activated (double slip, multiple slip), the net slip direction is the final orientation of the tensile axis. The net slip direction is the vector sum of all the slip directions. For the corner orientations $\langle 100\rangle, \langle 110\rangle, \langle 111\rangle$, the net slip direction is identical with the initial orientation. Single crystals of these orientations should maintain their orientation during tensile deformation. This is true for $\langle 111\rangle$ and $\langle 100\rangle$ oriented crystals. In contrast $\langle 110\rangle$ oriented crystals behave in an unstable manner, since they typically activate only two coplanar slip systems.

Within a stereographic triangle one slip system has the largest Schmid factor, therefore, single slip is activated. For single slip the slip direction is located in a different stereographic triangle than the initial orientation of the tensile axis (Fig. 6.42a). Therefore, the tensile axis moves first towards the $[001]$-$[\bar{1}11]$ symmetry line. If the tensile axis reaches the border of the standard triangle because of orientation change during deformation, a second (secondary or conjugate) slip system is also activated and the resulting slip direction changes from $[\bar{1}01]$ to $[\bar{1}12]$ ($[\bar{1}01 + [011] = [\bar{1}12]$). Correspondingly, the orientation of the tensile axis approaches $[\bar{1}12]$, which it theoretically would reach but only for infinitely large strains. If the initial orientation of the tensile axis is located on the great circle $[\bar{1}01] - [011]$, the orientation of the tensile axis reaches the $[\bar{1}12]$ direction directly upon arrival at the symmetry line and, therefore, does not change its orientation any more.

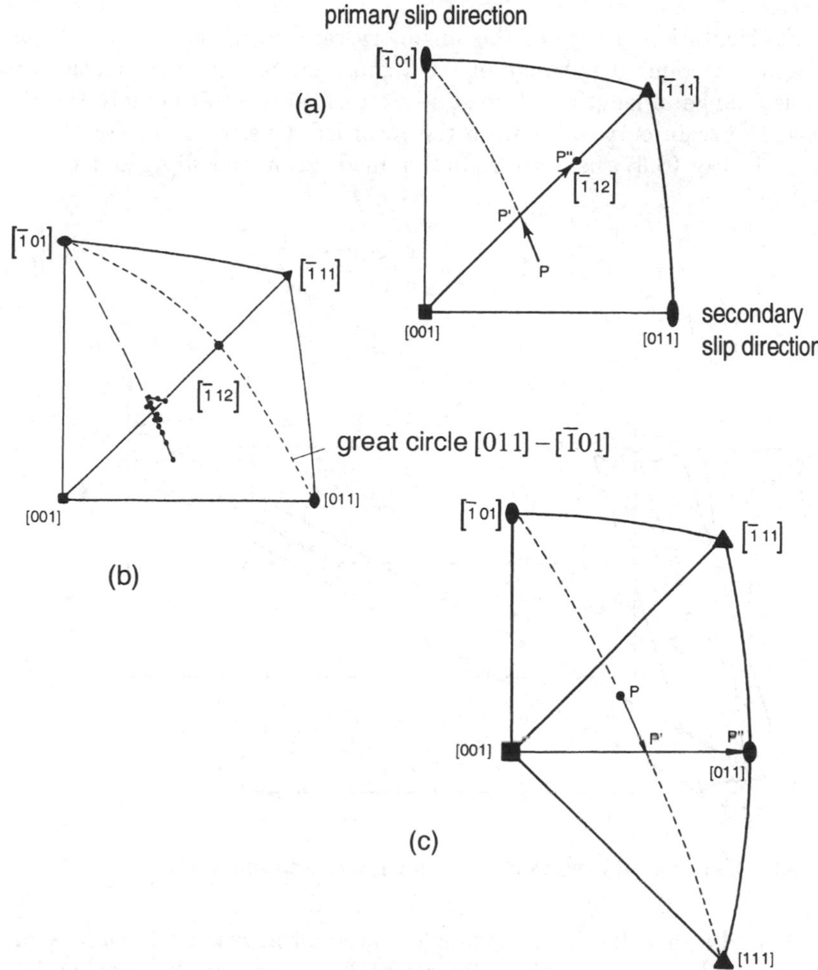

Fig. 6.42. Orientation change due to tensile deformation by single slip. (a) theoretical; (b) experimental (with "overshooting"); (c) orientation change in a compression test.

Usually slight deviations from this ideal behavior are observed, such that upon arrival at the symmetry line, the primary slip system continues to dominate with the result that the tensile axis "overshoots" the symmetry line (Fig. 6.42b). The closer the tensile axis approaches the primary slip direction (here [$\bar{1}$01]), the smaller the Schmid factor of the primary slip system. This eventually causes the conjugate slip system to be activated and the orientation of the tensile axis returns to the symmetry line. This "overshooting" has to be understood as latent hardening, which means that the secondary slip system is hardened more than the primary because of the activation of the primary slip system.

The orientation change of the single crystal during deformation requires to take into account the change of the Schmid-factor with deformation - besides the change of length and cross section - in order to calculate the shear stress-shear strain curve $\tau(\gamma)$ from the recorded σ vs. ε curve. For this $\lambda(\varepsilon)$ and $\kappa(\varepsilon)$ in Eq. (6.37) have to be determined. From the slip geometry (Fig. 6.43) it follows

$$\frac{\ell}{\ell_0} = 1 + \varepsilon = \frac{\sin \lambda_0}{\sin \lambda} = \frac{\cos \kappa_0}{\cos \kappa} \tag{6.69}$$

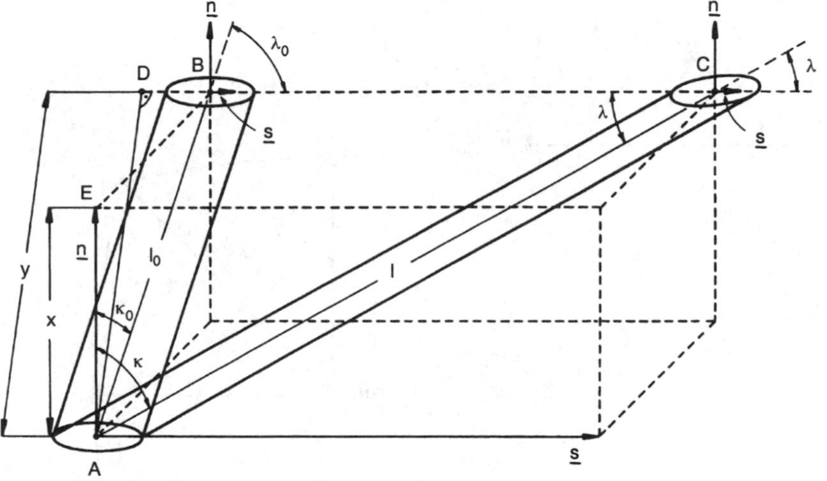

Fig. 6.43. The geometry of orientation changes due to single slip.

where λ_0 and κ_0 are the respective angles of the initial orientation and λ and κ the corresponding angles of the orientation after a strain ε. With this relation we obtain the shear stress

$$\begin{aligned}\tau = \sigma_t \cos\kappa \cos\lambda &= \sigma_t \cdot \frac{\cos\kappa_0}{1+\varepsilon} \cdot \sqrt{1 - \frac{\sin^2\lambda_0}{(1+\varepsilon)^2}} \\ &= \sigma_t \cdot \frac{\cos\kappa_0}{(1+\varepsilon)^2}\sqrt{(1+\varepsilon)^2 - \sin^2\lambda_0}\end{aligned} \tag{6.70}$$

The shear strain can be calculated because the work of deformation $\tau d\gamma = \sigma_t \cdot d\varepsilon_t$ and, therefore,

$$d\gamma = \frac{d\varepsilon_t}{\cos\kappa \cdot \cos\lambda} = \frac{\frac{d\varepsilon}{(1+\varepsilon)}}{\cos\kappa \cdot \cos\lambda} = \frac{(1+\varepsilon)d\varepsilon}{\cos\kappa_0\sqrt{(1+\varepsilon)^2 - \sin^2\lambda_0}} \tag{6.71a}$$

and by integration

$$\gamma = \int_0^\gamma d\gamma = \int_0^\varepsilon \frac{\frac{d\varepsilon}{(1+\varepsilon)}}{\cos\kappa \cdot \cos\lambda}$$
$$= \frac{1}{\cos\kappa_0}\left[\sqrt{(1+\varepsilon)^2 - \sin^2\lambda_0} - \cos\lambda_0\right] \qquad (6.71b)$$

The equations get more complicated if ε becomes large enough that the orientation hits the symmetry line and double slip is activated.

The success of Schmid's law (see Section 6.4.1) in explaining the orientation dependence of the yield stress has encouraged people to speculate that the orientation dependence of the single crystal hardening curve may disappear if $\tau(\gamma)$ instead of $\sigma(\varepsilon)$ is plotted. This "extended Schmid law" does not hold for fcc single crystals. The shear stress-shear strain curves of single crystals of different orientations are very different (Fig. 6.44) although similar in character.

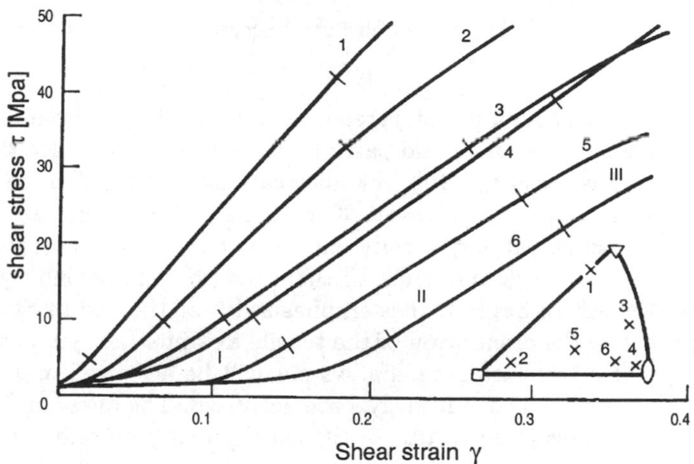

Fig. 6.44. Hardening curves of copper single crystals of different orientations under tensile deformation (after [6.10]).

6.5.2 Dislocation Models of Strain Hardening

Single crystals that deform by single slip reveal a typical hardening curve as plotted in Fig. 6.45. Apart from the elastic range we can distinguish three stages.

Stage I: (Easy glide regime) Very small hardening coefficient.

Stage II: Large linear increase of strength, $\theta_{II} = d\tau/d\gamma \approx G/300$.
Stage III: Decrease of the hardening rate $d\tau/d\gamma$ (dynamic recovery).

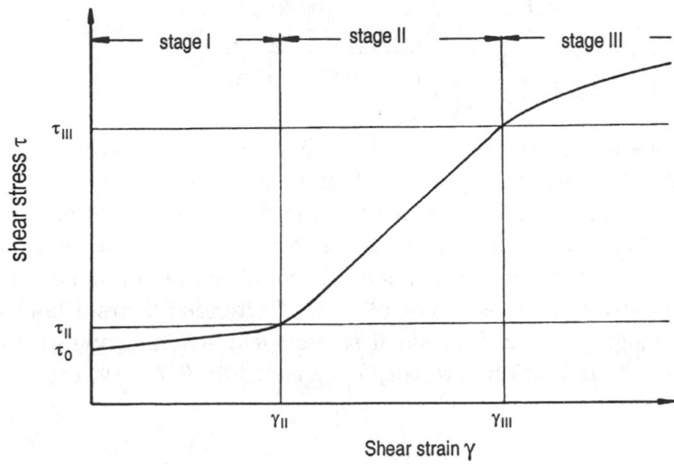

Fig. 6.45. Schematic hardening curve of fcc single crystals oriented for single slip.

The hardening curve can be interpreted as follows. During Stage I, dislocations can move long distances and partially leave the crystal, once the flow stress reaches and exceeds τ_0. Only few dislocations are stored in the crystal. Because of the long range stresses of the stored dislocations, secondary glide systems are activated only locally and do not contribute noticeably to the strain, but have a large influence on the strength. The secondary dislocations cause the end of Stage I. It is emphasized that the end of Stage I is attained long before the orientation of the tensile axis hits the symmetry line [100] – [111], where the conjugate slip system will be activated and double slip - i.e. two slip systems concurrently - are activated. The onset of Stage II is not caused by a massive activation of the conjugate slip system.

Stage II is caused by the reaction of primary dislocations with dislocations on secondary slip systems that generates a population/network of immobile dislocations (so-called Lomer-locks or Lomer-Cottrell locks, Fig. 6.46) which cannot be overcome by successive dislocations. Successive dislocations will get stuck at these locks in the crystal, i.e. they become immobilized, but they contribute to a further increase of the internal stresses and, therefore, create a larger activity of secondary slip systems. For each immobilized dislocation another mobile dislocation has to be generated to maintain the imposed strain rate, for instance during a tensile test. This causes a rapid increase of the dislocation density in Stage II, which, in turn, raises the passing stress τ_{pass} (Eq. (6.56)) and the cutting stress τ_c (Eq. (6.57b)) and, consequently, the stress necessary to maintain plastic flow, i.e. the flow stress $\tau = \tau_{pass} + \tau_c = \alpha G b \sqrt{\rho}$

which increases strongly. Thus the single crystal strain hardens (Fig. 6.47). The hardening rate θ_{II} in Stage II is almost independent of the crystal orientation or even crystal structure and amounts to approximately

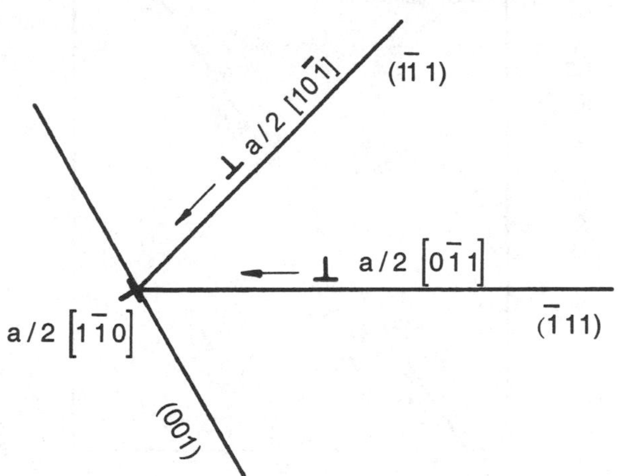

Fig. 6.46. A Lomer-lock. The dislocation in the (001) plane is immobile.

$$\theta_{\mathrm{II}} = \left.\frac{d\tau}{d\gamma}\right|_{\mathrm{II}} \approx \frac{G}{300} \tag{6.72}$$

This Eq. (6.72) can be understood from a simple calculation. Let us consider a small time interval during which dislocations of density $d\rho$ are generated and become immobilized after a slip length L. The slip length will scale with the spacing of the obstacles, i.e., $L = \beta/\sqrt{\rho}$. With Eq. (6.31b) and Eq. (6.59) we obtain

$$d\tau = \alpha G b \frac{d\rho}{2\sqrt{\rho}} \tag{6.73}$$

$$d\gamma = \frac{\beta b}{\sqrt{\rho}} d\rho \tag{6.74}$$

$$\frac{d\tau}{d\gamma} = \frac{\alpha}{\beta} \frac{G b \sqrt{\rho}}{b 2\sqrt{\rho}} = \frac{\alpha}{2\beta} G \tag{6.75}$$

Measurements have shown that the slip length of dislocations exceeds the average dislocation spacing by about a factor one hundred, i.e., $\beta \approx 100$. With $\alpha \approx 0.6$ we obtain $\theta_{II} \approx G/300$, the typical measured value.

The dislocation density strongly increases in Stage II as evident from the flow stress increase (Fig. 6.47). Where do these dislocations come from? Up to now we have only considered the case in which dislocations are generated

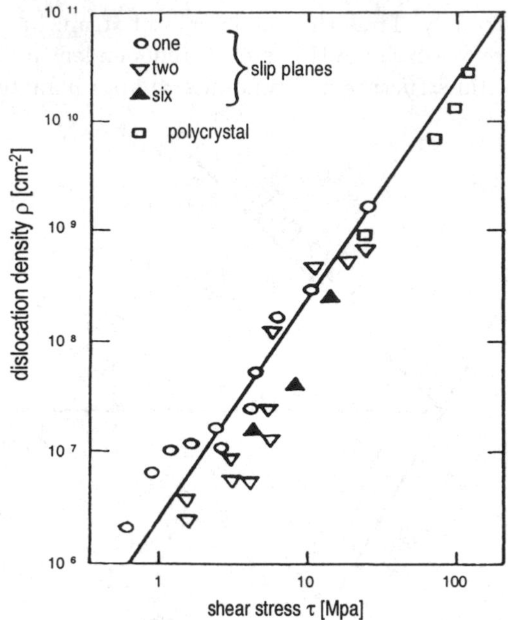

Fig. 6.47. Correlation of yield stress and dislocation density (after [6.11]).

on the surface. Because the slip length is usually small compared to the specimen thickness or the crystal size, there must also be mechanisms to generate dislocations in the interior of the crystal. These dislocations are provided by internal dislocation sources, the best known of which is the Frank-Read source (Fig. 6.48). The unstressed configuration of such a source consists of a mobile dislocation segment of length ℓ in the slip plane which, for instance, may have been generated by cross slip (see below). If an appropriate stress is applied, the dislocation segments bows out, and the radius of curvature R is related to the acting shear stress τ by

$$R = \frac{Gb}{2\tau} \tag{6.76}$$

The radius of curvature is smallest and, correspondingly, the required shear stress largest, when $R = \ell/2$. In this case the shape of the curved dislocation is a semicircle. Further motion of the dislocation increases the radius of curvature and, therefore, the necessary shear stress decreases. Correspondingly, if a shear stress

$$\tau_0 = \frac{Gb}{\ell} \tag{6.77}$$

is applied, the semi-circle configuration expands and eventually forms a closed loop with the original dislocation segment in its center. The loop continues

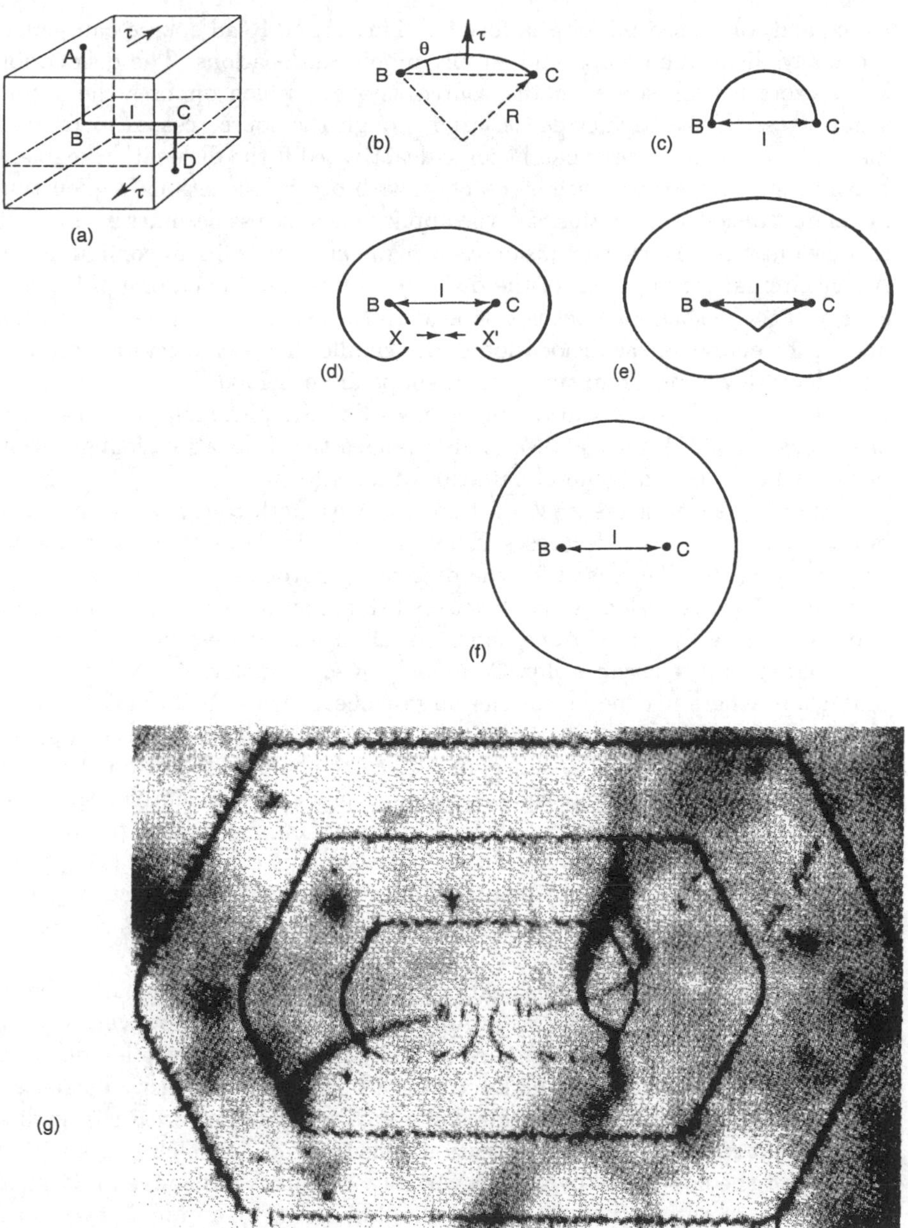

Fig. 6.48. Principle of the Frank-Read source: (a) free segment of the dislocation BC (edge dislocation) in a slip plane; (b) curvature of the dislocation under stress; (c) critical configuration (semicircle); (d)-(f) generation of a dislocation loop; (g) Frank-Read source observed in silicon [6.12].

to expand and a second loop is formed. This Frank-Read source can generate many dislocation loops and supply mobile dislocations. The dislocation loops exert a back stress on the source, though, which opposes the acting shear stress. If the back stress is large enough the source ceases to provide dislocations. Also a source can become deactivated if the dislocation segment becomes smaller, for instance by reaction with other dislocations, e.g. cutting by a forest dislocation. In this case the applied shear stress becomes insufficient to bow out the now shorter dislocation segment to its critical configuration. A semicircular arrangement of the dislocation line is obtained only if the line energy of the dislocation does not depend on the orientation of the dislocation line. If the energy of the dislocation line is smaller in specific crystallographic directions, deviations from the circular shape are observed, for instance in silicon (Fig. 6.48). The dislocation line is piecewise straight, namely along $\langle 110 \rangle$ directions on $\{111\}$ glide planes. This changes the dislocation configuration but does not affect the general behavior of a source.

After the shear stress τ_{III} is attained the strength continues to increase, but the hardening rate decreases. This Stage III is the longest stage of the hardening curve. The reason for the decreasing hardening rate is mainly the cross slip of screw dislocations. A screw dislocation does not have a defined glide plane (see Chapter 3) and thus, can change its glide plane (Fig. 6.49), which is referred to as cross slip. Commonly, a screw dislocation will move on that plane where it experiences the largest shear stress. If its motion, however, is impeded by an obstacle on the primary slip plane, it can change to another slip plane, the cross slip plane. Since the Schmid-factor on the cross slip plane is smaller than for the primary slip plane, a sufficiently large shear stress is necessary to make the dislocation move on the cross slip plane. In Stage III this is always the case. In fcc metals slip is confined to $\{111\}\langle 110 \rangle$ slip systems. Since two $\{111\}$ planes intersect along a $\langle 110 \rangle$ direction, there is exactly one slip plane onto which a screw dislocation can cross slip in fcc crystals (see Fig. 6.31).

Cross slip enables screw dislocations to circumvent obstacles and, therefore, increases the slip length of a dislocation. Most likely, however, a cross slip dislocation will meet an antiparallel dislocation on the new glide plane so that both dislocations are annihilated, and the dislocation density decreases. In a small time interval an additional slip $d\gamma_{cr}$ is caused by cross slip with a concomitant reduction of the dislocation density by $d\rho_{cr}$, which scales with a decrease of the flow stress by $d\tau_{cr}$. Invariably, the processes active in Stage II continue to operate and contribute to the flow stress by an amount $d\tau_h$ and to the shear strain by $d\gamma_h$. Correspondingly,

$$\left.\frac{d\tau}{d\gamma}\right|_{\mathrm{III}} = \frac{d\tau_h - d\tau_{cr}}{d\gamma_h + d\gamma_{cr}} < \frac{d\tau_h}{d\gamma_h} = \left.\frac{d\tau}{d\gamma}\right|_{\mathrm{II}} = \theta_{\mathrm{II}} \tag{6.78}$$

The decrease of the strain hardening rate is a recovery process. Since it occurs during deformation, it is also referred to as dynamic recovery, as opposed to static recovery during annealing.

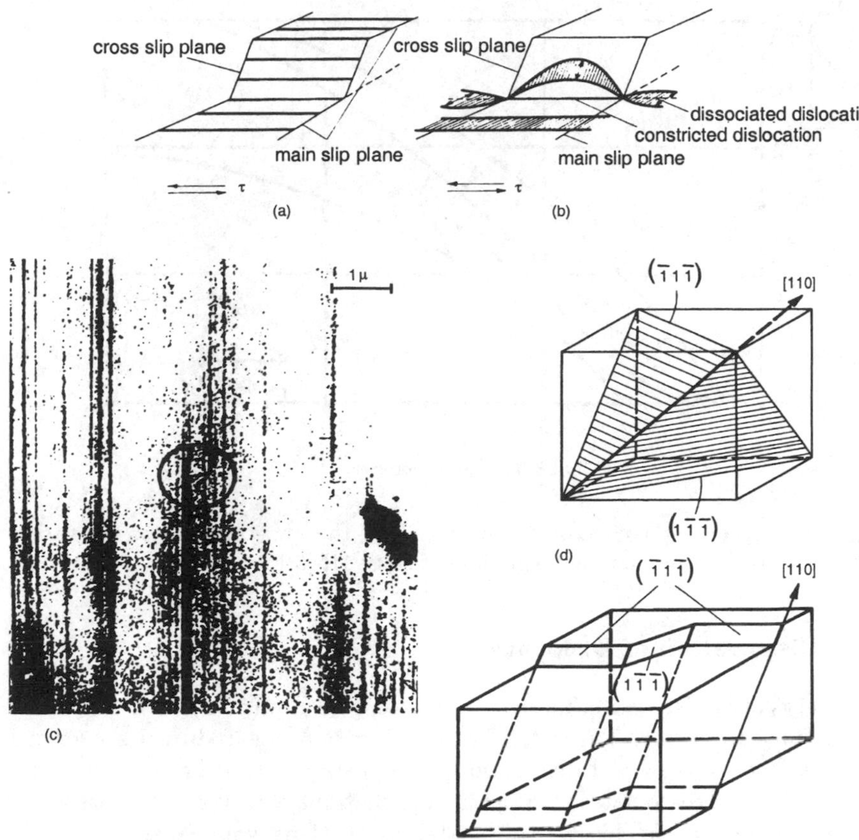

Fig. 6.49. Cross slip of (a) perfect and (b) dissociated screw dislocations; (c) cross slip trace on the surface of a copper single crystal; (d) geometry of cross slip in the fcc lattice.

The occurrence of cross slip in stage III has been proven experimentally. Still we have to understand why in most metals and alloys τ_{III} is much larger than the stress necessary to reach the critical shear stress in the cross slip system and also why τ_{III} depends so strongly on material (Fig. 6.50). Even fcc metals with very similar values of lattice parameter, shear modulus, and melting temperature such as silver and aluminum have very different values of τ_{III}. Besides, τ_{III} and thus the duration of stage II, depends strongly on temperature in such a way that τ_{III} decreases drastically with increasing deformation temperature (Fig. 6.51). The reason for this behavior is the dissociation of dislocations into partials.

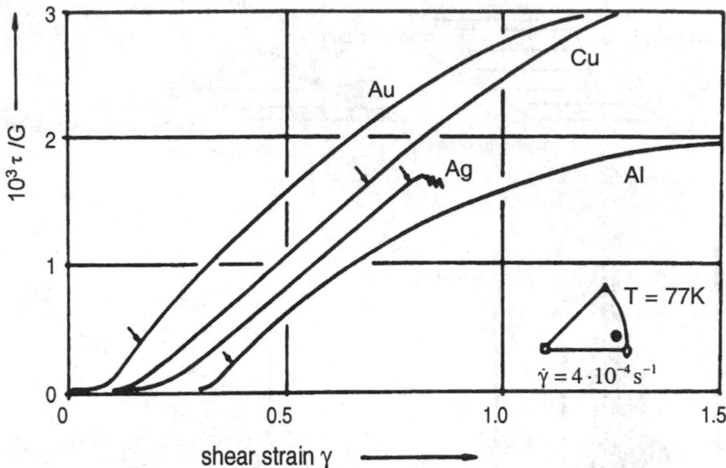

Fig. 6.50. Hardening curves of single crystals of the same orientation for various fcc metals. Arrows indicate the beginning of stage III [6.13].

6.5.3 Dissociation of Dislocations

The energy of a dislocation increases according to Eq. (6.52) with the square of the Burgers vector. Theoretically, a dislocation can reduce its energy if it decomposes into partial dislocations. For instance, if a dislocation with a Burgers vector b dissociates into partial dislocations each with Burgers vector $b/2$, the energy E_2 of this pair of partial dislocations would read

$$E_2 = 2 \cdot \frac{1}{2} G \left(\frac{b}{2}\right)^2 = \frac{1}{2} \cdot \left(\frac{1}{2} G b^2\right) = \frac{1}{2} E_1 \tag{6.79}$$

i.e. only half as large as the energy E_1 of a single dislocation. The approximation, however, makes the tacit assumption that both partial dislocations are far apart, such that R_0 is large and Eq. (6.52) can be applied. Indeed, two parallel dislocations on the same glide plane repel each other and try to maximize their spacing. The vector $b/2$, however, is not a translation vector of the crystal lattice. Between the two partial dislocations the crystal lattice on the slip plane is perturbed. The energy of this planar fault in most crystal structures is much larger than the energy gain by dissociation. Therefore, the dislocation remains undissociated, i.e. perfect.

In cubic and hexagonal crystals, however, the Burgers vector can dissociate into smaller vectors which are associated with a planar fault of small energy. The most important example is the existence of Shockley partial dislocations in fcc crystals. On the (111) slip plane a perfect dislocation with Burgers vector $\mathbf{b}_1 = a/2[1\bar{1}0]$ can dissociate into partial dislocations (Shockley dislocations) (Fig. 6.52) according to

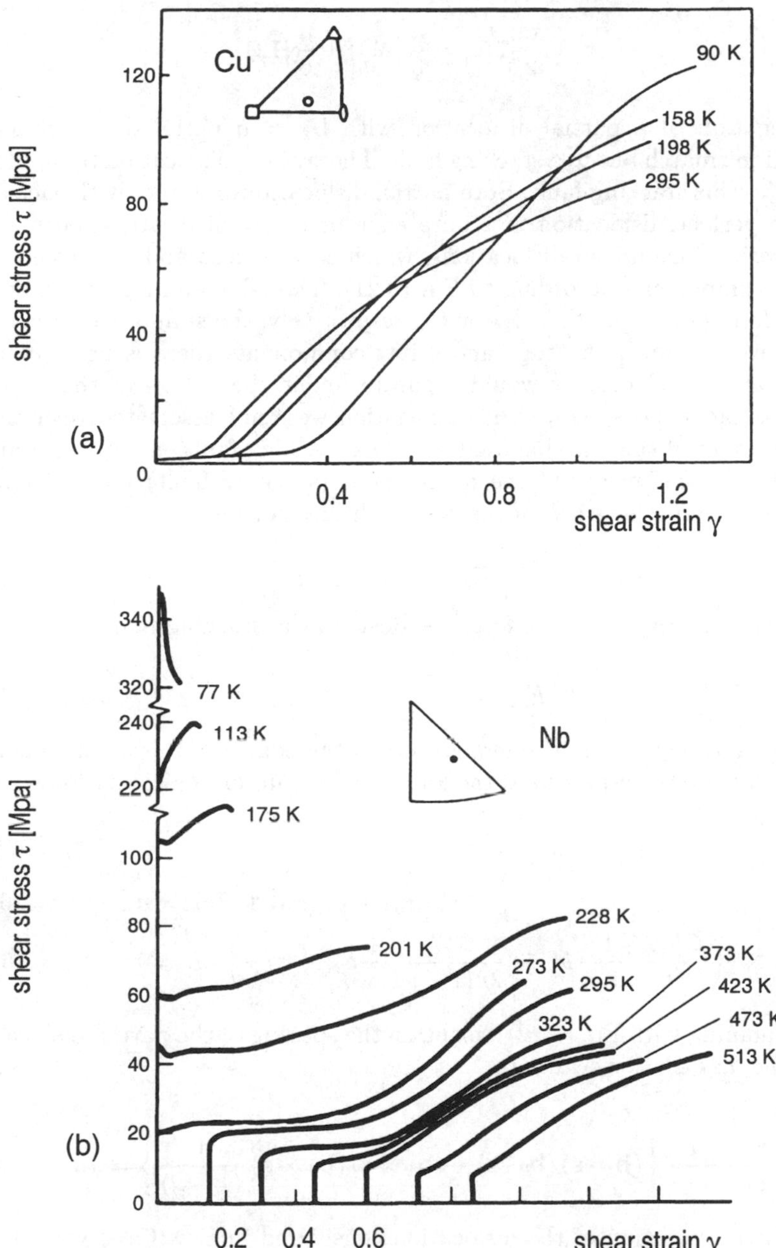

Fig. 6.51. Hardening curves of tensile deformed single crystals at various temperatures (a) fcc copper [6.14]; (b) bcc niobium [6.15].

$$\frac{a}{2}[1\bar{1}0] = \frac{a}{6}[2\bar{1}\bar{1}] + \frac{a}{6}[1\bar{2}1] \tag{6.80}$$

The motion of a partial dislocation with $\mathbf{b}_2 = a/6[2\bar{1}\bar{1}]$ does not lead to planar mismatch but to a stacking fault. The trailing dislocation $\mathbf{b}_3 = a/6[1\bar{2}1]$ removes this stacking fault. Both partial dislocations interact with each other. If the perfect dislocation is a pure edge or screw dislocation, both partial dislocations are mixed dislocations, which can be decomposed into edge and screw components according to Eq. (3.21). The edge and screw components exert forces on each other, $\mathbf{F}_e$ or $\mathbf{F}_s$, respectively, the sum of which is always repulsive. Between the edge and screw components there is no interaction. The partial dislocations would separate under the action of this repulsive force as far as possible if their dissociation were not associated with an area enlargement of the stacking fault. Given a stacking fault energy per unit area $\gamma_{SF}[\mathrm{J/m^2}]$, the energy of the planar fault (stacking fault) for a dislocation spacing x and a length L of the partial dislocations is

$$E_{SF} = \gamma_{SF} \cdot L \cdot x \tag{6.81}$$

Correspondingly, there is a force to constrict the stacking fault

$$F_{SF} = -\frac{dE_{SF}}{dx} = -\gamma_{SF} \cdot L \tag{6.82}$$

If b_e and b_s denote the Burgers vectors of the edge and screw components of the dislocations, respectively, the force equilibrium for a dissociation width x_0 is

$$F_e(x_0) + F_s(x_0) + F_{SF} = 0 \tag{6.83a}$$

$$\left(\frac{Gb_{1s}}{2\pi} \cdot \frac{1}{x_0} \cdot b_{2s} + \frac{Gb_{1e}}{2\pi(1-\nu)}\frac{1}{x_0}b_{2e} - \gamma_{SF}\right) \cdot L = 0 \tag{6.83b}$$

so combining with Eq. (3.22) we obtain the spacing of the partial dislocations, i.e., the dissociation width x_0

$$\frac{G}{2\pi} \cdot \frac{1}{\gamma_{SF}} \left\{ (\mathbf{b}_1 \cdot \mathbf{s})(\mathbf{b}_2 \cdot \mathbf{s}) + (\mathbf{b}_1 \times \mathbf{s})(\mathbf{b}_2 \times \mathbf{s}) \frac{1}{(1-\nu)} \right\} = x_0 \tag{6.84}$$

For the dissociation of the dislocation considered in Eq. (6.80) we obtain in case of an edge dislocation, with $\mathbf{s} = 1/\sqrt{6}[11\bar{2}]$ and $b = |\mathbf{b}| = a/\sqrt{2}$

$$x_0 = \frac{Gb}{\gamma_{SF}} \frac{b}{24\pi} \frac{2+\nu}{1-\nu} \tag{6.85}$$

The dissociation width depends primarily on the stacking fault energy, which can be very different for metals which seem otherwise much alike, for instance

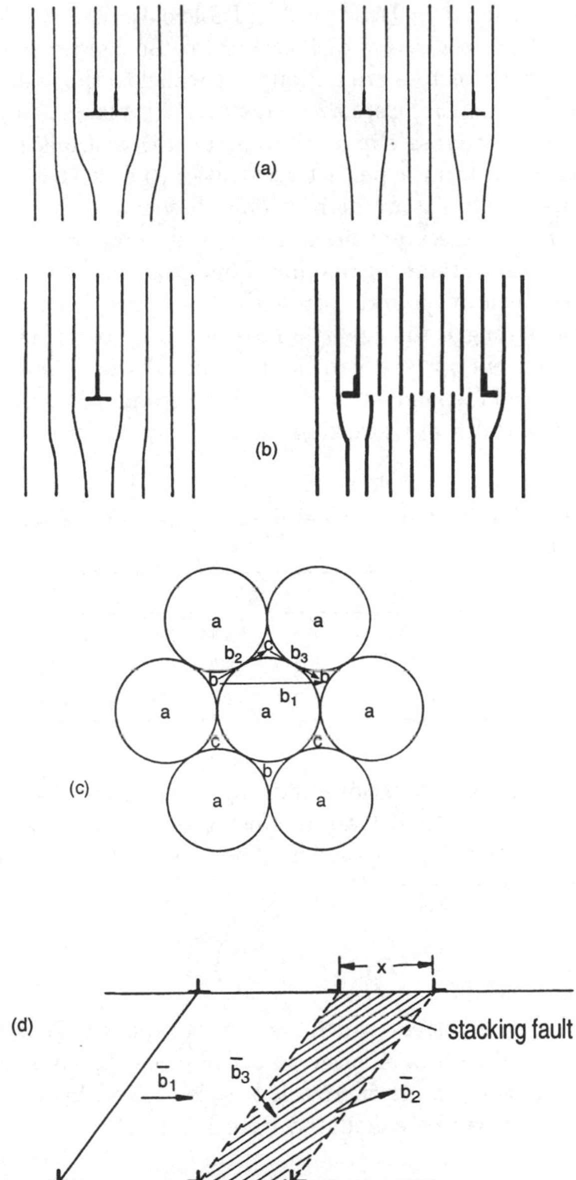

Fig. 6.52. Dissociation of dislocations. (a) A double dislocation splits into two single dislocations; (b) a single dislocation dissociates into two partial dislocations; (c) arrangement of successive planes in the fcc lattice; (d) dissociation of an edge dislocation in an fcc lattice into Shockley partial dislocations.

180 mJ/m^2 for Al and 20 mJ/m^2 for Ag (Table 6.4). The Shockley dislocations of a dissociated perfect screw dislocation are no longer screw dislocations, because the Burgers vectors are no longer parallel to the dislocation line. The partial dislocations, therefore, have a defined slip plane. A dissociated screw dislocation can only cross slip if the partial dislocations recombine over a certain length to re-form a perfect dislocation, i.e. if they "constrict" (Fig. 6.49b). Dislocations in Ag are more widely dissociated than in Al. Therefore, much higher τ_{III} stresses are necessary in Ag compared to Al to constrict the dissociated dislocations for making them cross slip (Fig. 6.50). Aluminum for this reason has a much lower strength than silver, because τ_{III} is a rough measure of the strength that can be introduced by deformation. Conversely, the stacking fault energy γ_{SF} can be determined from the start of stage III, namely τ_{III}, as an independent method of the standard microscopic approach of measuring the width of a stacking fault.

Table 6.4. Stacking fault energy and dissociation width of screw dislocations in various fcc metals.

Material	Ag	Cu	Ni	Al
γ_{SF} $[mJ/m^2]$	20	40	150	180
γ_{SF}/Gb $[10^{-3}]$	3.0	4.3	9.9	27.4
x_0/b	15	11	5	1

The constriction of dislocations and, therefore, cross slip of dissociated dislocations can be thermally activated, since the partial dislocations thermally vibrate about the equilibrium position. The cross slip frequency for an applied shear stress τ is given by

$$\Gamma_Q = \nu_D \left(\frac{\tau}{\tau_M}\right)^{A/kT} \tag{6.86}$$

Here τ_M is the shear stress which has to be applied to enforce cross slip without thermal activation. A is the so-called cross slip constant. Both A and τ_M depend on stacking fault energy. If τ_{III} is defined by a certain cross slip frequency Γ_{crIII}, it can be calculated from Eq. (6.86)

$$\tau_{\mathrm{III}} = \tau_M \left(\frac{\Gamma_{cr\mathrm{III}}}{\nu_D}\right)^{\frac{kT}{A}} \tag{6.87}$$

τ_{III} strongly depends on temperature which is reflected by a stage II shortening with increasing temperature (Fig. 6.51).

The mechanism of mechanical twinning in fcc crystals can be related to Shockley partial dislocations. The motion of a Shockley dislocation results in a stacking fault on the slip plane. This changes the ideal stacking $ABCA_{\uparrow}^{B}CABC$ to $ABCA_{\uparrow}^{C}ABCA$, if the dislocation moves on the second plane termed B. If on the next neighbor slip plane (now A) another Shockley dislocation moves with the same Burgers vector, the corresponding displacement modifies the stacking sequence to $ABCA'CB'CAB$, i.e., the planes CB represent a twin that is two layers thick. This twin grows in thickness by the motion of further partial dislocations on adjacent slip planes.

Conversely, a stacking fault can be considered as a limiting case of a crystal where the twin consists of only one atomic layer. Hence, a stacking fault is bounded by two twin boundaries, and, therefore, the stacking fault energy ought to correspond to about twice the energy of a coherent twin boundary. This is a good approximation for many metals.

6.6 Strength and Deformation of Polycrystals

In contrast to single crystals, crystallites in polycrystals are subject to constraints during plastic deformation, because the polycrystal has to deform as a whole, instead of responding as a set of individual crystals. This requires that every grain has to participate in deformation and each grain has to adjust its deformation to the shape change of the neighboring grains, to maintain a perfect fit of the crystals along their common grain boundaries. This seemingly trivial boundary condition has significant impact on the deformation and strength of polycrystals.

The grains of a polycrystal have different orientations, by definition. When an external tensile stress is applied, crystals that are favorably oriented, i.e. have a high Schmid factor, deform first while others which are less favorably oriented are still stressed below the critical resolved shear stress. Single slip deformation of an individual grain would lead to a shape change, which is not shared by its surrounding. Therefore, this shape change must be suppressed by elastic deformation which rapidly engenders high internal stresses, and eventually causes the next neighbor grains also to reach the critical resolved shear stress. Only when all grains in a polycrystal deform plastically the yield stress is reached.

Let us consider this problem as a dislocation concept. If a single slip system of a grain is activated, dislocations in this slip system are produced and move. The grain boundaries are obstacles which can not be overcome by the dislocations, because the Burgers vector has to be a translation vector of the crystal which - except for very special cases - does not hold for the next neighbor crystal due to its different orientation. Hence, in the next neighbor crystal the slip directions are differently oriented and usually not parallel to each other, for instance in fcc metals the $\langle 110 \rangle$ directions. If slip in one grain were to be continued into the next neighbor grain with the Burgers vector **b**, the crystal structure of the neighbor grain would be destroyed, but this is suppressed due to energetic reasons. Hence, the dislocations have to pile up at the grain boundaries as observed in TEM micrographs (Fig. 6.53). The piled-up dislocations exert a back stress on subsequent dislocations which is opposite to the applied shear stress. The succeeding dislocations assume positions such that the applied shear stress and the back stress balance. Since the back stress increases with increasing number of piled-up dislocations, the spacing between successive dislocations increases away from the pile-up tip (Fig. 6.53). The length of the pile-up in a grain, however, is limited, to half the grain size $D/2$ (Fig. 6.54) since a source in the center of a grain will generate pile-ups on both sides of a grain. For an applied shear stress τ, the maximum number of dislocations which fits in a pile-up of length $D/2$ can be calculated to yield (for edge dislocations)

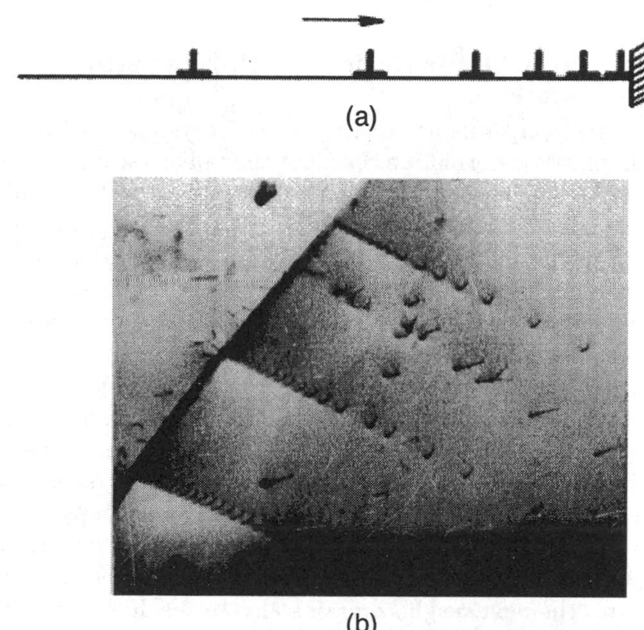

Fig. 6.53. Pile-up of dislocations at a grain boundary. (a) Schematic; (b) observed in stainless steel (TEM) [6.16].

$$n = \frac{\pi(1-\nu)}{Gb}\frac{D}{2}\tau \tag{6.88}$$

At the tip of the pile-up there act, in addition to the applied stress τ, the repulsive forces of the $(n-1)$ succeeding dislocations each with $F = \tau b$. Therefore, at the tip of the pile-up there is a stress

$$\tau_{\max} = n\,\tau \tag{6.89}$$

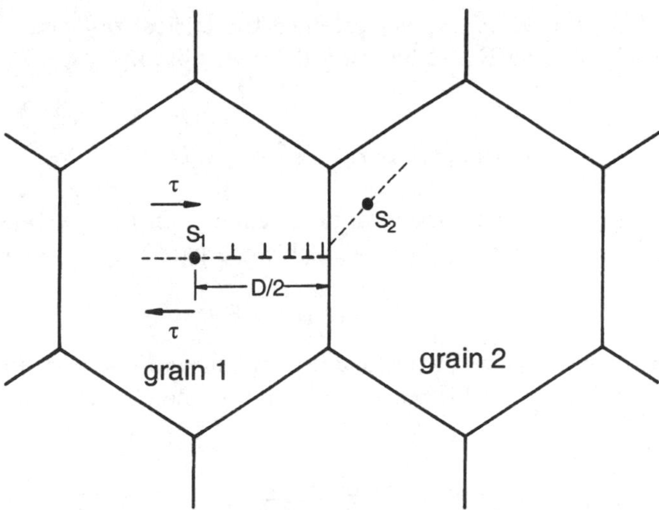

Fig. 6.54. Yield strength of single crystals. A pile-up in grain 1 activates a dislocation source S_2 in grain 2 (D = grain diameter).

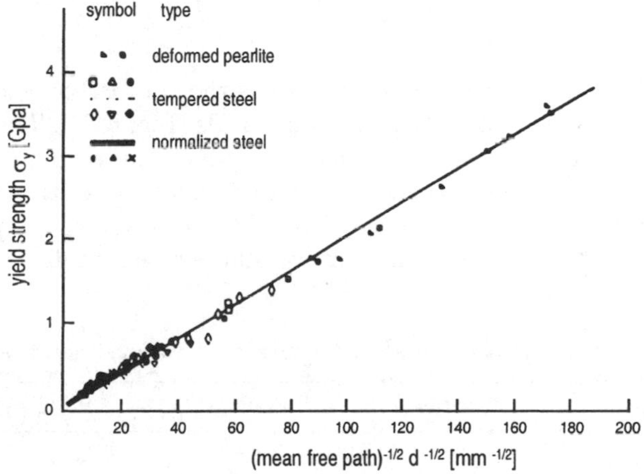

Fig. 6.55. Grain size dependence of yield strength in various steels (after [6.17]).

This internal stress $\tau_{\max}$ also affects the plastically undeformed next neighbor grain 2 and, therefore, increases the active stress in its slip systems. At a distance x from the grain boundary the shear stress in the neighbor grain is

$$\tau_2(x) = m_2 \cdot \sigma + \beta(x) \cdot \tau_{\max} \tag{6.90}$$

where $\beta(x)$ is an attenuation factor depending on location.

Plastic deformation in grain 2 will be initiated if at a distance x_0 from the grain boundary, the source S_2 has attained the critical resolved shear stress $\tau_2(x_0) = \tau_0$. Combining Eq. (6.88) and (6.89) we obtain

$$\tau_0 = m_2\sigma + \beta(x_0) \cdot \frac{\pi(1-\nu)}{2 \cdot G \cdot b} \cdot D \cdot \tau^2 \tag{6.91a}$$

Correspondingly, the shear stress τ in the original grain 1 has to be sufficiently high that (assuming that $m_2 \cdot \sigma$ is negligible compared to the second term)

$$\tau^2 \cdot D = \text{const.} = k'_y \tag{6.91b}$$

In the limit of very large D the critical resolved shear stress of a single crystal is the minimum necessary for plastic deformation, hence we obtain

$$\tau = \tau_0 + \frac{k'_y}{\sqrt{D}} \tag{6.92a}$$

or for the normal stress because of $\tau = m\sigma$

$$\sigma = \sigma_0 + \frac{k_y}{\sqrt{D}} \tag{6.92b}$$

where $k_y = k'_y/m$. Eq. (6.92b) is referred to as the Hall-Petch relation. It has been experimentally confirmed for many materials (Fig. 6.55). The Hall-Petch relation is the foundation for the strengthening by grain refinement, which is very important for the development of materials where a strength increase can not be obtained by changing chemistry. The constant k_y is referred to as the Hall-Petch constant and is different for different materials (Table 6.5).

Table 6.5. Parameters of the Hall-Petch-relation for various metals and alloys.

Material	Lattice	σ_0 $[MPa]$	k $[MPa \times \sqrt{m}]$
Cu	fcc	25	0.11
Ti	hex	80	0.4
low-carbon steel	bcc	70	0.74
Ni_3Al	$L1_2$	300	1.7

The deformation of grains does not occur independently of each other. If two neighboring grains were to deform by single slip, they would undergo a shape change different from each other because of the different spatial orientation of the slip systems. This would result in a separation of the crystallites (Fig. 6.56a), in contradiction to observations. Hence, other slip systems must also be activated to accommodate the shape change of their neighboring grains (strain compatibility). Since in a 3D polycrystal a grain is surrounded by many neighbor grains (14 on average), it must be capable of undergoing an arbitrary

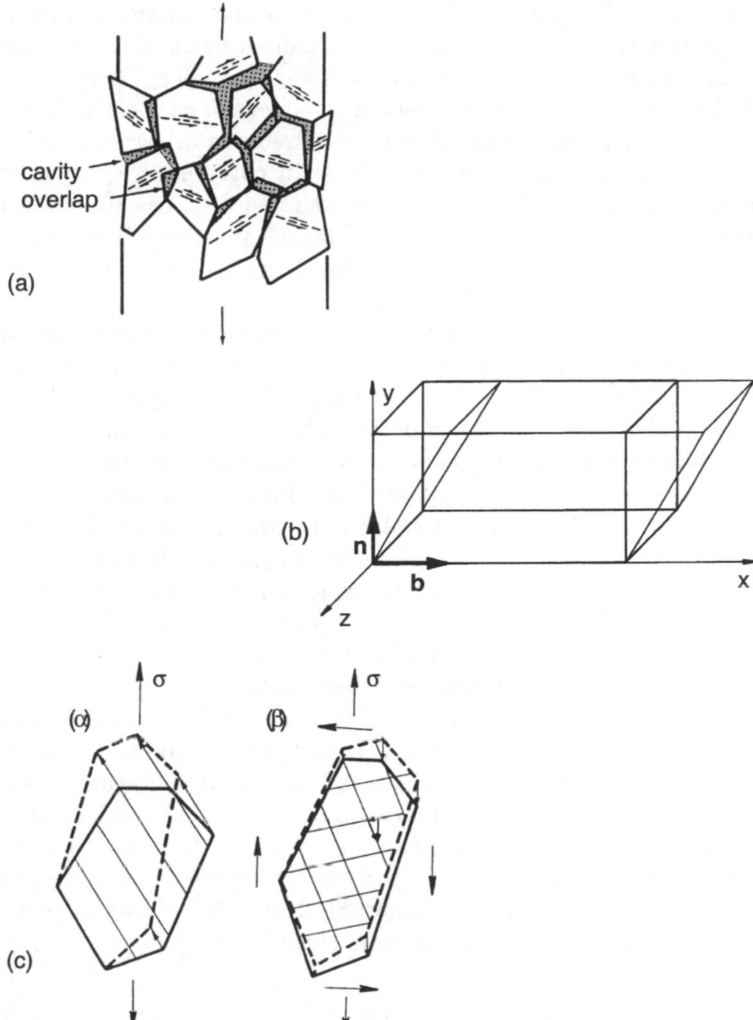

Fig. 6.56. (a) Change in shape of the grains of a polycrystal due to single slip; (b) illustration of the correspondence of shear γ_{xy} and slip on slip system set $\{\mathbf{n}\} < \mathbf{b} >$; (c) approximate restoration of the original shape by a second set of slip system.

shape change, in principle. For an arbitrary shape change the activation of five independent slip systems is necessary. This can be understood as follows:

Let us assume that the shape change consists of a simple shear γ_1 parallel to the slip plane with normal $n_1 \| \ y$ and the slip direction $b_1 \| \ x$ (Fig. 6.56b). In this case the strain tensor has only one independent component, i.e. $\varepsilon_{xy} = 1/2\gamma_1$. All other components of the strain tensor vanish. If the crystal is now additionally deformed parallel to another slip system by the amount γ_2, with for the sake of simplicity $n_2 || z, b_2 || x$, we obtain another component of the strain tensor $\varepsilon_{xz} = 1/2\gamma_2$. Obviously, ε_{xz} cannot be obtained by a shear in the first slip system, because γ_1 is already defined through ε_{xy} (Fig. 6.56b). On the other hand ε_{xy} and ε_{xz} are entirely independent of each other. Therefore, for such deformation two slip systems are required. An arbitrary strain tensor has five independent components. This is so in part because a symmetrical tensor like the strain tensor (see Eq. (6.8), Section 6.1), has six independent components. In addition, constant volume during plastic deformation means that $\varepsilon_{xx}+\varepsilon_{yy}+\varepsilon_{zz} = 0$ so one more component can be eliminated, for instance $\varepsilon_{xx} = -(\varepsilon_{yy} + \varepsilon_{zz})$.

So, in the two-dimensional case we need a minimum of two slip systems (Fig. 6.56c) and in 3D five slip systems, in order to satisfy an arbitrary shape change. More specifically one needs five independent slip systems. A slip system is independent of others if the deformation that it produces can not be replaced by a combination of shears on the other slip systems. For instance in hexagonal crystals there is a set of three slip systems, namely the basal plane with three slip directions, of which only two are independent, because the deformation on one of the three slip systems can be replaced by a (linear) combination of the deformation on the remaining two slip systems. Hexagonal metals which deform only by basal slip are brittle as polycrystals, but can be very ductile as single crystals, for instance zinc (Fig. 6.57). Mechanical twinning is, therefore, very important for the plastic deformation of hexagonal crystals (see Section 6.3.2). In cubic crystals there are five independent slip systems so that polycrystals of cubic materials are ductile if deformation is not constrained by other factors. Since there are twelve different slip systems in fcc crystals, there are 384 different combinations of five independent slip systems which will generate an arbitrary shape change. The selection of a particular set of five slip systems is unimportant for ductility but it is significant for strain hardening and texture formation. Taylor has shown that a set of slip systems is selected for which the total shear

$$d\Gamma = \sum_{s=1}^{5} d\gamma_s \tag{6.93}$$

is minimized. This allows the orientation change during deformation and the strain hardening behavior of polycrystals to be calculated. For this, first the average Schmid factor m_T has to be determined. By analogy to the relationship in single crystals (Eq. (6.71a))

$$d\varepsilon = m \; d\gamma$$

we obtain for polycrystals

$$d\varepsilon = m_T \; d\Gamma = m_T \sum_{s=1}^{5} d\gamma_s \tag{6.94}$$

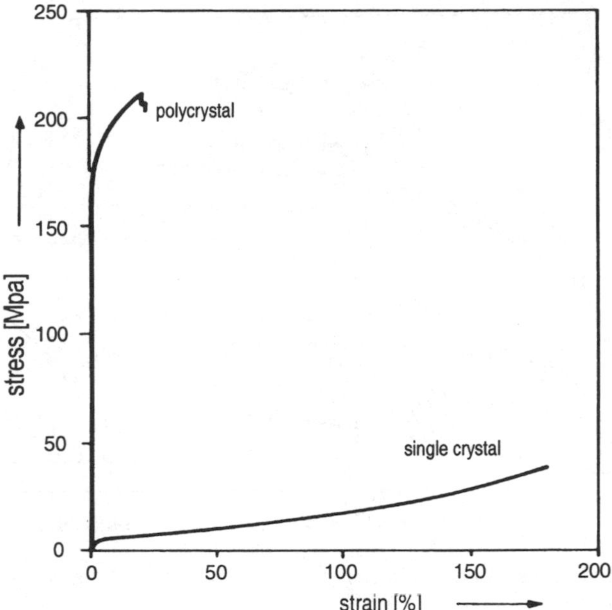

Fig. 6.57. Hardening curves of Zn single and polycrystals (after [6.18]).

For a random orientation distribution, Taylor calculated according to this method the so-called Taylor factor for tensile deformation

$$M_T = \frac{1}{m_T} = 3.06 \tag{6.95}$$

The average Schmid factor for polycrystals during tensile deformation is, therefore,

$$m_T = \frac{1}{3.06} = 0.327 \tag{6.96}$$

With Eq. (6.96) the stress-strain diagrams of polycrystals can be translated into shear stress-shear strain curves. To calculate the hardening curve, Taylor assumed that all grains harden like $\langle 111 \rangle$ oriented single crystals. This is reasonable since $\langle 111 \rangle$ crystals also deform by multiple slip (six slip systems). The computed hardening curves of polycrystals are similar to the measured curve (Fig. 6.58). If, instead, the average Schmid factor m_S is calculated by averaging the Schmid factor of a random distribution of crystallites for unconstrained deformation, i.e., as if they were single crystals, one obtains the so-called "Sachs factor"

$$M_S = \frac{1}{m_S} = 2.24 \tag{6.97}$$

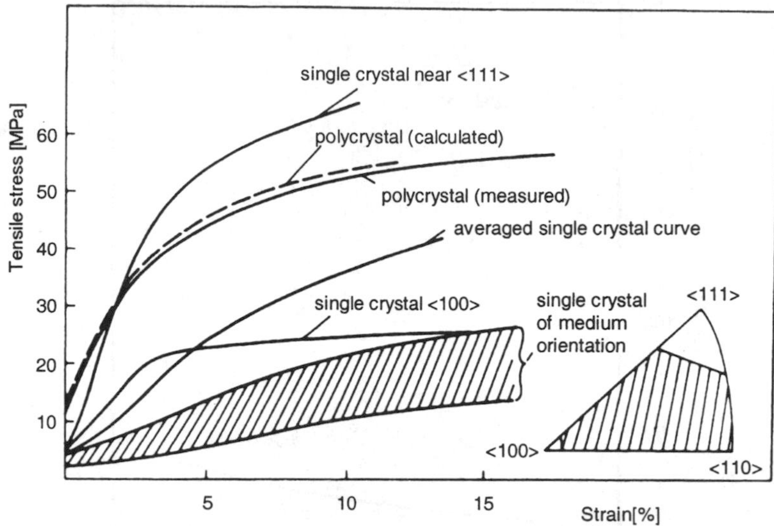

Fig. 6.58. Single crystal hardening curves of Al compared to measured and calculated polycrystal hardening curves.

and the respective hardening curve would be the average single crystal hardening curve. This curve, however, is distinctly different from the measured hardening curves, as evident from Fig. 6.58.

Plastic deformation by shear causes an orientation change of the crystals. Therefore, the selection of activated systems determines the evolution of the deformation texture, which is very important for many applications (see Chapter 2). The problem of computing the deformation texture is exacerbated by the fact that there is more than one combination or set of five slip systems which results in the same minimum shear $d\Gamma$. Each combination leads to a different grain rotation, however. These problems are the subject of current research.

6.7 Strengthening Mechanisms

6.7.1 Solid Solution Hardening

In the previous chapters we have already addressed two mechanisms of strengthening, namely by grain refinement (Section 6.6) and by plastic deformation (Section 6.5). However, there are even stronger means to increase the strength of materials, notably by alloying. If the alloy is a solid solution, we call the strength differential between the solid solution and the pure metal "solid solution hardening" as expressed in terms of the change of τ_0 (Fig. 6.59) or σ_y, respectively.

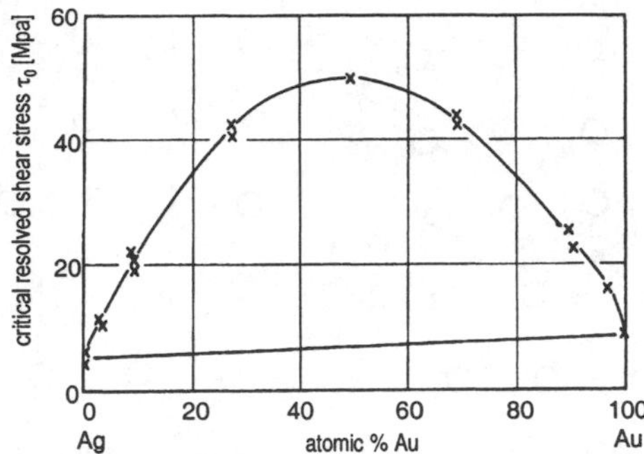

Fig. 6.59. Composition dependence of critical shear stress in Ag-Au single crystals (after [6.19]).

Solid solution hardening is caused by the interaction of an alloying element with the dislocations, which results in an increased glide resistance. There are three different ways in which solute atoms can interact with dislocations

(a) Parelastic interaction (lattice parameter effect)

(b) Dielastic interaction (shear modulus effect)

(c) Chemical interaction (Suzuki effect).

(a) Parelastic interaction: Solute atoms have an atomic size different from that of the matrix atoms. Their incorporation into the crystal lattice causes compressive or tensile stresses, depending on whether the solute atom is larger or smaller than the matrix atom. An edge dislocation has in its core dilated and compressed regions. Correspondingly, the energy associated with the elastic distortion by the solute atom is reduced if the solute atom segregates from the perfect crystal to a dislocation (Fig. 6.60). If such dislocation moves, dislocation and solute atoms have to separate. This requires the expenditure of elastic energy gained during segregation of the solute atom to the dislocation and corresponds to a (back stress) on the dislocation. Hence, an additional stress has to be applied to overcome this back-driving force, and the critical resolved shear stress of the solid solution crystal is, therefore, larger than that of the pure matrix. This stress differential can be calculated.

The edge dislocation has a hydrostatic stress field (see Eq. (6.42))

$$p = \frac{1}{3}\left(\sigma_{xx} + \sigma_{yy} + \sigma_{zz}\right) = -\frac{Gb}{3\pi r}\frac{1+\nu}{1-\nu}\sin\theta \tag{6.98}$$

If the volume changes by incorporation of a solute atom in the crystal this results in an interaction energy

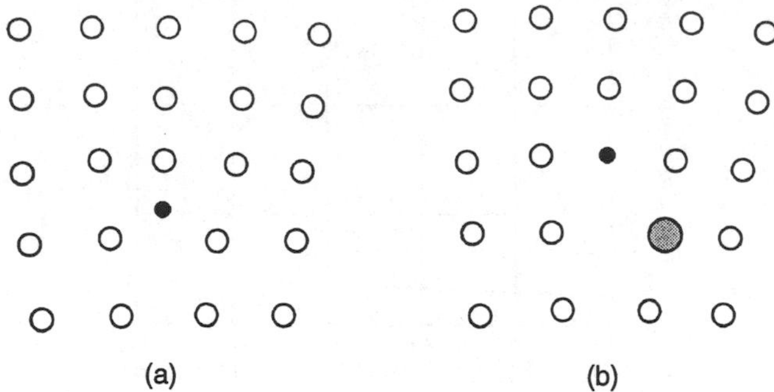

Fig. 6.60. Preferred positions of substitutional atoms at the core of an edge dislocation. (a) Interstitial atom; (b) smaller (filled circle) and larger (shaded circle) substitutional atom.

$$\Delta E^p = -p\Delta V \left(3\frac{1-\nu}{1+\nu}\right) \tag{6.99}$$

(The term in brackets reflects the energy in the volume ΔV). If y is the distance of the solute atom from the slip plane, the parelastic interaction force is given by

$$F^p = -\frac{d\Delta E^p}{dx} = \frac{Gb\Delta V}{\pi y^2}\frac{2\left(\frac{x}{y}\right)}{\left(1+\left(\frac{x}{y}\right)^2\right)^2} \tag{6.100}$$

F^p is at maximum for $x = y/\sqrt{3}$, where $y = b/\sqrt{6}$ (half of the slip plane spacing in fcc crystals). The magnitude of ΔV can be determined from the change of the lattice parameter during alloying. For an (atomic) concentration change by dc^a, the change of the lattice parameter is

$$dc \cdot \Delta V = dc^a \cdot \frac{\Delta V}{\Omega} = \frac{1}{a^3}\left\{a^3\left(1+\frac{da}{a}\right)^3 - a^3\right\} \cong 3\frac{da}{a} \tag{6.101}$$

$$\Delta V = 3\Omega\delta \tag{6.102a}$$

$$\delta = \frac{d\ln a}{dc^a} \tag{6.102b}$$

Here $\Omega \approx b^3$ is the atomic volume. This yields the maximum parelastic interaction force

$$F^p_{\max} = Gb^2|\delta| \tag{6.103}$$

We note that the interaction assumes the existence of a hydrostatic stress field around the dislocation. Screw dislocations do not have a hydrostatic stress field and, therefore, do not show parelastic interaction with solute atoms. However, if the solute atom has a non-isotropic distortion, for instance like C in α-Fe, then screw dislocations also contribute to parelastic interaction.

(b) Dielastic interaction: The dielastic interaction is caused by the fact that the energy density of a dislocation is proportional to the shear modulus. If the solute atom has a different shear modulus, the volume of the solute atom contributes to the total energy of the dislocation differently from atoms of the pure solvent crystal. This interaction energy can be calculated for the screw dislocation to be

$$\Delta E^d = \frac{Gb^2}{8\pi^2 r^2}\Omega\eta \tag{6.104}$$

$$\eta = \frac{d\ln G}{dc^a} \tag{6.105}$$

From this we can calculate the maximum dielastic interaction force

$$F^d_{\max} \approx \frac{1}{20}Gb^2|\eta| \tag{6.106}$$

In comparison to the parelastic force, ΔE^d decreases more rapidly than ΔE^p with increasing r, however, $|\eta|$ is usually much larger than $|\delta|$.

(c) Chemical interaction: This effect, named after Suzuki, is caused by the dependence of the stacking fault energy on chemical composition in such a way that the stacking fault energy decreases with increasing solute concentration. A decreasing stacking fault energy engenders an increase of the dissociation width of dislocations, which reduces the total dislocation energy. Solute atoms, therefore, segregate to the dislocations to decrease the stacking fault energy. If the dislocation moves, it has to leave the segregated solute atoms behind, which changes the composition of the dislocation core and, therefore, increases the energy of the dislocation. This makes itself felt as a back stress on the dislocation. It can be computed in a similar way as the previous two cases and will not be treated here.

The increase of the critical resolved shear stress for plastic flow by solid solution hardening can be calculated from the total back-driving force

$$F_{\max} = F^p_{\max} + F^d_{\max} \tag{6.107}$$

This maximum force must be compensated by an increase $\Delta\tau_c$ of the critical resolved shear stress, to make dislocations move in a solid solution. If the average free dislocation segment length is ℓ_F, the Peach-Koehler equation states that

$$\Delta\tau_c \cdot b\ell_F = F_{\max} \tag{6.108}$$

As a first approximation, the free segment length ℓ_F can be associated with the average spacing of solute atoms in the slip plane. In a more sophisticated approach, a flexible dislocation can be considered to bow out between the obstacles. Because of this curvature which depends on the applied shear stress according to Eq. (6.76), the probability of meeting another solute atom increases. The average free dislocation length ℓ_F at a shear stress $\Delta\tau_c$ is according to Friedel (the Friedel length)

$$\ell_F = \sqrt[3]{\frac{6E_d}{\Delta\tau_c\, c_A \cdot b}} \tag{6.109}$$

Here E_d is the dislocation energy (Eq. (6.52)) and c_A the number of solute atoms per unit area in the slip plane. Inserting in Eq. (6.108) yields

$$\Delta\tau_c \cdot b = F_{\max}^{3/2}\sqrt{\frac{c_A}{6E_d}} \tag{6.110}$$

or, because

$$c^a = c_A \cdot b^2 \tag{6.111}$$

and $E_d \cong \frac{1}{2}Gb^2$

$$\frac{\Delta\tau_c}{G} = \frac{1}{\sqrt{3}}\left(\frac{F_{\max}}{Gb^2}\right)^{3/2}\sqrt{c^a} \tag{6.112a}$$

and combined with Eqs. (6.103), (6.106), and (6.107) with a constant β

$$\frac{\Delta\tau_c}{G} = (|\delta| + \beta|\eta|)^{3/2}\sqrt{c^a/3} \tag{6.112b}$$

As a result the critical resolved shear stress of a solid solution increases with the square root of the concentration. This has been experimentally confirmed in many metallic and nonmetallic systems (Fig. 6.61).

The increase in strength depends on the nature of the alloying atoms as obvious from Eq. (6.112b). The hardening of copper alloyed with Sn or In is much more effective than with the same amount of Zn or Ni (Fig. 6.62).

So far we have assumed that only the dislocations are mobile, while the solute atoms are immobile. If the solute atoms can also diffuse, they will segregate to the dislocations (Fig. 6.63). To enforce plastic deformation, the dislocations have to be removed from their solute cloud and can then move at a lower stress on their glide planes through material that has the average solute concentration. This is the reason for the yield phenomena for instance in carbon steels. The mobility of carbon atoms in iron at room temperature

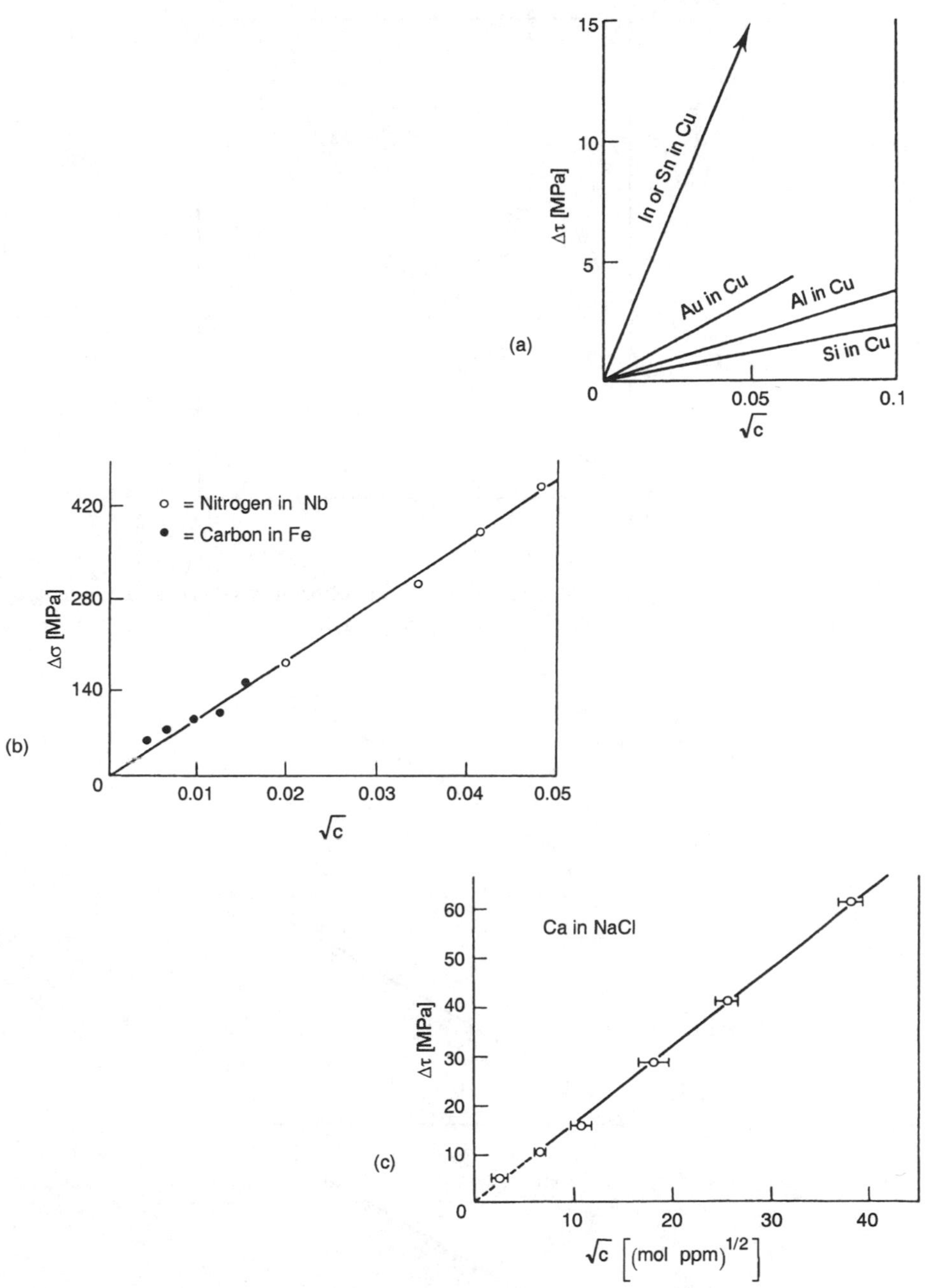

Fig. 6.61. The critical shear stress increases with the root of concentration (after [6.20]). (a) Substitutional solid solution; (b) interstitial solid solution; (c) doped ionic crystal.

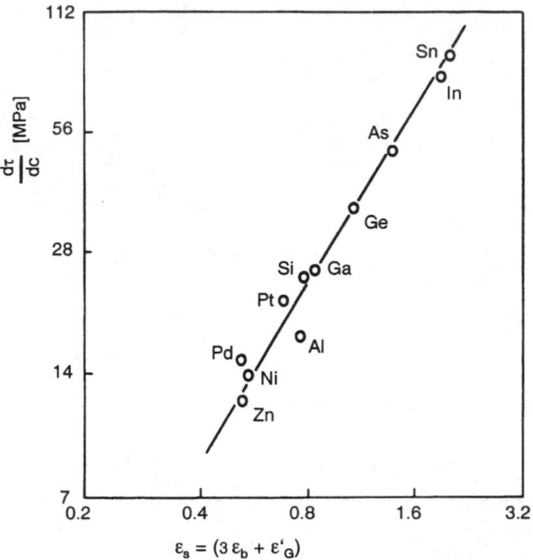

Fig. 6.62. The hardening effect in copper solid solution crystals is dependent on the alloying element ($\varepsilon_b = |\delta|$, $\varepsilon'_G = |\eta|$) (after [6.21]).

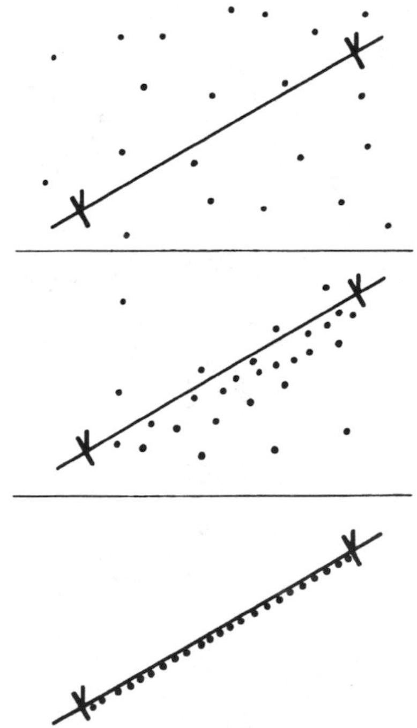

Fig. 6.63. Mobile solute atoms, e.g. C in α-Fe, segregate to the cores of dislocations.

is sufficiently large that they can diffuse to a stationary dislocation. During a short interruption of deformation, no yield point phenomena are observed (Fig. 6.64) since the solute atoms do not have enough time to segregate to the dislocations. Longer interruptions cause yield point phenomena to reappear, however, since segregation has time to occur.

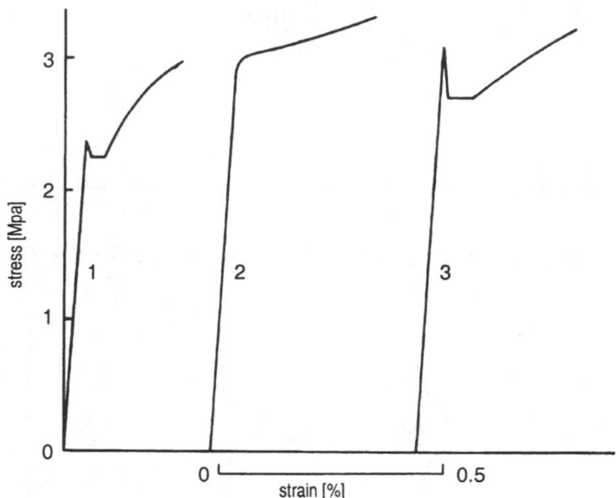

Fig. 6.64. Pronounced yield point in carbon steel (1). If the stress is relieved for a short time and then reapplied no pronounced yield point is observed (2). After a longer period of stress relief a pronounced point reappears (3) (after [6.22]).

At elevated temperatures the mobility of solute atoms can become large enough that they can follow a dislocation during its motion and segregate to its core while it has to wait in front of an obstacle. The repeated segregation and detachment processes cause the flow stress to oscillate. In such case serrations on the hardening curve are observed, which are referred to as dynamic strain aging or the Portevin - Le Chatelier effect (Fig. 6.65).

6.7.2 Dispersion Hardening

If a material contains nonmetallic particles, for instance oxides, carbides, or borides, the latter of which are frequently used for grain refinement during solidification, substantial strengthening can be obtained. The reason for this dispersion hardening is particle-dislocation interaction. The dislocations cannot cut or penetrate a hard particle so they have to circumvent the particle by bowing out between the particles (Fig. 6.66).

As shown for the Frank-Read source (Eqs.(6.76) (6.77)), there is a critical configuration when the shear stress attains a value

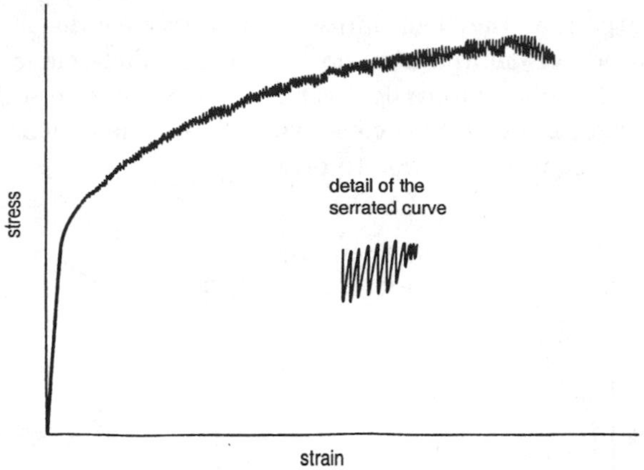

Fig. 6.65. Schematic hardening curve of a material exhibiting the Portevin-Le Chatelier effect.

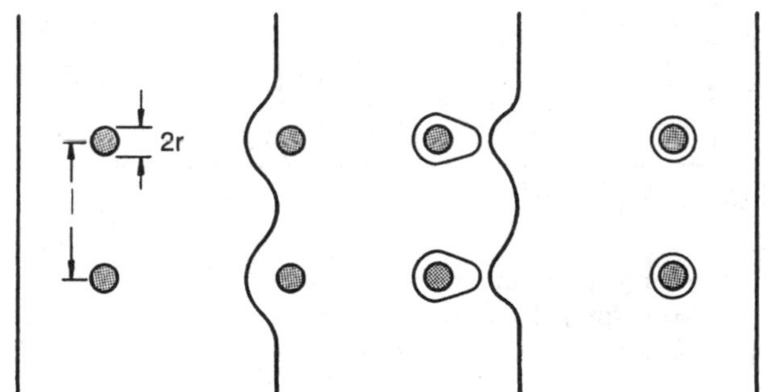

Fig. 6.66. Different stages of the Orowan mechanism for bypassing particles. A dislocation loop remains around the particles.

$$\tau = \frac{Gb}{\ell - 2r} \tag{6.113}$$

(Fig. 6.66), where $2r$ is the particle diameter and ℓ is the average particle spacing counted from particle center to particle center. Correspondingly, $\ell - 2r$ is the average free dislocation segment length. Any further motion of the dislocation will increase the radius of curvature which requires a lower stress for motion. Eventually, antiparallel dislocation segments will meet behind the particle and annihilate, which results in a dislocation loop around the particles and a free dislocation that can move on. This mechanism is referred to as the Orowan mechanism. Orowan loops have been observed in transmission electron micrographs (Fig. 6.67). The critical resolved shear stress of dispersion hardened alloys is given by Eq. (6.113). It increases with decreasing particle spacing ℓ. The average particle spacing is difficult to determine experimentally, but it is related to the volume fraction and the particle size by

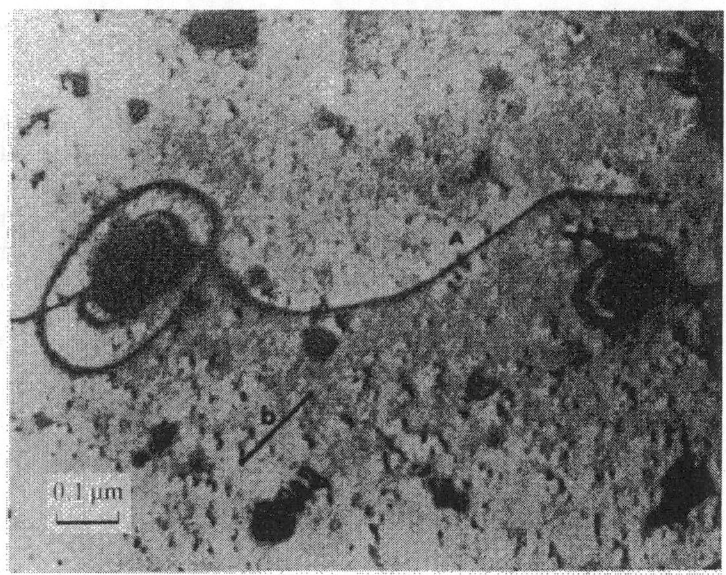

Fig. 6.67. Orowan loops around Al_2O_3 particles in Cu30%Zn [6.23].

$$\ell = \frac{r}{\sqrt{f}} \tag{6.114}$$

Eq. (6.114) can be understood from the fact that all particles which have a distance of less than the radius r from the slip plane will intersect the glide plane from above or below and, therefore, will be encountered by dislocations in the glide plane. If N is the number of particles per unit volume, then $N_E = 4rN$ is the number of particles per unit area in the slip plane, i.e. in a volume of thickness $4r$. For spherical particles

$$f = N \cdot \frac{4}{3}\pi r^3 \tag{6.115}$$

$$\ell = \frac{1}{\sqrt{N_E}} = \frac{1}{\sqrt{4rN}} = \frac{1}{\sqrt{\frac{4rf}{\frac{4}{3}\pi r^3}}} \tag{6.116}$$

which yields Eq. (6.114) with $\pi \approx 3$.

Since the particles are usually very small compared to their spacing (i.e., $r \ll \ell$), the Orowan stress is given by

$$\tau_{OR} \cong \frac{Gb\sqrt{f}}{r} \tag{6.117}$$

The flow stress in dispersion hardened alloys correspondingly strongly depends on the degree of dispersion f/r. It is most effective if the particles are very small.

The applied shear stress forces Orowan loops against the particle surface. On the other hand each loop exerts a back stress on the dislocation behind it. A second dislocation to pass a particle needs a higher stress than the first one. Therefore, the flow stress of dispersion hardened alloys increases strongly during deformation. This is indeed confirmed experimentally (Fig. 6.68). Finally, the stress on the dislocation loops becomes so large that they change their configuration by forming prismatic dislocation loops, or by generating dislocations on other slip systems to limit the stress field. In conclusion, besides their strong contribution to the yield stress particles also lead to strong strain hardening, which is much larger than strain hardening in pure metals or solid solutions (Fig. 6.68).

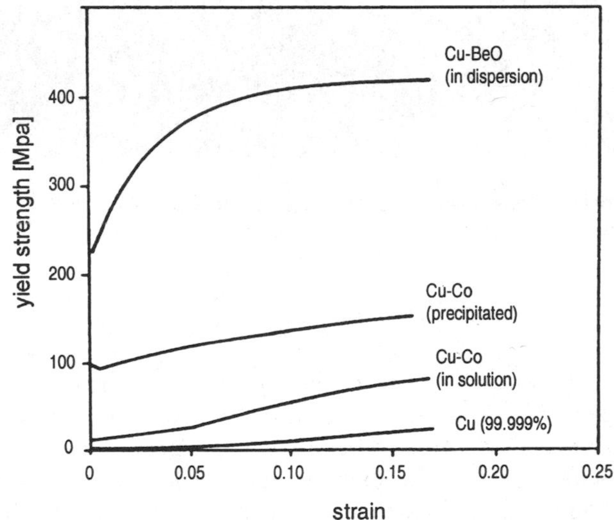

Fig. 6.68. Hardening curves of high purity copper and copper alloys. In CuCo particles are shearable, in BeO they are not (after [6.24]).

6.7.3 Precipitation Hardening

Precipitates which form during cooling of a homogeneous solid solution into the two-phase regime contribute to the strength of a material. They share a phase boundary with the matrix. In Chapter 3 we have learned to discriminate three types of phase boundaries, namely coherent, partially coherent, and incoherent phase boundaries. Incoherent phase boundaries act on dislocations like grain boundaries, they are unsurmountable obstacles. Dislocations can only circumvent incoherent precipitates by the Orowan mechanism. Therefore, incoherent precipitates have the same effect on the strength as particles. However, precipitates which are formed at high temperatures are usually very large and, therefore, contribute only very little to strength according to Eq. (6.117).

If a solid solution is quenched and subsequently tempered at lower temperatures, metastable phases can form with coherent or partially coherent phase boundaries (see Chapter 9). In this case all crystallographic planes and directions in the matrix continue into the precipitate with only a slight distortion. Dislocations are capable of cutting through such a particle. However, the precipitate will exert forces on the dislocation which have to be overcome. To begin with there will be parelastic and dielastic interactions as in solid solution hardening. However, the parelastic interaction increases with increasing size of the precipitates

$$F^p_{\max} \cong Gb|\delta|r \tag{6.118}$$

$$F^d_{\max} \cong Gb^2|\eta| \tag{6.119}$$

If a dislocation cuts through a coherent precipitate the precipitate will be sheared because the top part above the slip plane will be displaced by a Burgers vector with respect to the bottom part (Fig. 6.69). This causes the creation of additional phase boundary area, the energy of which has to be expended by the applied stress during the cutting process. The respective force on the dislocation - apart from a geometry factor - is related to the interface energy γ_p by

$$F^c = \gamma_p \cdot r \tag{6.120}$$

If the particle is long range ordered, such as the well-known $\gamma\prime$ phase Ni_3Al in a superalloy, the long range order will be destroyed along the slip plane and an antiphase boundary is generated. The energy of this antiphase boundary γ_{APB} has to be expended (Fig. 6.70)

$$F^{\mathrm{APB}} = \gamma_{\mathrm{APB}} \cdot r \tag{6.121}$$

Commonly, the precipitate has a stacking fault energy γ^P_{SF} different from that of the matrix γ^M_{SF}. Therefore, the width of the dislocation core in the precipitate is different from that in the matrix (Fig. 6.71). If $\gamma^P_{SF} < \gamma^M_{SF}$, the dissociation width in the particle is larger than in the matrix. When particle and dislocation are separated this additional dissociation has to be removed which requires a force

$$F^{\mathrm{SF}} = 2 \cdot \left(\gamma^P_{\mathrm{SF}} - \gamma^M_{\mathrm{SF}}\right) r \tag{6.122}$$

Conversely, if $\gamma^P_{SF} > \gamma^M_{SF}$, this force has to be exerted to move the dislocation into the particle, i.e. an obstacle of equal strength for dislocation motion. The magnitude of these various interaction forces and their significance for the total strength depends on the alloying elements.

In superalloys, the lattice parameter and shear modulus of Ni and Ni_3Al

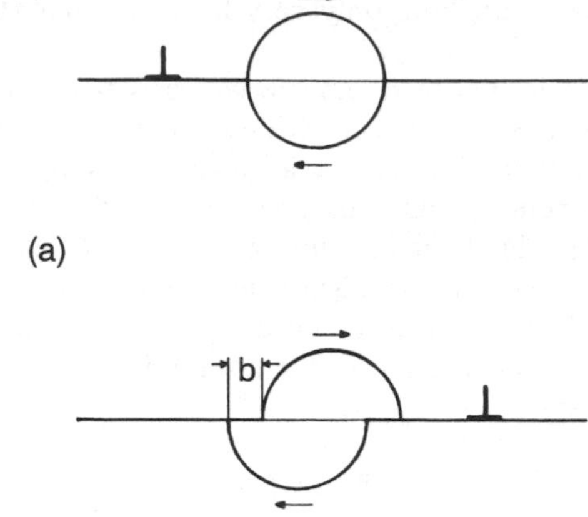

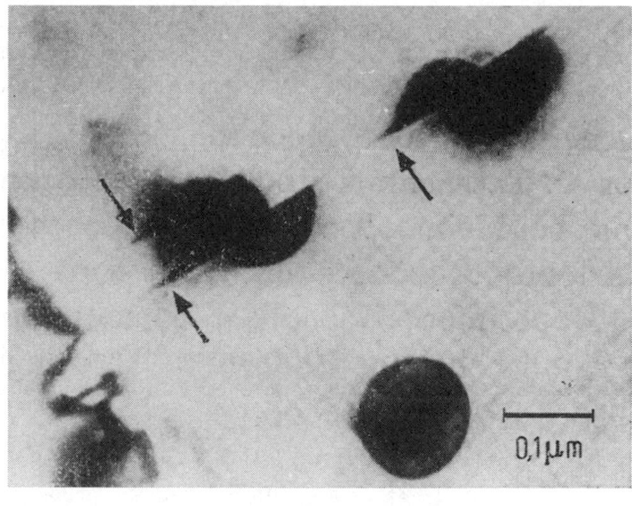

Fig. 6.69. If a dislocation cuts a particle, the particle shears off. (a) Schematic; (b) observed in Ni19%Cr6%Al (aged 540h at 750°C and deformed 2% [6.25]).

are very similar, but the energy of the antiphase boundary on {111} planes is very high. For disordered particles in decomposition zones, F^{APB} does not play a role. All the forces considered, except for the dielastic contribution, increase in proportion to the size of the precipitate, and the proportionality constant is always an interface energy. For sake of completeness we introduce Eq. (6.123) for the parelastic force

$$Gb|\delta| = \gamma_b \tag{6.123}$$

The sum of all forces $F_{\max}$ then reads

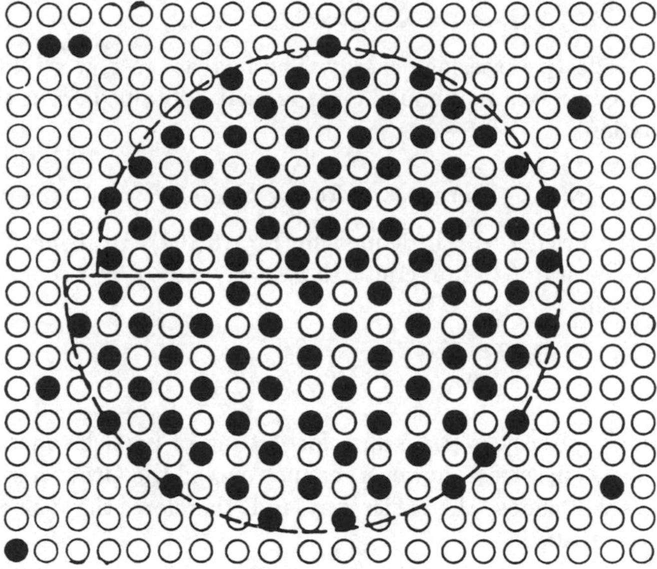

Fig. 6.70. Intersecting an ordered particle produces an antiphase boundary.

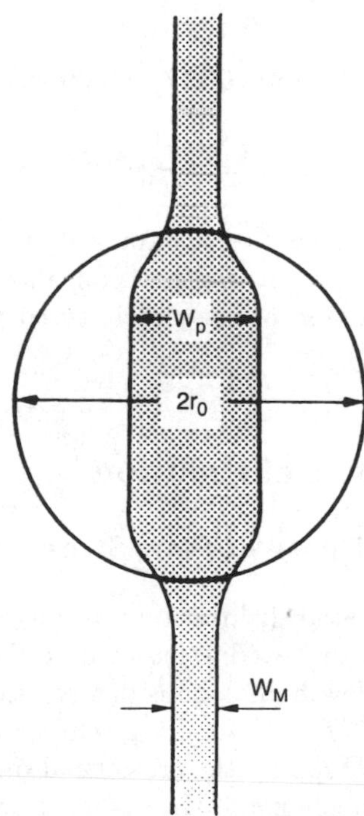

Fig. 6.71. Change in dissociation width in a particle of different stacking fault energy.

$$F_{\max} = \tilde{\gamma} \cdot r \tag{6.124}$$

where $\tilde{\gamma}$ is an effective interface energy. The increase in flow stress $\Delta\tau_c$ necessary to overcome this maximum force is again obtained from the Peach-Koehler equation

$$\Delta\tau_c \cdot b\ell_F\left(\Delta\tau_c\right) = F_{\max}$$

where ℓ_F is the Friedel length defined by Eq. (6.109). Utilizing Eq. (6.124) and the area density of the precipitates $c_F = 1/\ell^2 = f/r^2$ according to Eq. (6.114), results in

$$\Delta\tau_c \cdot b \cong \tilde{\gamma}^{3/2}\sqrt{f} \cdot \frac{\sqrt{r}}{\sqrt{6E_v}} \tag{6.125}$$

Correspondingly, the shear stress $\Delta\tau_c$ for cutting of particles increases with $\sqrt{r}$ (Fig. 6.72a). This stress, however, cannot exceed the Orowan stress, since nature always will choose the easier way and, therefore, the dislocation will circumvent the obstacle.

Therefore, there is a precipitate size r_0 which yields the maximum strength if

$$\Delta\tau_c = \tau_{OR} \tag{6.126}$$

A comparison of Eqs. (6.125) and (6.117) combined with Eq. (6.52) yields

$$r_0 = \frac{Gb^2}{\tilde{\gamma}}\sqrt[3]{3} \tag{6.127}$$

It is the aim of age hardening (see Chapter 9) to obtain the particle size r_0 that imparts the maximum strength to an alloy. Note that r_0 does not depend on the volume fraction of the precipitated phase. This is confirmed experimentally (Fig. 6.72b).

6.8 Time Dependent Deformation

6.8.1 Strain Rate Sensitivity of Flow Stress: Superplasticity

Except for thermally activated dislocation motion, we have so far tacitly assumed that deformation can be completely described by stress and strain. This is essentially true at low homologous deformation temperatures (homologous temperature $T^* = T/T_m$, T_m-melting temperature). For instance, if we interrupt a tensile test by stopping the cross head motion of the machine, the stress remains essentially unchanged. (In reality it decreases slightly, which is referred to as stress relaxation (see Section 6.8.3).) If the specimen is unloaded for a short time, yielding during reloading will occur at the same stress as before unloading. Therefore, each point on the stress-strain curve corresponds

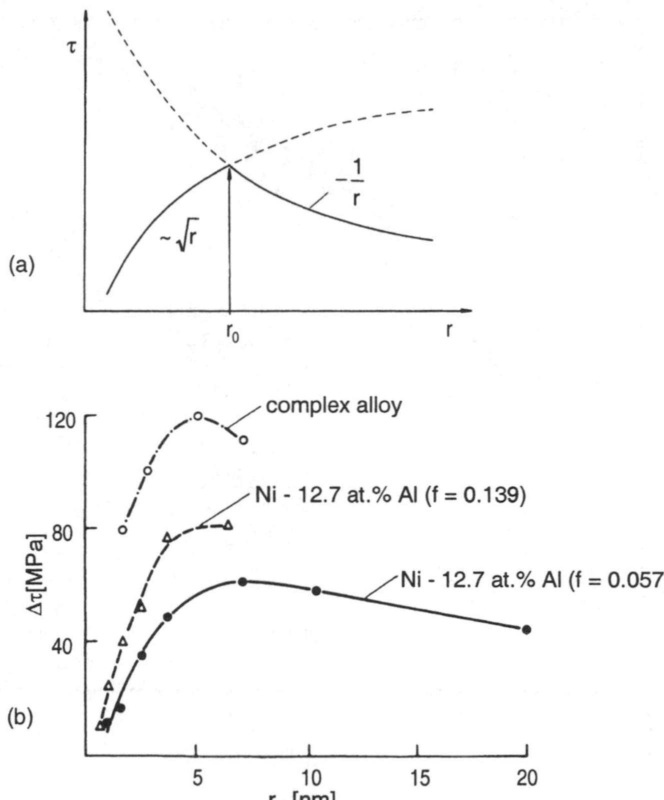

Fig. 6.72. Dependence of strengthening on particle size. (a) Schematic; (b) observed in a Ni-Al alloy (after [6.26]).

to the flow stress of the instantaneous state of deformation. Obviously, time dependent processes do not play an essential role at low deformation temperatures. This fact is reflected in the dependency of the flow stress σ on the strain rate $\dot{\varepsilon}$ which is expressed in terms of the strain rate sensitivity m

$$m = \frac{d\ ln\ \sigma}{d\ ln\ \dot{\varepsilon}} \tag{6.128}$$

At low temperatures in fcc metals $m \approx 1/100$, i.e. the flow stress is practically independent of the strain rate. The choice of Eq. (6.128) to define m results from the fact that traditionally the dependency $\sigma(\dot{\varepsilon})$ has been expressed by a power law

$$\sigma = K\dot{\varepsilon}^m \tag{6.129}$$

where $K = K(\varepsilon)$.

When the deformation temperature is increased to $T^* \geq 0.5$, m increases

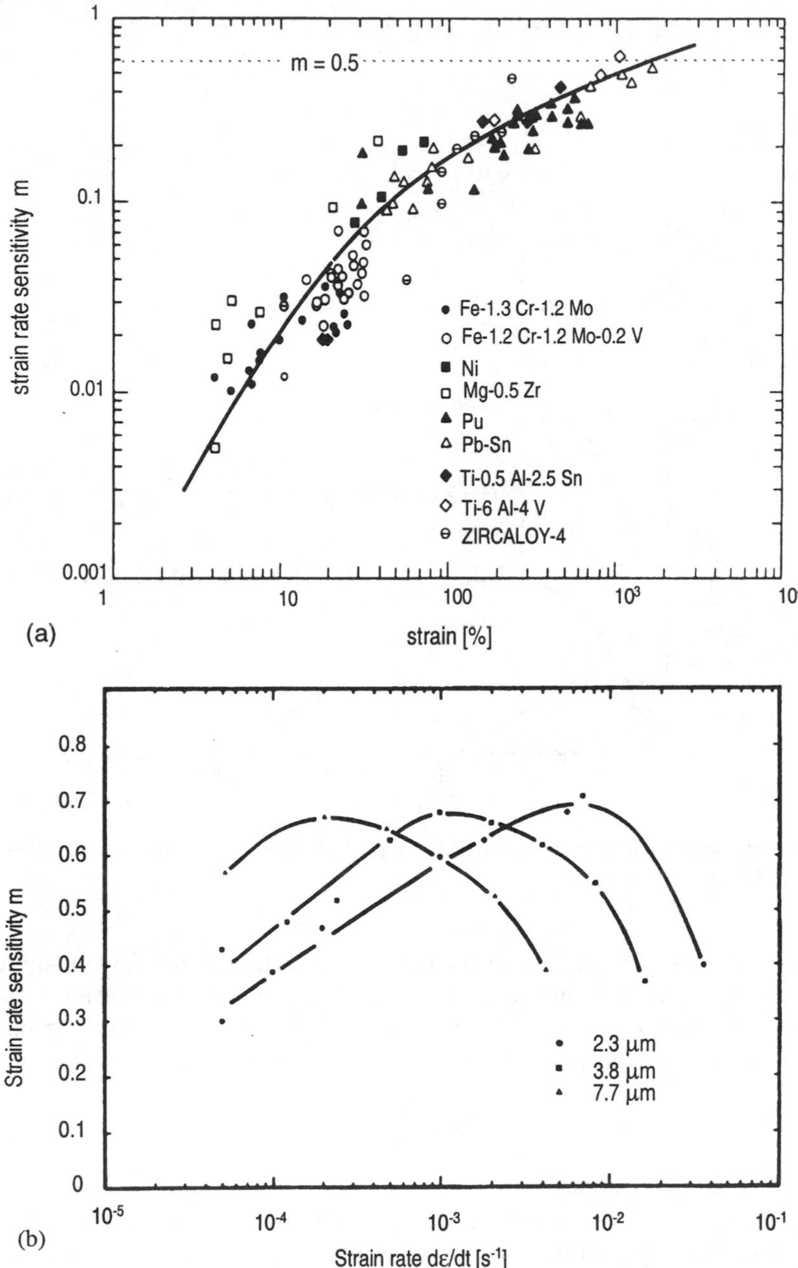

Fig. 6.73. (a) Elongation achieved in a tensile test as a function of strain rate sensitivity m [6.27]. (b) Dependence $m(\dot{\varepsilon})$ on grain size [6.28].

and typically attains values of about 0.2. In very fine grained materials $m \gtrsim 0.3$ is observed (Fig. 6.73) in specific regimes of the strain rate, typically for $\dot{\varepsilon} \approx 10^{-3}$/s. Such a high strain rate sensitivity is accompanied by an increase of ductility in a tensile test, and under these circumstances, strains to fracture of the order of 1000% or higher can be obtained. This phenomenon is referred to as superplasticity. The world record is currently about 8000% (Fig. 6.74b). The reason for the high strain to fracture is the high strain rate sensitivity. At low temperatures deformation becomes unstable at the Considère criterion, because physical hardening can no longer compensate for geometrical softening (see Section 6.2). If a local constriction of the cross section occurs, deformation becomes concentrated at that location, which drastically increases the strain rate at that location. If m is large, there will also be a consequent increase of flow stress according to Eq. (6.129). As a result, further deformation in the region of the constriction will be suppressed until the specimen has reestablished a uniform cross section. Consequently, any constriction is suppressed and very large strains are obtained in a tensile test. Superplastic behavior is principally defined by a large strain to fracture in a continuous tensile test. While Eq. (6.129) gives a phenomenological explanation for superplasticity, the physical reason for this phenomenon is the very fine microstructure and concurrently the high deformation temperature. This causes deformation to be accomplished by processes in the grain boundary, i.e. grain boundary sliding or grain boundary diffusion, while dislocation motion in the crystals is rendered unimportant. There is practically no dislocation storage in the crystallites and, therefore, negligible strain hardening during superplastic deformation (Fig. 6.74a).

A fine grained microstructure is the main prerequisite for the occurrence of superplasticity. The grain size should be less than 10 μm. Since in pure metals grain growth occurs at elevated temperatures and becomes faster with decreasing grain size, superplasticity practically does not occur in pure metals. Typically, superplastic materials are two-phase alloys, frequently at the eutectic composition. Superplasticity has also been observed, however, in single-phase materials with low grain boundary mobility, for instance in ordered alloys like Ni_3Al or in fine-grained ceramics, for instance in ZrO_2 doped with $3\%Y_2O_3$. The strains to fracture in ceramic superplastic materials are, however, substantially smaller than in metallic materials.

6.8.2 Creep

In contrast to low temperature deformation, materials are typically subjected to a continuous deformation at constant load, or constant stress, at elevated temperatures. This phenomenon is termed creep. A tensile test at constant load is also called a static tensile test, in contrast to the dynamic tensile test in a testing machine at constant strain rate. A typical creep curve $\varepsilon(t)$ (for $\sigma = \text{const.}$) comprises three stages (Fig. 6.75). Immediately upon loading, an instantaneous strain ε_0 is attained very rapidly. This is followed by stage I, or

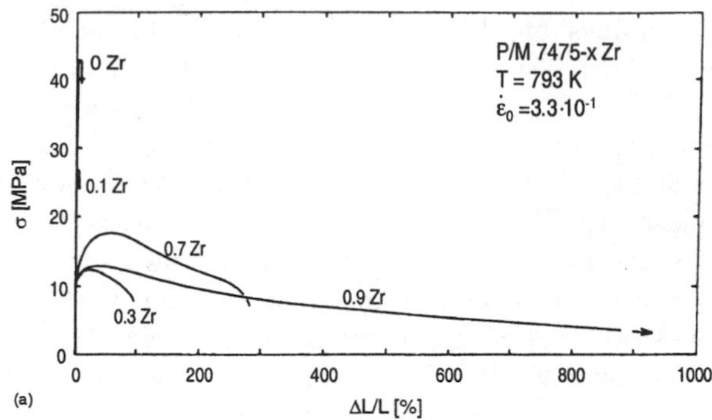

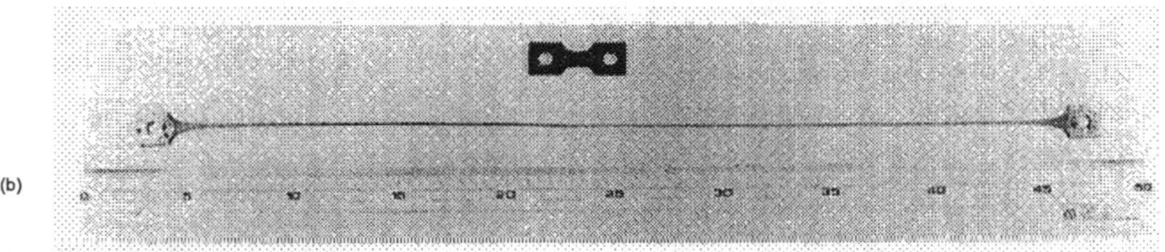

Fig. 6.74. (a) Stress-strain curve of a superplastic alloy (after [6.29]); (b) undeformed and deformed specimen of superplastic aluminum bronze (elongation approx. 8000% !) [6.30].

primary creep. During primary creep the creep rate continuously decreases. Eventually, the creep rate will become constant which defines the state of stationary creep or stage II of the creep curve, i.e., the strain increases linearly with time. Eventually, the creep rate increases again in the range of tertiary creep, stage III, until creep fracture occurs. The stationary creep rate strongly depends on deformation conditions, i.e., on the applied stress, on deformation temperature and on material, in particular on the diffusion coefficient and the stacking fault energy. In a phenomenological approach, the steady state creep rate can be expressed as a function of stress and temperature according to

$$\dot{\varepsilon}_s = A\left(\frac{\sigma}{G}\right)^n e^{\left(-\frac{Q}{kT}\right)} \tag{6.130}$$

The mechanism of creep is associated with self diffusion, because the activation energy for creep is found to be the same as the activation energy for self-diffusion. However, the vacancies that move during diffusion do not directly cause the shape change but rather they assist the edge dislocations to overcome obstacles on their slip planes (dislocation creep). If vacancies become attached to the dislocation core, they remove atoms from the core, and the dislocation climbs perpendicular to the glide plane on to the next

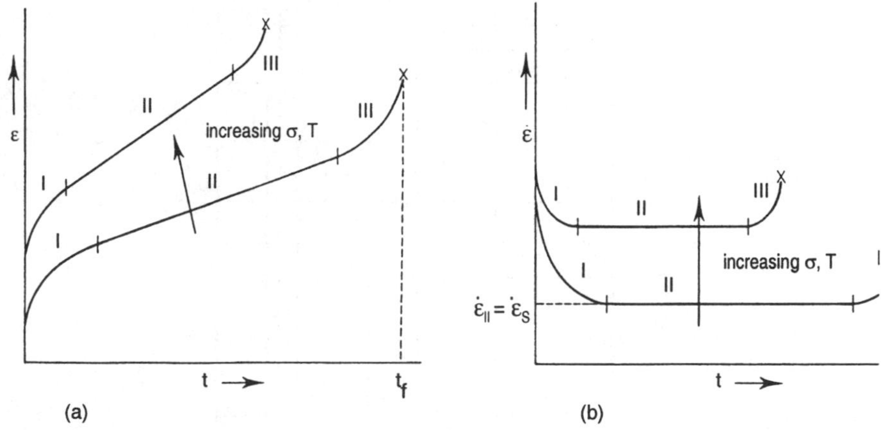

Fig. 6.75. (a) Schematic creep curve $\varepsilon(t)$. (b) Schematic diagram of the creep rate as a function of time.

neighbor glide plane (Fig. 6.78). Note that many vacancies are required to make a dislocation line climb because a single vacancy will only cause climb of a dislocation segment of length b, and to make a dislocation move on the adjacent glide plane, the climbed dislocation segment has to exceed a critical free length.

At high temperatures, the vacancy concentration in thermal equilibrium is sufficiently large that the diffusion controlled climb process can proceed continuously. In a simple model one can consider dislocation motion to be like the motion of a rod in a viscous medium, for instance in honey. The dislocation velocity v_D is then a drift velocity, which according to Eq. (5.9) under a force F, is given by

$$v_D = BF = \frac{D}{kT}\tau b \tag{6.131}$$

The strain rate is related to the velocity by Eq. (6.31a) $\dot{\gamma} = \rho b v$. And since $\rho \sim \tau^2$ according to Eq. (6.59) we obtain

$$\dot{\gamma} = \frac{\tau^2}{\alpha^2 G^2 b^2}\frac{Db^2}{kT}\tau = A_0 G\left(\frac{\tau}{G}\right)^3 \frac{D_0}{kT}\, e^{\left(-\frac{Q_{SD}}{kT}\right)} \tag{6.132}$$

or

$$\dot{\varepsilon} = A\left(\frac{\sigma}{G}\right)^3 e^{\left(-\frac{Q_{SD}}{kT}\right)} \tag{6.133}$$

Eq. (6.133) is very similar to the phenomenological equation (6.130), however, the stress exponent $n = 3$ in contrast to $n \approx 5$ and higher as experimentally observed. A theoretical justfication of $n > 3$ requires special assumptions on details of the mechanism of dislocation motion which also is affected by the width of the dislocation core. There is still no generally accepted theory to

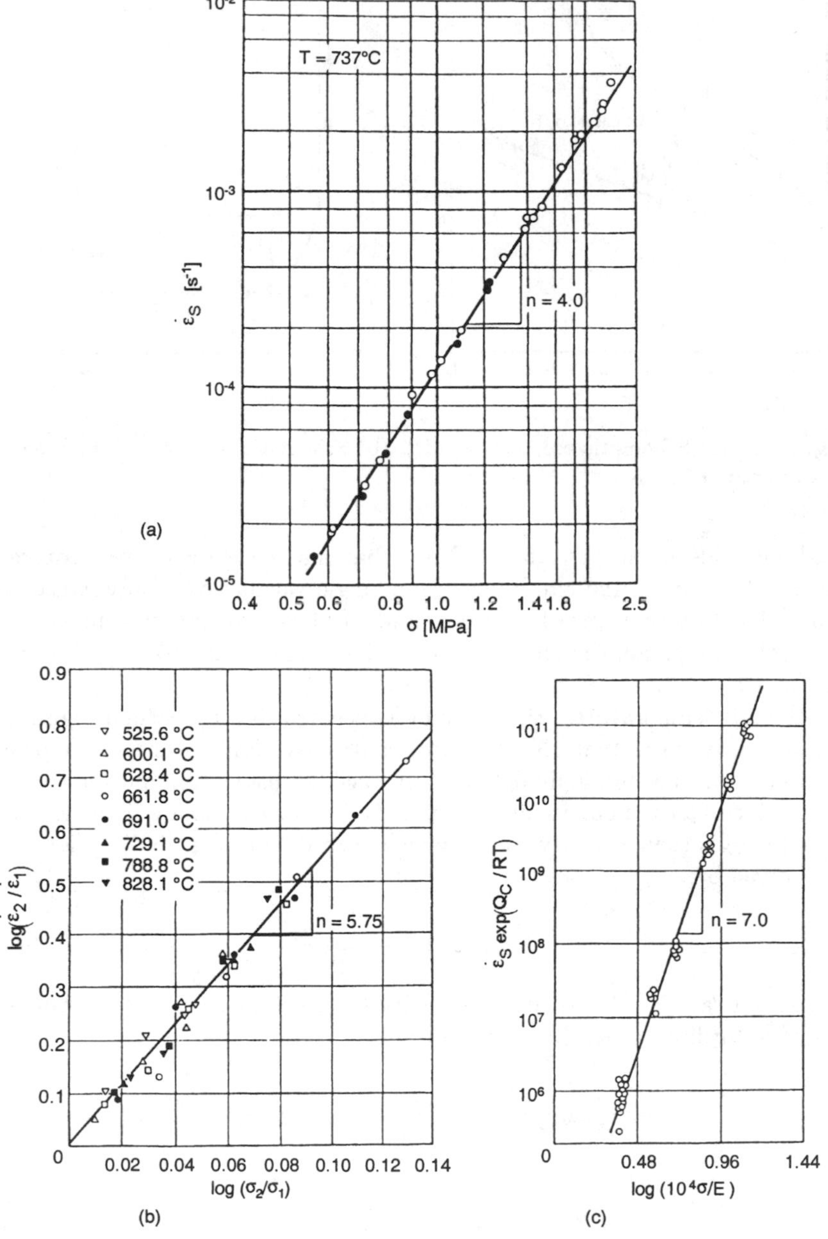

Fig. 6.76. Stress dependence of the stationary creep rate (n - stress exponent) [6.31]. (a) NaCl single crystal at 737°C; (b) FeSi solid solution, derived from load changes at different temperatures; (c) pure polycrystalline Ni at different temperatures.

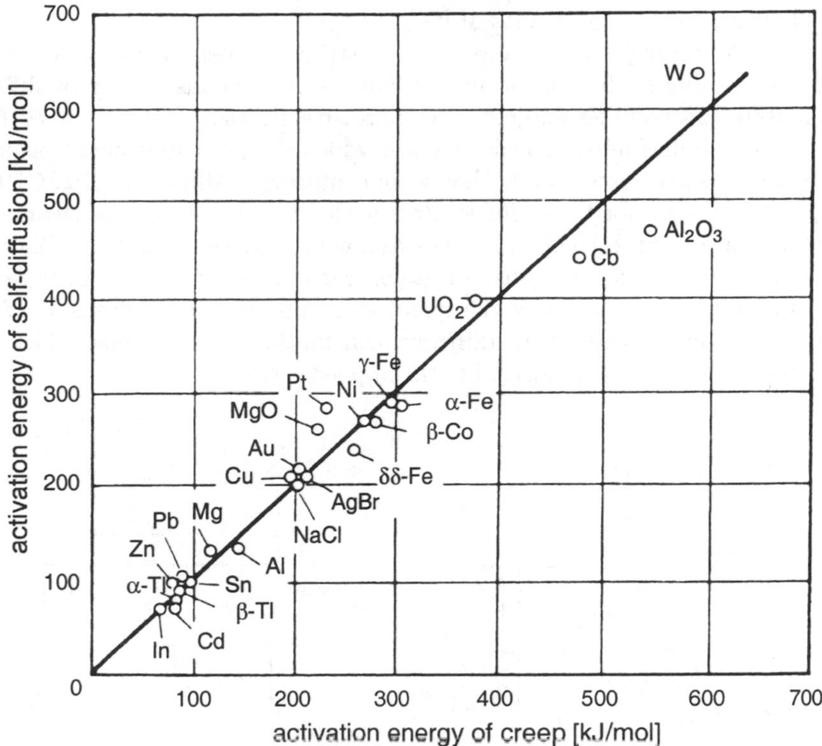

Fig. 6.77. Correlation of the activation energies of self-diffusion and stationary creep [6.32].

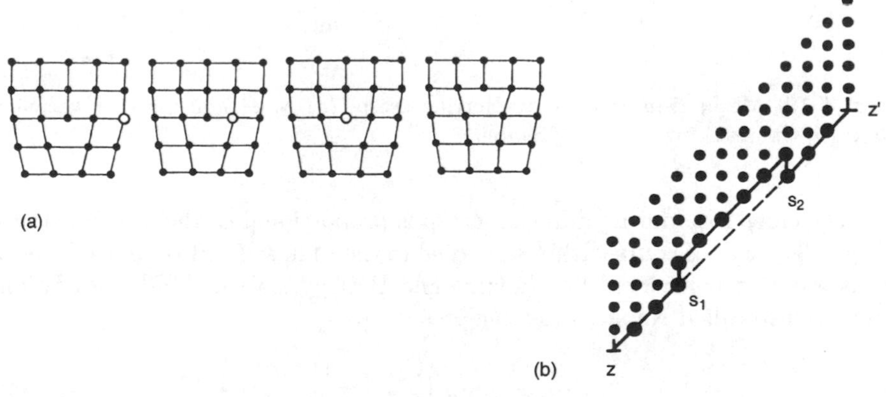

Fig. 6.78. (a) Mechanism of climb of an edge dislocation by absorption of vacancies. (b) Climb of a dislocation line requires the addition of many vacancies.

naturally account for high values of n.

At very high temperatures and very low stresses, creep phenomena occur which are not due to dislocation motion but which are caused only by diffusion. Material in volumes subject to compressive stresses, will be moved by diffusion to volumes under tensile stresses, which causes a specimen to elongate in the tensile direction or to shorten in compression direction (Fig. 6.79). The physical reason for this diffusion flux is the dependency of the chemical potential of atoms on the elastic stress state, which causes a flux from dilated to compressed volumes. The flux of atoms corresponds to a flux of vacancies in opposite direction. Since deformation is only carried by diffusion, the strain rate is determined by the diffusive drift motion of the atoms, which is proportional to the driving force, i.e. the applied stress.

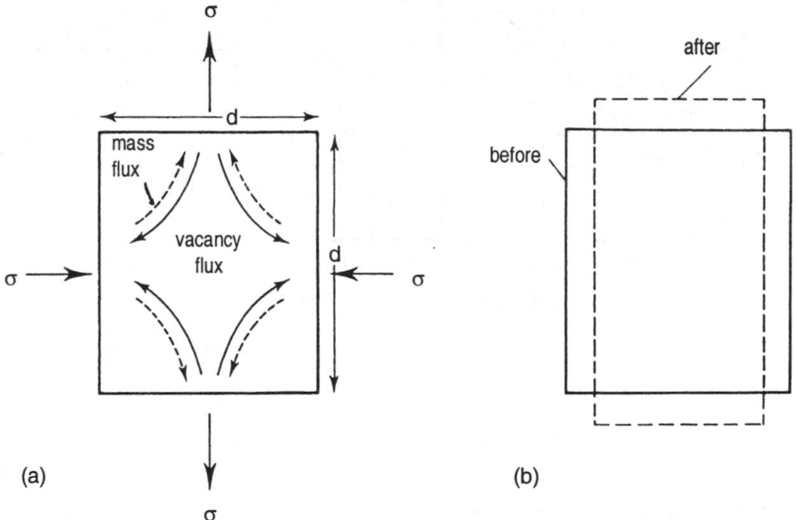

Fig. 6.79. Mechanism of Nabarro-Herring creep. (a) Mass transport or vacancy flux; (b) the resulting change of shape.

The creep rate during diffusion creep is proportional to the applied stress (Fig. 6.80) and, of course, is proportional to the bulk self-diffusion coefficient. This was first considered by Nabarro and Herring, so this kind of diffusion creep is also called Nabarro-Herring creep

$$\dot{\varepsilon}_{NH} = A_{NH} \left(\frac{D}{kT} \right) \sigma \frac{\Omega}{d^2} \tag{6.134}$$

Here $\Omega \approx b^3$ is the atomic volume and d is the grain size. A_{NH} is a constant.

The transport of material does not necessarily have to proceed through the volume. In particular, at not so high temperatures and in fine-grained materials, the rate of atomic transport through grain boundaries can exceed volume

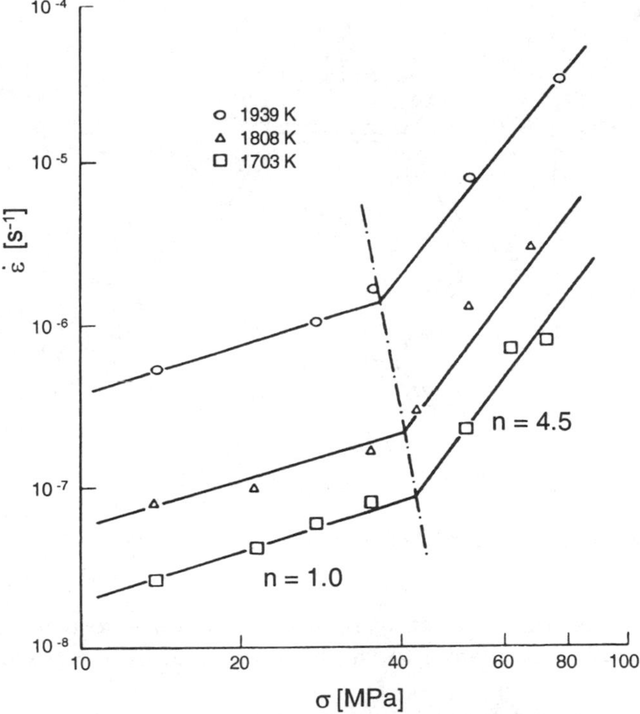

Fig. 6.80. Stress exponent of creep of UO_2 polycrystals with a grain size of 10 μm. Under small stress and high temperature diffusion creep with $n = 1$ is the dominant mechanism (after [6.33]).

diffusion and dominate diffusion creep (Fig. 6.81). This case of diffusion creep is referred to as Coble creep, and we obtain

$$\dot{\varepsilon}_C = A_C \left(\frac{D_{GB}\delta}{kT} \right) \sigma \frac{\Omega}{d^3} \tag{6.135}$$

where δ is the thickness of a grain boundary and D_{GB} is the coefficient of grain boundary diffusion. Both mechanisms of diffusion occur concurrently in polycrystals and contribute to the macroscopic creep rate. Therefore, both are usually subsumed under the term diffusion creep and the total creep rate reads

$$\dot{\varepsilon}_D = \dot{\varepsilon}_{NH} + \dot{\varepsilon}_C = A_D \frac{\sigma \Omega D}{d^2 kT} \left(1 + \frac{D_{GB}\delta}{d \cdot D} \right) \tag{6.136}$$

Diffusion creep is pronounced if dislocation creep is suppressed. This is typically the case in ceramic materials and less pronounced in metals. The dependency of the creep rate on the grain size demonstrates that not all strengthening mechanisms at low temperatures are also mechanisms for improved creep

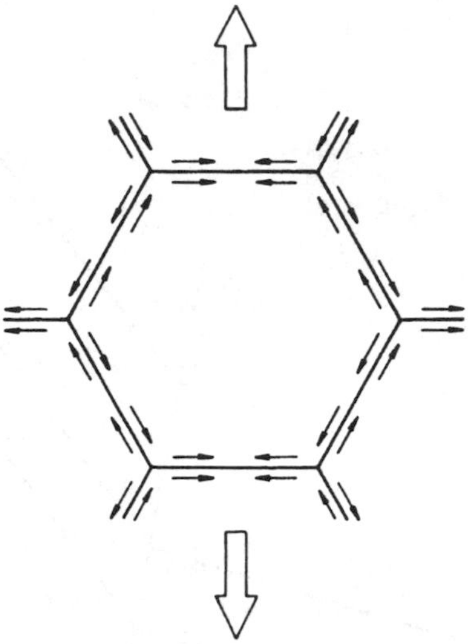

Fig. 6.81. Material flow due to Coble creep.

resistance. Strengthening by grain refinement at low deformation temperatures is obviously detrimental to the high temperature creep strength of a material.

The multitude of deformation mechanisms and their different dependencies on external deformation conditions (temperature, strain rate) and material properties (shear modulus, diffusion constant, grain size) engenders a confusing scenario for the prediction of material behavior. Materials engineers, however, need to know and to predict how a material behaves under service conditions, in order to select the right material and the optimum design. For this purpose the so-called deformation mechanism maps are helpful, which were invented by Ashby and co-workers. In these maps, fields are indicated as a function of stress and temperature where specific deformation mechanisms dominate (Fig. 6.82). Once the field is located that represents the service conditions, the respective equation of state derived for the mechanism in this field can be used to predict material behavior for intelligent processing and materials design.

6.8.3 Anelasticity and Viscoelasticity

At temperatures far below half of the melting temperature, time-dependent deformation in the elastic regime can also be observed, i.e. at stresses far below the yield stress. The total strain generally comprises a time-independent and a time-dependent contribution. The time-dependent component is usually very small and special experimental methods are required for its measurement. If the additional time-dependent deformation is removed on unloading, so that the shape of the specimen before and after unloading is the same, this is called

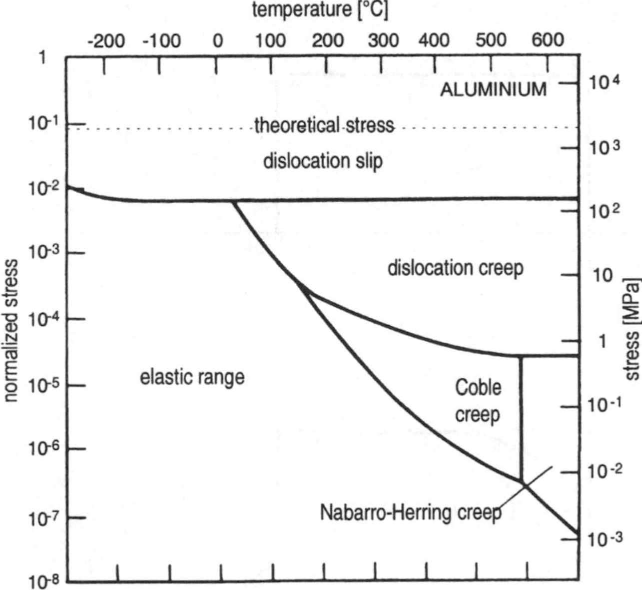

Fig. 6.82. The deformation mechanism map of aluminum (after [6.34]).

anelastic behavior, or anelasticity. If an anelastic body is loaded for a long time with a stress σ_0 (Fig. 6.83), where σ_0 is of course smaller than the yield stress, a spontaneous purely elastic strain ε_1 is instantaneously established, which is followed, however, by a time-dependent (anelastic) strain $\varepsilon_2(t)$, which approaches a maximum of ε_{20}. The anelastic part depends exponentially on time, hence

$$\varepsilon(t) = \varepsilon_1 + \varepsilon_{20}\left\{1 - e^{(-t/\tau)}\right\} \tag{6.137}$$

The time τ is referred to as the relaxation time. It is a measure for the time that has to pass until the anelastic steady state (the strain no longer changes with time) is attained. The relaxation time τ can be determined graphically as the point of intersection of the tangent of the curve $\varepsilon(t)$ for $t = 0$ with the maximum total strain $\varepsilon_1 + \varepsilon_{20}$ as sketched in Fig. (6.83). This is because for small times t

$$\varepsilon(t) - \varepsilon_1 = \varepsilon_{20}\left\{1 - e^{\left(-\frac{t}{\tau}\right)}\right\} \cong \varepsilon_{20} \cdot \frac{t}{\tau} \tag{6.138}$$

On unloading the strain is reversed. A spontaneous decrease by the instantaneous elastic strain $-\varepsilon_1$ is followed by a time-dependent strain $-\varepsilon_{20}\exp(-t/\tau)$, hence, after long times the initial state $\varepsilon = 0$ is restored. In contrast to plastic deformation there is no permanent shape change in anelasticity.

An important example of anelastic behavior is the Snoek effect in plain

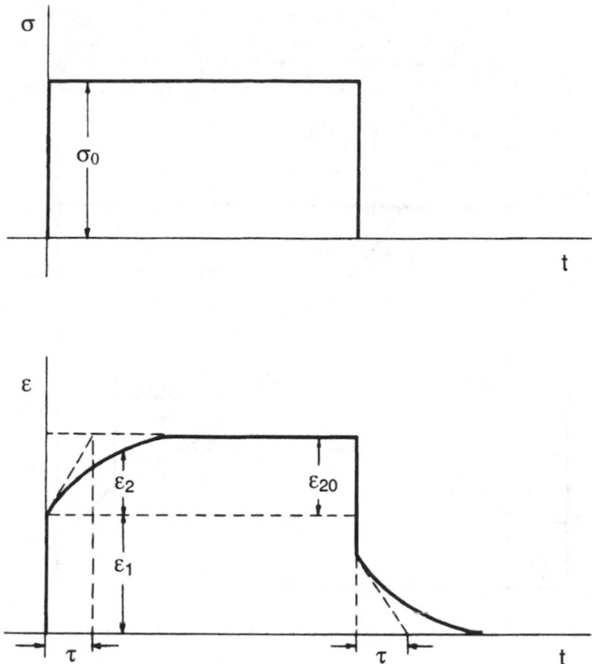

Fig. 6.83. Diagrammatic illustration of the elastic after-effect. Application of a constant stress σ_0 spontaneously generates an elastic stress ε_1 and additionally, over time an elastic after-effect ε_2, that tends to a limiting value ε_{20} with time constant τ.

carbon steel. The carbon atoms in α-Fe are positioned on the octahedral interstitial sites of the bcc lattice. At zero applied stress the carbon atoms are uniformly distributed over the x, y, and z sites of the bcc lattice (Fig. 6.84). Since the carbon atoms are larger than the size of the interstice, they cause an anisotropic elastic distortion of the lattice. If an elastic tensile stress is applied in z direction, the lattice will be slightly dilated in z direction and compressed in the xy plane due to Poisson contraction. Consequently, the z sites become energetically more favorable for the carbon atoms because the elastic distortion around a carbon atom is smaller on z sites than on x or y sites. However, if there are more z atoms on z sites, the lattice constant in z direction is larger than in the x or y direction. The reorientation from x and y sites to z sites occurs by diffusion. In the course of time a new equilibrium distribution of C-atoms on x, y, and z sites will be established such that on average more z sides are occupied. As the atoms re-arrange themselves, the specimen dimension parallel to the direction of the applied stress also changes, i.e. there is an associated time-dependent strain. After unloading, the uniform distribution over x, y, and z sites will be re-established in the course of time and the anelastic strain will reverse itself.

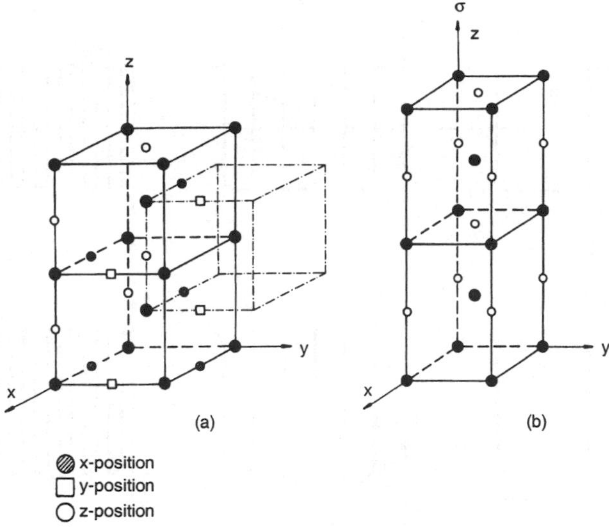

Fig. 6.84. Schematic representation of the distribution of C atoms in a bcc lattice in an unstressed state (a) and under tensile stress (b), illustrating the Snoek effect.

It is experimentally very difficult to measure this anelastic length change with sufficient accuracy, because the strains are extremely small. So instead, experimental methods utilize the attenuation of elastic vibrations. Instead of a constant stress, an elastic alternating stress is applied to the specimen (Fig. 6.85), so how much of the anelastic strain can develop depends on the frequency, or, equivalently, the time for each cycle. At very low frequencies the full anelastic strain will be attained for each cycle. With increasing frequency, the anelastic strain per cycle decreases until at very high frequencies, practically no measurable anelastic strain will appear. Since the Young's modulus is defined as

$$E = \frac{\sigma}{\varepsilon}$$

but $\varepsilon = \varepsilon(t)$ in a dynamic (vibration) experiment, the dynamic Young's modulus depends on frequency (Fig. 6.86). At very low frequencies the static or relaxed modulus (Fig. 6.86b)

$$E_r = \frac{\sigma_0}{\varepsilon_1 + \varepsilon_{20}} \tag{6.139a}$$

is obtained whereas at very high frequencies the unrelaxed modulus (Fig. 6.86a)

$$E_u = \frac{\sigma_0}{\varepsilon_1} \tag{6.139b}$$

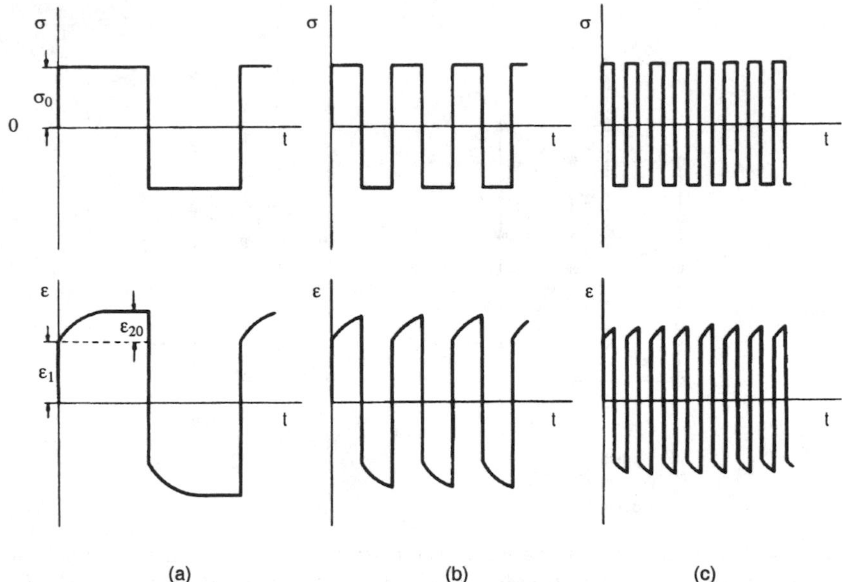

Fig. 6.85. Strain change due to a square shape alternating stress, for three different frequencies. (a) Low frequency: the full after-effect can develop; (b) medium frequency: the stress lasts exactly as long as the adaptation time τ; (c) very high frequency: the elastic after-effect can barely develop.

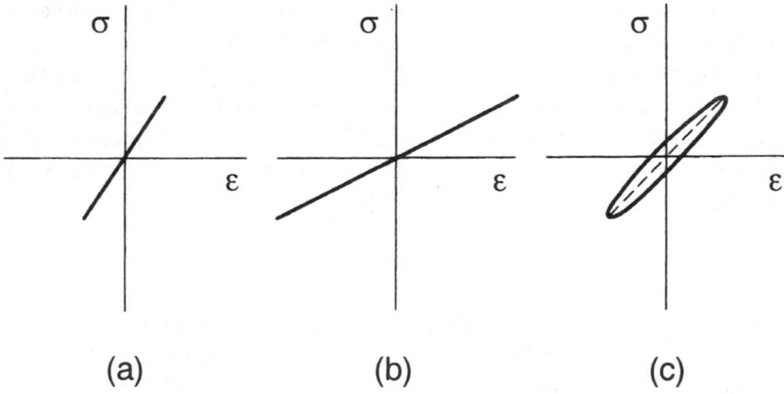

Fig. 6.86. Stress-strain-diagrams for three different frequencies, illustrating the elastic after-effect. (a) High frequency: the unrelaxed modulus of elasticity can be derived from the slope of the steeply inclined Hooke line; (b) low frequency: the relaxed modulus of elasticity can be derived from the gradually inclined Hooke line; (c) medium frequency: in addition to the elastic strain, there is an anelastic component, that is different for loading or unloading. This results in a hysteresis-like stress-strain dependence.

is measured. The stress is always out of phase with the strain, i.e. the stress tends to be ahead of strain. Thus, at intermediate frequencies, what would be the elastic line in the $\sigma - \varepsilon$ diagram is an elliptical hysteresis curve, and the area enclosed by the ellipse corresponds to the absorbed elastic energy per cycle. Elastic waves commonly do not have a rectangular shape but a trigonometric time dependence. For a vibration frequency ν, corresponding to a circular frequency $\omega = 2\pi\nu$, the time dependent strain and stress are

$$\varepsilon = \varepsilon_0 \sin \omega t \tag{6.140a}$$

$$\sigma = \sigma_0 \sin(\omega t + \delta) \tag{6.140b}$$

where δ is the phase shift between stress and strain, which reflects how much the stress is ahead of the strain. Mathematically, the handling of trigonometric functions is convenient with the use of complex functions so we replace Eq. (6.140) by

$$\varepsilon = \varepsilon_0 \; e^{i\omega t} \tag{6.141a}$$

$$\sigma = \sigma_0 \; e^{i(\omega t + \delta)} \tag{6.141b}$$

By definition $e^{i\omega t} = \cos \omega t + i \sin \omega t$. According to Eqs. (6.140a) and (6.140b), the measurable quantities, stress and strain in Eqs. (6.141a) and (6.141b), are described by the imaginary part. Consequently, we can define a complex modulus E^*

$$E^* = \frac{\sigma}{\varepsilon} = e^{i\delta}\frac{\sigma_0}{\varepsilon_0} = \frac{\sigma_0}{\varepsilon_0}(\cos\delta + i\sin\delta) \equiv E_1 + iE_2 \tag{6.142}$$

or

$$\sigma_0 \cos\delta = E_1 \varepsilon_0 \tag{6.143a}$$

$$\sigma_0 \sin\delta = E_2 \varepsilon_0 \tag{6.143b}$$

E_1 is referred to as the storage modulus, while E_2 is called the loss modulus, which will become clear in the following.
The elastic energy density Γ for a given stress σ is

$$\Gamma = \frac{1}{2}\sigma\varepsilon \tag{6.144}$$

or with Eqs. (6.141a) and (6.141b)

$$\Gamma = \frac{1}{2}\varepsilon_0\sigma_0 (\sin\omega t \cos\delta + \cos\omega t \sin\delta) \sin\omega t \tag{6.145}$$

Γ attains a maximum for $\omega t = \pi/2$. In this case

$$\Gamma_{\max} = \frac{1}{2}\varepsilon_0\sigma_0 \cos\delta = \frac{1}{2}E_1\varepsilon_0^2 \tag{6.146}$$

E_1 is called storage modulus because it indicates the elastically stored energy of a specimen at strain ε_0.

The energy loss per cycle is the area under the ellipse and, therefore,

$$\Delta\Gamma = \oint \sigma d\varepsilon = \int_0^{\frac{2\pi}{\omega}} \sigma \cdot \dot{\varepsilon} dt \tag{6.147}$$

Because

$$\dot{\varepsilon} = \frac{d\varepsilon}{dt} = \frac{d}{dt}(\varepsilon_0 \sin\omega t) = \varepsilon_0\omega\cos\omega t \tag{6.148}$$

we obtain

$$\begin{aligned} \Delta\Gamma &= \int_0^{\frac{2\pi}{\omega}} (\sin\omega t\cos\delta + \cos\omega t\sin\delta)\,\omega\varepsilon_0\cos\omega t\; dt \\ &= 2\pi\sigma_0\varepsilon_0\sin\delta = \pi E_2\varepsilon_0^2 \\ \Delta\Gamma &= \pi E_2\varepsilon_0^2 \end{aligned} \tag{6.149}$$

E_2 denotes the energy loss per cycle and, therefore, is referred to as the loss modulus.

The ratio of the moduli

$$\frac{E_2}{E_1} = \tan\delta \tag{6.150}$$

has a physical meaning.

The energy loss ratio per cycle is

$$\frac{\Delta\Gamma}{2\Gamma} = \frac{\pi E_2\varepsilon_0^2}{E_1\varepsilon_0^2} = \pi\tan\delta \tag{6.151}$$

Since $E_1 \gg E_2$ (typical values are $E_1 = 1GPa$, $E_2 = 10MPa$) tan $\delta \approx \delta$ or

$$\frac{\Delta\Gamma}{2\Gamma} \approx \pi\delta \tag{6.152}$$

The quantity δ is called the logarithmic decrement. The term decrement comes from the fact that δ can be deduced from the attenuation of an elastic vibration (Fig. 6.87). If A_n and A_{n+1} are the amplitudes of two consecutive vibration cycles, we obtain

$$\frac{\Delta\Gamma}{2\Gamma} = \ln\frac{A_n}{A_{n+1}} \approx \pi\delta \tag{6.153}$$

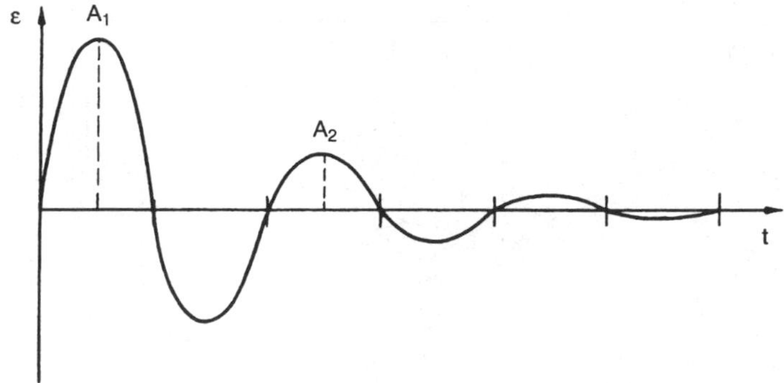

Fig. 6.87. An attenuating oscillation in a material with strong damping, like rubber or cast iron.

This relation can be easily derived from the equation of motion of a damped pendulum.

Commonly, the quantity tanδ is also termed Q^{-1} or internal friction. It depends on the frequency and temperature. For instance, for the diffusion-dependent Snoek effect, the anelastic strain is established more rapidly at higher the temperature so, correspondingly, the anelastic strain per cycle will change with temperature. Likewise, the respective energy loss, i.e. the internal friction, changes.

The temperature and frequency dependence of internal friction can be calculated from a simple model. The elastic and anelastic behavior of a solid corresponds to a coupling of an elastic body, in the most simple case of a spring, with a damping component, for instance a hydraulic dash pot.

These elements can be connected in very different ways, for instance in series (Fig. 6.88a). This arrangement is called the Maxwell model. The spring has an effective modulus E_M, hence

$$\sigma_1 = E_M \varepsilon_1 \tag{6.154}$$

The dash pot behaves like a viscous fluid, i.e. a stress causes a constant strain rate

$$\sigma_2 = \eta_M \cdot \dot{\varepsilon}_2 \tag{6.155}$$

where η_M is the viscosity of the Maxwell dash pot. If a constant stress is applied to the Maxwell element,

$$\sigma = \sigma_1 = \sigma_2 \tag{6.156a}$$

$$\varepsilon = \varepsilon_1 + \varepsilon_2 \tag{6.156b}$$

Combining Eqs. (6.154) and (6.155) with Eqs. (6.156a) and (6.156b) we obtain

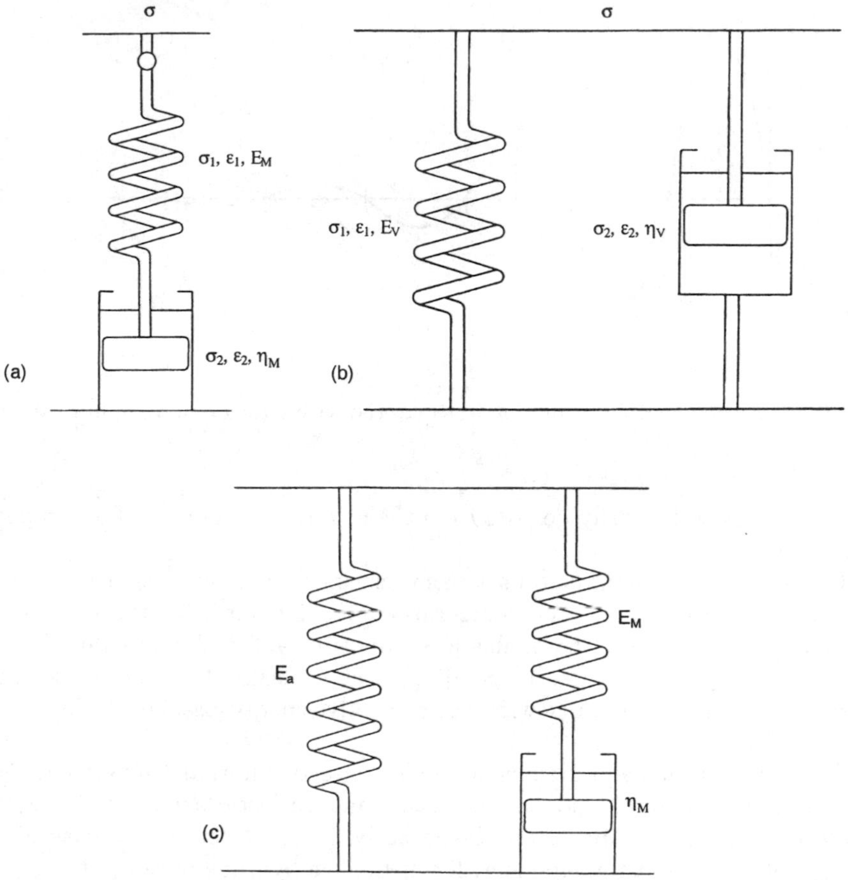

Fig. 6.88. (a) Maxwell model of the solid body. (b) Voigt-Kelvin model of the solid body. (c) Linear-elastic standard body.

$$\dot{\varepsilon} = \frac{\dot{\sigma}}{E_M} + \frac{\sigma}{\eta_M} \tag{6.157}$$

For a constant stress $\dot{\sigma} = 0$ and, therefore

$$\dot{\varepsilon} = \frac{\sigma}{\eta_M} \tag{6.158}$$

The Maxwell element does not show anelastic but rather viscoelastic behavior, i.e. a linearly increasing strain with time. If instead of the serial arrangement a parallel arrangement is used (Voigt-Kelvin model, Fig. 6.88b), the stress and strain are given by

$$\sigma = \sigma_1 + \sigma_2 \tag{6.159}$$

$$\varepsilon = \varepsilon_1 = \varepsilon_2 \tag{6.160}$$

with

$$\sigma_1 = E_V \cdot \varepsilon_1 \qquad \sigma_2 = \eta_V \cdot \dot{\varepsilon}_2 \tag{6.161}$$

In this case one observes anelastic behavior for σ = const., also in the case of creep, but a purely elastic behavior for $\dot{\varepsilon} = 0$, i.e., no stress relaxation, in contrast to the behavior of an anelastic solid.

A model with a more realistic behavior is obtained from a combination of these two basic models (Fig. 6.88c). This arrangement is referred to as a linear elastic standard body. With the relations

$$\begin{aligned} \varepsilon &= \varepsilon_1 = \varepsilon_2 \\ \varepsilon_2 &= \varepsilon_{21} + \varepsilon_{22} \\ \sigma &= \sigma_1 + \sigma_2 \end{aligned} \tag{6.162}$$

and

$$\begin{aligned} \sigma_1 &= E_a \varepsilon \\ \varepsilon_{21} &= \frac{\sigma_2}{E_M} \\ \sigma_2 &= \dot{\varepsilon}_{22} \cdot \eta_M \end{aligned} \tag{6.163}$$

we obtain the equation of state

$$\sigma + \tau\dot{\sigma} = E_a \varepsilon + (E_M + E_a)\,\tau\dot{\varepsilon} \tag{6.164}$$

where

$$\tau = \frac{\eta_M}{E_M} \tag{6.165}$$

For dynamical loading, σ and ε are again given by $\varepsilon = \varepsilon_0\ e^{i\omega t}$, $\sigma = \sigma_0\ e^{i(\omega t + \delta)}$ (Eqs. (6.141a) and (6.141b))

$$\begin{aligned} \sigma &= E^* \varepsilon \\ \sigma_0\ e^{i(\omega t+\delta)} &= (E_1 + iE_2)\,\varepsilon_0\ e^{i\omega t} \end{aligned} \tag{6.166}$$

The solution of Eq. (6.164) yields

$$E^* = E_1 + iE_2 = \frac{E_a + (E_a + E_M)\,\omega^2\tau^2}{1 + \omega^2\tau^2} + i\frac{E_M \omega\tau}{1 + \omega^2\tau^2} \tag{6.167}$$

According to Eq. (6.150), the internal friction reads

$$\tan\delta = \frac{E_M\omega\tau}{E_a + (E_M + E_a)\,\omega^2\tau^2} \tag{6.168}$$

Thus, $\delta \to 0$ for $\omega = 0$ and $\omega \to \infty$. δ is maximized for

$$(\omega\tau)^2 = \frac{E_a}{(E_a + E_M)} \tag{6.169}$$

The damping behavior passes through a maximum as function of $\omega\tau$. The elastic moduli E_a and E_M of the springs of the standard body correspond to the moduli defined in Eqs. (6.139a) and (6.139b)

$$E_a = E_r \tag{6.170a}$$

$$E_M = E_u - E_r \tag{6.170b}$$

Upon loading both springs are instantaneously strained to the same extent, while the dash pot is not yet active. Consequently,

$$\sigma = \sigma_1 + \sigma_2 = E_a \cdot \varepsilon + E_M \cdot \varepsilon = E_u \cdot \varepsilon$$

After a long time the dash pot compensates the spring E_M and

$$\sigma = \sigma_1 = E_a \cdot \varepsilon = E_r \cdot \varepsilon$$

This results in the following relations

$$\tan\delta_{\max}\left(\omega^2\tau^2 = \frac{E_r}{E_u}\right) = \frac{E_u - E_r}{2\sqrt{E_u \cdot E_r}} \tag{6.171a}$$

$$E_{2,\max}\left(\omega^2\tau^2 = 1\right) = \frac{E_u - E_r}{2} \tag{6.171b}$$

Also the anelastic strain is usually very small so that $E_r/E_u \approx 1$.

The damping maximum is, therefore, obtained at about

$$\omega\tau = 1 \tag{6.172}$$

Under these conditions the loss by damping and, according to Eq. (6.171b) also the loss modulus are maximized.

The storage modulus E_1, which is in phase with the stress, is according to Eqs. (6.167), and (6.170a) and (6.170b)

$$E_1 = \frac{E_r + E_u\omega^2\tau^2}{1 + \omega^2\tau^2} \tag{6.173}$$

and changes with increasing $\omega\tau$ at $\omega\tau \approx 1$ from E_r to E_u. The dependency of the functions $\delta(\omega\tau)$ and $E_1(\omega\tau)$ are plotted schematically in Fig. 6.89.

The damping maximum is physically caused by the fact that, for $\omega = 1/\tau$, the excitation frequency and relaxation time of the process, for instance the

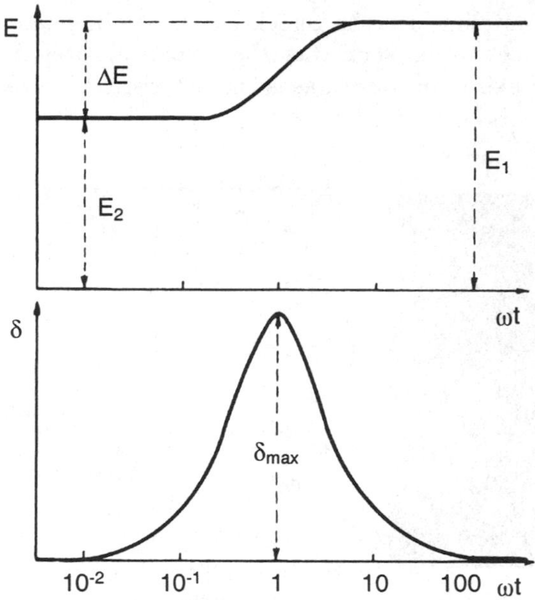

Fig. 6.89. Young's modulus and the logarithmic decrement δ as a function of the measurement frequency (schematic).

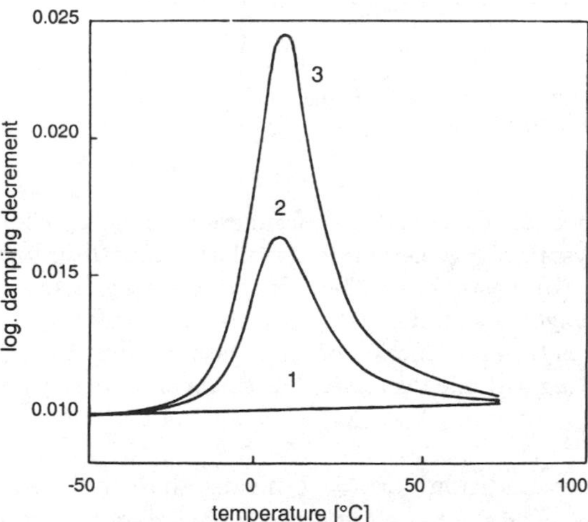

Fig. 6.90. Temperature dependence of the logarithmic decrement for α-Fe at different nitrogen concentrations. Damping increases with increasing concentration, but the temperature of the maximum is concentration independent [6.35].

characteristic jump time for the Snoek effect, are in phase and, therefore, generate resonance effects, such that the system absorbs maximum energy from the applied exitation source, as is characteristic of forced vibrations.

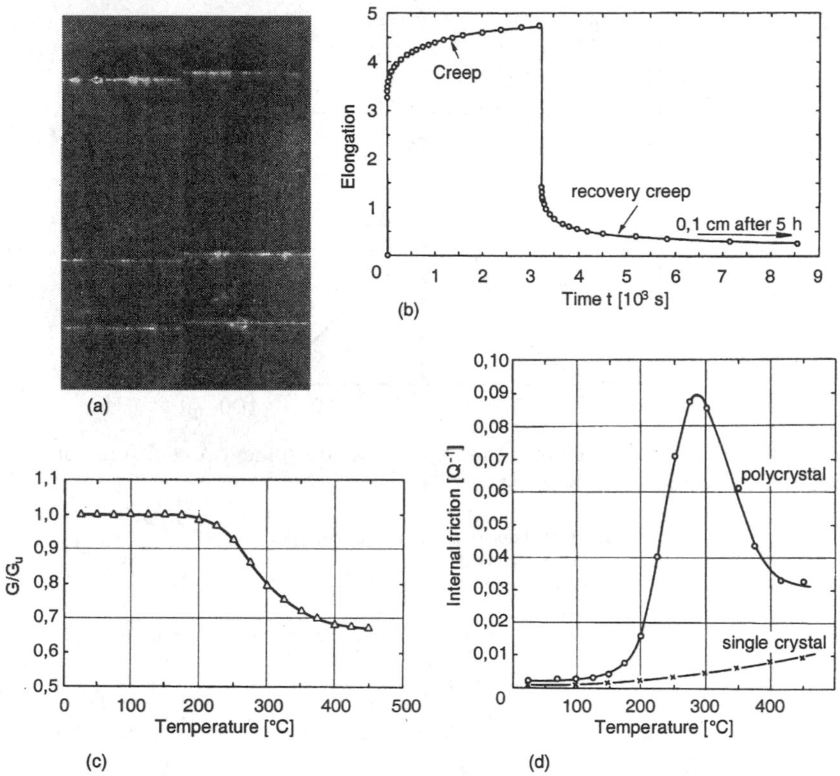

Fig. 6.91. The anelastic effect resulting from grain boundary sliding. (a) Displacement of two tin crystals along their grain boundary under shear stress (grain boundary sliding) [6.35]; (b) elastic after-effect (creep) following strain and strain release of an aluminum polycrystal (after [6.37]); (c) ratio of shear modulus G and unrelaxed shear modulus G_u as a function of temperature (after [6.38]); (d) temperature dependence of the logarithmic decrement for single and polycrystals (after [6.39]).

The resonance condition $\omega\tau = 1$ means that the resonance frequency changes if τ varies. For the Snoek effect τ is obviously the characteristic jump time for diffusion, during which the atoms change from energetically less favorable to more favorable sites. The diffusion rate depends exponentially on temperature (see Chapter 5)

$$\frac{1}{\tau} = \nu_D \; e^{\left(-\frac{G_m}{kT}\right)} \tag{6.174}$$

where G_m is the free energy of migration of the carbon atoms. Correspondingly, the resonance frequency $\omega = 1/\tau$ increases with increasing temperature, and the damping, therefore, depends on temperature (Fig. 6.90). At room temperature the resonance frequency of the Snoek effect is of the order of 1Hz, so that the oscillations of a torsional pendulum can be used to measure the effect. According to Eq. (6.174) the diffusion coefficient D can be very accurately determined with this effect because according to Eq. (5.17)

$$D = \frac{\lambda^2}{6\tau_S} \tag{6.175}$$

where τ_S is the time elapsed between two successive jumps of a diffusing atom, and λ is the jump distance that depends on the crystal structure. More accurately

$$\tau_S = \frac{3}{2}\tau \tag{6.176}$$

because, for the Snoek effect, only jumps from x and y sites to z sites are significant but not the jumps of C-atoms from z sites to other z sites although they do contribute to diffusion. One specific advantage of the anelastic measuring method is the possibility of measuring the diffusion constant at low temperatures because the measurement of the effect essentially captures a single diffusion jump, which occurs in a finite time even at low temperatures. All low temperature values of the diffusion constant of carbon in α-iron have been gained from these anelastic mesurements (see Fig. 5.6).

Anelasticity or damping behavior of a solid is not restricted to the Snoek effect and there is a large number of reasons that cause anelastic behavior in metals, for instance grain boundary sliding (Fig. 6.91) or dislocation damping, which, however, show very different resonance behavior and in different frequency ranges. Of particular importance is the phenomenon of internal friction in polymers. Their behavior is not anelastic but viscoelastic since under a constant load they do not show an anelastic additional strain but rather, in addition to the elastic strain, a constant strain rate, that is proportional to the applied stress. The equation of state of a viscoelastic body under a constant shear stress σ_{xy} is

$$\sigma_{xy} = G\gamma_{xy} + \eta\dot{\gamma}_{xy} \tag{6.177}$$

The behavior of a viscoelastic material corresponds to a mixture of an elastic body and a viscous fluid. Diffusion creep and grain boundary sliding in metals are actually viscoelastic phenomena, not anelastic.

Dynamic viscoelastic behavior can be described with the same formalism as anelastic behavior. In polymers, however, there is a large number of viscoelastic processes which leads to a very complex spectrum of the damping behavior (Fig. 6.92), for instance caused by the rotation of molecules, unfolding of molecule chains, and so on. Several complementary experimental methods are necessary to adequately describe the mechanisms of such complex phenomena.

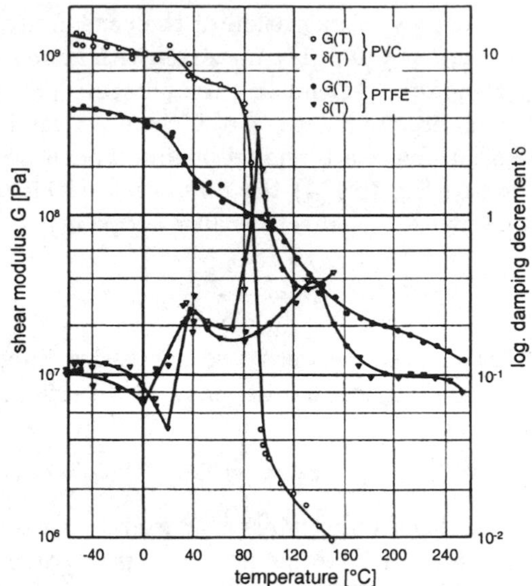

Fig. 6.92. Temperature dependence of the shear modulus G and the logarithmic decrement δ of PVC (polyvinyl chloride) and Teflon (PTFE) (after [6.40]).

7

Recovery, Recrystallization, Grain Growth

7.1 Phenomena and Terminology

Many materials undergo a heat treatment during their processing that strongly impacts properties. During heat treatment subsequent to plastic deformation in particular the mechanical properties and the microstructure are strongly affected, while other physical properties (e.g. electrical resistivity) are scarcely influenced (Fig. 7.1).

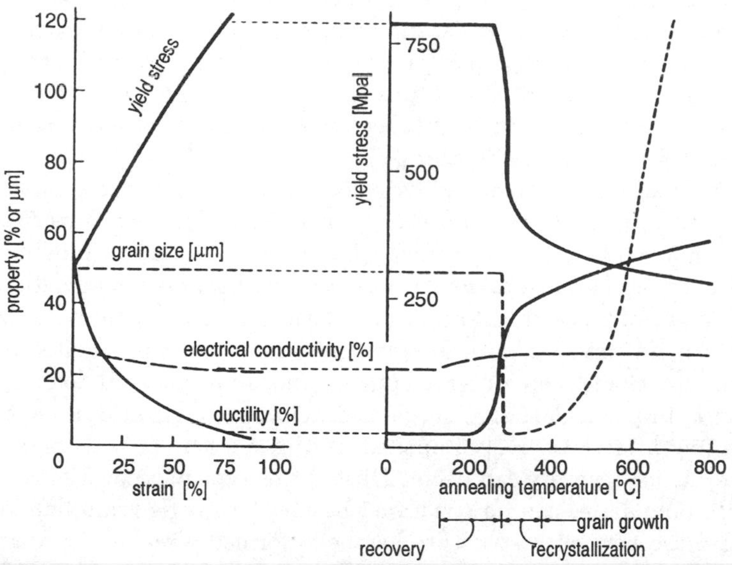

Fig. 7.1. The effect of cold forming and annealing on the properties of a Cu35%Zn alloy.

In the course of plastic deformation the strength of a material increases (strain hardening), while the strain remaining until fracture decreases. Conversely, during heat treatment the strength decreases and ductility improves. By successive deformation and heat treatment, large degrees of deformation can be imposed on a material.

The underlying physical reason for these phenomena are the dislocations, which are stored during plastic deformation and cause strain hardening. Their rearrangement and removal during annealing softens the material. In principle, there are two different reasons for the loss of strength during annealing, recovery and recrystallization.

Recrystallization is referred to as a reconstruction of the grain structure during annealing of deformed metals. It proceeds by generation and motion of high-angle grain boundaries that concurrently removes the deformed microstructure. In contrast, recovery comprises all phenomena that are associated with the rearrangement or annihilation of dislocations. More specifically, the term recrystallization as defined here relates to the most important among many recrystallization processes, namely primary "static" recrystallization. The term recrystallization, however, is commonly used in a much broader sense, by including all kinds of phenomena associated with grain boundary migration that lead to a lower free energy of the crystalline aggregate. These phenomena comprise in particular all processes of grain growth, granular restructuring during deformation, and special cases of strong recovery.

Generally, one distinguishes whether the processes occur during deformation (dynamic recrystallization, dynamic recovery) or subsequent to cold forming during annealing treatment (static recrystallization, static recovery).

If recrystallization occurs during heat treatment of a deformed material, at first the generation of small new grains is observed that grow at the expense of the deformed microstructure until they impinge and, eventually, completely replace the deformed microstructure (Fig. 7.2).

This process — characterized by nucleation and nucleus growth — is referred to as primary recrystallization. Since the dislocation density in the material is not removed homogeneously, but discontinuously at moving grain boundaries it is also termed discontinuous recrystallization in association with the terminology of phase transformations. Although primary recrystallization is by far the most important annealing phenomenon, sometimes quite different microstructural changes are observed during annealing of a cold deformed microstructure. In particular, after large cold deformation or in case that grain boundary migration is strongly impeded, for instance by dispersion of a second phase, strong recovery can occur. During this type of recovery even high-angle grain boundaries can be generated besides low-angle grain boundaries. In this case the new microstructure has been formed without the migration of high-angle grain boundaries, so this phenomenon is referred to as in-situ recrystallization. This process — as any recovery process — occurs homogeneously throughout the microstructure and is, therefore, also referred to as continuous recrystallization to discriminate it from discontinuous (primary)

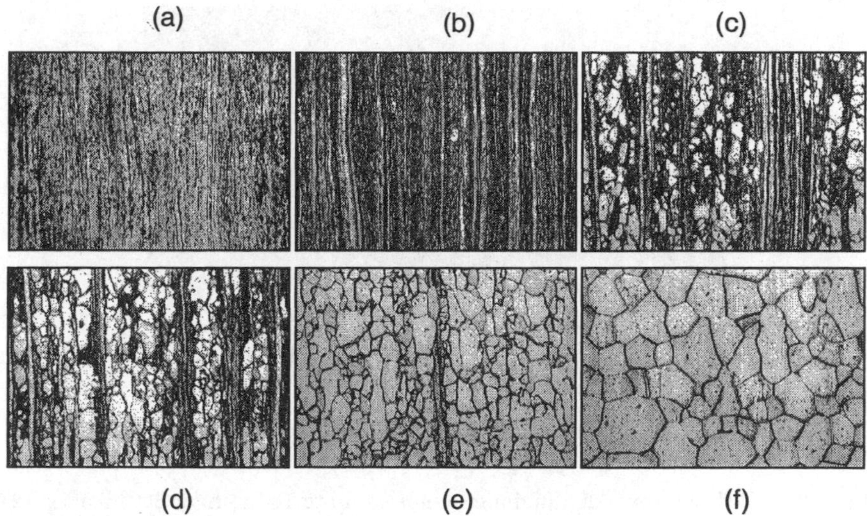

Fig. 7.2. Microstructural change during recrystallization of cold rolled Armco iron.

recrystallization.

If the degree of prior cold forming is small, it is observed that nucleation of new grains is suppressed. Specifically, existing grain boundaries migrate locally and remove the dislocation structure in the swept volumes (SIBM: strain-induced grain boundary motion). Figure 7.3 gives an example of this process in aluminum. At low degrees of deformation not all grains are deformed equally. During SIBM a less deformed crystal grows into the adjacent crystal of higher energy and removes the deformed microstructure of the consumed grain. The energetic reason for this grain boundary migration is the difference of stored deformation energy (i.e. dislocation density) in adjacent grains.

If the heat treatment is continued after complete primary recrystallization — or in other essentially dislocation-free microstructures, for instance in cast structures — the grain size usually increases. The respective phenomena are subsumed under the term grain growth, which occurs in two modes. Either the average grain size continuously increases, which is referred to as continuous or normal grain growth (Fig. 7.4), or only a few grains grow rapidly, while the other grains grow slowly or not at all. The latter process is called discontinuous or abnormal grain growth (Fig. 7.5). Because of the similarity of its microscopic appearance to primary recrystallization (nucleation and nucleus growth), discontinuous grain growth is also termed secondary recrystallization. Note that only discontinuous grain growth is referred to as secondary recrystallization. It leads to large grains and its occurrence is usually undesirable in commercial materials.

During grain growth not only the average grain size but the entire grain

Fig. 7.3. SIBM (c.f. text) of aluminum annealed up to 130 min. at 350° C after 12% rolling deformation. The original position of the grain boundary is still visible [7.1].

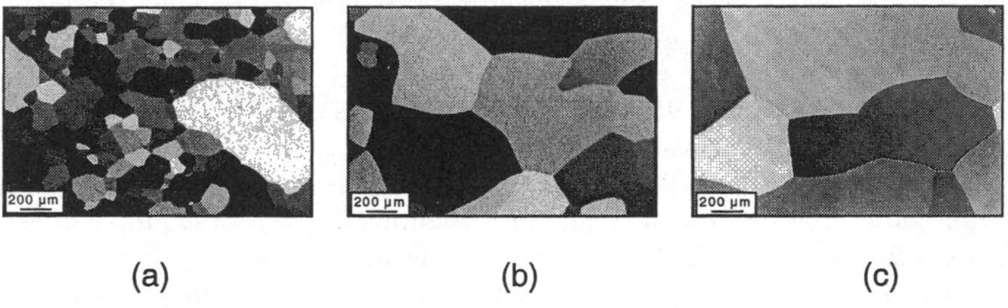

Fig. 7.4. Microstructural change during continuous grain growth of Al-0.1%Mn after 95% cold rolling. Annealing temperature 450° C.

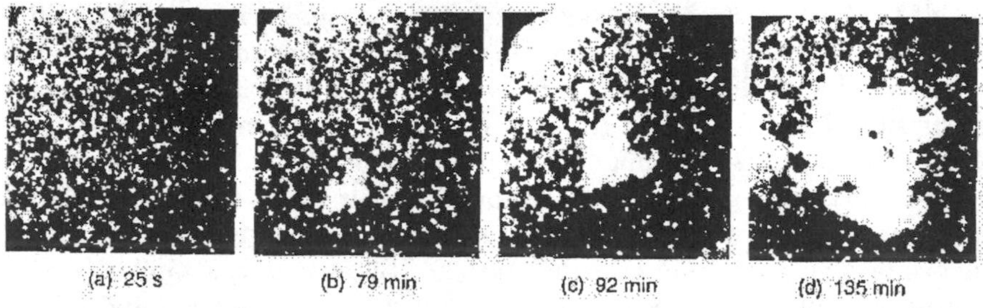

Fig. 7.5. Discontinuous grain growth of high purity zinc at 240° C in a hot stage of a microscope after 40% deformation. Annealed 25 s (a), 79 min. (b), 92 min. (c), and 135 min. (d).

size distribution changes, and this is characteristically different for continuous and discontinuous grain growth. During normal grain growth, the average (logarithmic) grain size lnD_m is shifted to larger values, but the height of the maximum and the standard deviation remain unchanged (Fig. 7.6a). Such behavior of the distribution is also referred to as self-similarity, i.e., if the distribution is plotted versus the normalized logarithmic grain size $ln(D/D_m)$, the distribution does not change during normal grain growth. Of course, this requires that the distribution is normalized as any probability distribution. Normalization means in this context that the integral of the distribution has a constant value, for instance unity. If this were not the case, the maximum of the distribution would decrease, since fewer and fewer grains remain during grain growth.

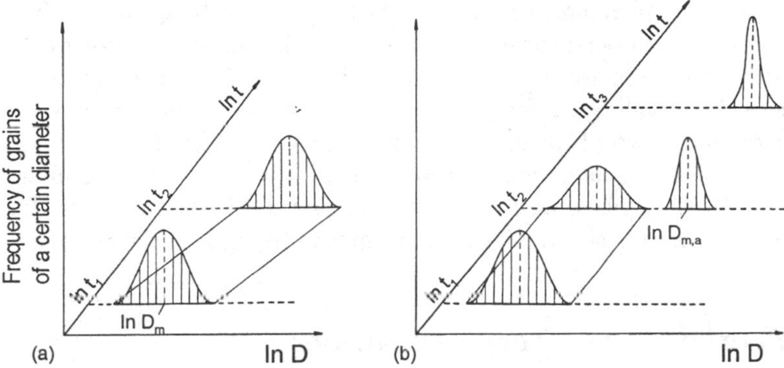

Fig. 7.6. Time dependence of grain size distribution for normal (a) and abnormal (b) grain growth(schematic).

During discontinuous grain growth, however, the grain size distribution does not remain self-similar. Rather, during incomplete secondary recrystallization a bimodal distribution develops; where one distribution relates to the slowly-growing grains and the other distribution represents the rapidly-growing grains. The distribution of the remaining slowly-growing grains becomes smaller and will disappear eventually, but its average grain size does not change. In contrast, the distribution of the few abnormally growing grains significantly changes, because the respective average grain size $lnD_{m,a}$ and the maximum frequency f_{max} increase with increasing annealing time until abnormal grain growth has gone to completion (Fig. 7.6b).

Normal grain growth usually ceases, if the grain size becomes comparable to the smallest specimen dimension, for instance the sheet thickness. In some cases, especially for thin sheet, discontinuous growth of a few grains is observed after continuous growth has come to an end. By choice of an appropriate gas atmosphere during annealing this process can be promoted,

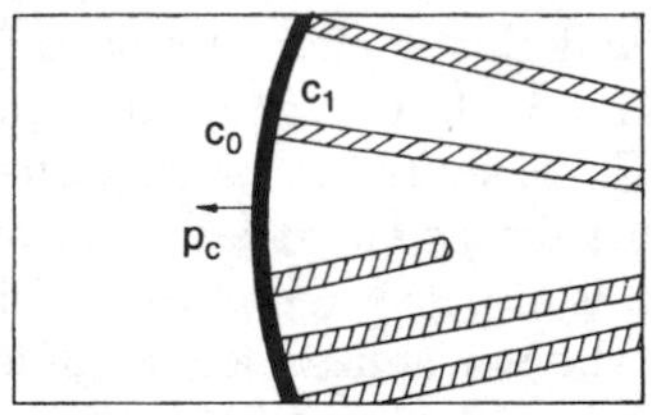

Fig. 7.7. Schematic representation of discontinuous precipitation. The supersaturated solution of concentration c_0 acts on the grain boundary as a chemical driving force p_c.

suppressed or even inverted. Because of its discontinuous appearance, and to distinguish it from discontinuous grain growth owing to different energetic reasons, this is referred to as tertiary recrystallization.

A particular phenomenon of recrystallization can be observed during recrystallization in supersaturated solid solutions if recrystallization occurs concurrently with the phase transformation. The otherwise impeded precipitation processes are accelerated by grain boundary diffusion in the moving grain boundary, and a two-phase microstructure appears behind the moving boundary (Fig. 7.7). This phenomenon is called discontinuous precipitation although according to its physical nature it is also a recrystallization process. The high driving forces for phase transformation can lead to high recrystallization rates.

7.2 Energetics of Recrystallization

In contrast to the atomistic details of recrystallization, the energetic reasons for recrystallization are well understood. Basically, there is always a driving force on a grain boundary if Gibbs free energy, G, of a crystal is reduced during motion of the boundary. If an area element dA of a grain boundary is displaced by an infinitesimal distance, dx, the free energy will be changed by

$$dG = -pdAdx = -pdV \tag{7.1}$$

where dV is the volume swept by the moving grain boundary. The term

$$p = -dG/dV \tag{7.2}$$

is referred to as driving force; it has the dimension of gained free energy per unit volume (J/m^3), but it can also be considered as a force acting per unit area on the grain boundary (N/m^2), i.e., as pressure on the grain boundary.

The driving force for primary recrystallization is the stored energy of the dislocations. If a recrystallized grain grows into the deformed microstructure, most of the dislocations in the swept volume are consumed by the boundary, and a volume with a substantially lower dislocation density is left behind (about $10^{10}[\mathrm{m}^{-2}]$ compared to $10^{16}[\mathrm{m}^{-2}]$ in heavily deformed metals).

The energy of a dislocation per unit length is given by (see Chapter 6)

$$E_d = \frac{1}{2}Gb^2 \tag{7.3}$$

(G - shear modulus, b - Burgers vector).

For a dislocation density, ρ, the driving force for primary recrystallization may be formulated (assuming that the remaining dislocation density is so small that it can be neglected)

$$p = \rho E_d = \frac{1}{2}\rho G b^2 \tag{7.4}$$

For $\rho \cong 10^{16}\mathrm{m}^{-2}$, $G \cong 5 \cdot 10^4$ MPa, and $b \cong 2 \cdot 10^{-10}$ m, the driving force amounts to about $p = 10$ MPa (10^7 J/m^3 ≈ 2 cal/cm^3), and compares well to values of stored energy measured by calorimetry.
The driving force for grain growth results from the boundaries themselves because the total grain boundary area of the crystalline aggregate is reduced during grain growth. If a large grain grows into an environment of small grains with stable grain size, i.e. for discontinuous grain growth (Fig. 7.8), the calculation of the driving force becomes simple. Assume that the consumed grains have a cuboidal shape. If their diameter is d and their grain boundary energy γ [J/m^2], then the grain boundary energy per unit volume and thus the driving force on a boundary sweeping such a volume equals

$$p = \frac{3d^2\gamma}{d^3} = \frac{3\gamma}{d} \tag{7.5}$$

The factor 3 results from the fact that each of the 6 faces of a cube is shared by two adjacent grains. From a quantitative assessment we obtain $p \cong 0.03$ MPa ($= 3 \cdot 10^4$ J/m^3) for a grain size of the consumed grains $d = 10^{-4}$ m typical for recrystallized microstructures and $\gamma \approx 1 J/m^2$. Obviously, the driving force even for discontinuous grain growth is smaller by orders of magnitude than for primary recrystallization. Therefore, it can be expected that grain growth phenomena proceed much more slowly or become noticeable only at much higher temperatures.

The derivation of Eq. (7.5) is based on the assumption that a large crystal grows into a fine-grained microstructure and thus liberates the energy of the consumed grain boundaries, i.e. the driving force is the same for any element of the moving grain boundaries. An arbitrary area element of a moving boundary, in general, does not "sense" the existence of the remote grain boundaries that provide the driving force. The force on this area element results from the fact that at the grain boundary junctions mechanical equilibrium has to be established, which always leads to a curvature of the boundary. A curved boundary, however, responds to a force by moving towards the center of curvature in order to become straighter and thus reduce its area. Therefore, the driving force on a boundary segment is given by the pressure on a curved

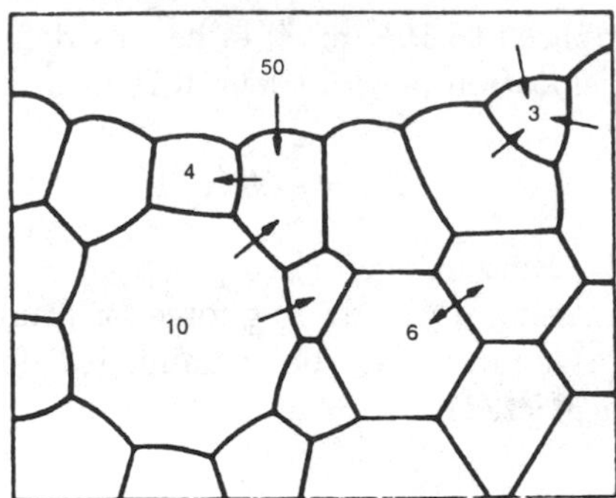

Fig. 7.8. Schematic representation of a primary recrystallized structure with different grain sizes. The numerals indicate the number of nearest neighbors of a grain. (Grain 50 is growing abnormally, grain 10 continuously, grain 3 is shrinking.)

surface. To calculate the magnitude of this driving force, let us consider the change of surface and volume of a spherical cap, or simply of an entire sphere

$$p = \frac{8\pi R\gamma dR}{4\pi R^2 dR} = \frac{2\gamma}{R} \tag{7.6}$$

The driving force in Eq. (7.6) resembles the one derived in Eq. (7.5) if the radius of curvature R is approximately the grain size. Usually, however, the curvature of a grain boundary is small, and correspondingly the radius of curvature is substantially larger than the grain size (by factors of 5-to-10). Therefore, the driving force for continuous grain growth (Eq. (7.6)) is about 5 to 10 times smaller than for discontinuous grain growth (Eq. (7.5)) so that continuous grain growth proceeds much more slowly than discontinuous grain growth or secondary recrystallization.

The driving force for tertiary recrystallization is caused by the orientation dependence of the free surface energy. A grain exposed to the surface is bound to grow at the expense of its neighbors if its surface energy γ_0 (because of the crystallography of its surface) is smaller than that of its neighbors. For a thin sheet of width B and a grain size that is large compared to the thickness of the sheet h, i.e. the grain boundaries extend through the entire sheet thickness, and they run perpendicular to the sheet plane (Fig. 7.9), one obtains for the driving force

$$p = \frac{2\left(\gamma_{02} - \gamma_{01}\right) B dx}{B h dx} = \frac{2\Delta\gamma_0}{h} \tag{7.7}$$

Again this driving force is transmitted to any boundary element in the interior by a curvature of the grain boundary due to its motion on the surface. For $\Delta\gamma_0 \cong 0.1$ J/m^2 and a sheet thickness $h \cong 10^{-4}$ m, the driving force amounts to $p \cong 2 \cdot 10^{-3}$ MPa ($= 2 \cdot 10^3$ J/m^3). Since the surface energy depends on the ambient atmosphere, $\Delta\gamma_0$ can be made larger, smaller, or even change

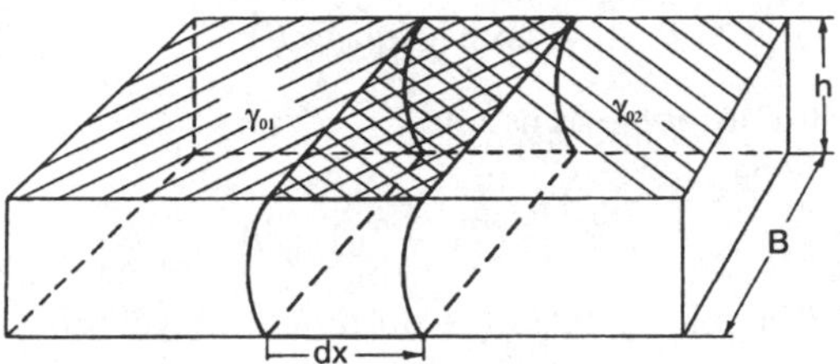

Fig. 7.9. Calculating the driving force of tertiary recrystallization for $\gamma_{01} < \gamma_{02}$.

sign by choice of an appropriate annealing atmosphere, and, therefore, tertiary recrystallization can be influenced correspondingly (see section 7.12). During discontinuous precipitation, recrystallization proceeds in a supersaturated solid solution with concurrent phase transformation. Therefore, besides the stored energy of cold work (Eq. (7.4)) the chemical driving force for phase transformation also contributes to the total driving force. Let us assume that the concentration of the supersaturated solid solution is c_0, the corresponding solvus temperature T_0 (Fig. 7.10) and the solubility limit is c_1 at temperature T_1. For a regular solution the chemical driving force at temperature T_1 results from the free energy of mixing

$$\begin{aligned} p_c = & \frac{Q_v}{\Omega} c_0 (1 - c_0) + \frac{kT_1}{\Omega} [c_0 \ln c_0 + (1 - c_0) \ln (1 - c_0)] \\ & - \left\{ \frac{Q_v}{\Omega} c_1 (1 - c_1) + \frac{kT_1}{\Omega} [c_1 \ln c_1 + (1 - c_1) \ln (1 - c_1)] \right\} \end{aligned} \tag{7.8}$$

where Q_v is the atomic heat of solution and Ω the atomic volume. Since Q_v can be obtained from the solvus line

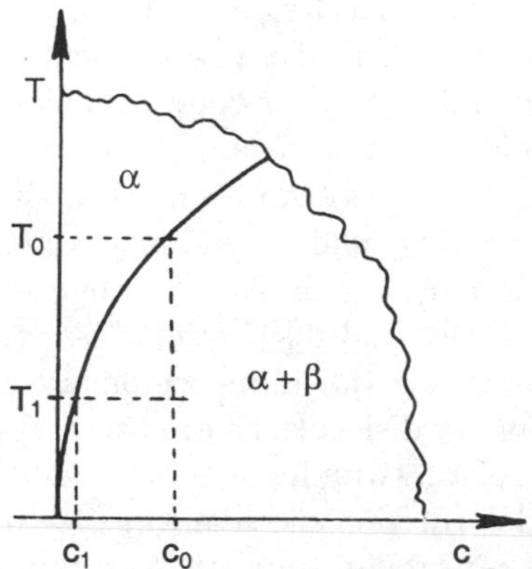

Fig. 7.10. Detail of a binary phase diagram with limited solubility.

$$c = \exp\left(-Q_v/kT\right) \tag{7.9}$$

we obtain for small concentrations and $c_1 \ll c_0$

$$p_c \cong \frac{k}{\Omega}\left(T_1 - T_0\right) c_0 \ln c_0 \tag{7.10}$$

For 5 at%Ag in Cu a solid solution is obtained above 780°C. If this solution is quenched and annealed at 300°C a driving force of $6 \cdot 10^2$ MPa is obtained, i.e., more than 10 times as much as for primary recrystallization.

In addition to the cases represented here, there are abundant examples of driving forces acting on grain boundaries. Any energy state that depends on orientation can be used to exert a driving force on the grain boundary and to force it to move. Examples are magnetic and elastic energy density owing to the orientational dependence of the magnetic susceptibility or Young's modulus. The respective driving forces, however, are much smaller than those for primary recrystallization and grain growth (Table 7.1), hence they play a minor role for the process of recrystallization.

7.3 Deformation Microstructure

Recrystallization always proceeds from a deformed microstructure by the formation of nuclei and their growth. Plastic deformation of metals is mainly caused by the motion of dislocations (see Chapter 6). The mechanisms of deformation and the obtained deformation microstructures depend on the availability and mobility of the dislocations in the microstructure. The mobility of dislocations is markedly influenced by obstacles that hinder their motion, such as solute atoms, precipitates, other dislocations, and, in particular, dissociation of the dislocation cores. The normalized stacking fault energy $\gamma_{SF}/Gb \approx \tilde{\gamma}_{SF}$ controls the dissociation width of the dislocations. The smaller $\tilde{\gamma}_{SF}$ of a material, the larger is the dissociation. With increasing width of dissociation cross slip of screw dislocations and climb of edge dislocations become more difficult. Therefore, obstacles can not be as easily circumvented and eventually work hardening increases. In materials with low stacking fault energy the flow stress can even reach the stress level required to activate mechanical twinning, and if twinning constitutes a major deformation mechanism, it will markedly impact deformation microstructure.

Besides internal obstacles and dislocation core structure, the deformation temperature also exerts a major influence on the deformation behavior, as both cross slip and climb of dislocations are thermally activated processes. At low temperatures, therefore, twinning can become favored over dislocation motion. In essence, the deformation mechanisms of a material can be different in different temperature regimes. Important examples are the deformation behavior of copper and its alloys.

After large degrees of deformation there are two major types of deformation microstructures in fcc metals. Depending on the magnitude of $\tilde{\gamma}_{SF}$

Table 7.1. The driving force of grain boundary migration.

Source	Equation	Approximate value of parameters	Estimated driving force in $[MPa]$
Stored deformation energy	$p = \frac{1}{2}\rho Gb^2$	ρ = dislocation density $\sim 10^{15}/m^2$ $\frac{Gb^2}{2}$ = dislocation energy $\sim 10^{-8} J/m$	10
Grain boundary energy	$p = \frac{2\gamma}{R}$	γ = grain boundary energy $\sim 0.5N/m$ R = grain boundary curvature radius $\sim 10^{-4}m$	10^{12}
Surface energy	$p = \frac{2\Delta\gamma_0}{d}$	d = sample thickness $\sim 10^{-5}m$ $\Delta\gamma_0$ = surface energy difference of two adjacent grains $\sim 0.1\ N/m$	$2{\cdot}10^{-2}$
Chemical driving force	$p = R(T_1 - T_0) \cdot c_0 \ln c_0$	c_0 = concentration $\widehat{=}$ max. solubility at T_0 T_1 ($< T_0$) annealing temperature	$6{\cdot}10^2$ (5% Ag in Cu at 300°C)
Magnetic field	$p = \frac{\mu_0 H^2}{2}(\chi_1 - \chi_2)$	material: Bismuth H = magnetic field strength ($10^7 A/m$) χ_1, χ_2 = magnetic susceptibilities of adjacent grains	$3{\cdot}10^{-5}$
Elastic energy	$p = \frac{\sigma^2}{2}\left(\frac{1}{E_1} - \frac{1}{E_2}\right)$	material: Bismuth H = elastic moduli of adjacent grains $\sim 10^5\ MPa$	$2.5{\cdot}10^{-4}$
Temperature gradient	$p = \frac{\Delta S \cdot 2a\, gradT}{V_m}$	ΔS = {difference in entropy between grain boundary and crystal (approx. equivalent to melting entropy) $\sim 8 \cdot 10^3 J/K \cdot mol$ $2a$ = grain boundary thickness $\sim 5 \cdot 10^{-10}\ m$ $gradT$ = temperature gradient $\sim 10^4\ K/m$ V_m = molar volume $\sim 10cm^3/mol$	$4 \cdot 10^{-5}$

and/or deformation temperature they are characterized by the appearance or absence of deformation twins.

Even at low degrees of deformation dislocations are not distributed homogeneously in the material. Rather dislocations cluster and eventually form a so-called cell structure with a distribution of cell sizes (Fig. 7.11). Such a cell structure is characterized by cell walls with high dislocation density that enclose cell interiors of relatively low dislocation density. The character and appearance of a cell structure depends on the material and is mainly determined by the normalized stacking fault energy ($\tilde{\gamma}_{SF}$), degree of deformation, and deformation temperature. With increasing temperature, respectively larger $\tilde{\gamma}_{SF}$, the thickness of the cell walls decreases until eventually sharp subgrain boundaries are formed, and concurrently the cell interior becomes

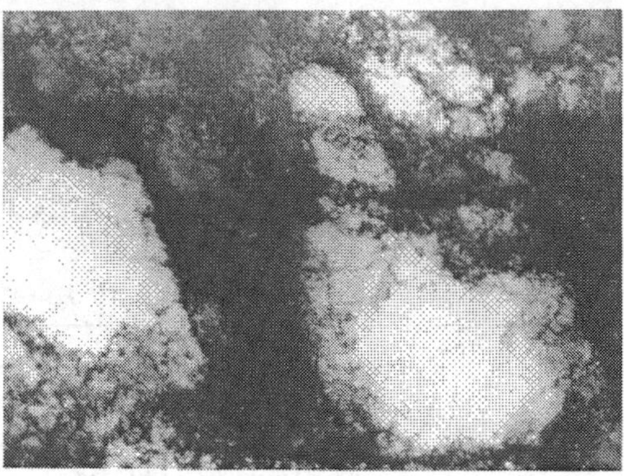

Fig. 7.11. Electron microscope image of the structure of a 10% rolled $\{112\}\langle 111\rangle$ copper single crystal with irregular cell size distribution. The image plane is perpendicular to the transverse direction.

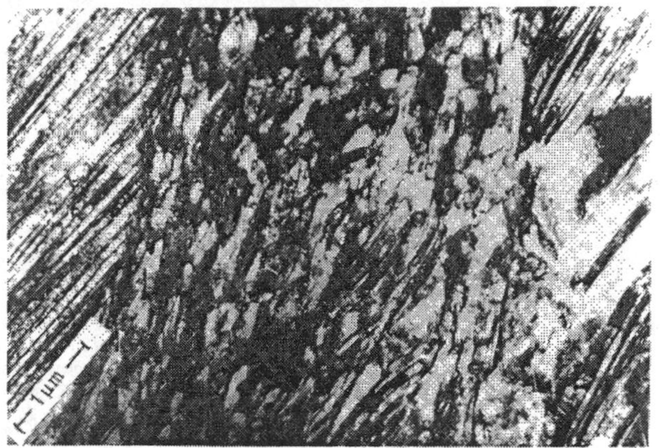

Fig. 7.12. Electron microscope image of a "brass-type" shear band in copper after 50% rolling deformation in liquid nitrogen. Twinning planes are parallel to the rolling plane.

depleted of dislocations. The degree of deformation affects the cell size and the misorientation between adjacent cells. With increasing degree of deformation the average cell size decreases, and the orientation difference between adjacent cells increases.

At large strains deformation inhomogeneities tend to form in a globular cell structure. They are referred to as bands, for instance kink bands during tensile deformation of single crystals, or shear bands during rolling (inclined 35° to the rolling plane) (Fig. 7.12, 7.13) and deformation bands (parallel to the rolling plane). These deformation inhomogeneities can contain orientations quite different from the matrix orientation, and frequently an orientation

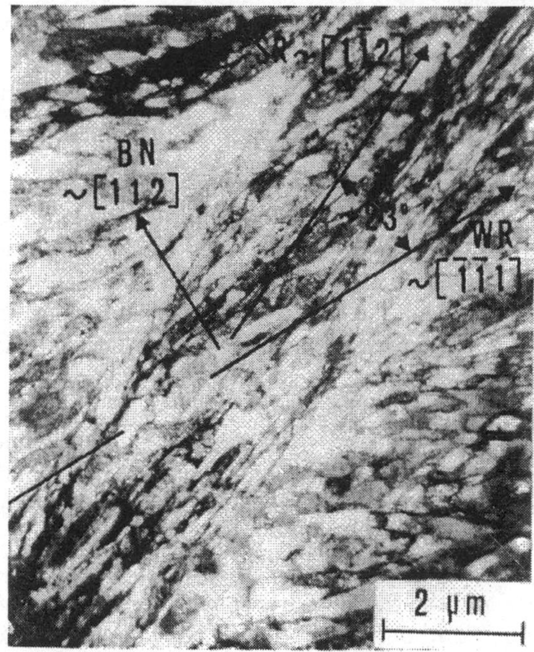

Fig. 7.13. Electron microscope image of a "copper-type" shear band in Cu 0.6% Cr after 95% rolling deformation.

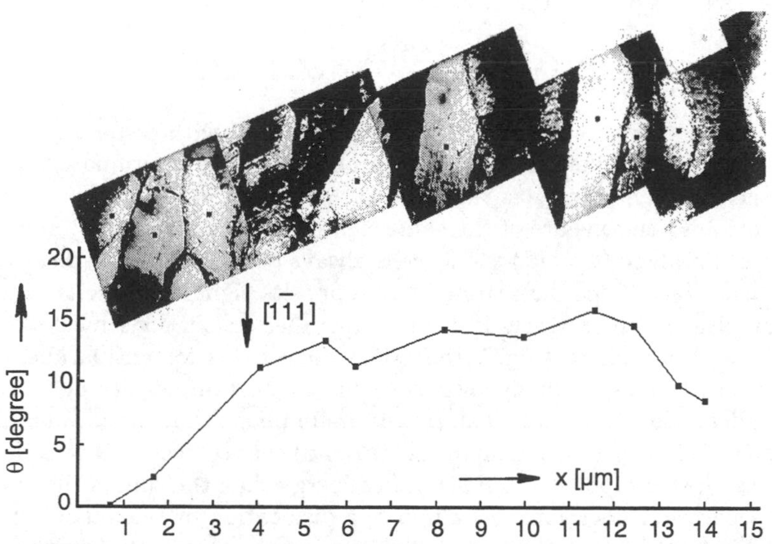

Fig. 7.14. Dislocation structure and orientation dependence in a kink band of a tensile deformed ⟨451⟩ copper single crystal.

gradient is observed at the transition from the matrix to the bands (see kink band, Fig. 7.14).

7.4 Recovery

The deformed state of a material is principally unstable, because the dislocation structure generated during deformation is not in thermodynamic equilibrium. At low deformation temperatures, however, the deformed state is conserved after deformation because the structure is mechanically stable, i.e., the dislocations of the microstructure are in a state of mechanical equilibrium. Upon increase of the temperature, however, this mechanical stability can be overcome by thermally activated processes, i.e., cross slip of screw dislocations and climb of edge dislocations by which the dislocations unlock and move to interact with other dislocations. Through climb, dislocations can leave their glide planes to arrange in energetically more favorable patterns, can mutually annihilate, or leave the crystal altogether. These processes are subsumed under the term recovery, which is always associated with a decrease of the dislocation density and the formation of special dislocation patterns, i.e., networks of low-angle grain boundaries, also referred to as polygonization.

Recovery is caused by the interaction of dislocations via their long-range stress fields. For instance, the interaction of an edge dislocation with Burgers vector b_1 with another parallel edge dislocation of Burgers vector b_2 (see Chapter 6.4) is given by the interaction force

$$F = \tau b_2 = \frac{G b_1 b_2}{2\pi r_d (1-\nu)} \cos\Phi \cos 2\Phi \tag{7.11}$$

where r_d and Φ denotes the position of dislocation 2 with respect to dislocation 1 (r_d - distance between the dislocations, Φ-angular coordinate with $\Phi = 0°$ on the slip plane) and ν is the Poisson ratio.

If both dislocations are of the same sign (parallel dislocations) and lie on the same slip plane ($\Phi = 0°$) the force is always positive, i.e., the dislocations repel each other. If the dislocations have opposite sign, the force is negative, and the dislocations attract. If such antiparallel dislocations meet they recombine and annihilate (Fig. 7.15a). The same holds for antiparallel screw dislocations. Correspondingly, over time the dislocation density decreases. If antiparallel dislocations are on an adjacent slip plane, they do not annihilate, but rather form a dislocation dipole (Fig. 7.15b) that corresponds to a chain of vacancies. Such a dipole has a much lower energy than the sum of the energies of two separate dislocations. By climb of a dislocation by one lattice spacing such a dipole can be annihilated, as observed in the electron microscope.

Even if the dislocations are several lattice spacings apart, they will still interact and, in case of attraction, they can annihilate by climb. If the angular coordinate $\Phi > 45°$, the sign of the interaction force changes (Eq. (7.11)). Now antiparallel dislocations repel each other, but parallel dislocations attract

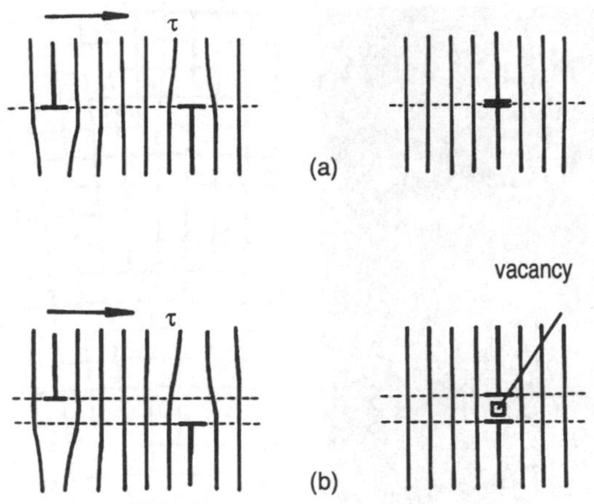

Fig. 7.15. Illustration of the principle of annihilation (a) and dipole generation (b) by edge dislocations.

each other along the glide plane. The equilibrium arrangement of two such parallel dislocations is an arrangement of one above the other. In this case $\Phi = 90°$, and according to Eq. (7.11) $F = 0$. Each displacement from this position results in a force to restore the equilibrium position. This arrangement is energetically most favorable. A large gain of energy is obtained if many dislocations polygonize to align in a periodic pattern in the plane $\Phi = 90°$, thus forming a low-angle symmetrical tilt grain boundary (LATB). The periodic arrangement of dislocations causes a superposition of the long-range stress fields in such a way that the range of the stress field is reduced to about the dislocation spacing r_d. This is associated with a substantial decrease of the energy of each dislocation in the arrangement. If there are Z_d dislocations per unit length in this arrangement, the energy per unit area is

$$\gamma_{\text{LATB}} = Z_d \left[\frac{Gb^2}{4\pi(1-\nu)} \ln \frac{r_d}{2b} + E_C \right] \tag{7.12}$$

(E_C - energy of dislocation core).

Such dislocation arrangement corresponds to a LATB as shown in Fig. 7.16, and γ_{LATB} in Eq. (7.12) reflects the specific energy of this low-angle grain boundary.

Since the orientation difference Θ of the adjacent grains is related to the dislocation spacing in the boundary

$$\Theta = \frac{b}{r_d} \tag{7.13}$$

and $1/r_d = \Theta/b = Z_d$ is the number of dislocations per unit length in the LATB, Eq. (7.12) can be rearranged to yield the specific energy of the LATB, namely

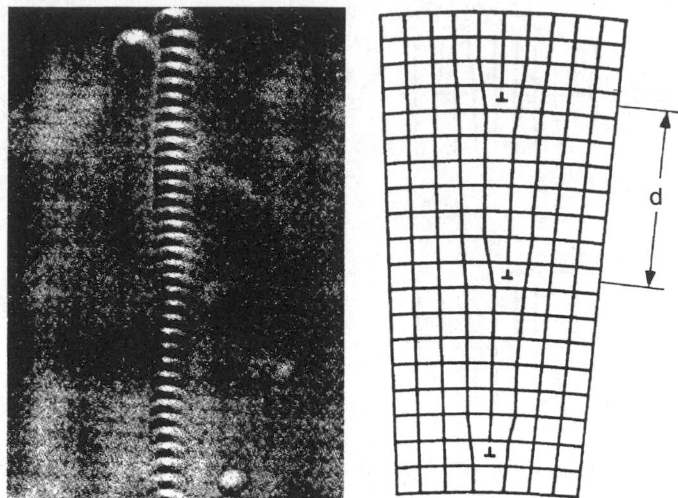

Fig. 7.16. A low-angle tilt boundary; right: schematic representation, left: etch pits of a low-angle grain boundary on the {100}-plane of germanium [7.2].

$$\gamma_{\text{LATB}} = \Theta\,(K_1 - K_2 \ln \Theta) \tag{7.14}$$

$$K_1 = \frac{E_c}{b} - K_2 \ln 2 \tag{7.15}$$

$$K_2 = \frac{Gb}{4\pi(1-\nu)} \tag{7.16}$$

A similar consideration for screw dislocations or mixed dislocations yields the result that such dislocations can also form low angle boundaries consisting of periodic patterns in the material (Fig. 7.17). Such boundaries generated by edge, screw, and mixed dislocations eventually form a spatial network of many low angle grain boundaries (subboundaries), the energy of which is much smaller than the energy of the same dislocations distributed randomly in the crystal.

The energy per dislocation decreases with the number of dislocations in the subboundaries, as r_d in Eq. (7.12) decreases. Therefore, low angle grain boundaries tend to combine, by which r_d decreases, whereas Θ increases according to Eq. (7.13). By combination of many subboundaries, eventually high-angle grain boundaries can be generated.

Recovery is controlled by climb and cross slip. Both processes depend on the normalized stacking fault energy, $\tilde{\gamma}_{SF}$, in such a way that climb and cross slip are promoted by increasing stacking fault energy. Therefore, materials with high $\tilde{\gamma}_{SF}$ are liable to suffer strong recovery phenomena, for instance fcc aluminum and most bcc metals; Ag, Cu, and their alloys, however, have low stacking fault energy and show little tendency to recovery.

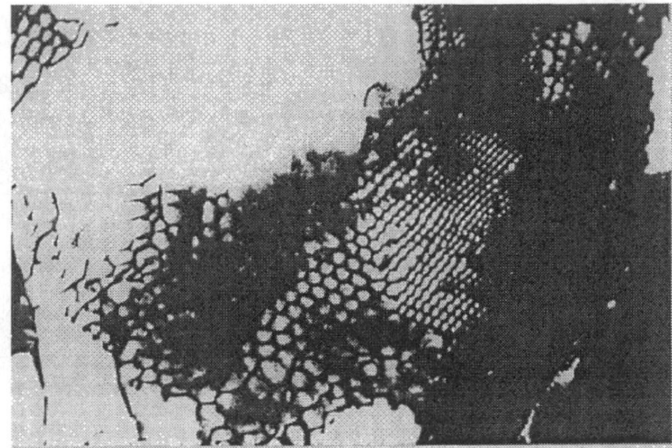

Fig. 7.17. A twist boundary formed by a network-like arrangement of screw dislocations in molybdenum [7.3].

The progress of recovery in a bent FeSi single crystal is shown in Fig. 7.18. After annealing at 650° for one hour the dislocations are still arranged along their slip planes (*a*). With increasing temperature at constant annealing time the dislocations rearrange perpendicular to the slip plane (polygonization (*e*)). After polygonization is complete at about 875°C, the structure coarsens, i.e. the average spacing of the low-angle grain boundaries increases (*h*).

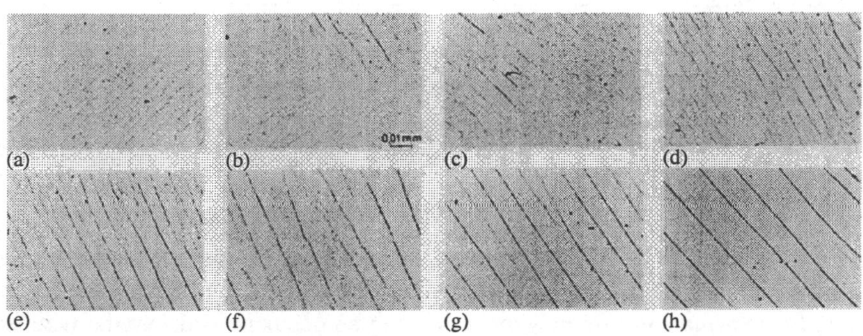

Fig. 7.18. Polygonization of edge dislocations in a bent iron-silicon single crystal, annealed for one hour at different temperatures.

Recovery processes are not confined to annealing after cold deformation (Fig. 7.19a,b), but can occur also concurrently with deformation. This is referred to as dynamic recovery and is manifested by a decrease of the work hardening rate. Dynamic recovery is caused by the arrangement of the dislo-

Fig. 7.19. TEM image of a deformed iron single crystal after annealing: (a) 20 min. at 400°C; (b) 5 min. at 600°C; (c) as in (b), but taken in the kink band, where subgrains were already generated during deformation [7.5].

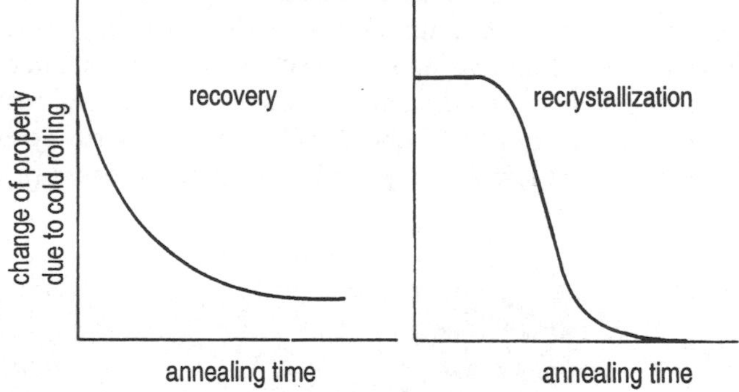

Fig. 7.20. Time dependence of recovery (a) and recrystallization (b) (schematically).

cations in cell walls, or in case of strong recovery, in subgrain boundaries. The extent of recovery depends on the type and prior arrangement of the dislocations. In deformation inhomogeneities, such as occur in kink bands in tensile deformed single crystals, a pronounced subgrain formation occurs during deformation (Fig.7.19c), while in other areas of the specimen the dislocation arrangement remains disordered (Fig. 7.19a).

Since recovery is only controlled by thermal activation it proceeds spontaneously and does not need an incubation time. Therefore, its kinetics are different from the kinetics of recrystallization. Hence, recovery makes itself felt particularly at small annealing times and attenuates with increasing annealing time, whereas recrystallization starts only after some incubation time and then usually progresses rapidly until completion (Fig. 7.20). Generally,

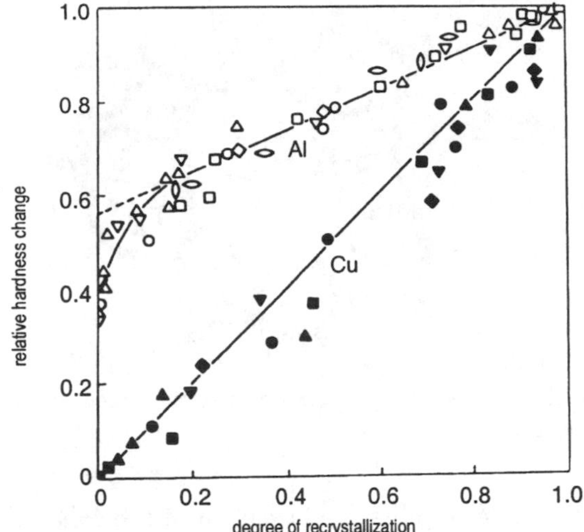

Fig. 7.21. Relative change in hardness as a function of the degree of recrystallization for copper and aluminum (after [7.6]).

recovery leads to similar property changes (for instance softening) as recrystallization. Therefore, one has to make sure to identify correctly the processes that cause a property change, especially if this property is used to determine recrystallization kinetics. This becomes clear from Fig. 7.21 where the hardness change and the recrystallized volume fraction, as determined by metallography, were measured on the same specimens. For Cu, which hardly recovers, a strict proportionality is found between the change of hardness and the recrystallized volume fraction. In the case of Al, the hardness decreases almost solely due to recovery. Only after some annealing time does recrystallization commence and cause the hardness to change linearly with the recrystallized volume fraction, X.

In some materials, under special circumstances, recovery can be strong enough to suppress completely recrystallization (recrystallization in-situ). Basically, recovery competes with, but also promotes, recrystallization. Recovery at the same time leads to the formation of nuclei for primary recrystallization.

7.5 Nucleation

In order to generate a recrystallization nucleus successfully from the deformed microstructure, three criteria have to be met, which are also referred to as the three instability criteria. These conditions are schematically sketched in Fig. 7.22.

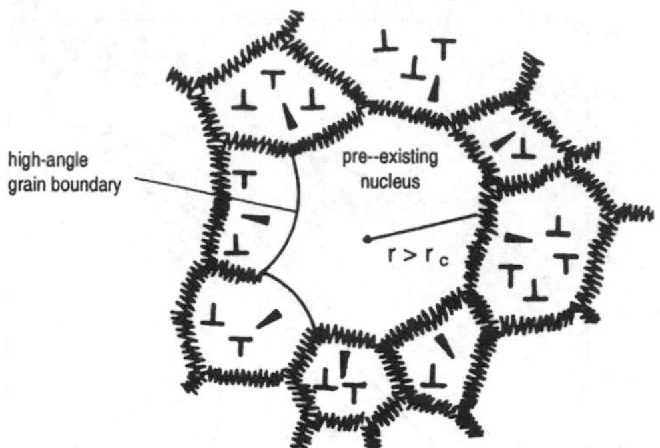

Fig. 7.22. Schematic illustration of a recrystallization nucleus with growth potential in a deformed structure.

i) Thermodynamic instability As in nucleation of the solid state from a melt (Chapter 8) a viable recrystallization nucleus has to exceed a critical size, owing to the fact that the growth of a nucleus requires a decrease of Gibbs free energy, even though the surface of the nucleus and, therefore, the total surface energy increases. The critical nucleus radius may be expressed with the driving force given by Eq. (7.4), namely

$$r_c = \frac{2\gamma}{p} = \frac{4\gamma}{\rho G b^2} \tag{7.17}$$

Because of the low driving force for recrystallization, the nucleation rate due to thermal fluctuations is by far too small to initiate recrystallization. Therefore, one must assume that a supercritical nucleus is already present in the deformed microstructure (pre-existent nucleus), for instance in form of a dislocation cell or subgrain. Additional recovery processes are necessary, however, to activate such a cell as a recrystallization nucleus.

ii) Mechanical instability The grain boundary of the nucleus has to move in a defined direction, which requires a local imbalance of the driving forces. This condition can be met by an inhomogeneous dislocation distribution or by a local imbalance of subgrain sizes that develops during the incubation period by recovery processes.

iii) Kinetic instability The surface of a nucleus, the grain boundary, must be mobile to make the nucleus grow. Only high angle boundaries usually have a sufficiently high mobility. The generation of a mobile high angle grain boundary from a deformed microstructure is one of the most difficult steps of nucleation of recrystallization. There are a variety of mechanisms that are currently discussed, e.g. discontinuous subgrain growth, nucleation at prior grain boundaries, deformation inhomogeneities or large particles, formation of recrystallization twins, to name just a few.

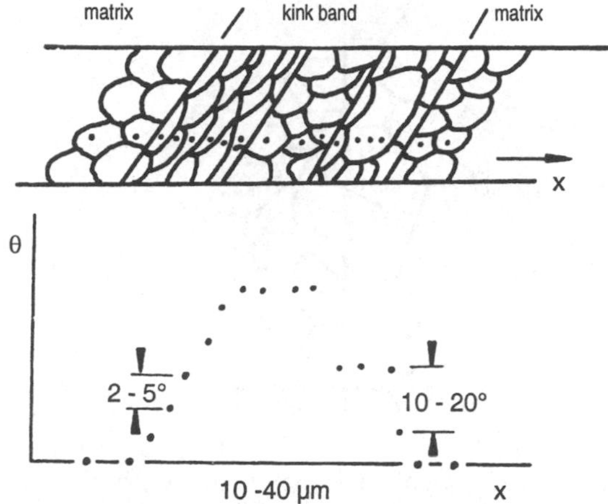

Fig. 7.23. Schematic illustration of the generation of a grain boundary by subgrain growth in deformation inhomogeneities.

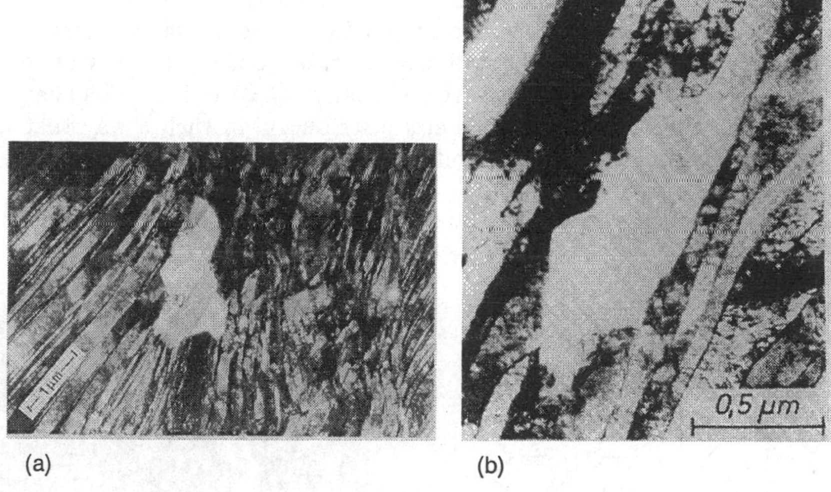

Fig. 7.24. TEM images of recrystallization nuclei that developed near the edge of a shear band and are growing into the deformed structure.

The constraint to comply with all three instability criteria concurrently favors nucleation in specific regions of the deformed microstructure, especially in deformation inhomogeneities and at prior grain boundaries. Deformation inhomogeneities usually contain orientations with a large misorientation to the matrix. By subgrain growth, a grain boundary with high misorientation and, thus, high mobility can be generated (Fig. 7.23 and Fig. 7.24). Across prior grain boundaries there is already a large orientation difference and, therefore, a potential high mobility. Nucleation at grain boundaries usually is preceeded

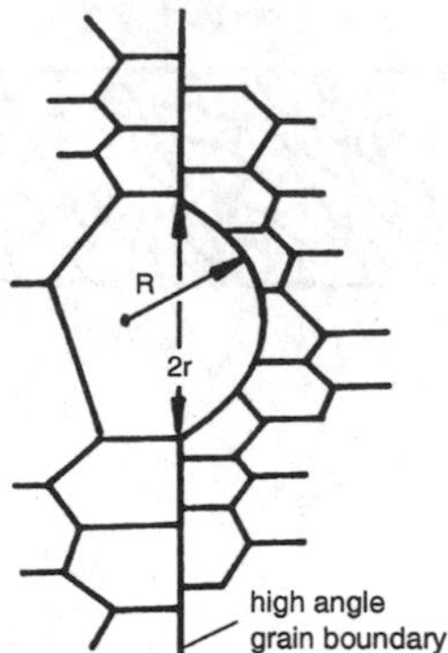

Fig. 7.25. Schematic illustration of nucleation at an existing high angle grain boundary.

by grain boundary bulging. Also, in this case, a thermodynamically critical subgrain nucleus size has to be exceeded, which is also given by Eq. (7.17) (Fig. 7.25). The difficulty for grain boundary bulging is the mechanical instability criterion (ii), that is the imbalance of driving forces. Usually the driving forces on both sides of a grain boundary are essentially equal, and an imbalance can only be introduced by the subgrain size distribution in such a way that the subgain size on one side of the boundary is smaller than that in the adjacent grain (Fig. 7.26).

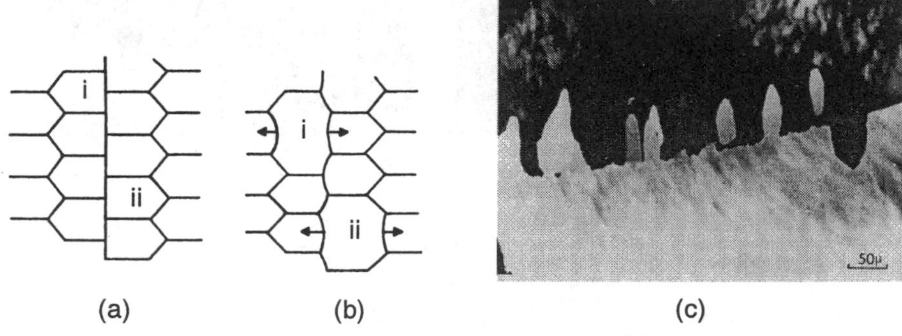

Fig. 7.26. "Strain induced boundary migration" (SIBM), left (a,b) schematic illustration, right (c) SIBM in weakly tensile strained aluminum [7.7].

Even without bulging of the boundary the area adjacent to a grain boundary is favored for nucleation because usually the dislocation density is higher and more inhomogeneous due to compatibility constraints arising from joint deformation of adjacent grains (Fig. 7.27). The same argument holds for large particles. Heterogeneous nucleation at a particle surface and a high and inhomogeneous dislocation density distribution promote rapid nucleation in two-phase alloys with a coarse dispersion (particle stimulated nucleation, PSN) (Fig. 7.28).

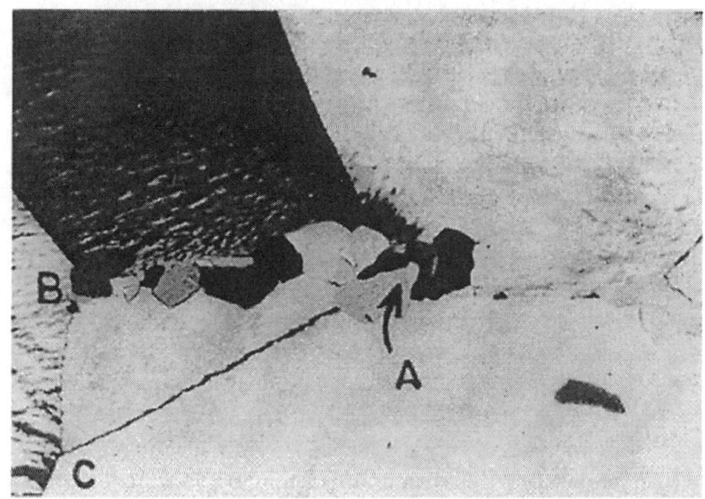

Fig. 7.27. Nucleation of recrystallization at grain boundaries and triple junctions in zone-refined aluminum [7.8].

In metals with low stacking fault energy, annealing twin formation can promote nucleation of recrystallization. By twin formation a different orientation is engendered and, therefore, a mobile high angle grain boundary is generated. Frequently, even twin chains are observed, i.e. twins of twins and twins of second order twins, and so on, which lead to highly mobile grain boundaries (Fig. 7.29).

All these processes require a local rearrangement of the deformed dislocation structure, i.e., nucleation is always connected to recovery processes. This is the reason for the so-called incubation time for recrystallization. On the other hand recovery and recrystallization are competing processes, because recovery degrades the driving force for recrystallization. In materials with strong recovery, i.e., with high $\tilde{\gamma}_{SF}$, for instance aluminum, (discontinuous) recrystallization can be delayed or even suppressed when softening progresses rapidly by (continuous) in-situ recrystallization.

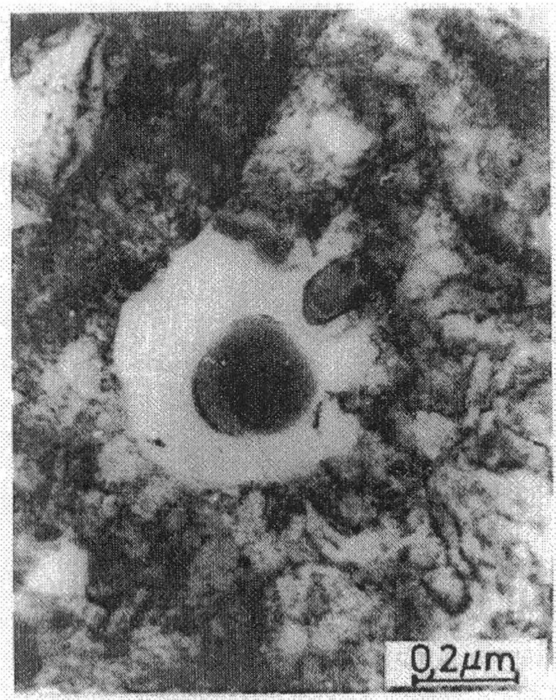

Fig. 7.28. Recrystallization nucleation at a TiC particle in high-strength microalloyed steel after 90% cold forming and annealing for 650h at 550°C [7.9].

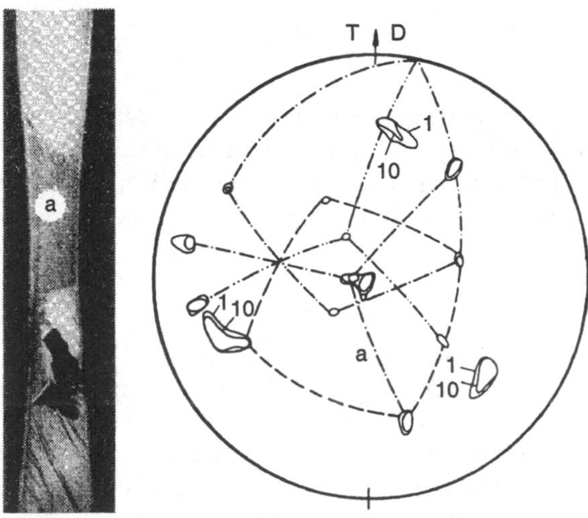

Fig. 7.29. Micrograph and {111} pole figure of a copper single crystal after dynamic recrystallization during a tensile test at 1103 K. (—1st, -·-2nd, -··-3rd. generation twin).

7.6 Grain Boundary Migration

If a grain boundary moves under the action of a driving force p [J/m^3] (see section 7.2) each atom gains the free energy pb^3, where b^3 is the atomic volume, if it detaches from the shrinking grain and subsequently attaches to the growing grain. The velocity of a grain boundary is the displacement per unit time from the difference of thermally activated diffusional jumps from the shrinking to the growing grain, and vice versa (Fig. 7.30)

$$v = b\nu_0 c_{vg} \left\{ \exp\left(-\frac{G_m}{kT}\right) - \exp\left(-\frac{G_m + pb^3}{kT}\right) \right\} \tag{7.18}$$

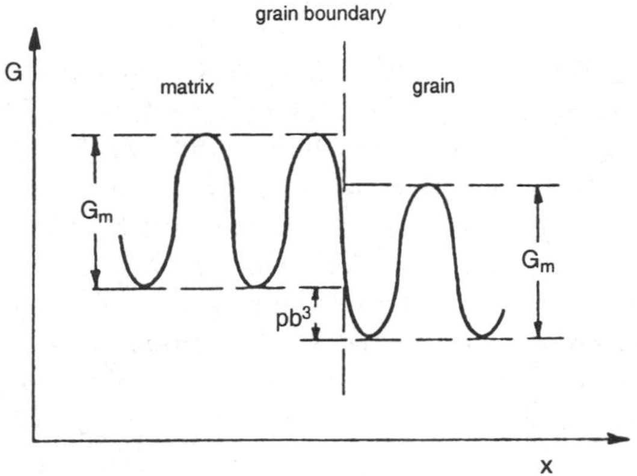

Fig. 7.30. Diagrammatic free energy curve at a grain boundary due to the action of a driving force p.

Here ν_0 is the atomic vibration frequency ($\approx 10^{13}\text{s}^{-1}$), G_m the free activation energy for a diffusional jump through the grain boundary, and c_{vg} the vacancy concentration in the grain boundary since, as in self diffusion, only a jump to an empty site in the grain boundary is possible.

Because of the small driving forces of recrystallization, at typical recrystallization temperatures ($T > 0.4T_m$)

$$pb^3 \ll kT \tag{7.19}$$

(for instance for strongly deformed Cu at half of its melting temperature (400°C) $pb^3 \cong 10^{-22}$ J, $kT \cong 1/20$ eV $\cong 10^{-20}$ J, $pb^3/kT \cong 0.01$). Hence, the exponential term in Eq. (7.18) can be linearly expanded as

$$
\begin{aligned}
v &\cong b\nu_0 c_{vg} \exp\left(-\frac{G_m}{kT}\right)\left\{1 - 1 + \frac{pb^3}{kT}\right\} \\
&= b^4 \nu_0 c_{vg} \frac{1}{kT} \exp\left(-\frac{G_m}{kT}\right) \cdot p
\end{aligned}
\tag{7.20a}
$$

or

$$
v = mp \tag{7.20b}
$$

The relation between mobility, m, and diffusion coefficient, D_m, for jumps through the grain boundary with the activation energy, Q_m, is given with the Nernst-Einstein relation as

$$
m = \frac{b^2 D_m}{kT} = \frac{b^2 D_0}{kT} \exp\left(-Q_m/kT\right) = m_0\, e^{-Q_m/kT} \tag{7.21}
$$

By comparison of Eqs. (7.21) and (7.20a) it is obvious that $Q_m = H_m$, if the vacancy concentration, c_{vg}, is not thermally activated. The experimental determination of grain boundary mobility is difficult, because only in special cases a constant driving force or a constant grain boundary geometry is obtained. Also the influence of other factors such as surfaces, specimen purity, etc., are difficult to account for properly. For an accurate determination of grain boundary mobility, experiments on specially grown bicrystals are most appropriate. From such experiments the proportionality between grain boundary velocity and driving force was confirmed (Fig. 7.31) and the grain boundary mobility could be determined according to Eq. (7.20b).

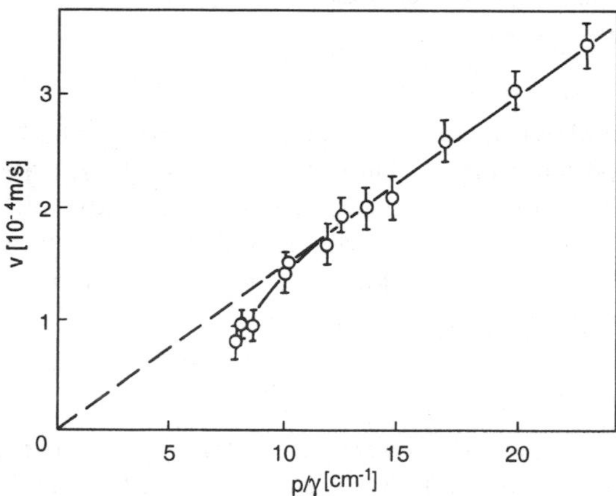

Fig. 7.31. Grain boundary migration rate as a function of the reduced driving force p/γ (γ grain boundary energy) in an aluminum bicrystal (after [7.10]).

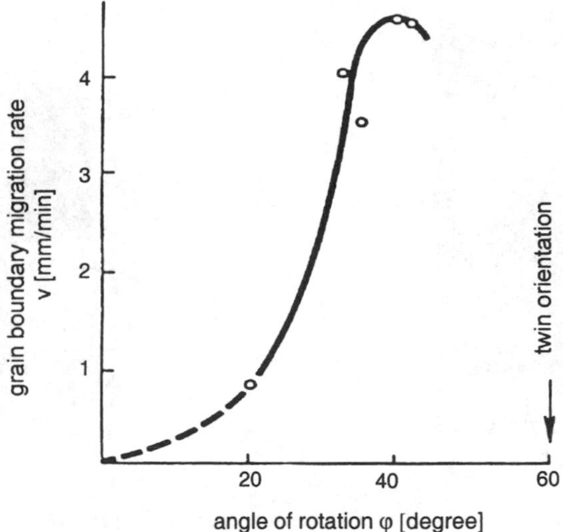

Fig. 7.32. Dependence of the grain boundary migration rate on the angle of rotation of ⟨111⟩ tilt boundaries in aluminum (after [7.11]).

Even a small bulk concentration of impurities strongly influences grain boundary mobility, since impurities tend to segregate at the grain boundary and exert a drag force on the boundary upon its motion (see Chapter 7.9). In such cases, high activation energies Q_m are observed.

The mobility of a grain boundary also depends on the orientation relationship between the adjacent grains $\omega\langle hkl\rangle$ (ω - rotation angle, $\langle hkl\rangle$ - rotation axes) (Fig. 7.32)

$$m = m\,(\omega < \text{hkl} >) \tag{7.22}$$

Low-angle grain boundaries have low mobilities, whereas grain boundaries with an orientation relationship 40°⟨111⟩ show high mobility in aluminum. Depending on material and crystal structure, other orientation relationships were also found for rapidly moving boundaries, for instance 30°⟨0001⟩ for Zn or 27°⟨110⟩ for Fe-3%Si. Also, a special orientation of the grain boundary plane may play a role regarding its mobility. For instance, in aluminum, a high mobility is found for ⟨111⟩ tilt boundaries (Fig. 7.33) whereas in Fe3%Si the ⟨100⟩ twist boundaries seem to move especially fast.

The orientation dependence, in particular the dependence of mobility on misorientation angle, is usually understood in terms of impurity segregation to grain boundaries. Highly ordered "coincidence" boundaries (see Chapter 3) are mobile, since they are less prone to segregation (Fig. 7.34). So, the orientation dependence of the mobility is commonly interpreted in such a way that the grain boundaries are affected by impurity drag when they absorb more or less impurity atoms, depending on their misorientation. Coincidence bound-

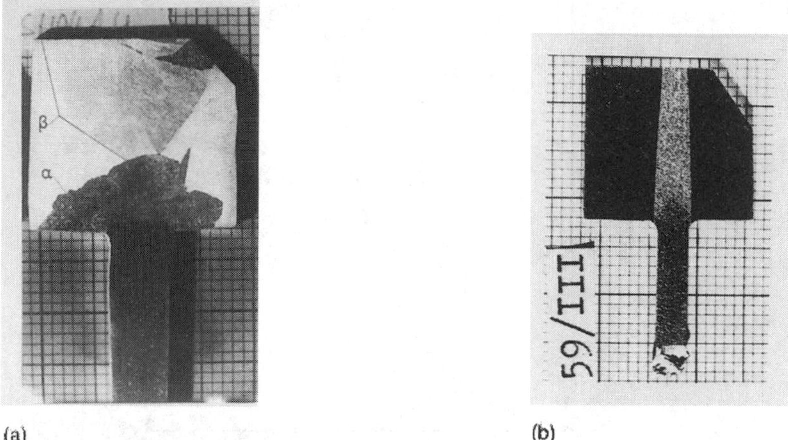

Fig. 7.33. Examples of isotropic and anisotropic grain growth in aluminum. Prior to the start of annealing the bicrystal consisted of a recrystallized grain in the handle and a slightly deformed grain in the blade. During annealing the grain boundary moved into the deformed structure isotropically in (a), i.e. at approximately the same rate for all directions for a ⟨100⟩ rotation axis, and very anisotropically in (b) for a ⟨111⟩ rotation axis. The long straight contrast lines in (b) are traces of {111} twist boundaries, which apparently are not very mobile [7.12].

aries segregate least and, thus, have the highest mobility. With an increasing content of solute atoms, grain boundary migration is drastically reduced. In contrast, at high purity for some boundaries (⟨100⟩ tilt boundaries in aluminum) the misorientation dependence of grain boundary mobility may even disappear (see Fig. 7.34).

7.7 Kinetics of Primary Recrystallization

Because of its high dislocation density the deformed state is thermodynamically unstable at all temperatures. Its removal by recrystallization is an irreversible process, because it represents a transition from a metastable equilibrium to a more stable state, and the initial state can not be reestablished. Therefore, recrystallization is sometimes also referred to as a phase transformation without a true equilibrium temperature. Nevertheless, the kinetics of recrystallization are frequently described in terms of a recrystallization temperature, which is defined by the condition that recrystallization at this temperature comes to completion in a time that can be commercially realized (typically 1 hour). Such a definition is a reasonable one, because recrystallization is a thermally activated process and, therefore, strongly dependent on temperature via a Boltzmann factor ($\exp(-Q/kT)$). Hence, small changes of temperature cause large changes of the recrystallization time (see Sec. 7.13);

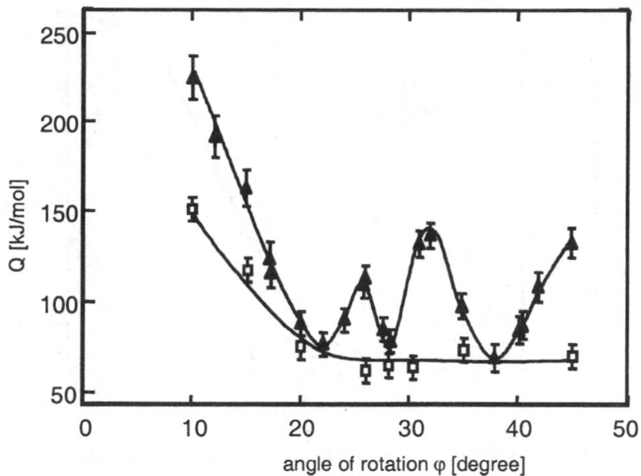

Fig. 7.34. Mobility activation energy of ⟨100⟩ tilt boundaries in aluminum of different purity as a function of the angle of rotation. The minima occur at coincidence orientations ($\Sigma < 17$) (after [7.13]).

conversely, a definition of the recrystallization time of 0.5 h, 1 h, or 2 h leads only to a small change of the recrystallization temperature.

The kinetics of primary recrystallization are determined by thermal activation of its mechanisms, nucleation, and nucleus growth, all of which control the progress of recrystallization with annealing time. For a quantitative treatment we define the nucleation rate $\dot{N}$ and the growth rate v, respectively, as

$$\dot{N} = \frac{\frac{dz_N}{dt}}{1 - X} \tag{7.23}$$

$$v = \frac{dR}{dt} \tag{7.24}$$

Here, $X = V_{RX}/V$ is the recrystallized volume fraction, t the annealing time, R the radius of a grain, z_N is the number of observed nuclei per unit volume. $\dot{N}$ is the number of nuclei generated per unit of time and volume in the unrecrystallized volume. The growth of nuclei is simply assumed to proceed isotropically, i.e. the nuclei grow like spheres, where R is the radius of the sphere.

Fig. 7.35 shows the measured recrystallized volume fraction X as function of annealing time t for cold deformed aluminum. For a quantitative description one often uses the Johnson-Mehl-Avrami-Kolmogorov equation

$$X = 1 - \exp\left\{-\left(\frac{t}{t_R}\right)^q\right\} \tag{7.25}$$

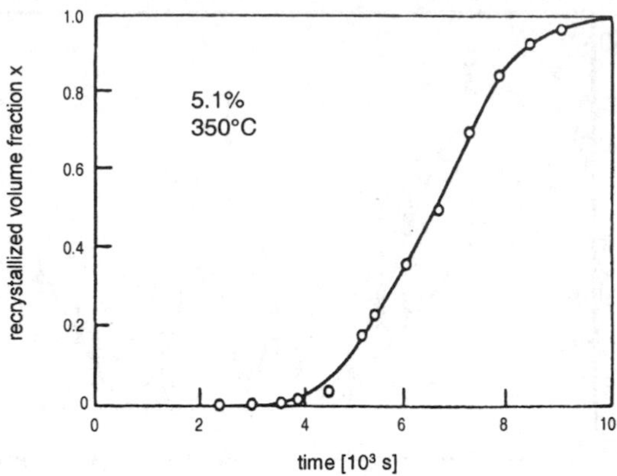

Fig. 7.35. Recrystallized volume fraction as a function of annealing time following 5.1% tensile deformation (after [7.14]).

Here t_R is the characteristic time for recrystallization. Intuitively, the recrystallization time is associated with the time necessary for recrystallization to go to completion. According to Eq. (7.25) this is mathematically obtained only after an infinitely long annealing time. Practically, it is more reasonable to define the recrystallization time as the time necessary to reach a certain volume fraction recrystallized. Most convenient mathematically (and also physically sensible) is a definition $X(t_R) = 1 \text{-} (1/e) = 0.63$. This is the definition of t_R in Eq. (7.25). One could also use $X = 0.99$ to define the characteristic time, but its dependency on deformation and annealing conditions would be similar. However, at values of X close to 100% deviations from Eq. (7.25) are usually observed. Hence, an intermediate value of X is most sensible for the definition of t_R.

Under simplified assumptions the recrystallization time, t_R, and the primary recrystallized grain size, d, the recrystallized volume fraction, X, and the time exponent, q, can all be derived if $\dot{N}$ and v are known. If the grains grow as spheres until impingement (isotropic growth), nucleation occurs homogeneously throughout the microstructure (homogeneous nucleation) and v and $\dot{N}$ remain constant during the entire process (i.e., no pronounced recovery during recrystallization). The recrystallized volume fraction can then be derived as a function of time to be

$$X(t) = 1 - \exp\left(-\frac{\pi}{3}\dot{N}v^3t^4\right) \tag{7.26}$$

By comparison of Eq. (7.26) and (7.25) the recrystallization time and the primary recrystallized grain size can be immediately derived

$$t_R = \left(\frac{\pi}{3}\dot{N}v^3\right)^{-1/4} \tag{7.27}$$

and

$$d \cong 2vt_R = 2\left(\frac{3}{\pi}\frac{v}{\dot{N}}\right)^{1/4} \tag{7.28}$$

Of course, the calculation of the recrystallized grain size according to Eq. (7.28) is only an approximation, because in a real microstructure the recrystallized grains will eventually impinge and, therefore, cannot expand freely anymore. Eq. (7.28), however, gives the right order-of-magnitude and the correct dependency on deformation and annealing conditions.

Grain boundary mobility and nucleation rate are thermally activated processes (see Sec. 7.6 and 7.3). With the respective activation energies as Q_v and $Q_{\dot{N}}$ we can express v and $\dot{N}$ by

$$v = v_0 \exp\left(-Q_v/kT\right) \tag{7.29}$$

$$\dot{N} = \dot{N}_0 \exp\left(-Q_{\dot{N}}/kT\right) \tag{7.30}$$

where the pre-exponential factors v_0 and $\dot{N}_0$ are independent of the temperature.

From Eq. (7.27) in combination with Eq. (7.29) and Eq. (7.30) it is obvious that also the recrystallization time depends exponentially on temperature in such a way that the recrystallization time decreases exponentially with increasing temperature.

$$t_R = \left(\frac{3}{\pi \dot{N}_0 v_0^3}\right)^{1/4} \cdot \exp\left(\frac{Q_{\dot{N}} + 3Q_v}{4kT}\right) \tag{7.31}$$

This has been confirmed by measurements on various materials (Fig. 7.36). The slope of an Arrhenius plot of t_R versus $1/T$ yields the apparent activation energy of primary recrystallization: $(Q_{\dot{N}} + 3Q_v)/4$.

As previously mentioned, one can define the recrystallization temperature as the temperature where recrystallization goes to completion in a defined time — about 1 h in commercial practice — ($T_R = T(t_R = \text{const.})$). In pure metals, the recrystallization temperature after large strain deformation approximately scales with the melting temperature ($T_R \cong 0.4T_m$). The recrystallization temperature can be much higher in alloys depending on the alloying elements or the dispersion of a second phase, which impedes recrystallization.

The temperature dependence of the recrystallized grain size can be derived from Eq. (7.28) through (7.30) to yield

$$d = \left(\frac{48v_0}{\pi \dot{N}_0}\right)^{1/4} \exp\left(\frac{Q_{\dot{N}} - Q_v}{4kT}\right) \tag{7.32}$$

Eq. (7.28) and Eq. (7.32) demonstrate that the recrystallized grain size is determined by a competition between nucleation rate and growth rate. An

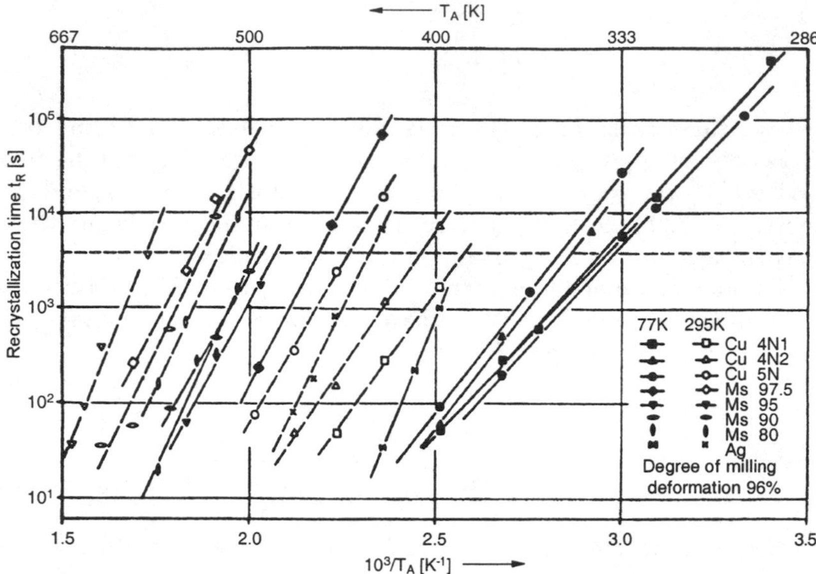

Fig. 7.36. Dependence of recrystallization time on annealing temperature T_A for different materials and rolling temperatures (after [7.15]).

increase of the growth rate leads to a larger grain size, whereas an increase of the nucleation rate at a constant growth rate yields a much finer grain size.

Very often both activation energies $Q_{\dot{N}}$ and Q_v are approximately equal. In this case the primary recrystallized grain size ought to be independent of temperature according to Eq. (7.32). In fact, this is usually observed. In some cases, however, $Q_{\dot{N}}$ is much larger, as in the case of aluminum. In this case a finer grain size is observed with increasing temperature.

Both $\dot{N}$ and v grow with increasing degree of deformation, which also shortens the recrystallization time. Since $\dot{N}$ mostly increases more strongly with degree of deformation than v, the recrystallized grain size becomes smaller according to Eq. (7.28) with increasing strain. A smaller grain size prior to deformation has a similar effect.

The influence of alloying elements is important. If the alloying elements are in solid solution, usually v decreases (see Sec. 7.9). Often $\dot{N}$ is also reduced upon alloying in particular at low concentrations, approximately to the same extent as v. This becomes obvious from the fact that the recrystallized grain size and, therefore, the ratio $v/\dot{N}$ according to Eq. (7.28) changes only a little whereas the absolute change of $\dot{N}$ and v, and, accordingly, the recrystallization time, can vary by many orders-of-magnitude.

The simplified assumptions for the derivation of Eqs. (7.26) and (7.28) are in most cases at variance with reality. For instance, the nucleation rate frequently tends to alter with increasing annealing time, as obvious from di-

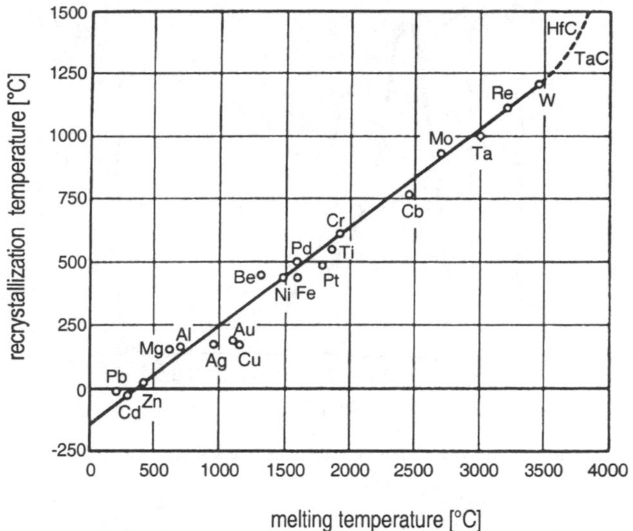

Fig. 7.37. Recrystallization temperatures of various metals as a function of melting temperature (after [7.16]).

rect metallographic measurements (Fig. 7.38) and in virtually no case is it completely isotropic. Also nucleation does not occur homogeneously throughout the volume. These deviations from ideal behavior become obvious in a so-called Avrami plot $lgln(1/1 - X)$ versus lgt. According to Eq. (7.26) one ought to obtain straight lines with a slope $q = 4$. In reality, for engineering materials $q \cong 2$. This indicates that either $\dot{N}$ or v or both are not constant. Rather these quantities subside with time (for instance by recovery), or the growth does not proceed equally well in all three dimensions (for instance during nucleation at grain edges). However, even if there are deviations from the simple assumptions of the quantitative analysis, the derived general dependencies remain valid.

7.8 The Recrystallization Diagram

The recrystallization diagram represents the dependencies between recrystallized grain size, degree of deformation, and annealing temperature. It is particularly helpful for industrial applications. The grain size usually increases with increasing degree of deformation and increasing annealing temperature (Fig. 7.39). It is stressed, however, that the recrystallization diagram does not reveal the grain size at terminal primary recrystallization, rather after a constant time of annealing.

At low temperatures, recrystallization has not yet occurred or has not gone to completion at the chosen annealing time. In this case it is impossible

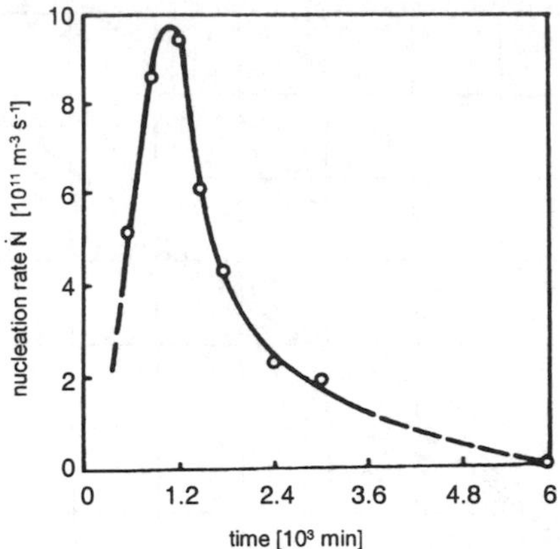

Fig. 7.38. Dependence of nucleation rate on annealing time for tensile deformed aluminum (after [7.17]).

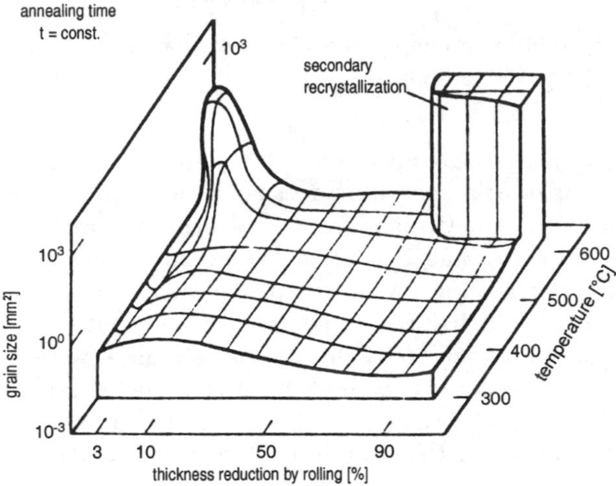

Fig. 7.39. Recrystallization diagram of high purity aluminum (after [7.18]).

to define a recrystallized grain size, i.e., the recrystallization diagram only begins above a certain temperature which depends on the degree of deformation.

At high temperatures and with the chosen annealing time, primary recrystallization usually has gone to completion, and grain growth has also progressed. Since the grain size increases during grain growth (Fig. 7.11), the recrystallization diagram reveals a larger grain size with increasing annealing temperature, whereas after primary recrystallization the grain size ought to be essentially independent of temperature, or even to decrease with rising temperature. Hence, to obtain a small recrystallized grain size, annealing has to be terminated immediately after complete recrystallization.

If secondary recrystallization is set off (see section 7.10), the recrystallization diagram will predict large grains. Moreover, it is important to realize that the heating-up time is not taken into account in this diagram, although it is of technical importance. A low heating rate corresponds to a prior recovery treatment at low temperatures and frequently decreases the number of recrystallization nuclei, hence affecting recrystallization kinetics and recrystallized grain size.

7.9 Recrystallization in Homogeneous Alloys

The addition of alloying elements has little influence on the nucleation rate, but depending on the character of the alloying element it may have a strong influence on the grain boundary migration rate (Fig. 7.40). The reason is a preferential segregation of solute atoms to grain boundaries to lower the free energy.

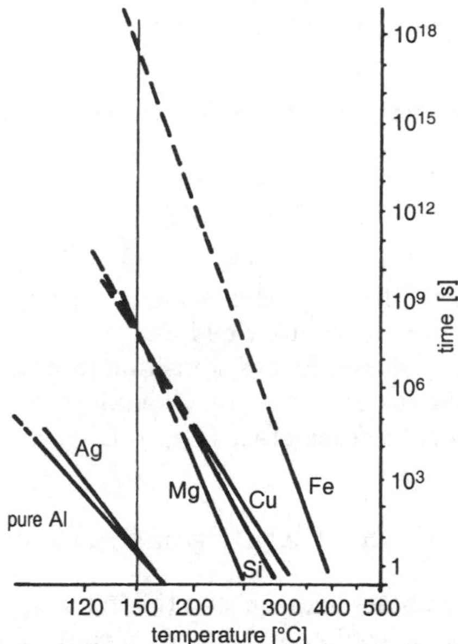

Fig. 7.40. Temperature dependence of the recrystallization time of binary alloys of high purity aluminum and 1/100 atomic percent of a second metal (after [7.19]).

During motion of the grain boundary the solute atoms have to stay in the boundary to maintain their low energy state, i.e., they have to move along with the boundary via diffusion and thus exert a back-driving force p_R on

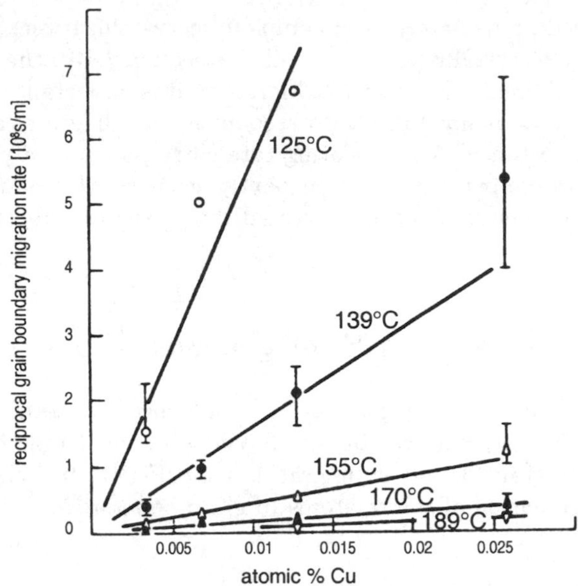

Fig. 7.41. Dependence of reciprocal grain boundary migration rate on copper concentration in zone refined aluminum (after [7.20]).

the boundary, which depends on the boundary migration rate, v, and solute concentration, c, according to

$$v = m\,(p - p_R(c, v)) \tag{7.33}$$

At low velocities the grain boundary is loaded with solute atoms and the back driving force grows with increasing velocity. At high velocities the grain boundary can detach from its solute cloud and move freely. At intermediate velocities there may be a discontinuous transition from the loaded to the free state of the boundary. In this state the grain boundary velocity and the driving force are not proportional to each other (Fig. 7.41 - 7.44).

7.10 Recrystallization in Multiphase Alloys

The presence of second phases has a substantial influence on recrystallization. It can accelerate recrystallization or even completely suppress it. Generally, recrystallization kinetics are sped up by coarse particles, whereas a fine dispersion of particles strongly hinders recrystallization.

This complex influence of particles on recrystallization is due to the fact that particles influence both the deformation structure, recovery during nucleation, and eventually grain boundary migration. Hard particles cannot be deformed and geometrically necessary dislocations will generate an inhomogeneous dislocation structure around those particles. If the particles are large,

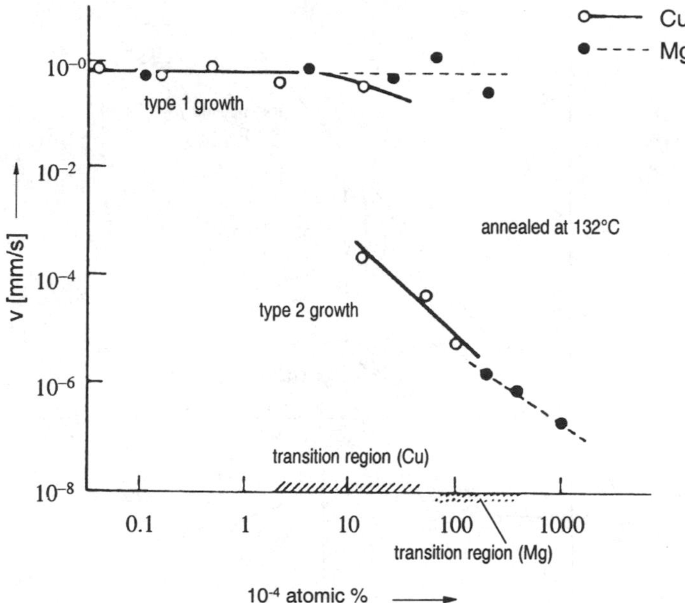

Fig. 7.42. Growth rate of recrystallized grains in rolled aluminum with added Cu or Mg (after [7.21]).

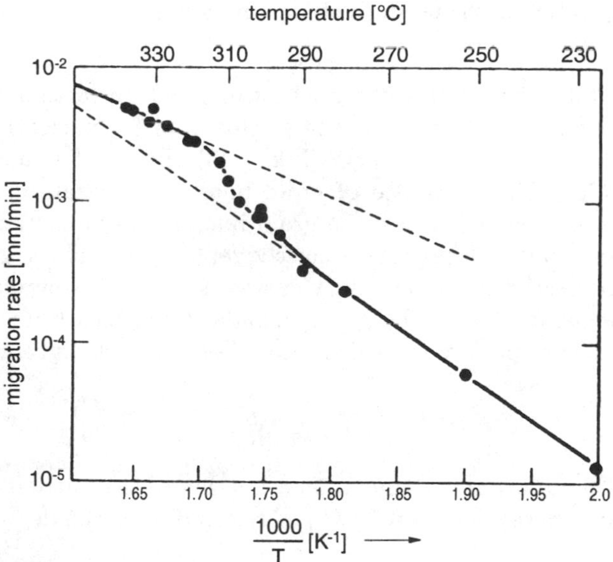

Fig. 7.43. Migration rate of a recrystallization front in rolled gold with 20 ppm iron (after [7.22]).

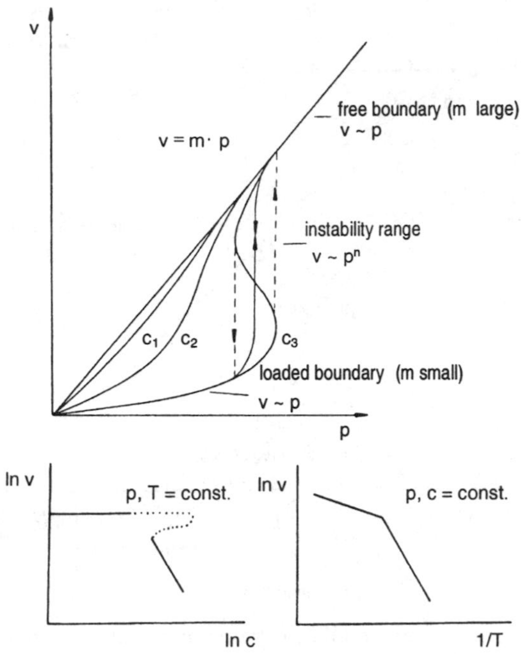

Fig. 7.44. Theoretical dependence of the grain boundary migration rate on driving force and temperature in the presence of impurity atoms.

this can facilitate nucleation by particle-stimulated nucleation (PSN) (Fig. 7.45). Finely dispersed particles are less influential for nucleation but they hinder dislocation motion (recovery) (Fig. 7.46) and grain boundary migration (Fig. 7.47a). The hindrance of grain boundary migration by particles is due to a back driving force on the grain boundary. On contact of a grain boundary with a particle the particle surface replaces part of the grain boundary and the respective grain boundary energy is reduced. However, this area has to be regenerated when the grain boundary detaches from the particle (Fig. 7.47b). The respective retarding force (Zener force) is given by

$$p_R = -\frac{3}{2}\gamma\frac{f}{R_p} \tag{7.34}$$

where R_p is the radius and f the volume fraction of the particles, and γ the grain boundary energy. The ratio f/R_p is referred to as the degree of dispersion.

Only the effective driving force $p_{eff} = p + p_R$ is available for grain boundary migration and since p_R is negative, p_{eff} may be small. In heterogeneous alloys with a high degree of dispersion, the backdriving force can become large and recrystallization is retarded (higher recrystallization temperatures). For commercial applications the particles are important to stabilize the grain size after primary recrystallization, because the Zener force affects grain growth even more strongly and frequently completely suppresses it. This can be easily seen from a comparison of driving force and Zener force. For instance, for $f = 1\%$, $r_p = 1000\text{Å}$ and $\gamma = 0.6\ \text{J/m}^2$ a Zener force of about 0.1 MPa is obtained which is of approximately the same order of magnitude or even larger than the driving force for grain growth.

7.11 Normal Grain Growth

Primary recrystallization goes to completion if the growing recrystallized grains fully impinge and the entire deformed microstructure has been consumed. The result is a new strain-free polycrystalline structure with a Gibbs free energy much lower than the energy of the deformed state.

A polycrystalline microstructure consists of an arrangement of grains, the morphology of which can be described by polyhedra that remain in contact at their faces, edges, and corners. The shape of the grains results from a compromise to satisfy two requirements: namely, complete filling of space and mechanical equilibrium of surface tensions at edges and corners. The mathematical treatment of this three-dimensional problem is difficult, and even difficult to visualize. The fundamental physical processes, however, can be easily realized in a two-dimensional model. In a two-dimensional structure the boundaries are only in stable equilibrium if all grains have 6 corners, because in this case all boundaries are straight and all contact angles at the junctions are equilibrium angles (Fig. 7.48).

The equilibrium of contact angles due to the grain boundary surface tensions at the junctions is given (if torque terms resulting from anisotropy are neglected, see Chapter 3) (Fig. 7.49) by

$$\frac{\gamma_1}{\sin\alpha_1} = \frac{\gamma_2}{\sin\alpha_2} = \frac{\gamma_3}{\sin\alpha_3} \tag{7.35}$$

In single phase metals the grain boundary surface tension of most grain boundaries is similar, therefore, the contact angle is about 120°. In multiphase alloys there can be major deviations from 120°, because of different surface tensions at the interface boundaries. For instance in α/β brass the equilibrium angle at a triple junction of one β and two α grains is 95°.

If in a granular microstructure a grain has a number of edges different from six, for instance in Fig. 7.48 the five-sided grain, then the force equilibrium at the junctions can be attained according to Eq. (7.35) only if at least one grain boundary is curved. On a curved boundary there acts a force to move it towards its center of curvature, to mininize the grain boundary area. If the grain boundary is displaced, it disturbs the equilibrium angle of 120° at its terminal junctions. To reestablish equilibrium the other grain boundaries at the junctions have to readjust by migration, which again causes a grain

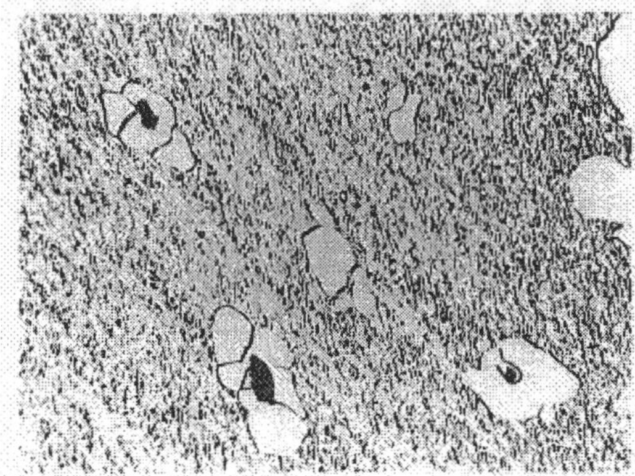

Fig. 7.45. Nucleation at oxide particles in 60% rolled iron (2 min. 540°C) (TEM image) [7.23].

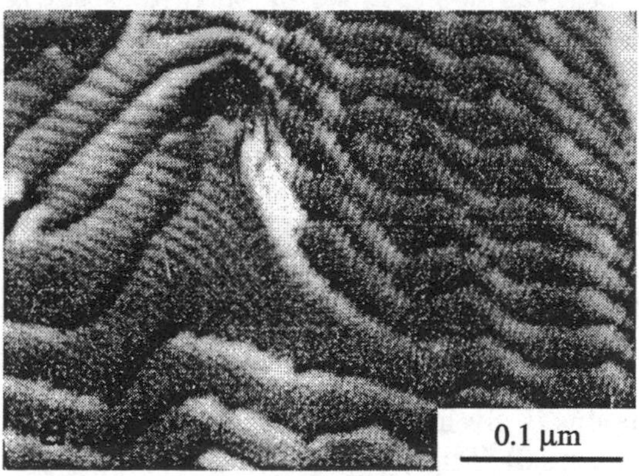

Fig. 7.46. Small particles (aluminum oxide) impeding the motion of dislocations at a subgrain boundary [7.24].

boundary curvature, further movement, and so on. As a consequence a stable equilibrium can never be reestablished.

It is known from morphological investigations that grains with more than six sides are bordered by grain boundaries with mainly concave curvature, whereas grains with less than six sides have convex curved boundaries to comply with the equilibrium constraints at the junctions (Fig. 7.50). Since large grains are surrounded by many smaller grains, they usually have more than six sides, whereas small grains usually have less than six sides. The motion of the boundaries towards their centers of curvature in order to reduce their grain boundary area causes grains with more than six sides, i.e. the large grains, to grow while grains with less than six sides, i.e. the small grains, shrink.

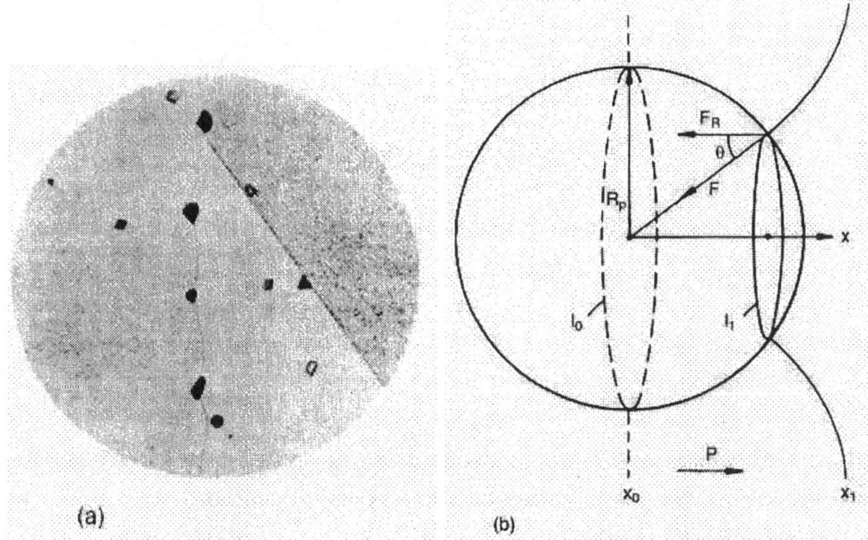

Fig. 7.47. Particles of a second phase impede the motion of a high angle grain boundary; (a) pinning of a grain boundary by inclusions in α-brass [7.25]; (b) schematic illustration (to explain the derivation of equation (7.34)).

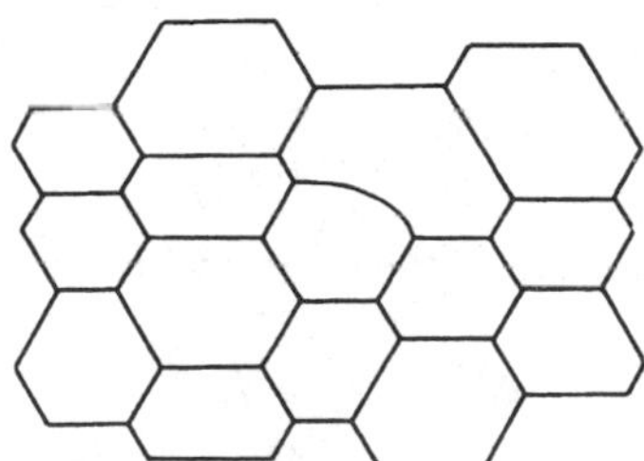

Fig. 7.48. Two-dimensional equilibrium structure that, except for one defect, consists only of hexagons with 120 °-contact angle.

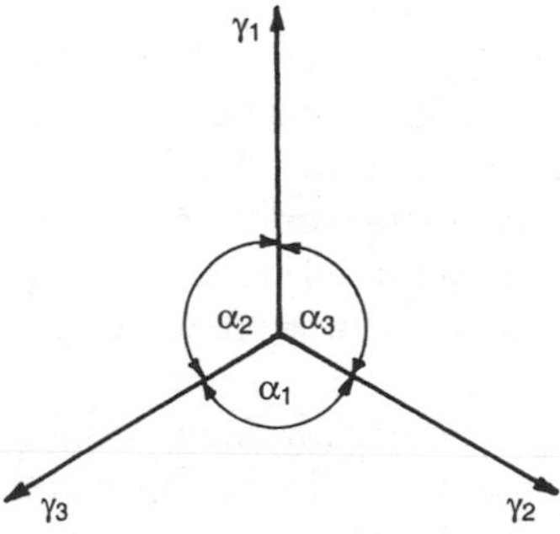

Fig. 7.49. Schematic illustration showing the forces and contact angles at a triple junction formed by grain boundaries.

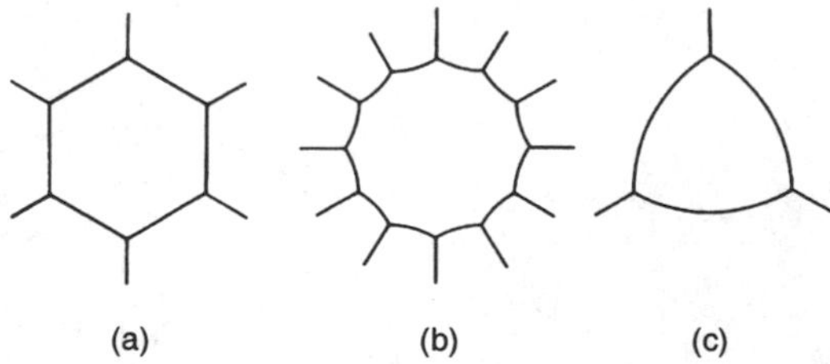

Fig. 7.50. Curvature of the sides of regular polygons with different numbers of corners and a contact angle of 120°.

In the three-dimensional case, complete filling of space and simultaneously planar grain boundaries can be obtained only if all grain boundaries meet at a contact angle of approximately 120° along a common edge and at a contact angle of approximately 109° in a quadruple point. Therefore, in a 3D granular structure it is never possible to establish interfacial equilibrium.

The increase of the average grain size during isothermal annealing can be predicted from a simple model. Under the assumption that the average radius of curvature of a grain boundary R is proportional to the grain size D, and that the average grain boundary migration rate v is proportional to the temporal change of the grain size (dD/dt) one obtains from Eqs. (7.20b) and (7.6)

$$\frac{dD}{dt} = mK_1 \frac{\gamma}{D} \tag{7.36}$$

and by integration, assuming that the constant K_1 does not change with time,

$$D^2 - D_0^2 = K_2 t \tag{7.37}$$

D_0 is the grain size at time $t = 0$, i.e., immediately after primary recrystallization.

If $D_0 \ll D$, Eq. (7.37) can be simplified to

$$D \cong Kt^n \tag{7.38}$$

with $n = 0.5$, i.e., the average grain size ought to increase in proportion to the square root of the annealing time. The exponent $n = 0.5$, however, is only confirmed for high purity metals and for annealing temperatures close to the melting temperature. In metals of commercial purity, typically n-values fall between 0.2 and 0.3 depending on purity, temperature, and texture (Fig. 7.51).

Grain growth will substantially slow down if the average grain size reaches the magnitude of the smallest specimen dimension. If a grain boundary touches a free surface, a triple junction between the grain boundary and two external crystal surfaces is generated that requires an equilibrium of surface tensions. The result is the formation of a (thermal) groove along the grain boundary (Fig. 7.52), which has to be dragged along or left behind upon further grain boundary migration. This causes a retarding force independent of the depth of the groove

$$p_{GR} = -\frac{\gamma_{GB}^2}{h\gamma_S} \tag{7.39}$$

(h - specimen thickness, γ_S - surface energy, γ_{GB} - grain boundary energy).

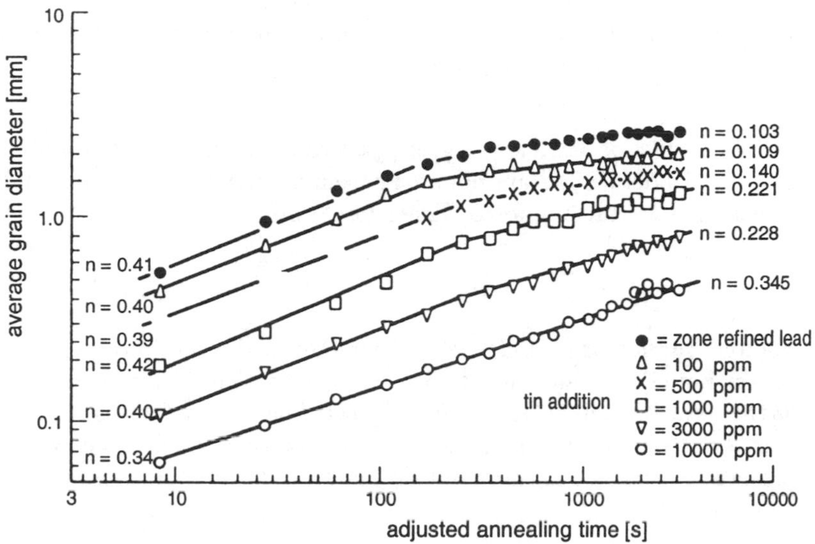

Fig. 7.51. Grain growth in zone refined lead and lead with different tin content. The deviation of the exponent from the ideal value 0.5, and the increase of this deviation with increasing tin content are apparent (after [7.26]).

This retarding force reduces the growth rate of grains in contact with the surface. As long as the grain size is much smaller than the specimen thickness this influence on grain growth is negligible, but it becomes increasingly significant as more grains touch the surface. Even if all grains extend through the thickness of a sheet (columnar structure) there is still a 2D curvature of the grain boundaries in the sheet plane and thus, a driving force for grain growth, which diminishes according to Eq. (7.6). Grain growth finally ceases due to grooving, if the average grain size corresponds to about 2 times the sheet thickness.

If a material also contains precipitates that do not dissolve during annealing, continuous grain growth is limited eventually by Zener drag and occurs only up to a maximum grain size that depends on the dispersion of the second phase. The final grain size would be smaller than in a comparable single phase

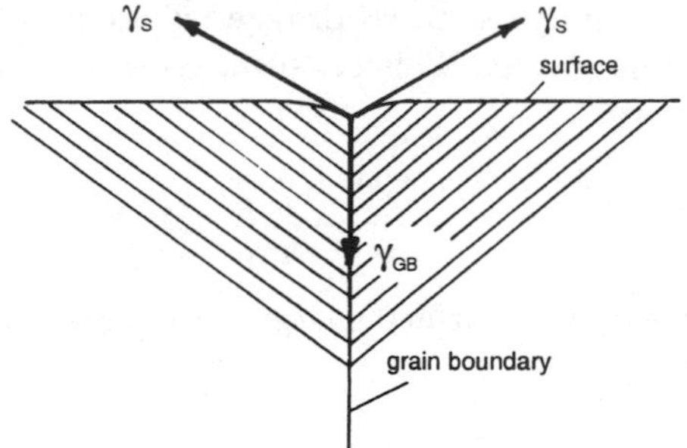

Fig. 7.52. The equilibrium of the forces of surface tension and grain boundary tension causes the formation of a groove.

material. The effect of particles on grain growth is due to the Zener force (see section 7.10). To detach a grain boundary from a particle the grain boundary has to overcome a retarding force (Zener force) p_Z

$$p_Z = -3\gamma \frac{f}{d_p} \tag{7.40}$$

(f = volume fraction of particles, d_p = diameter of the precipitates).

A grain boundary can only detach from a particle if the driving force exceeds the retarding force. During grain growth the driving force continuously decreases, since the curvature of the grain boundaries, which scales with the grain size, diminishes. Grain growth terminates, if driving force and retarding force compensate, i.e., $p = -p_Z$

$$\frac{2\gamma}{\alpha \cdot d} = 3\gamma \frac{f}{d_p} \tag{7.41}$$

(α is a constant relating radius of curvature of grain boundaries and grain size).

This yields the terminal grain size

$$d_{\text{max}} = \frac{2}{3} \cdot \frac{1}{\alpha} \frac{d_p}{f} \tag{7.42}$$

Since particles undergo Ostwald ripening (coarsening, see Chapter 9)at higher annealing temperatures (d_p in Eq. (7.41) increases) d_{max} usually increases with annealing temperature. The grain sizes calculated according to Eq. (7.42) are usually larger than observed because of simplifying assumptions in the derivation of Eq. (7.42), in particular regarding the particle size distribution.

7.12 Discontinuous Grain Growth (Secondary Recrystallization)

If the microstructure remains essentially stable except for the growth of a few grains, this is referred to as discontinuous grain growth. It requires that normal grain growth is globally suppressed in the specimen, but is triggered at a few locations. A technically important example is discontinuous grain growth in two-phase materials during annealing close to the solvus temperature. During such an annealing process, precipitates locally dissolve and initiate grain growth, while the remaining granular microstructure remains stabilized by the precipitates. Some of the locally growing grains gain such an overwhelming size advantage that they can overcome Zener pinning and, therefore, become extremely large grains by consuming the nearby small grains pinned by Zener drag.

Fig. 7.53 gives an example for discontinuous grain growth in an aluminum manganese alloy. Annealing at temperatures where the manganese is completely in solid solution causes continuous grain growth to proceed, which ceases if the grain size reaches about two times the sheet thickness. During annealing at temperatures below the solvus temperature, when the manganese is completely precipitated, there is no grain growth even after long annealing times. If the material is annealed close to the solvus temperature, strong secondary recrystallization is observed after some time.

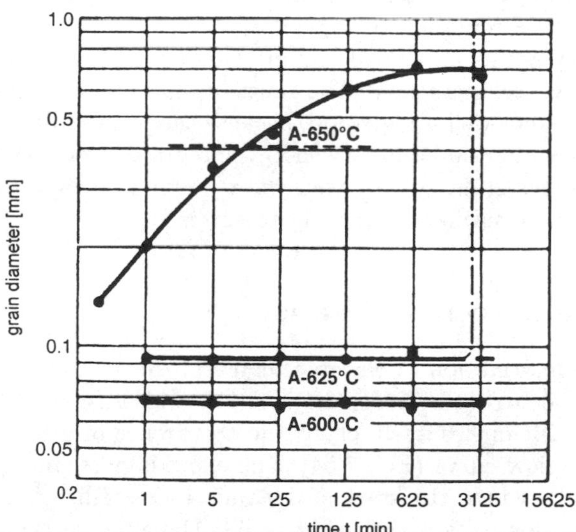

Fig. 7.53. Grain growth in an Al 1.1%Mn alloy. The horizontal dashed line indicates the sheet thickness, the vertical dotted-dashed line the beginning of discontinuous grain growth. Solubility temperature of the Al 1.1%Mn alloy is 625°C (after [7.27]).

The critical grain size for a grain to grow discontinuously into a matrix with precipitates can be derived to be

$$d > \frac{\bar{d}}{1 - \frac{\bar{d}}{d_{\max}}} \tag{7.43}$$

where $\bar{d}$ is the average grain size of the pinned grain structure and d_{max} is the maximum grain size according to Eq. (7.42).

For secondary recrystallization to occur $\bar{d}$ must be smaller than d_{max}. An increase of annealing temperature results in faster particle coarsening. Correspondingly, d_{max} increases according to Eq. (7.42) and a larger number of grains comply with Eq. (7.43). If several grains grow concurrently they finally impinge and the grain size after secondary recrystallization becomes smaller, correspondingly. Therefore, discontinuous grain growth is most pronounced upon annealing slightly above the critical temperature where secondary recrystallization is initiated.

Also a strong recrystallization texture can give rise to secondary recrystallization. Since the pronounced preferred crystallographic orientation results in a grain boundary character distribution with many low-angle grain boundaries, which have low mobility, the few grains that are bordered by high-angle grain boundaries can grow and eventually consume the stagnating granular microstructure.

Secondary recrystallization is used in some cases to obtain special material properties. Examples are grain oriented electrical steels, consisting of Fe3.5%Si, to produce soft magnetic materials (see Chapter 10.5.2), or ODS (Oxide Dispersion Strengthened) superalloys where the preferred orientations of the secondary grains impart special mechanical properties to the material for service at high temperatures. The cause of discontinuous grain growth in these materials is not well understood, because the recrystallization texture is weak. Anisotropy of grain boundary energy and grain boundary mobility may cause discontinuous grain growth as shown recently in computer simulations. These issues remain subjects of current research.

7.13 Dynamic Recrystallization

During plastic deformation at elevated temperatures ($T > 0.5T_m$), recrystallization is observed to occur during deformation. This is referred to as dynamic recrystallization. It makes itself felt by the occurrence of a single peak or oscillations of the flow curve (Fig. 7.54). The effect is particularly pronounced in single crystals, where the start of dynamic recrystallization is associated with a dramatic loss of strength (Fig. 7.55). Dynamic recrystallization can also occur during creep, usually recognized by a sudden increase of the creep rate (Fig. 7.56). The critical values of stress and strain at which dynamic recrystallization is initiated depend on material and deformation. Many concentrated alloys and dispersion strengthened materials do not recrystallize dynamically (Fig. 7.57). The flow stress to set-off dynamic recrystallization decreases with increasing deformation temperature (Fig. 7.58) and decreasing strain rate (Fig. 7.59).

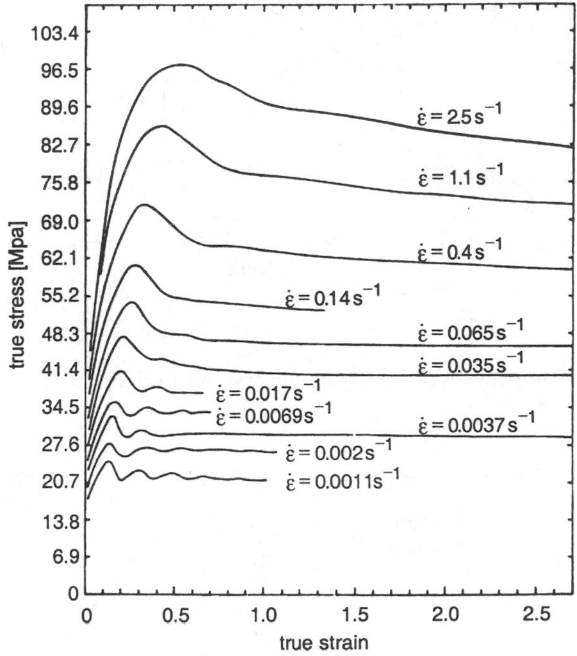

Fig. 7.54. Flow curves (torsion) of a carbon steel at 1100°C and various strain rates (after [7.28]).

The process of dynamic recrystallization is important for hot forming. It keeps the flow stress level low, i.e., requires only low deformation forces, and substantially improves ductility (Fig. 7.60). Moreover, the dynamically recrystallized grain size is directly correlated with the flow stress in such a way that with increasing flow stress the grain size decreases (Fig. 7.61). This allows one to adjust the recrystallized grain size by an appropriate choice of deformation conditions.

The flow stress behavior can be explained and formulated in a phenomenological theory by the assumption that a static recrystallization process is superimposed to deformation. A flow curve with oscillating flow stress is observed, if the cycles of recrystallization occur separately. If the recrystallization cycles overlap only a single flow stress peak is observed (Fig. 7.62).

This approach allows one to calculate the grain size development during hot rolling (Fig. 7.63). The predictions reveal that the recrystallized grain

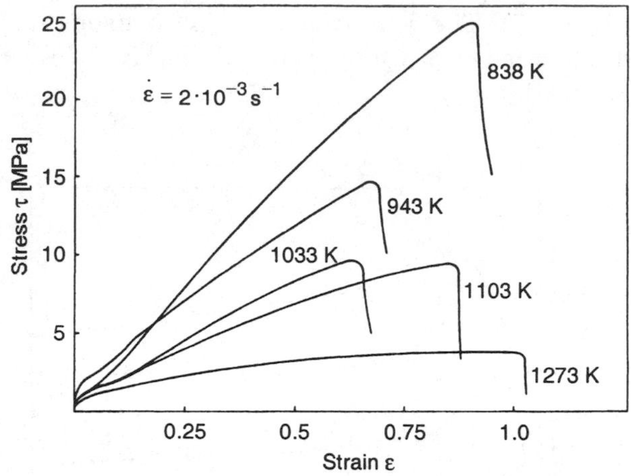

Fig. 7.55. Hardening curves of copper single crystals under tensile strain at different temperatures.

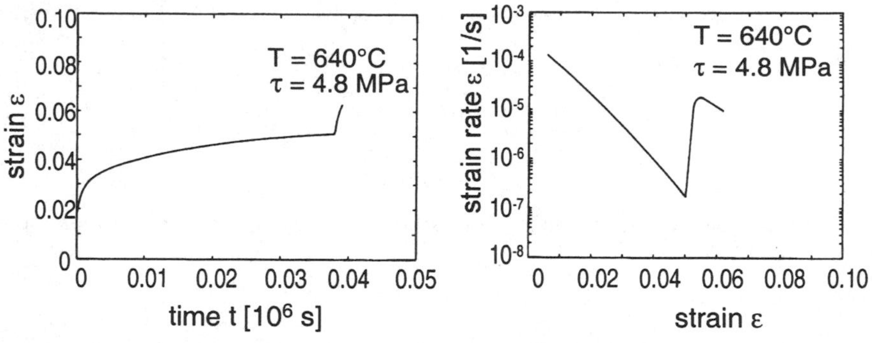

Fig. 7.56. Tensile creep deformation: $\varepsilon(t)$ and $\dot{\varepsilon}(\varepsilon)$ of a copper single crystal.

size does not depend on the initial grain size, but is strongly affected by the deformation schedule.

7.14 Recrystallization Textures

The distribution of orientations in a polycrystal is referred to as crystallographic texture (see Ch. 2.5.4). In the case of large strain deformation, for instance rolling or wire drawing, a pronounced texture is generated. Depending on the deformation mechanisms, whether crystallographic slip or twinning

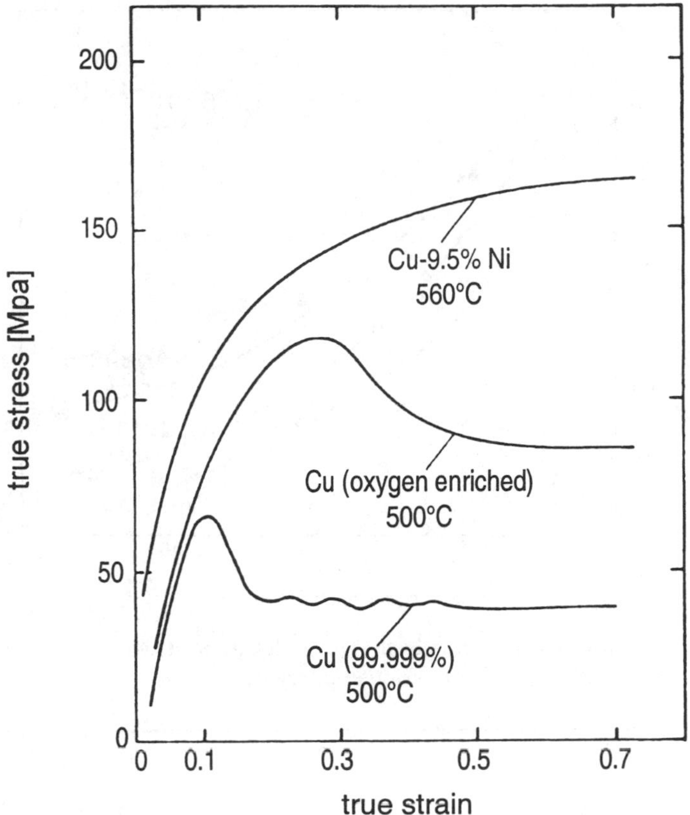

Fig. 7.57. Torsion flow curves of copper and copper alloys at a strain rate of $\dot{\varepsilon} = 2 \cdot 10^{-2} s^{-1}$ (after [7.29]).

predominate, a specific and reproducible texture develops. For the most important commercial forming process, rolling deformation, one obtains in fcc metals and alloys either a so-called copper-type rolling texture, or a brass-type rolling texture or a mixture of both. During recrystallization the texture changes drastically, but depending on the rolling texture specific types of recrystallization textures evolve: namely, either the so-called cube texture from the copper-type rolling texture, or the brass-recrystallization texture from the brass-type rolling texture (Fig. 7.64). Recrystallization textures are sensitive to alloying elements, which also affect the rolling textures, but to a much lower extent.

As indicated in Section 7.12 the texture will also change during grain growth, both during continuous grain growth and during secondary recrystallization. The desired Goss texture in FeSi electrical transformer sheet is caused by secondary recrystallization (Fig. 7.65).

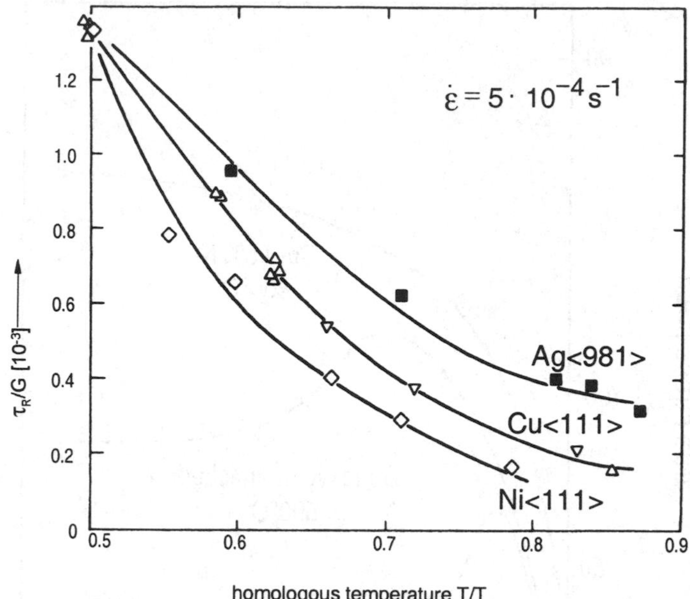

Fig. 7.58. Dependence of normalized recrystallization stress on the homologous temperature for single crystals of various materials.

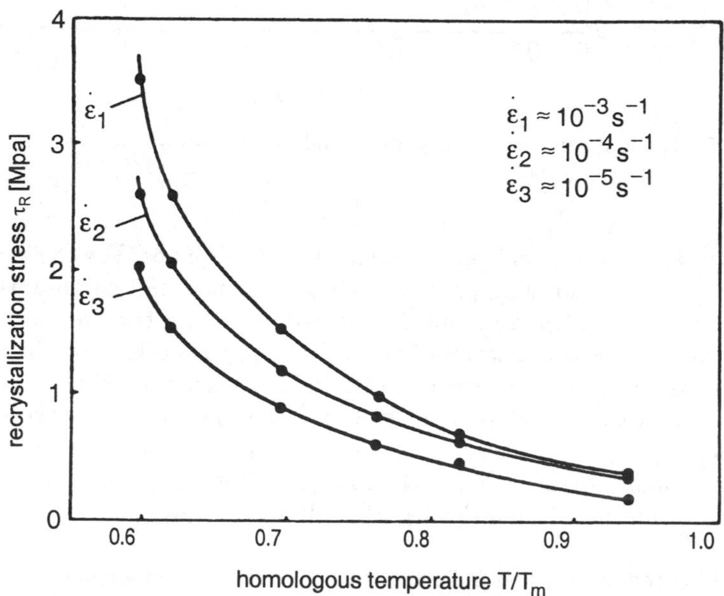

Fig. 7.59. Dependence of recrystallization stress of copper single crystals on deformation temperature for different deformation rates.

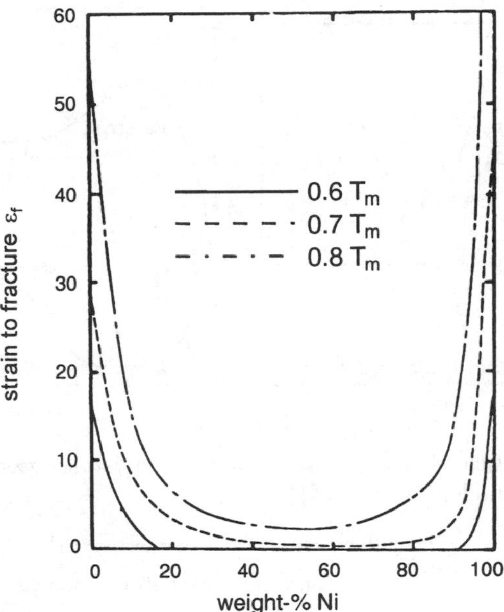

Fig. 7.60. The influence of solute content on ductility of a CuNi alloy at $0.6T_m$, $0.7T_m$, and $0.8T_m$ (after [7.30]).

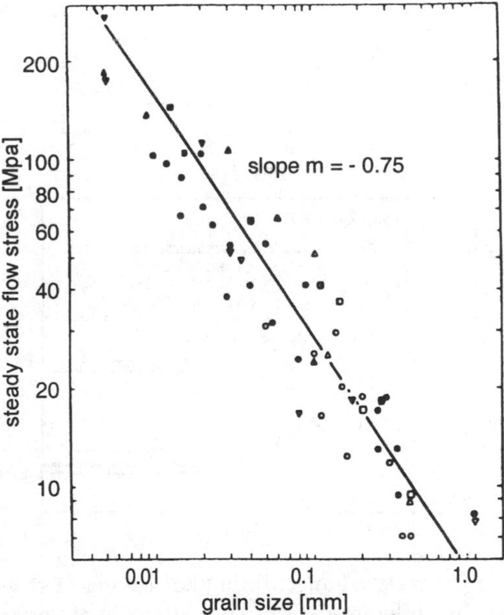

Fig. 7.61. Dependence of dynamically recrystallized grain size on steady state flow stress for Cu-Al alloys (after [7.32]).

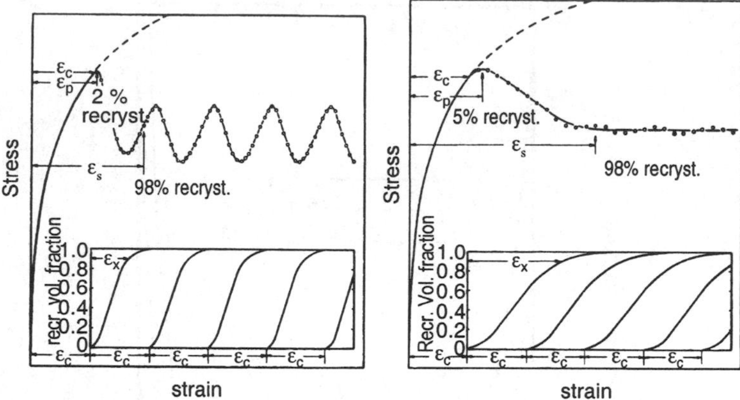

Fig. 7.62. Theoretical flow curves for dynamic recrystallization according to the Luton and Sellars model (after [7.32]).

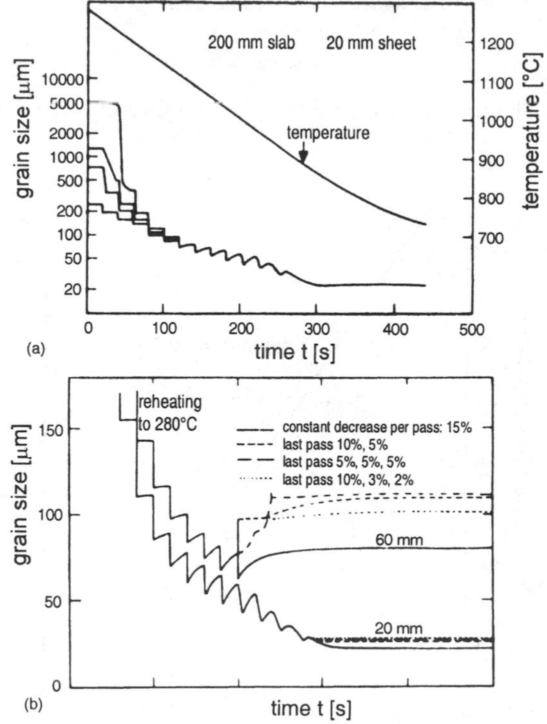

Fig. 7.63. Calculated grain size change during hot rolling of sheet metal. (a) Equal deformation per pass; (b) different deformation in the final passes but equal degree of total rolling deformation (after [7.33]).

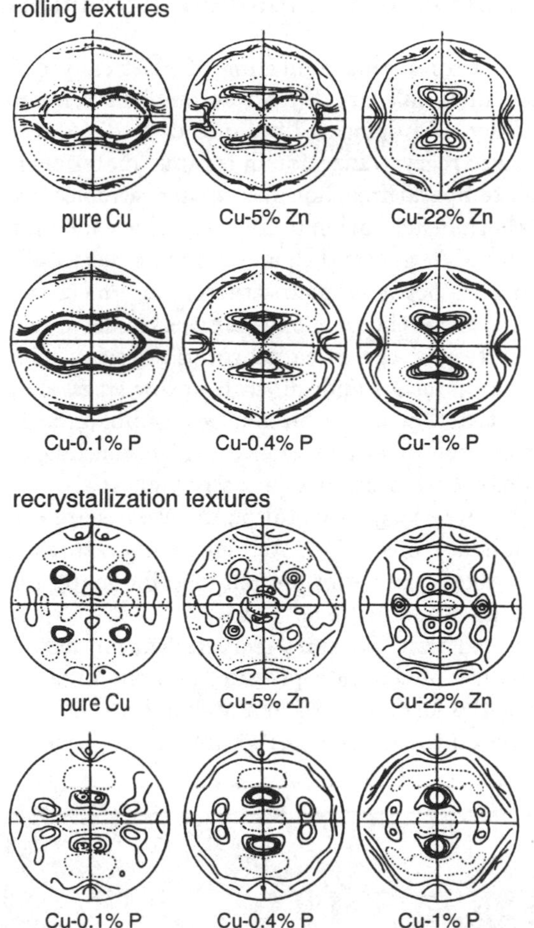

Fig. 7.64. Pole figures of several typical rolling and recrystallization textures (degree of rolling: 95%).

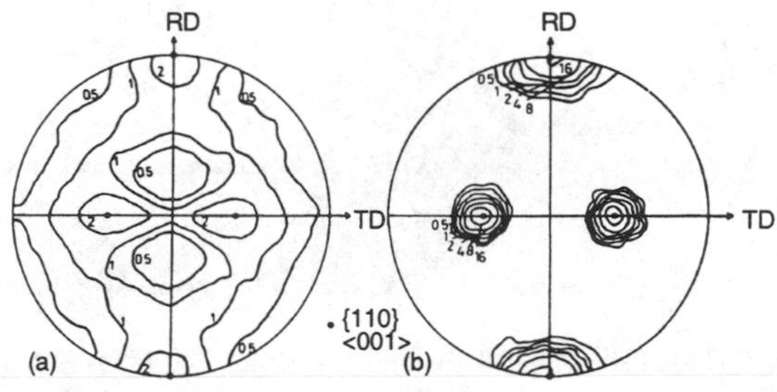

Fig. 7.65. Quantitative {100} pole figures of (a) primary, (b) secondary recrystallization textures of FeSi.

7.15 Recrystallization in Nonmetallic Materials

Recrystallization occurs during annealing of plastically deformed materials. Most nonmetallic materials are brittle at ambient temperature and, thus, cannot be plastically cold deformed because their dislocation mobility is too low. Therefore, static recrystallization in ceramic materials is of negligible importance. At high temperatures, however, many ceramics become increasingly ductile, because thermal activation facilitates dislocation motion. In such cases dislocations are stored in the crystal, and dynamic recrystallization can occur. The resulting relations between microstructure and mechanical properties are similar to metals. Besides the phenomena observed for dynamic recrystallization in metallic materials, the so-called rotation recrystallization is reported to occur in ceramic materials and minerals at low stresses. This phenomenon relies on the fact that low-angle boundaries (subboundaries) are generated during deformation, the misorientation of which continuously increases during progressing deformation, until eventually high-angle grain boundaries are produced. In such a way large orientation differences are created without noticeable grain boundary migration. At high temperatures and high stresses, however, dynamic recrystallization in non-metallic materials proceeds by the same mechanisms as in metals (Fig. 7.66).

Finally it is noted that dynamic recrystallization also occurs in geological materials. Because of the high pressure and the high temperature in the earth crust during the formation of mountains dynamic recrystallization has essentially contributed to the microstructure evolution of many rocks.

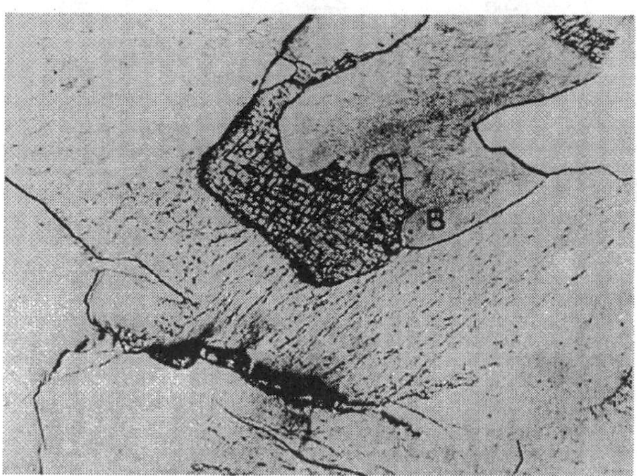

Fig. 7.66. Dynamic recrystallization of NaCl under compressive strain (480° C, $\sigma =$ 3.4MPa, $\varepsilon = 71\%$); grain B grows at the expense of grain A, which is strongly polygonized [7.34].

8

Solidification

8.1 The liquid state

In order to melt a metallic solid a certain amount of heat must be expended, the heat of melting. Different metals melt at different temperatures. This is related to the binding forces between the atoms. At the melting temperature, T_m, the thermal energy per mole (RT_m) has to be of the same order of magnitude as the binding energy (heat of melting per mole, H_m): $H_m \cong RT_m$ (Richards rule).

At the melting temperature the liquid and solid state coexist in thermal equilibrium, which requires (for constant pressure) that they have the same Gibbs free energy:

- $G_{liquid} = G_{solid}$
- $H_{liquid} - T_m \cdot S_{liquid} = H_{solid} - T_m \cdot S_{solid}$
- $H_{liquid} - H_{solid} = T_m(S_{liquid} - S_{solid})$
- $H_m = T_m \cdot S_m$

where H_m denotes the heat of melting, and S_m is the melting entropy. The quantities H_m and T_m can be measured (Table 8.1). For most metals the increase of entropy during melting is about 2 cal/mol/K (= 8.37 J/mole/K) $\approx$ R (R - gas constant).

Most metals crystallize in close-packed crystal structures (fcc and hcp). During melting the crystalline state breaks down and, therefore, the average packing density becomes smaller than in the solid state. Correspondingly, the volume increases during melting, as shown in Fig. 8.1 for Cu. This volume change is reversible and, therefore, restored during solidification by a volume contraction during solidification. In non-close packed crystalline materials the crystalline state has a low density because of special bonding arrangements. In this case the volume decreases upon melting, for instance for silicon, which crystallizes in the diamond cubic (dc) structure to satisfy the oriented covalent bonds. The dc crystal structure is "open" and has a lower density than

Table 8.1. Heat of melting and melting entropy of several metals.

Element	Heat of melting $[J/mol]$	Melting temperature $[K]$	Melting entropy $[J/mol \cdot K]$
Mn	8422	1517	5.45
Fe	11523	1812	6.29
Na	2640	371	7.12
K	2353	337	7.12
Mg	7333	923	7.96
Pb	4860	600	7.96
Cu	11187	1356	8.25
Ca	9344	1118	8.38
Ag	10685	1233	8.80
Ni	15880	1725	9.22
Cd	5782	594	9.64
W	33730	3683	9.22
Au	13282	1336	10.06
Al	9679	933	10.48
Zn	7123	692	10.48
Pt	22207	2046	10.89
Sn	7123	505	14.25
Bi	9972	544	18.44
Sb	19567	903	23.88

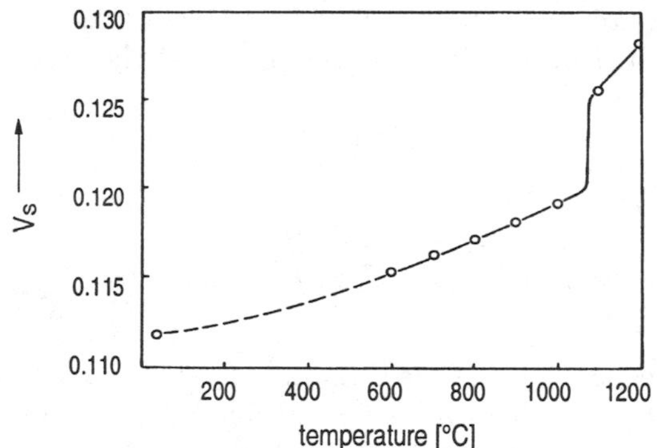

Fig. 8.1. Volume change due to thermal expansion and volume increase during melting of copper.

fcc and hcp (Table 8.2).

The atomic arrangement of atoms in a melt is not entirely random. Rather the interatomic forces keep trying to establish a locally dense arrangement, which is counteracted, however, by the thermal motion of the atoms. Therefore, liquid atomic arrangements are not as densely packed as in a close-packed crystalline solid. This becomes obvious from Fig. 8.2 in terms of the scattered X-ray intensity of liquid zinc at a temperature of 460°C, i.e. about 40° above

Table 8.2. Number of nearest neighbors in crystal and melt, and volume change during melting.

Element	Lattice	Nearest neighbors		ΔV[%]
		Crystal	Melt	
Al	fcc	12	10.6	+6.26
Au	fcc	12	11	+5.03
Cu	fcc	12		+5.03
Ag	fcc	12		+3.4
Pb	fcc	12	8	+3.38
δ-Fe	bcc	8		+3.0
K	bcc	8	8	+2.5
Zn	hex.	12	10.8	+4.7
Cd	hex.	12	8.3	+4.72
Ti	hex.	12	8.4	
In	tetr.	4	8.4	
Sn	tetr.	4	10	+2.6
Sb	rhomb.	3	4	−0.95
Ga	rhomb.	1	11	−3.24
Bi	rhomb.	3	7.5	−3.3
Si	diamond	4		−10
AlSb	diamond	4		−1.5

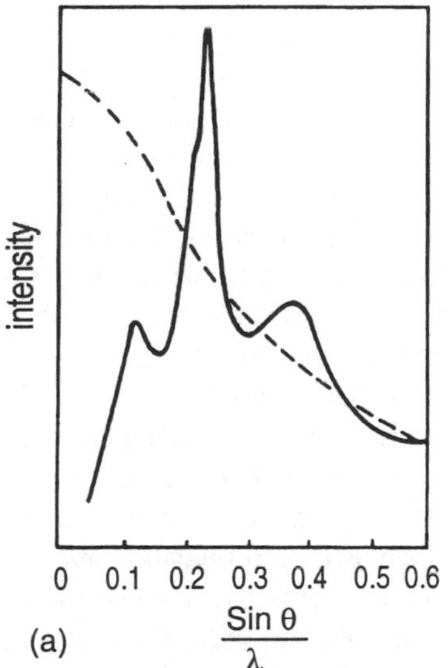

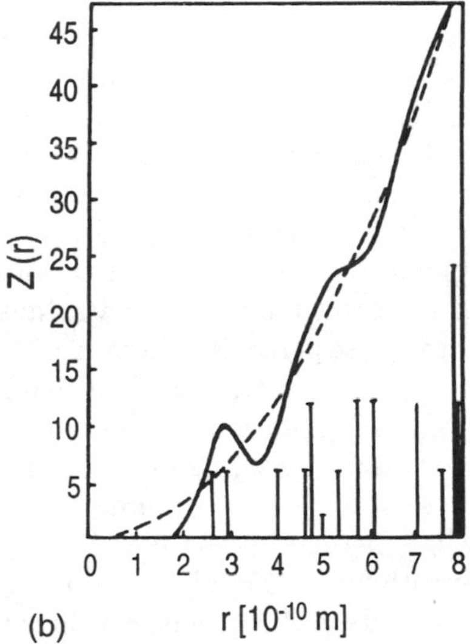

Fig. 8.2. (a) X-ray scattering intensity of liquid zinc at 460°C as a function of scattering angle; (b) atom density calculated from (a) at distance r from center atom (- - - disordered distribution) (after [8.2]).

the melting temperature (419.4°C). The broken line would be obtained for a completely random atomic distribution, as in a gas. The observed maxima of the measured curve indicate a higher density of atoms at specific interatomic spacings. In Fig. 8.2b the number of atoms is plotted as function of their spacing as calculated from Fig. 8.2a. Again, the broken line reflects a random distribution. The vertical lines indicate the spacing of atoms in crystalline zinc. It is apparent that for small radial distances, r, atoms in the melt and in the crystal have similar spacings, and in some cases even a similar number of next neighbors (Table 8.2). The results prove that the melt is by far more akin to the crystalline state than to a completely random state as a gas. There is a pronounced short-range order but no long-range ordered atomic structure. The concept that the crystal structure breaks down during melting and the atomic arrangement randomizes due to thermal motion is an oversimplification. Rather, the state of the melt is dominated by the formation and decomposition of ordered dense atomic arrangements with polytetrahedral symmetry, due to thermal fluctuations. This dynamical formation and destruction of locally ordered structures affects the nucleation of the solid state upon cooling below the melting temperature.

8.2 Nucleation of the Solid State

At a given temperature (and pressure) a material assumes the state with the lowest Gibbs free energy. For $T < T_m$ the free energy of the crystalline state is the lowest, but for $T > T_m$ it exceeds the free energy of the liquid state. At the melting temperature both free energies are equal, because at $T = T_m$ the crystalline and liquid state remain in equilibrium. Under the assumption that the free energy of both phases decreases approximately linearly with temperature one obtains the situation plotted in Fig. 8.3, where the solid lines represent the free energy in equilibrium. If the crystal is superheated or the melt is supercooled there is a driving force (free energy per unit volume) $\Delta g_u = (G_{liquid} - G_{solid})/V$ to change the state of the aggregate.

The solid state is not obtained spontaneously at the melting temperature rather a certain supercooling is necessary. For nucleation of the solid state to occur a small volume of crystalline atomic arrangement has to form through thermal fluctuation. As mentioned in Sec. 8.1 such fluctuations occur frequently in the liquid state due to motion of the atoms. Above the melting temperature such an "embryo" is principally unstable and will decompose. At a temperature, T, below the melting temperature an embryonic nucleus is stable only if it exceeds a critial size. This is due to the solid-liquid surface energy, which is always positive and thus, increases the free energy. If, for simplicity, we assume a spherical shape of the nucleus with radius r then the change of the free energy of the system by formation of the nucleus, ΔG_N, is given as the sum of two terms. First, one gains the volume energy $4/3\pi r^3 \cdot (-\Delta g_u)$, but

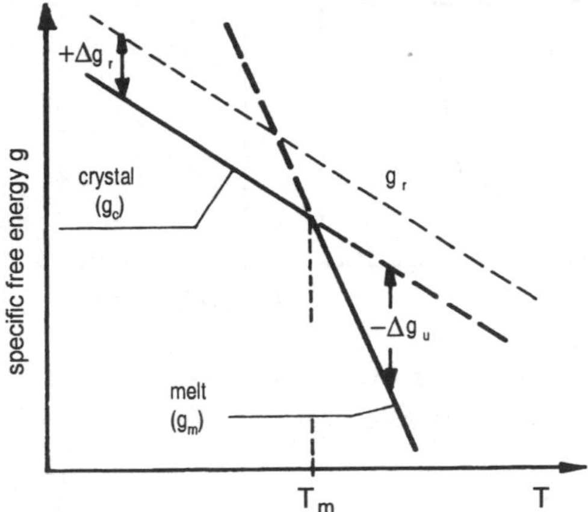

Fig. 8.3. Temperature dependence of the specific Gibbs free energy of a crystal (g_c) and melt (g_m) (—— equilibrium, - - - - specific free energy g_r of a crystal nucleus of radius r).

then the surface energy, $4\pi r^2\gamma$, has to be expended, where γ is the solid-liquid surface energy per unit area. Thus,

$$\Delta G_N = -\frac{4\pi}{3}r^3\Delta g_u + 4\pi r^2\gamma \tag{8.1}$$

For $T \geq T_m$ the change of volume free energy $\Delta g_u \leq 0$ and, therefore, ΔG_N is always positive. Any embryonic nucleus decomposes with a gain of free energy (Fig. 8.4a). For $T < T_m$ the free energy of a nucleus decreases only if the nucleus exceeds the size $r \geq r_0$. At $r = r_0$ the work of nucleation ΔG_N, attains a maximum (Fig. 8.4b), and r_0 can be calculated from Eq. 8.1 by differentiation: $d(\Delta G_N)/dr = 0$. The result is

$$r_0 = \frac{2\gamma}{\Delta g_u} \tag{8.2}$$

The free energy of the nucleus is indicated in Fig. 8.3 as a broken line. The work of nucleation, ΔG_0, is obtained by inserting Eq. 8.2 into Eq. 8.1 to yield

$$\Delta G_0 = \Delta G\left(r_0\right) = \frac{16}{3}\pi\frac{\gamma^3}{\left(\Delta g_u\right)^2} = \frac{1}{3}A_0\gamma \tag{8.3}$$

where A_0 is the surface area of the critical nucleus. For formation of a viable nucleus the required gain of volume energy does not have to compensate fully the surface energy of the nucleus, but only 1/3 of it. This is because the system will decrease its free energy by growth of a nucleus of size $r > r_0$, even though the absolute value of the free energy is larger than in a system without the nucleus.

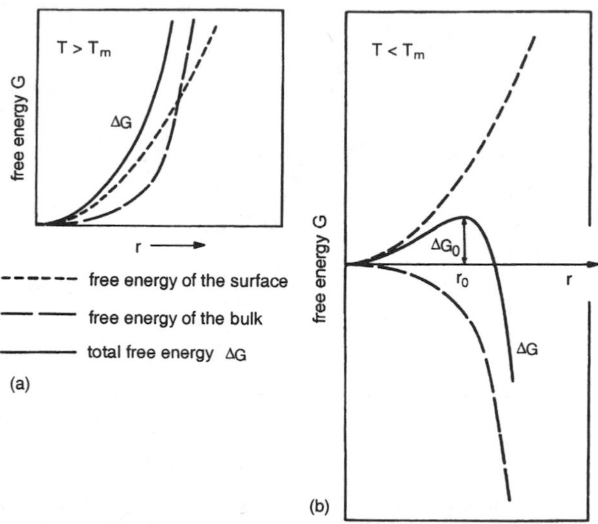

Fig. 8.4. Dependence of the free energy of a spherical crystal nucleus on its radius r.

Inasmuch as nucleation occurs through thermal fluctuations, the nucleation rate, $\dot{N}$, i.e. the frequency of nucleus formation per unit volume and unit of time is given as

$$\dot{N} \sim \exp\left(-\frac{\Delta G_0}{RT}\right) \tag{8.4}$$

From Fig. 8.3 and Eq. 8.3 we learn that

$$\lim_{\substack{T \to T_m \\ T \to 0}} \frac{\Delta G_0}{RT} = \infty$$

For $T \to T_m$ the term Δg_u tends to zero and, therefore, ΔG_0 becomes infinitely large.

In the limit $T \to 0$ the driving force Δg_u remains finite, namely $H_{liquid} - H_{solid}$ and, therefore, also ΔG_0 assumes a finite value, and $\Delta G_0/\mathrm{RT} \to \infty$. Thus, $\dot{N}(0) = 0$. Since the exponential function is always positive, $\dot{N} > 0$ for $0 < T < T_m$. $\dot{N}$ has to pass through a maximum (Fig. 8.5), which is observed (Fig. 8.6). As $\dot{N}$ depends exponentially on ΔG_0, and ΔG_0 is a function of temperature (via Δg_u, see Fig. 8.3 and Eq. 8.3) small changes of supercooling result in large changes of the nucleation rate (Fig. 8.7).

The nucleation rate calculated according to Eq. 8.4 is much smaller, however, than that observed in reality. The reason is that nucleation does not occur homogeneously within the volume as assumed (Fig. 8.8a), but occurs heterogeneously, i.e. at pre-existing surfaces, for instance on the surface of a

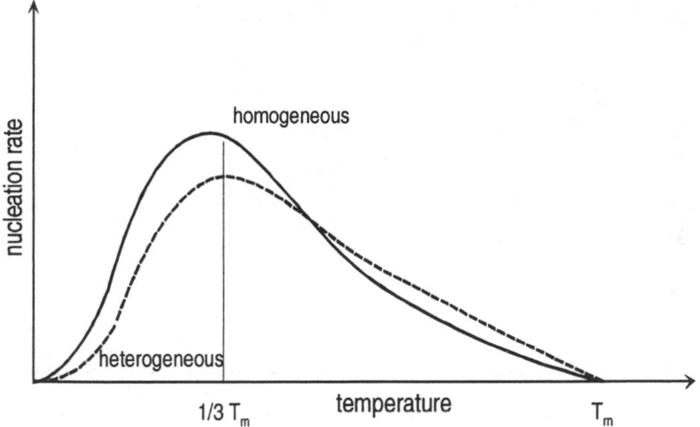

Fig. 8.5. Temperature dependence of the homogeneous and heterogeneous nucleation rate (schematic, T_m - melting temperature).

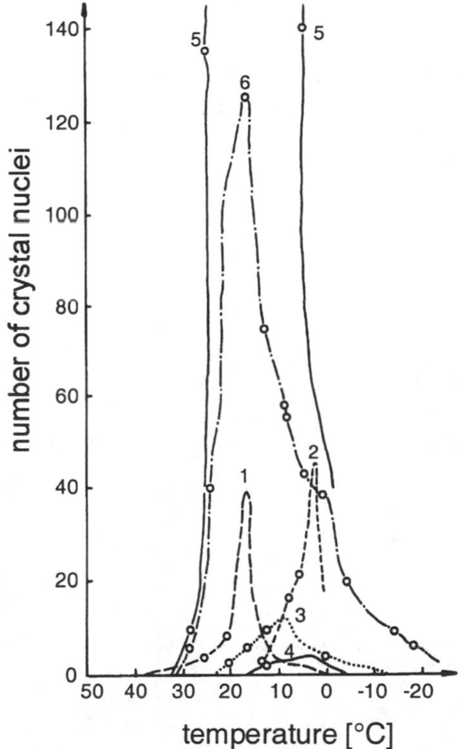

Fig. 8.6. Temperature dependence of the number of crystal nuclei in betol. Melting temperature is 91°C. (Line 1 - crystallized three times, line 4 - crystallized once, all other lines with organic additives) (after [8.3]).

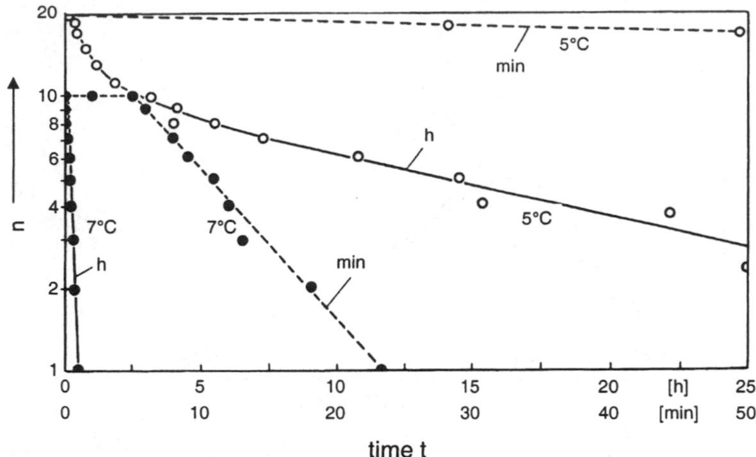

Fig. 8.7. How undercooling affects nucleation (number n of samples not solidified) in tin. The large scale is for the dashed lines.

mold or on particles in the melt (Fig. 8.8b). In such cases part of the nucleus surface is provided by the already existing surface of the particle, which reduces the work of nucleation. For heterogeneous nucleation on a flat surface the work of nucleation may be expressed as

$$\Delta G_{\text{het}} = f \cdot \Delta G_0 \tag{8.5}$$

$$f = \frac{1}{4}(2 + \cos\Theta)(1 - \cos\Theta)^2 \tag{8.6}$$

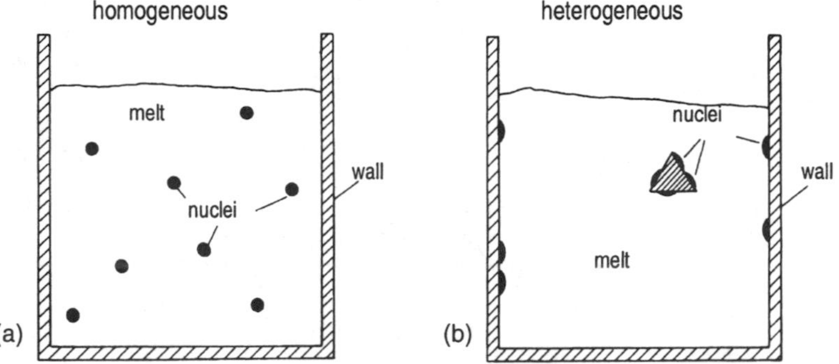

Fig. 8.8. Illustration of the principle of (a) homogeneous and (b) heterogeneous nucleation in a melt.

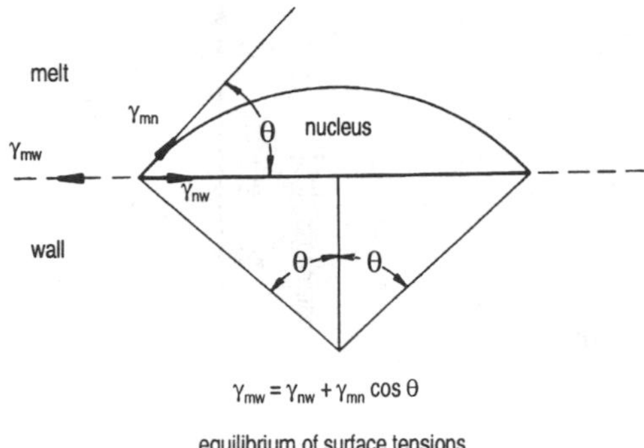

Fig. 8.9. Equilibrium shape of a drop of liquid on a plane surface.

where Θ is the contact angle (Fig. 8.9), with $0 \leq \Theta \leq \pi$ and thus $0 \leq f \leq 1$. Because of the strong (exponential) dependence of $\dot{N}$ on the work of nucleation the heterogeneous nucleation rate is much larger than the homogeneous nucleation rate, expecially at small supercooling.

In commercial ingot making processes insoluble particles are added to the melt prior to solidification. For instance, TiB_2 particles are added to aluminum to increase the number of nuclei and, therefore, to produce a cast microstructure with a finer grain size. The nucleation rate also depends on the superheating of the melt prior to solidification. This is due to the fact that particles in the melt that act as heterogeneous nucleation centers dissolve with increasing temperature. This influence strongly depends on the alloy composition (Fig. 8.10).

In commercial alloys the typical supercooling encountered during casting is only a few degrees, because nucleation occurs almost exclusively by heterogeneous nucleation. To obtain homogeneous nucleation special measures must be taken to avoid heterogeneous nucleation. One successful method involves the solidification of small metallic droplets. In large droplets the frequency of contaminating particles is always high enough to induce heterogeneous nucleation (Fig. 8.11). Crystallization of very small and, therefore, particle-free droplets occurs only at large supercooling. This is why pure metallic melts can be substantially undercooled by about 15% of the absolute melting temperature (Table 8.3). Inasmuch as the nucleus size can be measured from these droplet experiments, the surface energy, γ, can be calculated according to Eq. 8.2. The measured values agree reasonably well with surface energies calculated from first principles.

Large supercooling can be obtained in organic materials, because their more complicated molecular structures substantially impede crystallization, as shown in Fig. 8.6 for betol. The melting point of pure betol is 91°C. For noticeable nucleation to occur, a supercooling of about 50K is needed, but the nucleation rate can be markedly improved by organic additions.

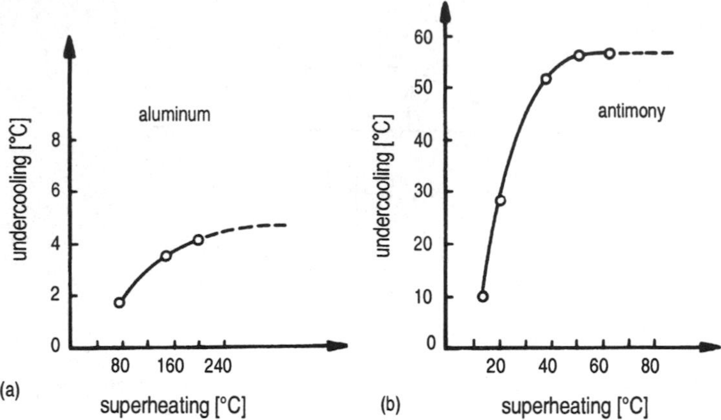

Fig. 8.10. Effect of superheating a melt on undercooling necessary for nucleation in (a) aluminum and (b) antimony (after [8.5]).

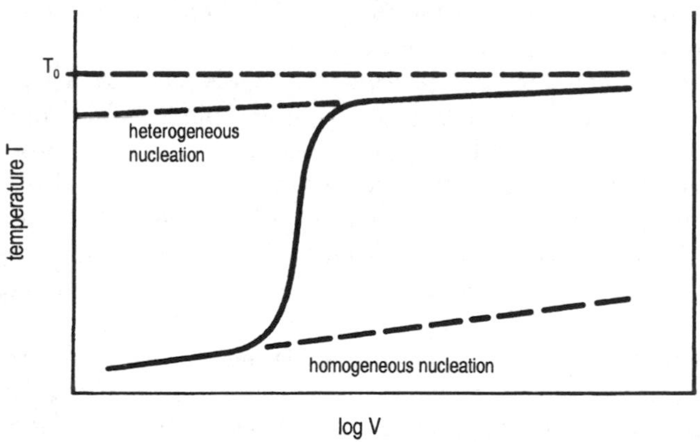

Fig. 8.11. Dependence of the undercooling necessary for solidification of a droplet on the droplet's volume.

Table 8.3. Maximum undercooling of metals melts.

Metal	Melting temperature T_m $[K]$	Maximum undercooling ΔT $[K]$	$\Delta T/T_m$ [%]
Au	1336	190	14.2
Co	1768	310	17.5
Cu	1356	180	13.3
Fe	1807	280	15.5
Ge	1210	200	16.5
Ni	1726	290	16.8
Pd	1825	310	17.0

8.3 Crystal Growth

8.3.1 Shape of a crystal

The surface represents an interruption of the perfect crystal. Thus, surfaces always increase the total energy of a crystal. One is tempted to assume that the equilibrium shape of a crystal in the absence of other forces is a sphere, because in this case the ratio of surface to volume is the smallest. This is true only, however, if the energy of the surface is independent of its crystallographic orientation. In crystalline materials the surface energy does depend on orientation. To minimize its free energy a crystal always tends to keep its surfaces with high energy as small as possible (Fig. 8.12). Under such circumstances the equilibrium shape of a crystal is given by a polyhedron, which complies with Wulff's theorem

$$2\gamma_i/\lambda_i = K_w \tag{8.7}$$

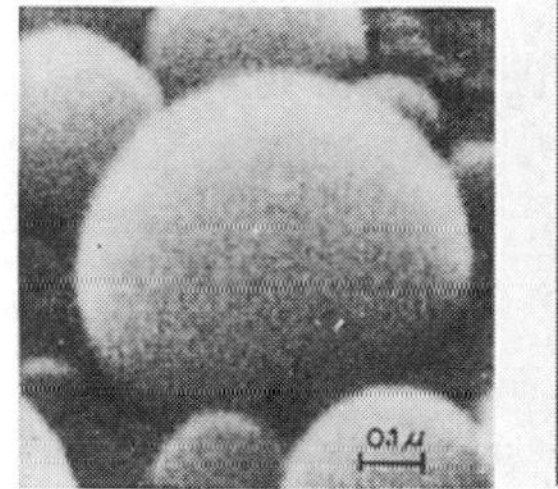

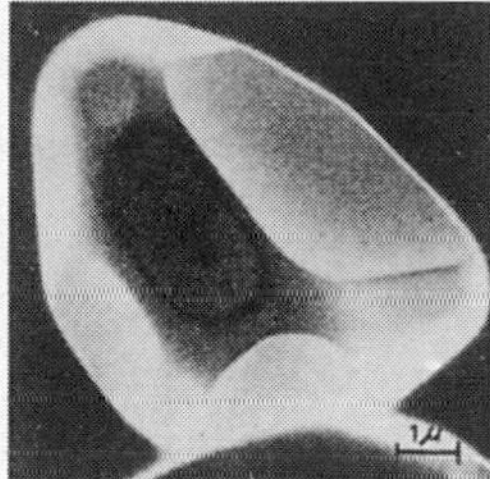

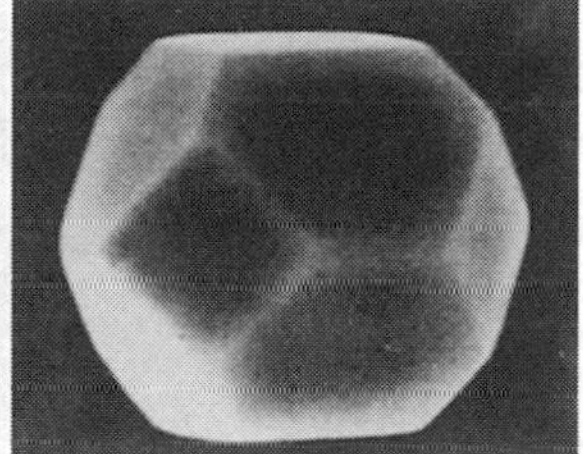

Fig. 8.12. Crystal shapes observed with a scanning electron microscope: (a) tin particles vapor deposited on a NaCl substrate (0.8 T_m); (b) iridium crystal after 50 h at 1700°C in a helium atmosphere; (c) vapor grown NiO crystal, surface facets are {111}, {001}, {011} [8.6].

Here γ_i is the specific energy of the surface, λ_i the distance of the surface from the center of the crystal and K_w Wulff's constant. According to the theorem the equilibrium shape of a crystal is obtained if a normal vector **i** of length γ_i is drawn in all spatial directions **i**. The inner envelope is the equilibrium shape of the crystal, i.e. a polyhedron (Fig. 8.13). Because of the small magnitude of the driving forces for this process and the slow nature of the kinetics of shape change, though, the crystal will obtain its equilibrium shape only after annealing for a long time at high temperatures in the absence of other influences (Fig. 8.12).

Therefore, the shape of a crystal observed during solidification is usually not the equilibrium shape, rather it is determined by the kinetics of the growth process. A spherical shape is obtained only if the crystal grows equally fast in all directions and is stable. Otherwise, a polyhedron is obtained where the surface is composed of the most slowly growing crystallographic planes because the fast growing planes disappear in the course of growth (Fig. 8.14), or a dendrite (tree-like) crystal forms because of instabilities in the shape caused by transport processes.

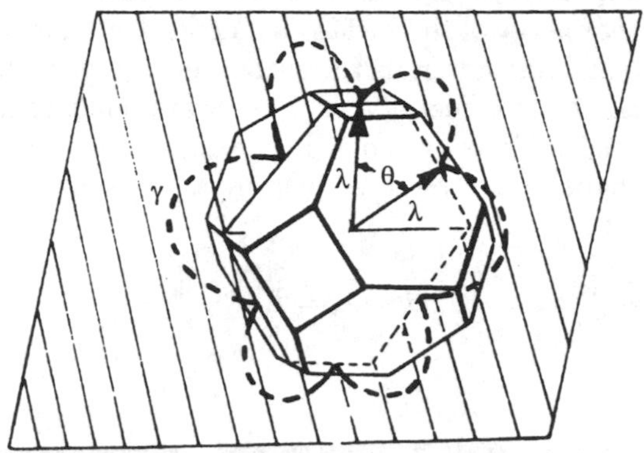

Fig. 8.13. Equilibrium shape of a crystal after Wulff. The dashed line corresponds to the surface energy. Plane surfaces are formed at energy minimum.

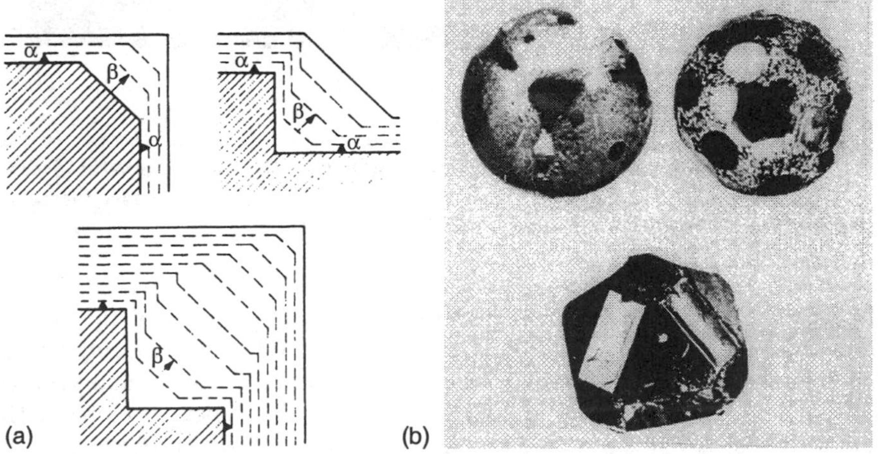

Fig. 8.14. (a) Transition stages of a slowly (α) and rapidly (β) growing surface. (b) Growth of an alum sphere from an aqueous solution [8.7].

8.3.2 Atomistics of Crystal Growth

The atomistic processes of crystal growth from the vapor phase were extensively studied by Kossel and Stranski. According to their theory, atoms adsorbed on a flat surface will meet to generate a (surface) nucleus which grows by the attachment of further atoms until a complete new layer forms, which requires the generation of a new nucleus for further growth, and so on. During deposition of atoms from the vapor phase onto a solid surface a variety of atomic configurations on the surface can be established, which are schematically shown in Fig. 8.15. If the surface energy is to first-order associated with the number of broken bonds (in comparison to the crystal interior) the so-called kink-ledge configuration is the energetically most favorable position for attachment. Because the atoms do not always touch the surface at their optimal position during the deposition process, other atomic configurations can be observed such as single atoms (adatoms), two atoms (double-adatoms), or even the formation of surface vacancies. Since the growth of a layer proceeds much faster than the formation of a nucleus, nucleation becomes the rate controlling process in this model. Experimentally it is observed that the growth rate increases with increasing supercooling, at rates much faster than predicted by the model (Fig. 8.16). The reason for this is the existence of crystal defects, e.g. screw dislocations, which allow for a continuous attachment of atoms around the dislocation line to form growth spirals (Fig. 8.17). Nucleation on the surface is not necessary in this case. This explains the high growth rates measured.

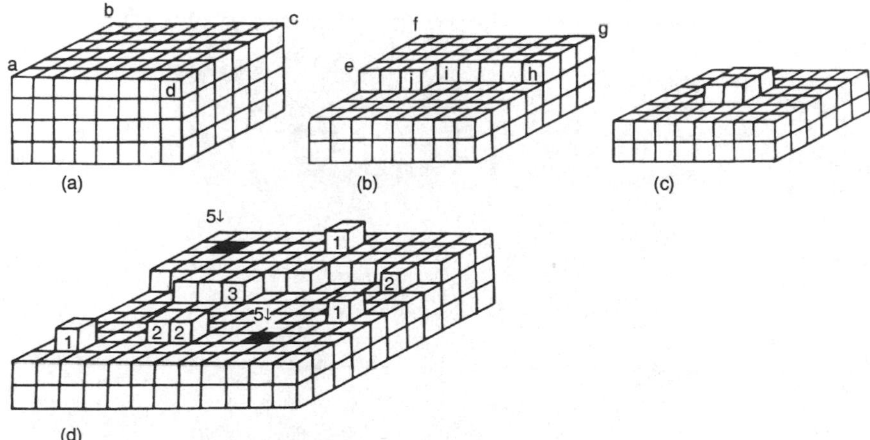

Fig. 8.15. Schematic illustration of the atomistic processes during crystal growth after Kossel and Stranski. (a) An atomically flat plane; (b) addition of atoms to an incomplete layer; (c) nucleation of a surface; (d) various crystal surface defects; the numbers indicate the number of bonds.

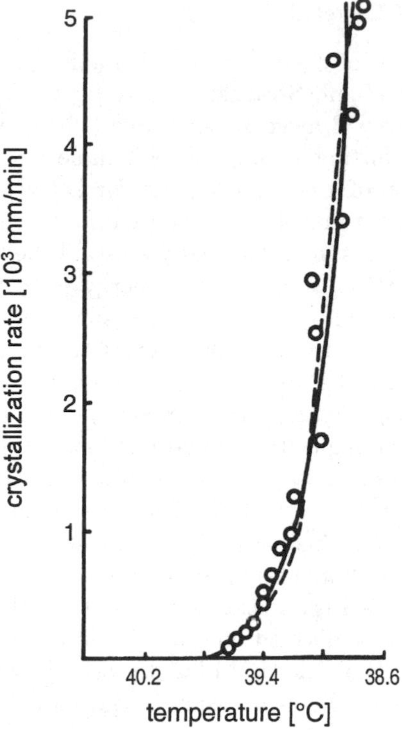

Fig. 8.16. Temperature dependence of the growth rate of alum crystals (after [8.8]).

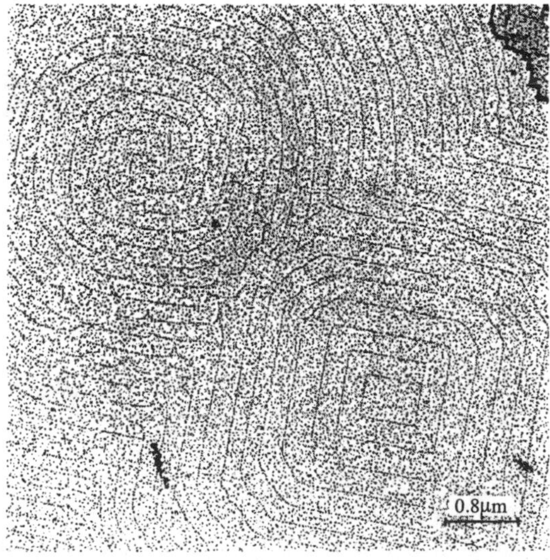

Fig. 8.17. Growth spiral around two screw dislocations, marked by vapor deposited gold particles [8.9].

8.3.3 Crystal growth in the melt

8.3.3.1 Solidification of pure metals

The shape of grains in a solidified metal is primarily determined by the heat flux distribution during solidification. For solidification to progress the generated heat of solidification has to be removed either through the solid or through the melt. The latent heat source located at the solidification front requires a discontinuity in the temperature gradient at this location (Fig. 8.18). A displacement of the solidification front by dx in a time interval dt generates the heat flux $h_S \cdot (dx/dt) = h_S \cdot v$ per unit of cross section and time. Here h_S denotes the heat of solidification per unit volume. If the latent heat flows through the solid, the total heat removed is the sum of the heat flux from the liquid and the latent heat of solidification. Since the heat flux density $j = \lambda(dT/dx)$, and the thermal conductivities are λ_L and λ_C in melt and crystal, respectively, the balance of heat fluxes may be expressed as

$$\lambda_C \left(\frac{dT}{dx}\right)_C - \lambda_L \left(\frac{dT}{dx}\right)_L = h_S v \tag{8.8}$$

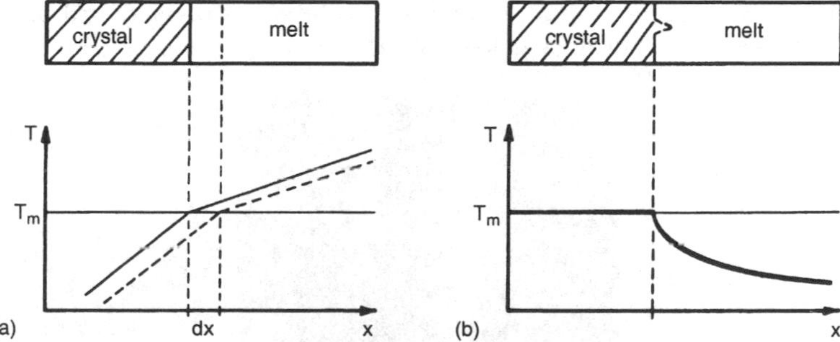

Fig. 8.18. Temperature distribution during growth of a crystal from the melt. Heat flux through (a) the crystal, (b) the melt.

Note, if the heat flow occurs through the crystal (Fig. 8.18a) then the temperature gradient in the crystal must be larger than that in the melt. If a region of the crystal at the solidification front grows a little faster and extends into the melt, the extended region will "feel" a higher temperature and dissolve. Correspondingly, the perturbed solidification front remains flat and grows in a stable fashion. In this case equiaxed grains develop.

If the melt is strongly supercooled during solidification the heat flow can also occur through the melt. In this case one obtains a temperature profile as depicted in Fig. 8.18b. Any irregularity that develops at the solidification front, will extend into a volume of undercooled melt, where it can grow faster.

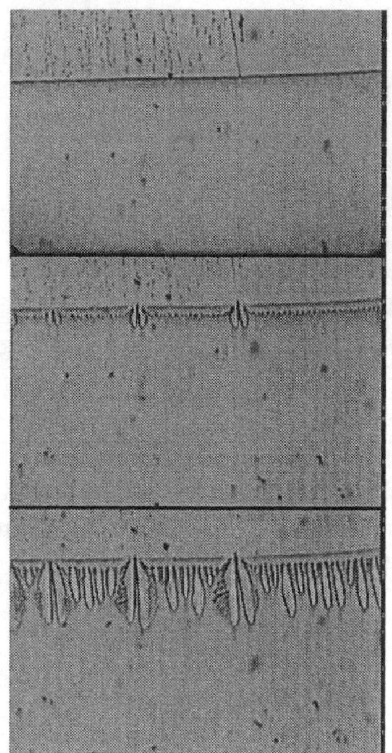

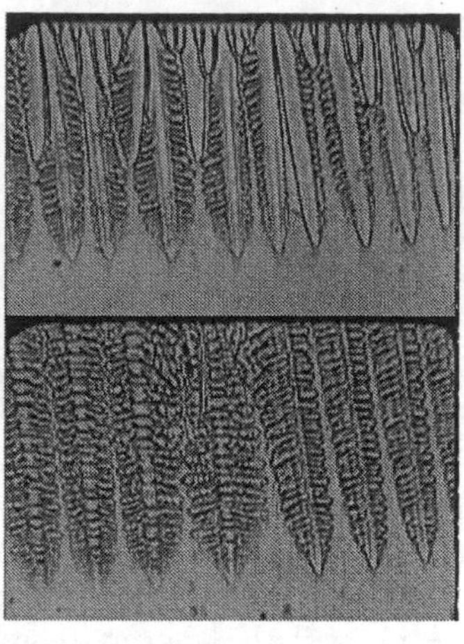

Fig. 8.19. Directional solidification of a solid solution of succinonitril in acetone. Dendrites caused by minor fluctuations [8.10].

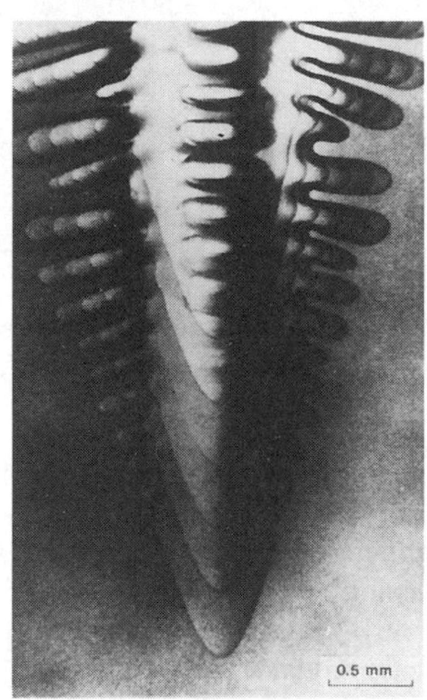

Fig. 8.20. The tip of a growing dendrite in a mildly undercooled melt of high purity succinonitril [8.11]

In such cases slender crystals form, which branch in perpendicular directions for the same reasons. The results are crystals that resemble pine trees and, thus, are called dendrites (Greek dendron: a tree). In this case, the solidification front does not move in a stable way. The process of dendritic growth can be easily observed (Fig. 8.19 - 8.20) in transparent organic materials.

During casting into a cold mold, the latter case occurs, because the melt becomes strongly supercooled at the mold wall (broken line in Fig. 8.21). When solidification starts and proceeds away from the mold wall, the crystal rapidly heats up to the melting temperature. The temperature profile in the melt results in a temperature minimum developing ahead of the solidification front (solid line in Fig. 8.21). This situation is favorable for dendrite formation, as is commonly observed (Fig. 8.22).

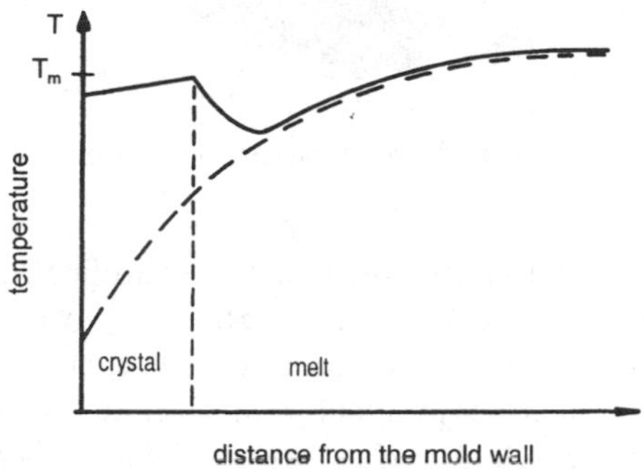

Fig. 8.21. Temperature distribution at the wall of a metal mold.

8.3.3.2 Solidification of alloys

Apart from some special cases, alloys solidify over a temperature range, that is over the temperature range between the solidus and liquidus temperature. Correspondingly, during solidification melt and crystal coexist in equilibrium but have different compositions (see Chapter 4). In such cases the heat flow problem as considered for pure metals in the previous section, is complicated by the diffusion problem, as sketched in Fig. 8.23.

For the sake of simplicity we now consider a binary system with complete solubility in the liquid and solid state, and assume that heat is removed through the crystal. In such case we obtain a temperature profile as plotted in the right-hand figures of Fig. 8.23 as solid lines. The broken line in the same figure reflects the temperature of the solidification front at that location. The dotted curve denotes the liquidus temperature of the melt that changes with composition according to the phase diagram. The solid lines in the left-hand side figures reveal the concentration profile developed in the crystal and melt at subsequent positions of the solidification front. The broken lines correspond

Fig. 8.22. Dendrites in the solidification cavity of an iron melt.

to the composition of crystal and melt at the solidification front that moves through the crucible from the left to the right.

If diffusion is fast in both the crystal and melt then both crystal and melt establish their temperature-dependent equilibrium concentrations through time (Fig. 8.23a). After solidification the solid has a homogeneous composition, c_0, equal to the initial melt. If, by contrast, diffusion in the crystal proceeds slowly while remaining sufficiently fast in the melt (Fig. 8.23b), the composition homogenizes only in the melt but not in the solid. Accordingly, as solidification progresses the melt becomes enriched with solute atoms, even beyond the terminal equilibrium concentration c_2. After complete solidification there remains a residual concentration gradient in the crystal.

If diffusion proceeds slowly both in the crystal and melt, the composition can not homogenize in any of the phases (Fig. 8.23c). In such a case, the solute atoms which are rejected from the crystal at the solidification front according to equilibrium at that temperature, remain in the melt close to the solidification front (i.e. within a small distance, δ). Remote from the solidification front the composition of the melt remains unchanged at the initial concentration, c_0. The maximum concentration of the melt at the solidification front can only be c_2. In this case, the crystal solidifies with a composition c_0 and enriches the melt to a concentration c_2. The composition of the melt changes rapidly in the vicinity of the solidification front, i.e. from c_2 to c_0 within a distance δ. Correspondingly, the respective liquidus temperature, that is the temperature where a melt with the corresponding composition would solidify — as predicted from the phase diagram — increases in the same range from T_2 to T_1 (dotted curve in Fig. 8.23c, right hand figure). If this temperature increase is larger than the actual temperature gradient in the melt (solid line)

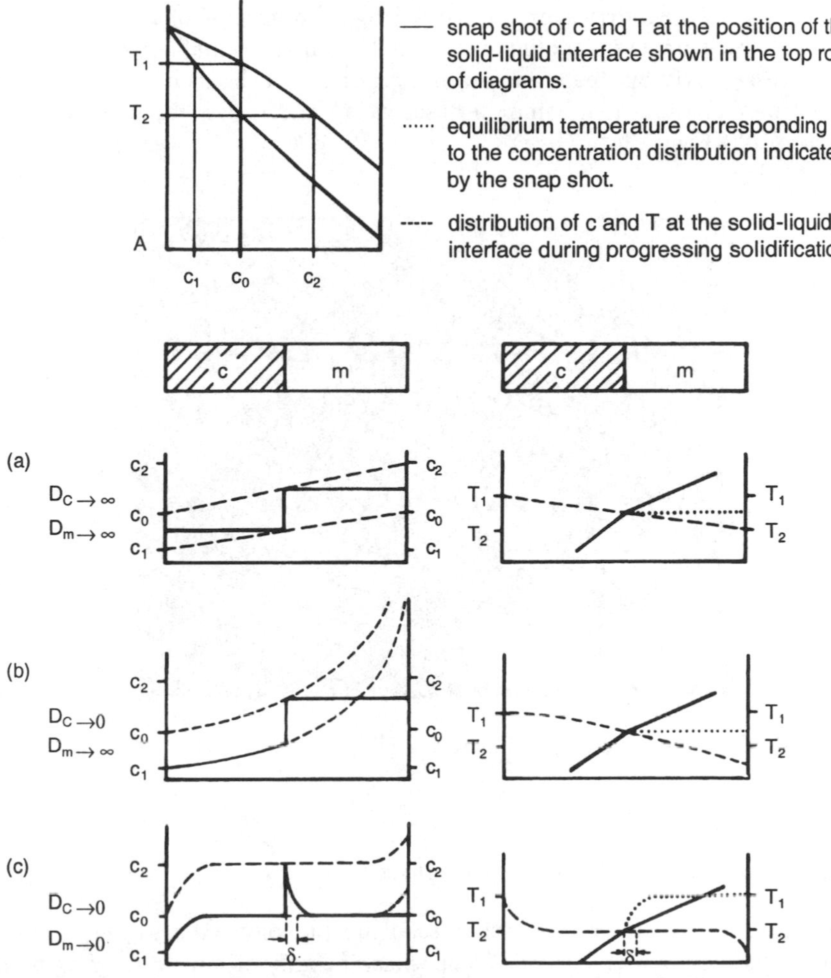

Fig. 8.23. Diagrammatic illustration of the solidification of a solid solution with concentration c_0. The left column depicts concentration distribution, the right column the temperature distribution (c - crystal, m - melt).

the temperature in front of the solidification front in the melt will be lower than the liquidus temperature of the melt with concentration c_0. This is the case of "constitutional" supercooling, because the supercooling is associated with the local constitution caused by sluggish diffusion of the alloy's components.

If during such a case of solidification an irregularity develops at the solidification front, a "bump" or protuberance will extend into a region of the melt that is colder than the respective equilibrium temperature corresponding to its composition. Rapid growth in the constitutionally supercooled region follows. In this way, dendrites are formed during solidification of alloys, although there is no thermal supercooling of the melt. Constitutional supercooling can be suppressed only by steep temperature gradients in the melt. The formation of dendrites in alloys is a common observation and caused by constitutional supercooling (Fig. 8.24).

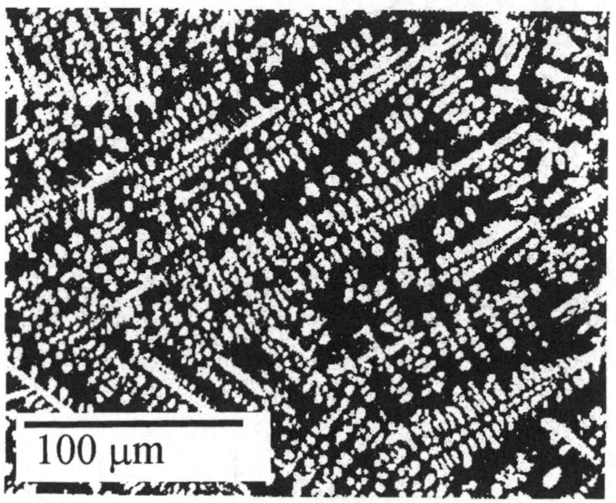

Fig. 8.24. Dendrites in a Fe-24%Cr casting alloy.

8.3.3.3 Solidification of eutectic alloys

Solidification microstructures with special morphology are also found in eutectic alloys. Alloys with eutectic composition solidify at a fixed freezing point by concurrent formation of two solid phases with different composition. The required composition change is achieved by mutual exchange of atoms, i.e. is diffusion controlled and, because of limited time, the composition change occurs over a spatially limited region that results in a lamellar microstructure (Fig. 8.25b). If the alloy composition is different from the eutectic composition, primary solid solution crystals develop first during solidification. The primary phase is usually dendritic because of constitutional supercooling. When the melt reaches the eutectic composition, the melt remaining between the primary crystals solidifies in a lamellar microstructure (Fig. 8.25c). When solidification of a eutectic melt proceeds in a temperature gradient, both phases may form continuous lamellae under appropriate conditions. The lamellae spacing ℓ is primarily determined by the cooling rate R via the relationship

$$\ell^2 \cdot R = \text{const.} \tag{8.9}$$

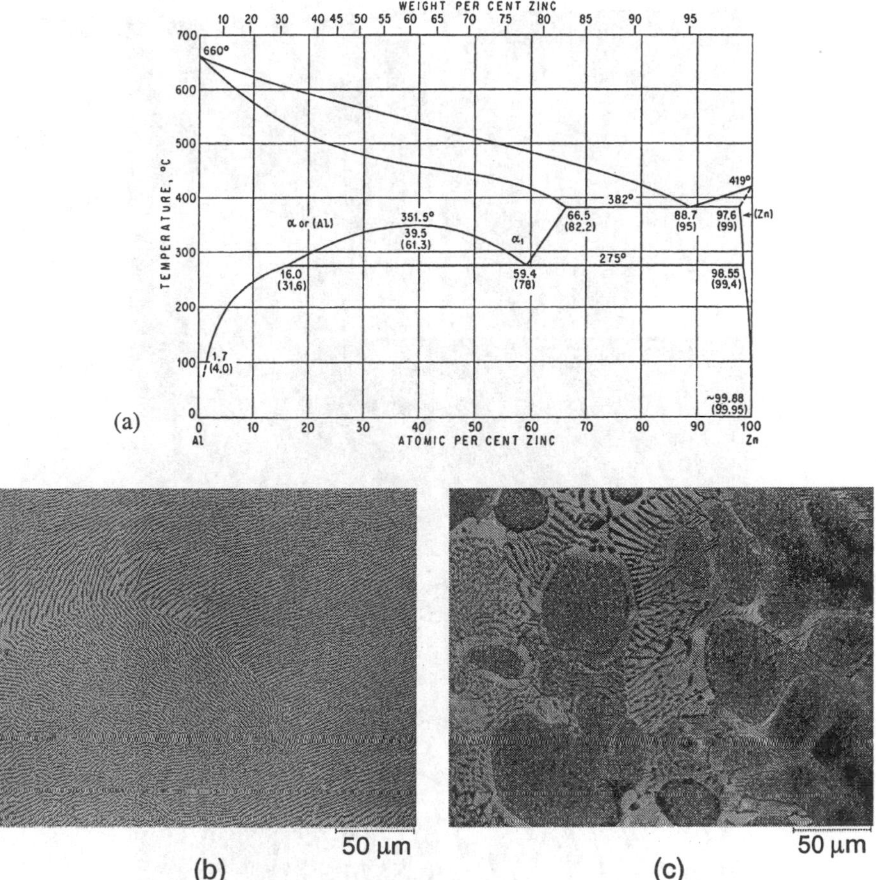

Fig. 8.25. Solidification microstructure of the eutectic system Al-Zn. (a) Phase diagram; (b) microstructure at 90%Zn (eutectic composition); (c) microstructure at 75% Zn. Note lamellar eutectic microstructure between primary solidified crystals.

High cooling rates correspond to small lamellae spacings. A directionally solidified eutectic microstructure corresponds to a fiber or lamellar reinforced composite (Fig. 8.26). Since this is not produced synthetically but occurs during solidification it is referred to as in-situ composite. Important examples are alloys for high temperature applications, which are used in gas turbine blades where excellent mechanical properties in specific spatial directions are required at elevated temperatures. Prominent examples are the COTAC alloys that contain Ta and C plus Co, Ni, and Cr. During solidification TaC forms as fibers embedded in a ternary solid solution (Fig. 8.26).

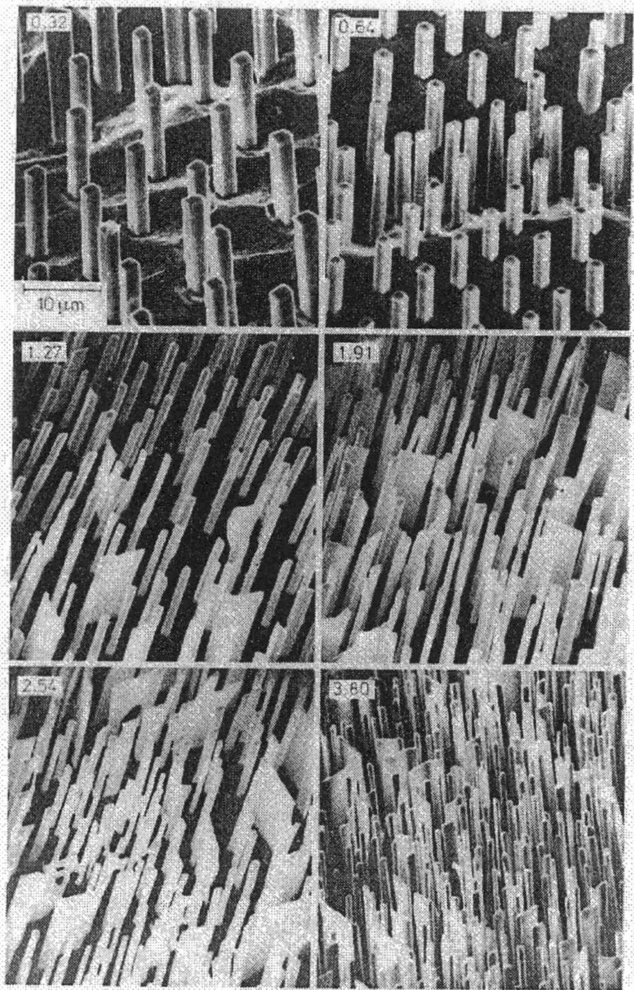

Fig. 8.26. Directionally solidified eutectic microstructure in the system CoNiCrTaC (COTAC 3), at different solidification rates (in cm/h). The TaC fibers were exposed by chemical removal of the matrix [8.14].

8.4 Microstructure of a Cast Ingot

The typical microstructure of a cast ingot (Fig. 8.27) consist of three zones (a) the fine-grained "chill" zone at the mold surface, consisting of randomly textured crystals, (b) the "columnar" zone of elongated crystals with strong crystallographic texture, and (c) the "equiaxed" zone consisting of round grains with random texture.

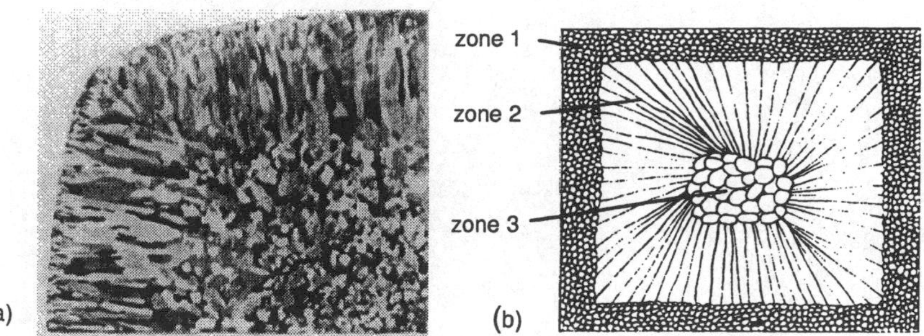

Fig. 8.27. The microstructure formed in a casting. (a) Microstructure of a block of cast Fe-4%Si; (b) schematic illustration: 1 - fine grained chill zone, 2 - columnar zone, 3 - equiaxed zone.

Fig. 8.28. TA casting of high purity aluminium.

In the beginning of solidification many randomly oriented grains are nucleated at the wall of the mold by heterogeneous nucleation. The nuclei grow subject to competition, where only those grains survive that have the highest growth rate along the direction of solidification (toward the center of the mold). Since the growth rate of a crystal depends on its crystallographic orientation, the surviving grains exhibit a preferred crystallographic direction of the grain's long axis, i.e. a crystallographic solidification texture results. The third zone in the center of the ingot comprises equiaxed grains that are caused by impurities, mostly with a high melting temperature. These components are rejected by the growing crystals and thus remain in the melt during solidification and are swept toward the casting's center. Here they accumulate and serve as nucleation centers for the residual melt, so that a fine-grained equiaxed microstructure is obtained in the center of the ingot. Correspondingly, this innermost equiaxed zone is missing in ingots of pure metals (Fig. 8.28). The width of the individual casting zones depends on solidification conditions (Fig. 8.29). If the temperature of the melt is high, the impurity particles dissolve, and because of a lower cooling rate a pronounced

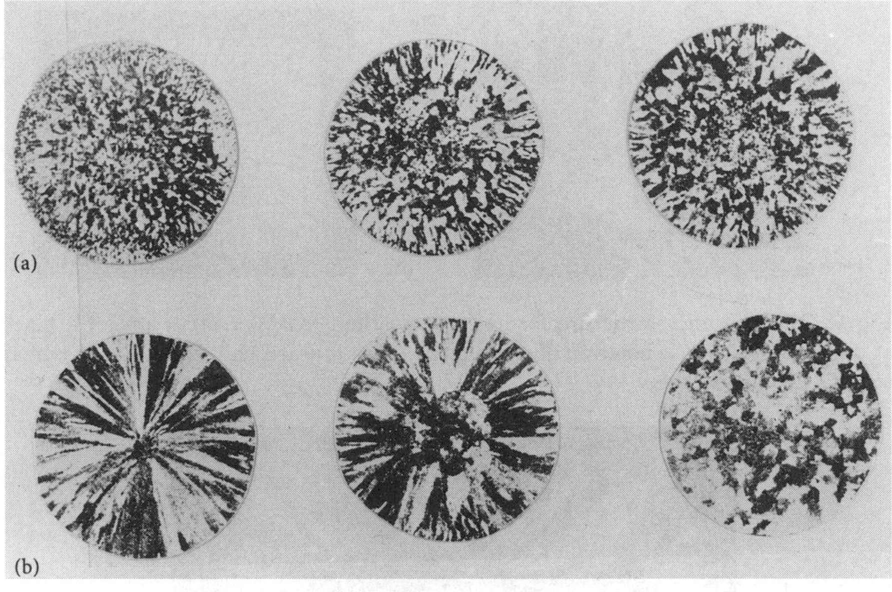

Fig. 8.29. Dependence of the microstructure of cast commercially pure aluminium on the temperature of the mold. (a) Casting temperature 700°C; (b) casting temperature 900°C.

columnar microstructure forms. Lower casting and mold temperatures induce a higher nucleation rate and, correspondingly, the globular "equiaxed" center zone increases.

8.5 Solidification Defects

Most metals and oxides have a larger volume in the liquid state than in the solid state. Correspondingly, upon solidification a volume contraction occurs. A consequence is the formation of a hole in the center of the cast ingot, which is referred to as shrinkage hole or "pipe" (Fig. 8.30). Besides the macroscopic shrinkage hole, microscopic cavities throughout the casting also form by contraction. This results in a porous casting.

Commonly, a liquid metal can dissolve a much larger amount of gas than the crystalline solid (Fig. 8.31). During solidification the gas molecules saturate, and can precipitate and combine to form gas bubbles that either rise in the liquid metal to cause mixing of the melt, or become trapped in the ingot to form permanent pores (gas porosity) (Fig. 8.32).

Strong segregation may result if the solidified crystals have a density different from the density of the liquid. Less dense crystals are lighter, and enrich

Fig. 8.30. A shrinkage hole in a zinc casting.

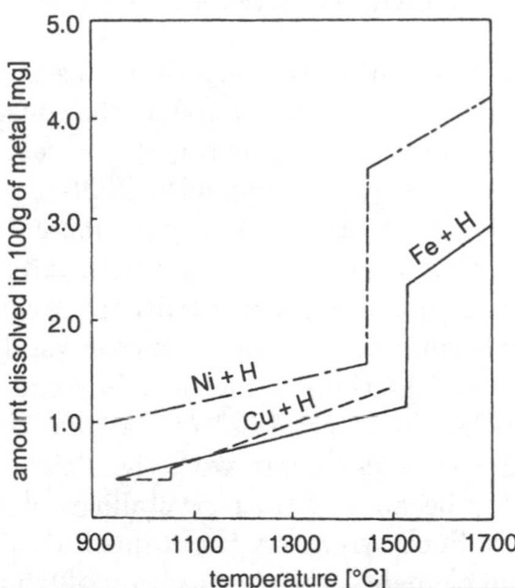

Fig. 8.31. Temperature dependence of the solubility of hydrogen in metals. Solubility changes sharply at the melting temperature (after [8.15]).

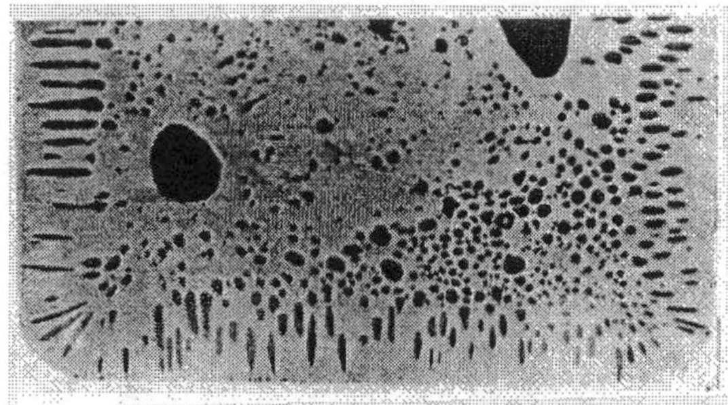

Fig. 8.32. Generation of gas bubbles in a rimming steel casting. Release of CO during the refining reaction causes gas bubbles in the solidified casting [8.16].

the top of the ingot, whereas heavier crystals sink to the bottom (Fig. 8.33, gravity segregation). Also, impurities can accumulate at various locations in the cast ingot (macrosegregation) and concentration gradients can arise in solid solution crystals (microsegregation) because of insufficient equilibration of composition differences during solidification of the crystal as discussed in Sec. 8.3.3.2. Microsegregation can be removed by a subsequent homogenization heat treatment. By contrast, it is practically impossible to remove macrosegregation by heat treatment (the length scales are too long for diffusion over practical time intervals), so measures have to be taken to try to minimize them, for instance by stirring.

8.6 Rapid Solidification of Metals and Alloys

An increasing cooling rate during solidification causes strong morphological changes of the solidified microstructure, impacts the constitution of alloys and affects even the crystalline state. In general, the as-cast grain size decreases with increasing cooling rate. In systems with limited solubility the solubility limit is shifted to higher concentrations with increasing cooling rate (supersaturated solid solutions), and in systems with intermetallic compounds metastable phases are frequently observed usually exhibiting simpler crystal structures. At high solidification rates microcrystalline microstructures are generated (grain size less than 1 micron). In some special systems (for instance Al-Mn) so-called "quasicrystals" can appear at high cooling rates. Their crystal structures do not comply with the strict principles of translational crystal symmetry because of their crystallographically unallowed five-fold rotation symmetry. Such symmetry is obtained usually from a mixture of two different structural elements, the composition of which does not generate a long-range ordered atomic arrangements but a five-fold rotation symmetry (so-called "Penrose" patterns, Fig. 8.35a). Such structures have been actually imaged in atomic resolution TEM (Fig. 8.35b). The potential of quasicrystalline materials for applications has yet to be shown, but some systems

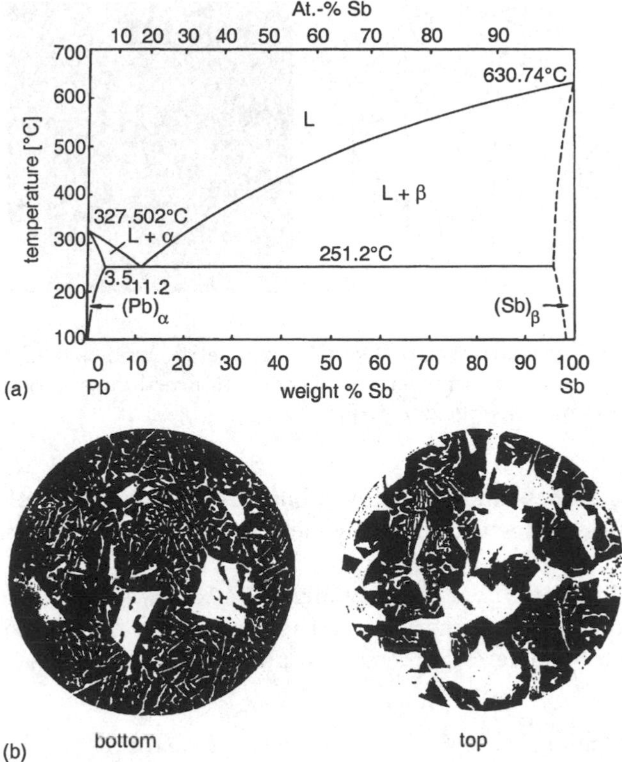

Fig. 8.33. Gravitational segregation in the lead-antimony system. (a) Phase diagram; (b) micrographs of a casting at the lower and upper limit of a hyper-eutectoid alloy (the less heavy antimony-rich solid solution appears lighter colored) [8.17].

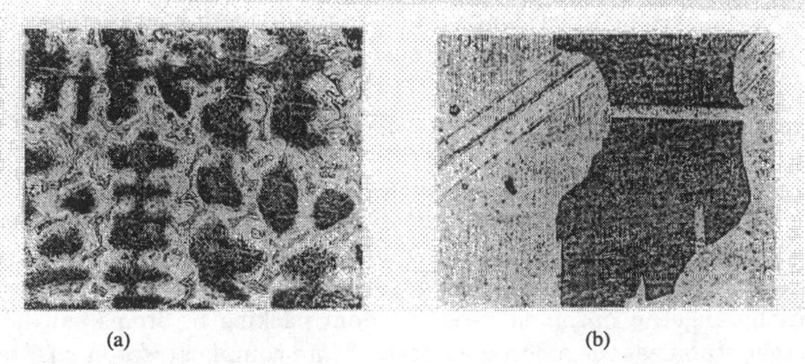

Fig. 8.34. Dendritic zone crystal of a casting bronze (90%Cu, 10%Sn). The difference in composition is obvious, due to the concentration dependent chemical attack. (b) After 30 min. homogenizing at 650°C the composition has become more uniform. The differences in brightness here are due to grain orientation contrast [8.18].

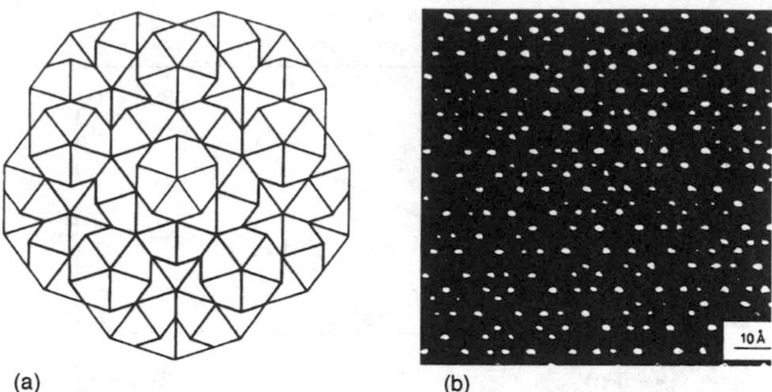

Fig. 8.35. (a) Quasi-lattice with five-fold symmetry, composed of two different rhombi (Penrose-sun): (b) arrangement of atoms in a real quasi-crystal (high resolution electron microscope image) [8.19].

exhibit high hardness and limited wetability that makes them of interest for special coatings, for instance in frying pans. Also unusual electrical properties are expected from these materials.

If the cooling rates are extremely high (10^5-10^6 K/s) alloys and even pure metals will solidify as an amorphous structure, usually referred to as metallic glasses. Typical commercial processes to produce metallic glasses are

- roller quenching, casting of the melt between cold rolls
- melt spinning, deposition of the melt on a rotating cold plate
- vapor deposition onto a cold substrate

More recently, metallic glasses can be obtained also by interdiffusion of thin films in specific systems with complicated crystal structure (for instance Au-La or Co-Zr). By roller quenching or melt spinning metallic glass tapes can be produced at an efficient rate. In the past, the application of metallic glasses was limited by the fact that only thin tapes could be fabricated. More recently, bulk metallic glasses can be produced in special alloy systems that provide new application potential for this class of materials, such as excellent elastic energy storage for the faces of golf clubs.

Typical commercial amorphous alloys consist of mixtures of transition metals (T) like Mn, Fe, Co, Ni, Sc, Ti, V and metalloids (M) like B, C, P, Si, Ge, As, Sb in a composition T_{1-X}, M_X with $0.15 < X < 0.25$. In this concentration range an optimum dense random packing of atoms can be produced that stabilizes the amorphous state. More complicated systems contain alkaline-earth and noble metals. To obtain a metallic glass by solidification, the melt must be rapidly cooled below the solidification temperature. Low melting temperatures are advantageous in accomplishing this because low-temperature eutectics are already close to the eutectic composition and much nearer to their glass transition temperatures.

In contrast to silicate glasses, metallic glasses are much less stable during annealing. During a temperature increase a structural relaxation occurs, which is caused by small atomic shuffles allowing more stable arrangements. This is frequently accompanied by the embrittlement of the amorphous material. At higher temperatures, crystallization (not recrystallization!) occurs. The service temperature of metallic glasses is therefore limited by their devitrification. For pure metals, the crystallization occurs at only a few K's. Much more stable are eutectic alloys based on transition metal-metalloid alloys for which the crystallization temperature can exceed several hundred K's.

Most metallic glasses exhibit predominantly metallic bonding and, therefore, are ductile even at high strength. Very often good corrosion resistance is also obtained. Of particular interest for commercial applications are ferromagnetic metallic glasses. These glasses are soft magnetic materials and, therefore, cause extremely low hysteretic losses during changing magnetization that makes them interesting for use as core materials in electrical transformers. The unique magnetic properties of metallic glasses are also exploited as theft-protection devices in department stores.

8.7 Solidification of Glasses and Polymers

8.7.1 Ionic crystals and glasses

The crystallization behavior of most ionic crystals is similar to metals. Some of them, however, can easily form permanently supercooled glasses. This is the case for common silicates, for instance Na-K silicates known as window glass. More accurately, the glassy state is distinguished from the supercooled melt (Fig. 8.36). If the temperature of a melt is decreased the state of the melt continuously changes corresponding to thermodynamic equilibrium at a given temperature, i.e. towards a continuous increase of the ordered state. At a temperature T_E easy thermal motion essentially ceases, and the arrangement of molecules will not change any further. Below this temperature the material is referred to as glass if it has not crystallized.

An amorphous or vitreous material can crystallize or transform into the state of a supercooled melt as, for instance, silicate materials. By adding nucleation agents to the melt like TiO_2, Cr_2O_3, P_2O_2, or by applying specific annealing treatments silicate glasses can be crystallized. These crystallized glasses are termed "vitroceramics" or "glass ceramics". The crystallite size and the degree of crystallization can be controlled by thermal processing to influence optical transparency, thermal expansion and its suitability for machining.

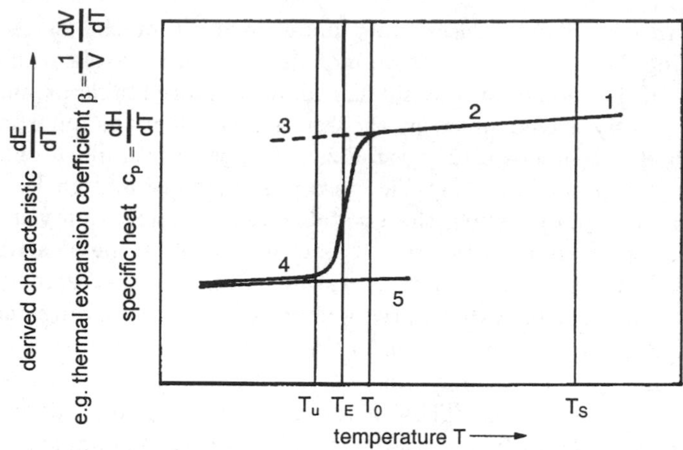

Fig. 8.36. Schematic characteristics-temperature diagram of high-polymers and engineering silicates. 1 - melt; 2 - state of the undercooled melt; 3 - equilibrium state of the undercooled melt; 4 - glassy state; 5 - crystalline state.

8.7.2 Polymers

The same physical principles that control the crystallization of metals or ionic crystals also hold for the crystallization of polymers. The macromolecules of polymers, however, have a morphology that is distinctly different from a spherical shape. This results in their special behavior during crystallization. The simpler and the more symmetrical the molecular chain structure, the more pronounced is the alignment of next neighbor molecules.

Asymmetrical molecular structures or entanglements of the molecule chains impede an ordered structure. For crystallization, both an orderly structure of the molecule chains and the type and magnitude of the intermolecular interactions are important. For instance, polyisobutylene $[\text{-}CH_2\text{-}C(CH_3)_2]_n$ can only be crystallized in a stretched condition, otherwise the intermolecular forces, (dispersion forces) are insufficient to accomplish crystallization.

As in other materials, a polymer nucleus attempts to assume a spherical shape that has the smallest surface for a given volume. Since the heat of crystallization Δg_u is only sufficiently large if the molecules are straight and parallel, the molecules in a polymer nucleus must be folded. Such nuclei of folded chains already exist in the melt in form of lamellae (Fig. 8.37). Like in metals or ionic crystals homogeneous and heterogeneous nucleation can be discriminated. Heterogeneous nucleation (also referred to as secondary nucleation) is generally enforced by the addition of inoculation agents, such as alkaline salts in polyamide.

The crystal growth rate, v_c, markedly varies among the polymers, since it depends strongly on the structure of the molecules. Highly symmetric molecules crystallize quickly (for instance polyethylene: $v_c = 5 \cdot 10^{-3}$ m/min),

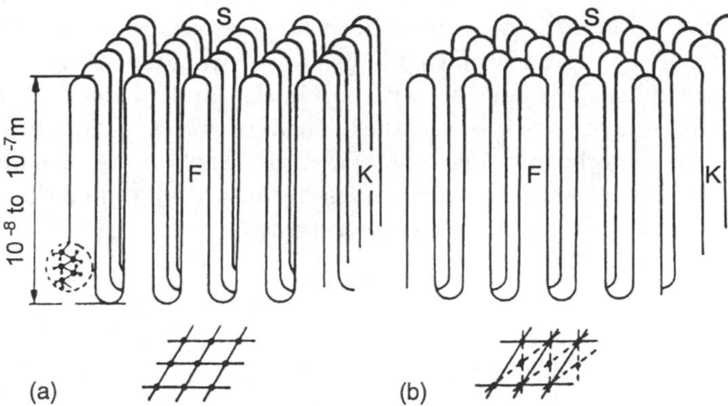

Fig. 8.37. Schematic illustration of the formation of lamella by the folding of macro-molecules (idealized folding). Side view and projection from above. (a) folding in overlapping sequence (e.g. polyamide); (b) folding in alternating sequence (e.g. polyethylene).

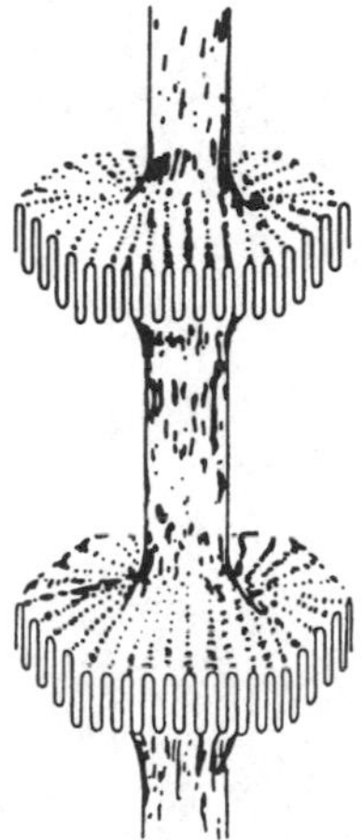

Fig. 8.38. Schematic illustration of a sheaf-of-wheat-structure (linear polyethylene).

whereas more complicated molecular structures grow slowly (for instance PVC: $v_c = 1 \cdot 10^{-8}$ m/min). Of course, the magnitude of v_c is determined by the supercooling.

The crystalline volumes are small in polymers relative to the length of the molecules. A macromolecule, therefore, usually belongs to several crystallites, separated by regions of amorphous structure. Commercial polymers are either partially crystalline or amorphous. The degree of crystallization typically does not exceed 40 - 60%, although some reach even 90%, and in very special cases even 95%. Completely crystalline polymers would behave very brittle, so the amorphous regions are of importance for their mechanical properties.

Under favorable nucleation conditions large blocks of folded chains can develop during cooling to form even larger polycrystalline assemblies. In this case so-called spherulites develop, with a diameter up to 1 mm. By branching or lateral attachment to fibrils, so-called "sheaf of wheat" structures develop, for instance in polyethylene (Fig. 8.38). Crystalline spherulites usually are disadvantageous for commercial applications because they tend to form cracks more easily than do fine-grained microstructures.

9

Solid State Phase Transformations

9.1 Pure Metals

The crystal structure of a metal is not necessarily stable at all temperatures below the melting temperature. This is due to the fact that the solid always assumes the crystal structure with the lowest Gibbs free energy, even if there are other crystal structures with a slightly higher free energy. This holds, in particular, for metals since the binding energy E_0 of a metal depends relatively little on its atomic arrangement. For instance, the change of heat of fusion of sodium from the bcc to the hexagonal structure amounts to only $E_0/1000$ (at 36 K). The major contributions to bonding are determined by the electronic structure, and small changes can cause an instability of the crystal structure, for instance by internal fields in ferromagnetic materials. The latter is the cause for the ferromagnetic bcc structure of iron (α-Fe) at low temperatures.

The different crystal structures of an element in its solid state are referred to as allotropic modifications. Depending on the element the change of crystal structure can occur at markedly different temperatures. Frequently there are even several of such phase transformations in the solid state (Table 9.1). If the same crystal structure exists in different temperature regimes some physical properties are continuous between these two regimes, such as the coefficient of thermal expansion of iron (Fig. 9.1). In both bcc phases of Fe, the thermal expansion of α and δ follows the same temperature dependence, which is distinctly different from the behavior in the γ phase.

9.2 Alloys

9.2.1 Diffusion Controlled Phase Transformations

9.2.1.1 General classification

Akin to the liquid to solid transition a solid state phase transformation can lead to different reactions, which are sketched in Fig. 9.2.

Table 9.1. Allotropic modifications of selected elements#).

Element	a.n.	Phase	Structure	a/c [Å]	t-temp. [°C]
Calcium (Ca)	20	α	fcc	5.58	$\alpha \xrightarrow{464} \beta$
		β	bcc	4.48	
Cobalt (Co)	27	α	hcp	2.51/4.07	$\alpha \xrightarrow{450} \gamma$
		γ	fcc	3.54	
Iron (Fe)	26	α	bcc	2.87	$\alpha \xrightarrow{909} \gamma$
		γ	fcc	3.67	$\gamma \xrightarrow{1388} \delta$
		δ	bcc	2.93	
Samarium (Sm)	62	α	hcp	3.62/26.25	$\alpha \xrightarrow{917} \beta$
		β	bcc	4.07	
Tin (Sn)	50	α (grey)	cub	6.49	$\alpha \xrightarrow{13.2} \beta$
		β (white)	tetr	5.83/3.18	
Strontium (Sr)	38	α	fcc	6.09	$\alpha \xrightarrow{225} \beta$
		β	hcp	4.32/7.06	$\beta \xrightarrow{570} \gamma$
		γ	bcc	4.85	
Titanium (Ti)	22	α	hcp	2.95/4.68	$\alpha \xrightarrow{882} \beta$
		β	bcc	3.31	
Uranium (U)	92	α	orthor		$\alpha \xrightarrow{662} \beta$
		β	tetr	10.76/5.66	$\beta \xrightarrow{775} \gamma$
		γ	bcc	3.53	

#) a.n. - atomic number; t-temp. - transition temperature

1. Dissolution or precipitation of a second phase on crossing of one phase boundary: $\alpha + \gamma \rightarrow \alpha$ (e.g. $a - b, a' - b'$)
2. Transformation of a crystal structure into another crystal structure of the same composition on crossing two phase boundaries: $\gamma \rightarrow \beta$ (e.g. $c - d, c' - d'$)
3. Decomposition of a phase into several new phases on crossing three phase boundaries: $\gamma \rightarrow \alpha + \beta$ (e.g. $e - f, e' - f'$). A special case is the eutectoid decomposition ($e'' - f''$).

9.2.1.2 Thermodynamics of decomposition

To begin we consider the case that a new phase β is formed from a homogeneous phase α, i.e. the reaction: $\alpha \rightarrow \alpha + \beta$. This can proceed in two ways

1. β has the same crystal structure as α, but with a different composition. In this case the phase transformation is referred to as decomposition.
2. β has a crystal structure and composition different from α. This is the general case of precipitation.

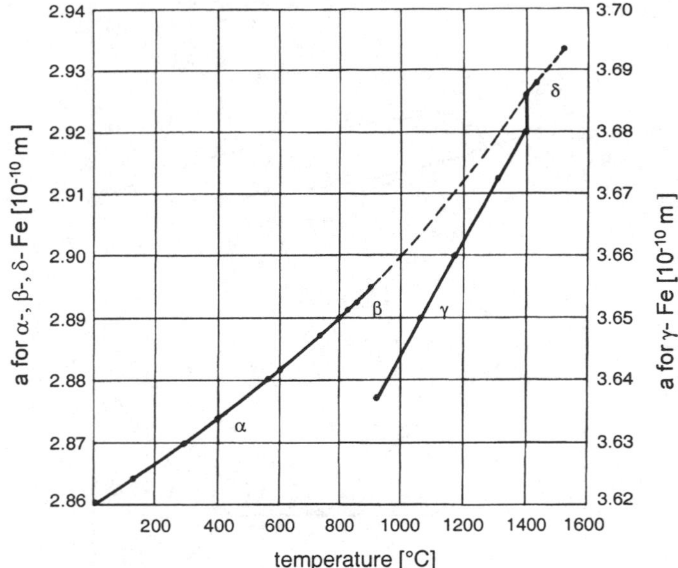

Fig. 9.1. Temperature dependence of the lattice constant of iron (after [9.1]).

Under simplified assumptions the thermodynamics of decomposition can be solved analytically in closed form. This quasi-chemical model of a regular solution will now be introduced.

At a given temperature and pressure the thermodynamic equilibrium is determined by the minimum of the Gibbs free energy, G.

$$G = H - TS \tag{9.1}$$

The enthalpy H represents in this model only the contributions of the binding energies between next neighbor atoms and is given by the binding enthalpies H_{AA}, H_{BB}, H_{AB} between next neighbor AA-, BB-, or AB-atoms. Correspondingly the total binding enthalpy, also referred to as the enthalpy of mixing H_m, is

$$H_m = N_{AA}H_{AA} + N_{BB}H_{BB} + N_{AB}H_{AB} \tag{9.2}$$

where N_{ij} is the total number of bonds between atoms i and j (i = A,B; j = A,B). For this model it is assumed that H_{ij} and, therefore, also H_m is positive. The enthalpy in this model is not the gain of enthalpy ($H < 0$) of an atom during its transition from the free to the bonded state, rather it is the difference between the bonded state and a reference state with absolute minimum enthalpy. Hence, a small value of H refers to a more strongly bonded state than a large enthalpy. The choice of reference state is unimportant for the result. Only the differences between the enthalpy levels determine the physical behavior of the system.

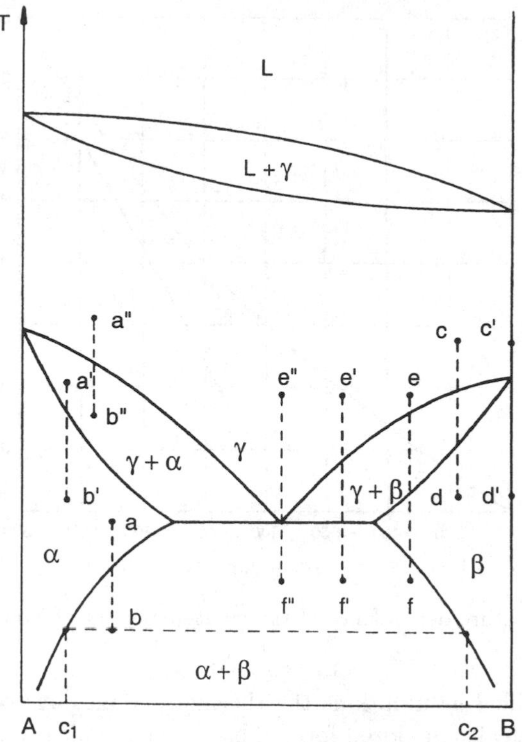

Fig. 9.2. Phase diagram of a binary alloy with phase transitions in the solid state.

If the total number of atoms is N and the number of next neighbors is z (coordination number) then the (atomic) concentration c of B-atoms is $N_{BB} = 1/2Nzc^2$, because there are $N \cdot c$ B-atoms and each B-atom has on average $z \cdot c$ B-atoms as neighbors. The factor 1/2 results from the fact that in this consideration all B-atoms in BB-bonds are counted twice. Respective considerations for N_{AA} and N_{AB} yield

$$\begin{aligned} H_m &= \frac{1}{2} N \cdot z \left[(1-c)^2 H_{AA} + 2c(1-c) H_{AB} + c^2 H_{BB}\right] \\ &= \frac{1}{2} N \cdot z \left[(1-c) H_{AA} + c H_{BB} + 2c(1-c) H_0\right] \end{aligned} \tag{9.3}$$

with $H_0 = H_{AB} - 1/2(H_{AA} + H_{BB})$. H_0 is referred to as exchange energy, which is gained ($H < 0$) or lost ($H > 0$), if two AB-bonds are generated from an AA- and BB-bond, respectively. The special case $H_0 = 0$ is called an "ideal solution". In this case the binding enthalpy does not depend on the arrangement of atoms.

The entropy, S, is composed of the vibration entropy, S_v, and the entropy of mixing , S_m. The vibration entropy is of the order of the Boltzmann constant k and nearly independent of the atomic arrangement. Hence, for alloys

$$S \cong S_m \tag{9.4}$$

According to Boltzmann the entropy of mixing reflects the multiplicity of potential atomic arrangements of N_A A-atoms and N_B B-atoms on N lattice sites.

$$S_m = k \ln \omega_m \tag{9.5a}$$

The number of possible distinctly different arrangements is

$$\omega_m = \frac{N!}{N_A! \cdot N_B!} \tag{9.5b}$$

which can be easily recognized for $N_B = 1,2,...n$. Using Stirling's formula, which is a good approximation for large x

$$\ln x! \cong x \ln x - x \tag{9.5c}$$

and $N_A = N(1\text{-}c)$ and $N_B = Nc$ one obtains for Eq.(9.5a) in a binary system

$$S_m = -Nk \cdot [c \ln c + (1-c)\ln(1-c)] \tag{9.6}$$

S_m is always positive, symmetric with regard to $c = 0.5$, and approaches the limits $c = 0$ and $c = 1$ with infinite slope. The latter behavior provides the reason for the difficulty of obtaining pure elements from alloys, because in the limit as $c \to 0$ the gain dG of free energy for a concentration change by dc (of impurities) approaches infinity (in case of infinitely large systems).

The free energy of a regular solution G_m (Gibbs free energy of mixing) may be expressed with equations 9.3 and 9.6 as

$$\begin{aligned} G_m = {} & 1/2Nz \cdot [(1-c)H_{AA} + cH_{BB} + 2c(1-c)H_0] \\ & + NkT \cdot [c \ln c + (1-c)\ln(1-c)] \end{aligned} \tag{9.7}$$

Enthalpy, entropy and the Gibbs free energy are sketched in Fig. 9.3. We have to distinguish two fundamentally different cases. If $H_0 \lesssim 0$, then $G_m(c)$ exhibits a single minimum. In case of $H_0 > 0$, the free energy of mixing $G_m(c)$ reveals two minima at sufficiently low temperatures, at c_1' and c_2' in Fig. 9.3. This is due to the fact that the entropy of mixing tends to infinity upon approaching the pure elements A and B, whereas the change of enthalpy remains finite for all compositions. Therefore, the free energy will always decrease for small concentrations, whereas for larger concentrations the enthalpy of mixing may exceed the entropy term. In the range $c_1 < c < c_2$, the system can reduce its free energy if it decomposes into a mixture of two phases with the compositions c_1 and c_2, respectively. The free energy of the mixture is given by the common tangent that intersects the $G(c)$ curve at c_1 and c_2 (rule of common tangent).

$$G_g = G_m(c_1) + \frac{c - c_1}{c_2 - c_1} \cdot (G_m(c_2) - G_m(c_1)) \tag{9.8}$$

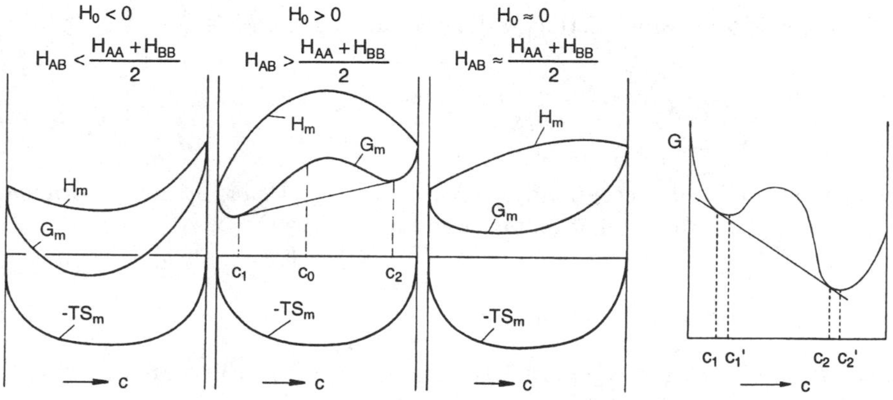

Fig. 9.3. Temperature dependence of free energy of a solid solution for different heat of fusion H_0.

Note that the points of tangency are usually not the minima of the $G_m(c)$-curve. This would only be true if the tangent were a horizontal line, i.e. $G_m(c_1) = G_m(c_2)$. In contrast, if $H_0 < 0$, the atoms of each element favor unlike partners as next neighbors. In this case long-range ordering is preferred over decomposition (see Chapter 4).

For $c < c_1$ and $c > c_2$ the solid solution has always a lower free energy than a phase mixture, i.e., the alloy is always single phase. Correspondingly, c_1 and c_2 represent the solubility limits of the solid solutions at temperature T on both sides of the phase diagram. The dependencies $c_1(T)$ or $c_2(T)$, respectively, reflect the terminal solubilities (solvus line) in the phase diagram of the binary system $A - B$. For the most simple case $H_{AA} = H_{BB}$, G_m is symmetrical with regard to $c = 0.5$ and has minima at c_1 and $c_2 = 1 - c_1$. Hence, the common tangent intersects $G_m(c)$ at the minima and the solubility limit $c_1(T)$ is then obtained by differentiation of eq. (9.7)

$$\left.\frac{dG_m}{dc}\right|_{c=c_1} = 0 \tag{9.9}$$

and we obtain

$$c_1(T) \cong \exp\left(-\frac{zH_0}{kT}\right) \tag{9.10}$$

This dependency is plotted in Fig. 9.4. It corresponds to the terminal solubility of a solid solution in a phase diagram of a system with limited solubility, for instance the system Au-Ni (Fig. 9.5). Usually the maximum of the curve is not attained, because the alloy becomes liquid at a lower temperature. Depending on the behavior of the solidus temperature (monotonic or with a minimum) one obtains a peritectic or eutectic phase diagram (Fig. 9.4), as discussed in more detail in Chapter 4.2.

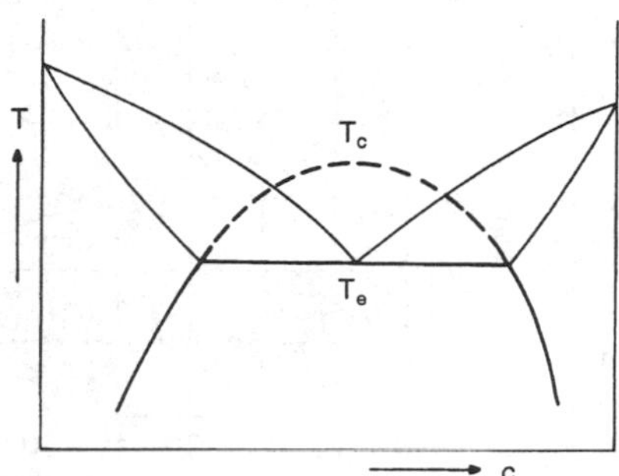

Fig. 9.4. Theoretical phase diagram of a solid solution with a miscibility gap.

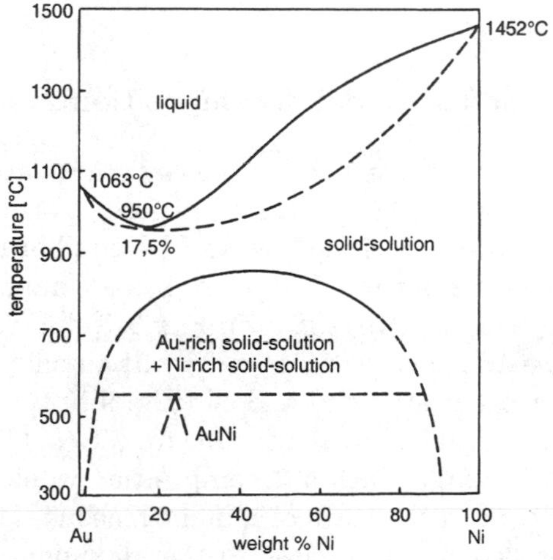

Fig. 9.5. Phase diagram of AuNi (after [9.2]).

If the two phases α and β have different crystal structures, there is no common curve $G_m(c)$, rather there are separate curves for each phase $G_\alpha(c)$ and $G_\beta(c)$ that intersect at a certain concentration. Since only the phase or phase mixture with the smallest free energy is thermodynamically stable, however, one is left with the same situation as for decomposition, i.e. a homogeneous α or β solid solution for $c \leq c_1$ or $c \geq c_2$ respectively, and a phase mixture of α and β for $c_1 < c < c_2$, where c_1 and c_2 denote the points of intersection of the $G(c)$-curves with the common tangent (Fig. of 9.6a).

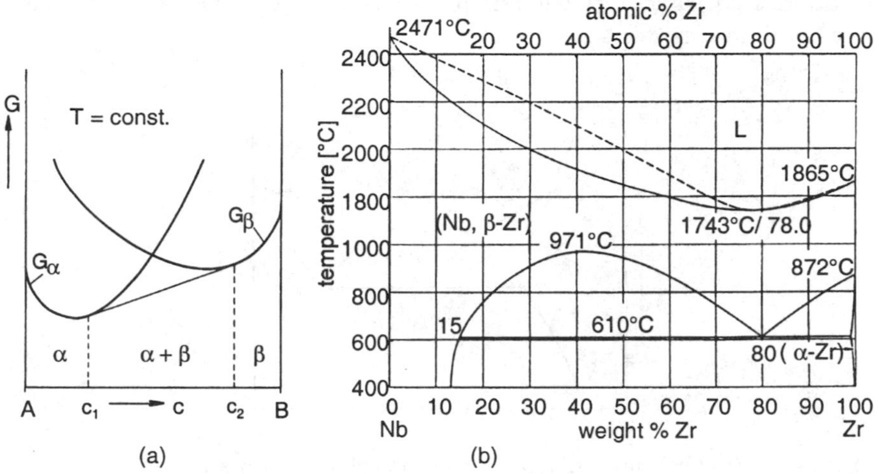

Fig. 9.6. (a) Composition dependence of the free energy in a system with two different phases α and β; (b) NbZr phase diagram (after [9.2]).

9.2.1.3 Nucleation and spinodal decomposition

The process of decomposition can occur either by a nucleation process, where a nucleus is formed with equilibrium concentration c_2, or by spontaneous decomposition (spinodal decomposition), where the equilibrium composition is attained only in the course of time. If a phase has a composition close to the minimum of the free energy curve, e.g. c_1 in Fig. 9.7, fluctuations of composition will always cause an increase of free energy. If an alloy with composition c_1 decomposes into concentrations c_1' and c_1'' (Fig. 9.7) the free energy of the phase mixture G_E is determined by the tie line from $G(c_1')$ to $G(c_1'')$, such that $G_E(c_1) \equiv G_{1E} > G(c_1) \equiv G_1$. Such a decomposition would be unstable, and the system would return to the state of a homogeneous solid solution. If the phase, however, has a composition close to the maximum of the $G(c)$ curve (e.g. c_3 in Fig. 9.7) any decomposition is accompanied by a free energy gain which grows with progressing decomposition. In such case any fluctuation of concentration will be amplified, the system decomposes spontaneously. This process is referred to as spinodal decomposition. Because the diffusion flux during spinodal decomposition runs from low to high concentrations (up hill diffusion) — in contrast to normal diffusion — i.e. opposite to the concentration gradient, this behavior corresponds to a negative diffusion coefficient, owing to a negative thermodynamic factor (see Chapter 5).

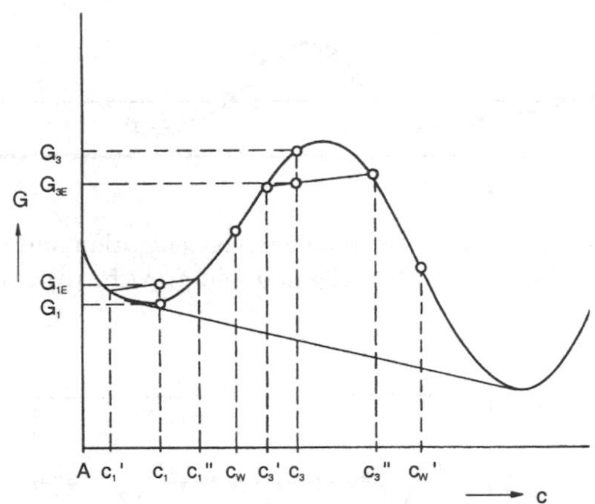

Fig. 9.7. How decomposition of a solution affects free energy. The curve's points of inflection are at c_w and c'_w.

The difference between spinodal decomposition and regular nucleation is demonstrated in Fig. 9.8. During nucleation a nucleus of the β phase is formed by thermal fluctuation with the correct composition c_β. In the course of time it grows by diffusion while its composition remains unchanged. Nucleation during solidification and solid state phase transformations has to exceed a critical nucleus size for stable growth. Therefore, nucleation is preceded by an incubation time. In contrast, spinodal decomposition proceeds spontaneously (without incubation time), and decomposition is amplified until equilibrium is attained. Thermodynamically the terminal states are the same, but the intermediate morphologies of the phase mixture are totally different. During spinodal decomposition a periodic wave-like pattern develops during the early stages, whereas during nucleation the shape of the precipitated phase is determined by interfacial and elastic energy as will be shown below. The wave length of spinodally decomposed structures is usually small, for instance 50 Å in Al-37%Zn at 100° (Fig. 9.9). Such fine lamellae composites have favorable properties, and in the case of ferromagnetic alloys yield high coercive forces making excellent permanent magnets.

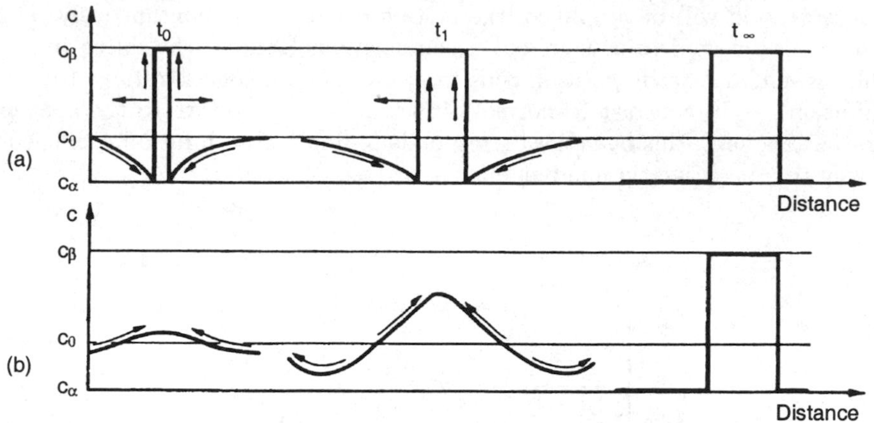

Fig. 9.8. Schematic diagram of the change in concentration and dimensions during decomposition of a solution by (a) nucleation and growth, (b) spinodal decomposition.

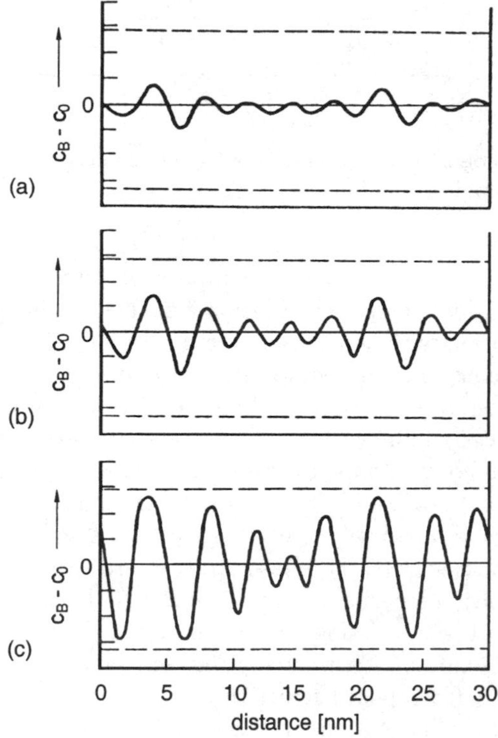

Fig. 9.9. Numerically calculated concentration profiles for spinodally decomposed Al-37%Zn aged at 100°C for (a) 8 min., (b) 15 min., (c) 23 min. The dashed lines indicate the equilibrium concentration of decomposition (after [9.3]).

A reduction of free energy during decomposition is only possible if the tie line between two points on the $G(c)$ curve falls below the curve itself, i.e. if the curve has a concave curvature ($d^2G/dc^2 \leq 0$). This is true between the points of inflection of the curve $G(c)$. The points of inflection are defined by $d^2G/dc^2 = 0$. They are indicated in Fig. 9.7 as c_w and c'_w and can be calculated from the second derivative of Eq. 9.7. The dependency $c_w(T)$ defines the spinodal curve. For the simple case $H_{AA} = H_{BB}$ we obtain

$$c_w \cdot (1 - c_w) = \frac{kT}{2zH_0} \tag{9.11}$$

This parabolic spinodal curve is included in Fig. 9.10. Within the two phase regime one can discriminate two regions: namely the center, which is limited by the spinodal curve where spinodal decompositon can occur, and the region between the spinodal curve and the solubility limit on both sides, where precipitation can only occur by nucleation and nucleus growth.

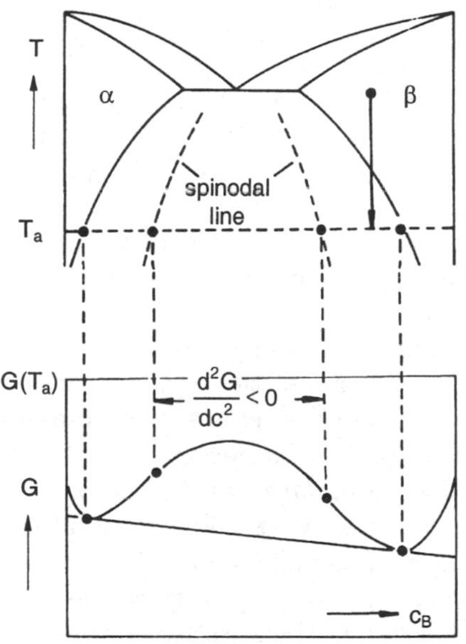

Fig. 9.10. Schematic phase diagram and free energy curve of a solid solution when a miscibility gap occurs; $G(T_a)$ = free energy at aging temperature T_a.

The nucleation of precipitates can principally be treated like nucleation during solidification (see Chapter 8.2) but there is an additional complication. In contrast to solidification the differences of molar volumes of mother phase and precipitates play an important role in the solid state. The volume difference has to be compensated by an elastic distortion with the elastic energy E_{el} that increases with increasing nucleus volume V: $E_{el} = \varepsilon_{el} \cdot V$ (ε_{el} - distortion energy per unit volume = elastic energy density).

Therefore, the increase of free energy due to the formation of a spherical nucleus with radius r, in analogy to Eq. 8.1 (chapter 8) is given by

$$\Delta G(r) = (-\Delta g_u + \varepsilon_{el}) \cdot 4/3\pi r^3 + \gamma \cdot 4\pi r^2 \tag{9.12}$$

γ is the specific energy of the $\alpha-\beta$ phase boundary, Δg_u is the free energy gain per unit volume during phase transformation. Correspondingly, the radius is given by the maximum of the curve $\Delta G(r)$:

$$r_0 = \frac{2\gamma}{\Delta g_u - \varepsilon_{el}} \tag{9.13}$$

The energy of the interface boundary and the elastic distortion energy play a major role since small changes of r_0 result in large changes of the nucleation rate (Chapter 8.2). The elastic distortion energy (per unit volume) for a hard precipitate β in a soft matrix α can be calculated to yield

$$\varepsilon_{el} = \frac{E_\alpha \delta^2}{1-\nu} (c_\beta - c_\alpha)^2 \cdot \varphi\left(\frac{c}{b}\right) \tag{9.14}$$

The concentrations c_α and c_β represent the composition of matrix and precipitate, respectively, E_α and ν denote Young's modulus and Poisson ratio of the α phase, $\delta = d(lna)/dc$ (a - lattice parameter) the atomic size factor (essentially the distortion) and φ the form factor. If one assumes that the precipitate has the shape of an ellipsoidal solid of revolution, where c is the ellipsoidal radius parallel to the rotation axis and b the respective perpendicular elipsoidal radius one obtains the dependency of φ on the axis ratio c/b as plotted in Fig. 9.11. Correspondingly, the distortion energy is largest for a sphere. The sphere, however, has the smallest surface for a given volume. The shape of a precipitate, therefore, is a compromise between a minimum of distortion energy and total surface energy. If both precipitate and matrix have about the same lattice parameter ($\delta \approx 0$ in Eq. 9.14) as in the system Al-Ag, the elastic energy plays a negligible role and the precipitates have a spherical shape. However, if the difference of the lattice parameters is large ($\delta \gg 0$) plate-shaped precipitates are preferred, as in the system Al-Cu. It is noted, though, that not only the area of the interface but also its specific energy plays a role. If the interfacial energy is anisotropic, i.e., different interfacial planes have different energies, then also plate-shaped precipitates are preferred, although the elastic energy is minimal. For instance, in the system Al-Ag when semicoherent interfaces with large energy anisotropy are generated spherical precipitates are observed in the beginning that change to a plate shaped-morphology in the course of annealing (Fig. 9.19).

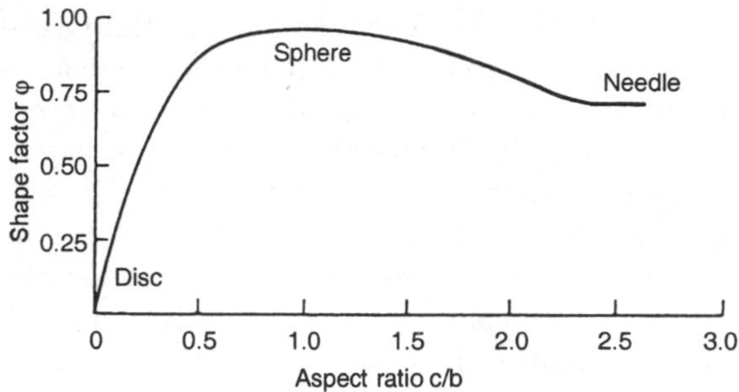

Fig. 9.11. Elastic shape factor φ for a rotational ellipsoid with aspect ratio c/b (after [9.4]).

9.2.1.4 Metastable phases

The nucleation rate $\dot{N}$ is by analogy to Eq. 8.4 given by

$$\dot{N} \sim \exp\left(-\frac{\Delta G_0}{RT}\right) \tag{9.15}$$

with

$$\Delta G_0 = \Delta G\left(r_c\right) = \frac{16}{3}\pi\frac{\gamma^3}{\left(\Delta g_u - \varepsilon_{el}\right)^2}$$

Correspondingly, $\dot{N}$ strongly depends on interface energy γ. The energy of an interface is essentially determined by its structure. We discriminate three types of interfaces: coherent, partially coherent, and incoherent interface boundaries (Fig. 9.12) (see Chapter 3.5).

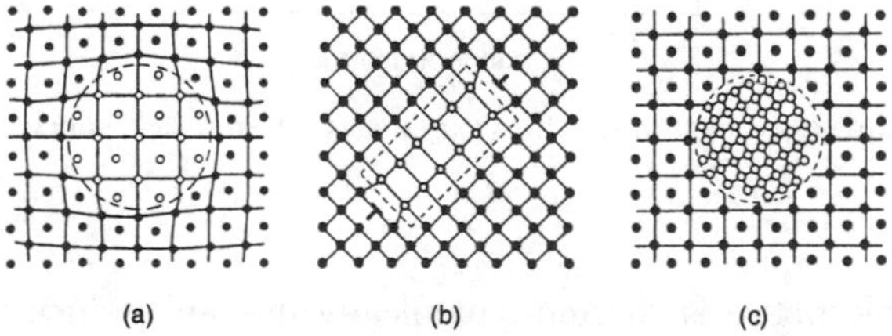

Fig. 9.12. Structure of phase boundaries: (a) coherent; (b) semi-coherent; (c) incoherent.

For coherent precipitates all lattice planes of the matrix continue through the precipitate (Fig. 9.12a). Because of differences of the lattice parameter in precipitate and matrix there will always be slight elastic distortions. If the difference in lattice parameter is too large then edge dislocations in the interface can compensate the elastic distortions (partially coherent interface, Fig. 9.12b), and consequently most lattice planes of the matrix are continued through the precipitate but a few terminate in the interface, and create the interfacial edge dislocations. If the crystal structure of both phases is different, or if the orientations of matrix and precipitate are different, both phases are separated by an incoherent interface (Fig. 9.12c). Commonly, precipitates have a crystal structure different from that of the mother phase. Therefore, their interface boundary is incoherent. The respective large interfacial energy renders the work of nucleation high and thus strongly delays nucleation, in particular at low temperatures, where diffusion proceeds slowly. In such case metastable phases are formed although the gain in free energy is not as large as for the equilibrium phase (e.g. the θ phase Fig. 9.13), but the energy of the interface boundary is low due to its coherent or partially coherent structure, which yields a high nucleation rate.

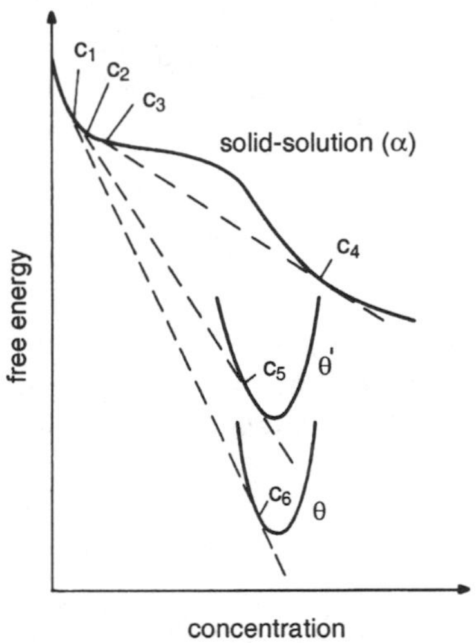

Fig. 9.13. Schematic diagram of the free energy change for successive metastable phases.

A metastable phase is thermodynamically unstable, since its free energy is higher than that of the equilibrium phase. Frequently, several metastable phases are successively formed until the equilibrium phase is attained. In such a case consecutive metastable phases have a continuously lower free energy (Fig. 9.13). The first phases to form are coherent phases or decomposition zones with a size of a few atomic layers. These layers are usually referred to as Guinier-Preston zones. Since thermal activation tends to destroy these zones the number of Guinier-Preston zones decreases with rising temperature until finally above a critical temperature further decomposition zones cease to form

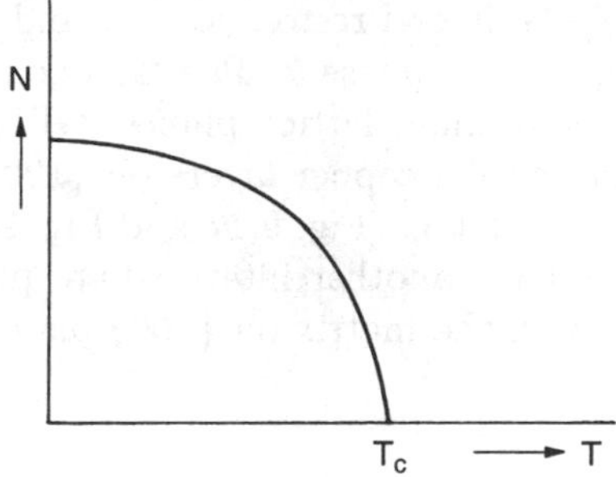

Fig. 9.14. Temperature dependence of the equilibrium density of Guinier-Preston zones (existence curve). Due to insufficient diffusion, equilibrium cannot be reached at low temperatures.

and only incoherent phases appear (Fig. 9.14). At low temperatures diffusion effectively ceases, and decomposition cannot take place. Consequently, there is a certain temperature range in which these metastable phases exist. This is the temperature range used for the thermal treatment of age hardened alloys, which are technically of special importance.

9.2.1.5 Age hardening

Age hardening is one of the most important processes for strengthening of alloys. Age hardened aluminum alloys are the base material of the aerospace industry, because they have a high specific strength (strength/density).

The principle of age hardening is the strengthening by precipitates of a second phase during annealing of a supersaturated solid solution. A typical example is the system Al-Cu. The aluminum-rich side of the phase diagram is eutectic (Fig. 9.15a). The maximum solubility of copper in aluminum is 5.65w.% at 548°C. For temperatures below 300°C the solubility decreases to less than 1%. For age hardening a composition is used that constitutes a solid solution at elevated temperatures, but yields a two-phase regime $\alpha + \theta$ at low temperatures (Fig. 9.15b). Here θ denotes the intermetallic phase Al_2Cu that has a crystal structure (tetragonal) different from the fcc Al-Cu solid solution. If the homogeneous alloy is quenched to room temperature, a supersaturated solid solution is obtained. Upon heating to slightly higher temperatures a considerable increase of strength is obtained as shown in Fig. 9.16. At low annealing temperatures (100°C) the hardness increases slowly but continuously until a plateau is reached. At slightly higher temperatures another rise of hardness is observed after the plateau. This second hardness increase passes through a maximum. At even higher temperatures the plateau and the maximum will be reached earlier but the plateau hardness decreases. Eventually, at 300°C the plateau is not obtained anymore, rather only the second hardness regime occurs. This material behavior can be interpreted in terms of the phase transformation $\alpha \rightarrow \alpha + \theta$. During the heat treatment only coherent metastable Guinier-Preston zones I and II are formed, rather than the incoherent equilibrium phase θ. The GPI zones are single layers of copper atoms on {100} aluminum lattice planes. GPII zones (or θ'' phase) are comprised of several parallel copper layers on {100} planes that cause a tetragonal distortion of the matrix (Fig. 9.17 and Fig. 9.18). The equilibrium phase θ is finally attained via another intermediate phase θ' (CaF_2 crystal structure, also coherent with the matrix on {100} planes).

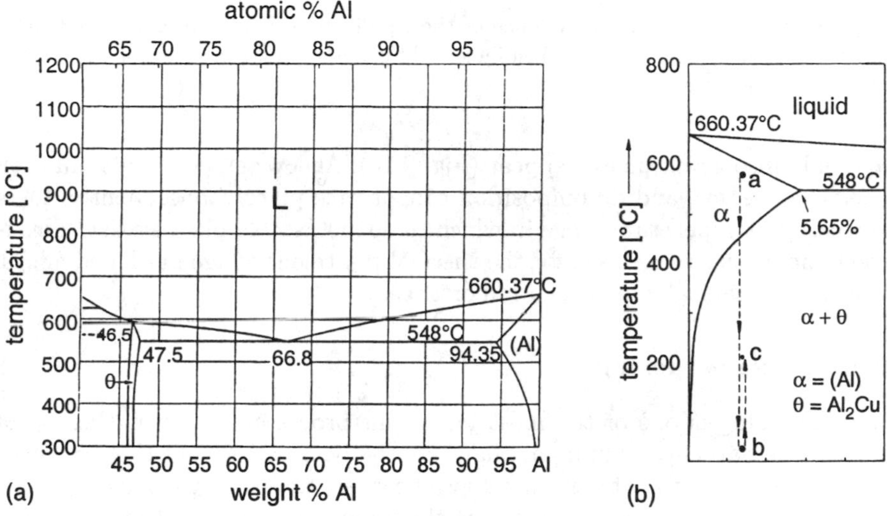

Fig. 9.15. (a) Al-Cu phase diagram, (b) detail of the Al-Cu phase diagram (Al-rich side) indicating how the tempering is managed for age hardening (after [9.5]).

The first hardening step, the plateau can be explained by the formation of GPI and GPII zones. GP zones are decomposition products, the number of which decreases with rising temperature and, correspondingly, the hardness of the plateau decreases at higher temperature. If age hardening remains confined to only this first hardening stage because of low temperature, this is referred to as room temperature age hardening. The second hardening stage is obtained only at higher annealing temperatures. This is called artificial aging. The latter is essentially caused by the formation of the θ' phase. Due to precipitate coarsening at longer annealing times (Ostwald ripening, see Section 9.2.1.6) the precipitate spacing increases and correspondingly, the strength decreases, because the precipitates can be more easily circumvented by dislocations via the Orowan mechanisms (see Chapter 6). It is the aim of age hardening to attain maximum strengthening. The undesired decrease of strength at long annealing times is called "overaging". The equilibrium phase θ does not play

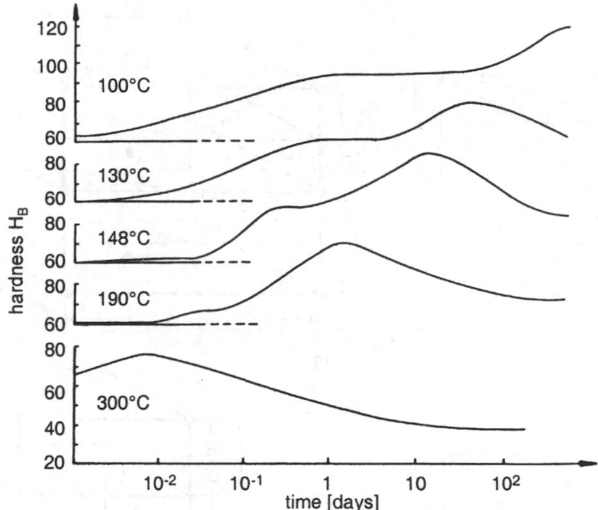

Fig. 9.16. Age hardening curves of Al-4%Cu-1%Mg (after [9.6]).

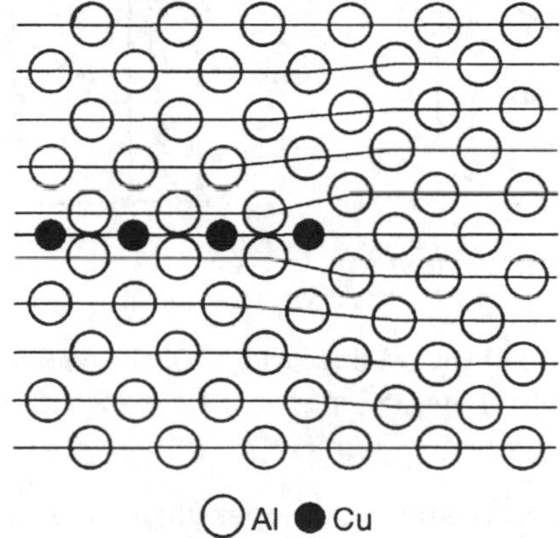

Fig. 9.17. Section through a GPI-zone in AlCu.

a role in age hardening and is obtained in most cases only after lengthy (severe) overaging.

Besides Al-Cu (mostly Al-4%Cu-1%Mg) also Al-Mg-Si (Al-1%Mg-1%Si, equilibrium phase Mg_2Si) and Al-Mg-Zn (Al-4.5%Zn-1.5%Mg, equilibrium phase $MgZn_2$) are age hardening alloys of commercial importance.

A second important group of age hardening alloys consists of Ni, Cr, and Co alloys with Al, Si, Ti, Nb, or W. They are used for high temperature applications (e.g. turbine blades) and are referred to as superalloys. The typical system here is Ni-Al where the intermetallic phase γ', Ni_3Al, is in equilibrium with the nickel-rich solid solution γ. Both phases have a cubic crystal structure with almost the same lattice parameter, thus coherent precipitates

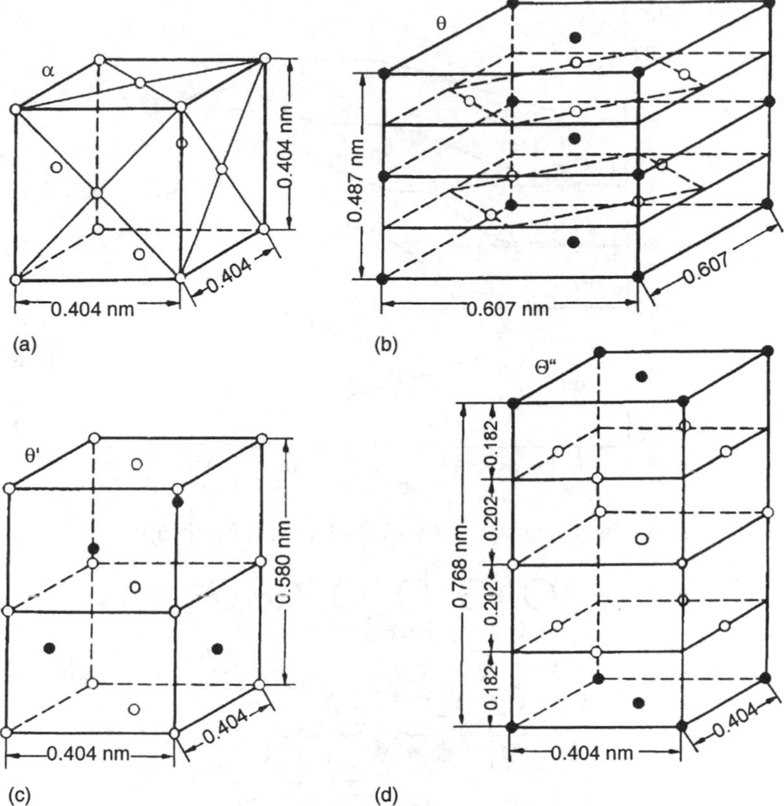

Fig. 9.18. Crystal lattice of the solid solution (a), the equilibrium phase Θ (b) and the metastable phases Θ' (c) and Θ'' of the Al-Cu system (Cu •, Al ∘) (after [9.7]).

can form easily. Modern commercial superalloys contain up to 80% volume fraction of γ' phase, and the γ matrix is confined to narrow channels between the precipitates.

For age hardening of Cu alloys most frequently Be is used (up to 3w.%). After annealing of the supersaturated solid solution at 300 - 400°C the hardness strongly increases and can be further improved by cold forming. The alloys are commercially used as paramagnetic materials or for non-sparking tools.

Age hardening becomes optimal if the precipitate spacing is small, i.e. if precipitation occurs homogeneously throughout the volume, and the precipitates are homogeneously distributed within the crystal (continuous precipitation). This is mostly the case for metastable decomposition zones. Incoherent precipitates preferentially form heterogeneously at crystal defects like dislocations, subgrain boundaries, or grain boundaries (Fig. 9.20). If precipitation is

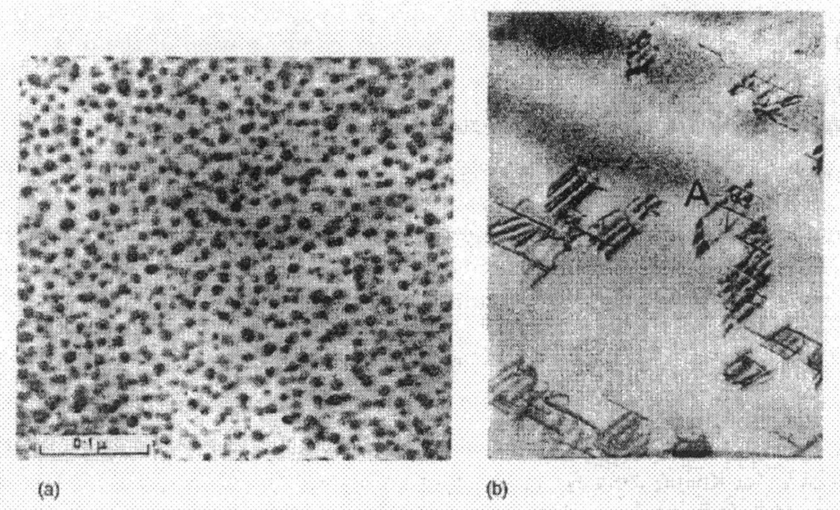

Fig. 9.19. TEM image of metastable phases in Al-32%Ag [9.8], (a) spherical coherent GP-zones after 300 min. at 160°C; (b) plate-shaped α-phase after 7200 min. at 160°C [9.9].

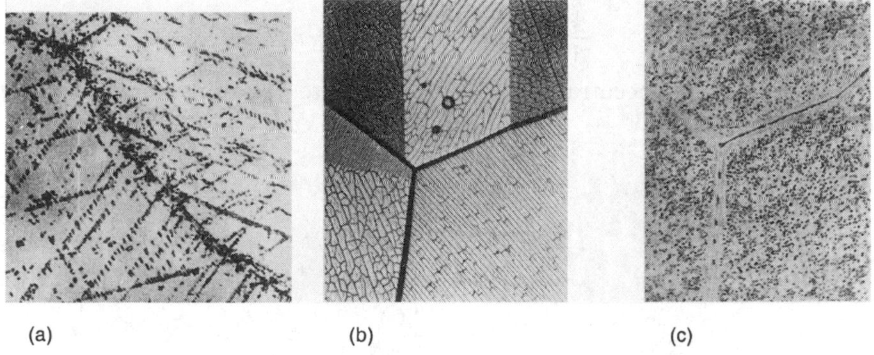

Fig. 9.20. Preferential precipitation at lattice defects. (a) At dislocations, i.e. along slip lines; (b) at subgrain boundaries (c) at grain boundaries. The precipitation-free zone around the grain boundary is caused by solute depletion due to grain boundary precipitation [9.10].

confined only to the volume in the vicinity of lattice defects, the mechanical properties of the material are commonly not optimal.

9.2.1.6 Growth kinetics of precipitates

In the course of annealing the nuclei of a precipitating phase grow until equilibrium is attained, i.e. when both phases have achieved their equilibrium concentration (Fig. 9.8). The growth kinetics are diffusion controlled and can be calculated in closed form under simplified assumptions. In a binary alloy AB with initial concentration c_0 the average concentration of the matrix $\bar{c}_B$ changes with time since a continuous flux of B atoms, j_B, is partitioned to the precipitates of radius r_0:

$$\frac{4}{3}\pi R^3 \cdot \frac{d\bar{c}_B}{dt} = j_B\,(r_0) \cdot 4\pi r_0^2 \tag{9.16}$$

The average matrix volume per precipitate is represented by a sphere of radius R. For steady state growth of the precipitate the concentration profile in front of the particle remains unchanged and is given by the solution of the stationary diffusion equation

$$c_B(r) = c_0 - (c_0 - c'_B) \cdot (r_0/r) \tag{9.17}$$

where c'_B is the equilibrium concentration at $r = r_0$. Fick's first law requires that

$$j_B\,(r_0) = -D_B \left.\frac{\partial c_B}{\partial r}\right|_{r=r_0} = -D_B \frac{c_0 - c'_B}{r_0} \tag{9.18}$$

and conservation of the number of B atoms imposes the relationship

$$\frac{4}{3}\pi R^3\,(c_0 - \bar{c}_B) = \frac{4}{3}\pi c_K r_0^3 \tag{9.19}$$

where c_K is the concentration in the precipitate. The precipitated volume fraction is found to be

$$X(t) \equiv \frac{c_0 - \bar{c}_B}{c_0 - c'_B} \tag{9.20}$$

for short times

$$X(t) = \left(\frac{2t}{3\tau}\right)^{3/2} \tag{9.21}$$

and after a long time when adjacent areas compete for the remaining excess B atoms

$$X(t) = 1 - 2\ \exp\left(-\frac{t}{\tau}\right) \tag{9.22}$$

The time constant τ is mainly determined by the diffusion constant

$$\frac{1}{\tau} = \frac{3D_B\,(c_0 - c'_B)^{1/3}}{c_K^{1/3} R^2} \tag{9.23}$$

This result has straight forward physical interpretation. Except for long annealing times, the radius of the precipitate grows with $r_0 \sim \sqrt{D_B \cdot t}$. This occurs because the B atoms that are needed to make the precipitate grow have to be supplied by diffusion from increasingly larger distances $a \sim \sqrt{D_B \cdot t}$. Since $X \sim r_0^3$, the dependency of $X \sim t^{3/2}$.

This simple model shows good agreement with experiments (Fig. 9.21) despite the fact that the influence of elastic distortion was ignored. There are cases, however, where less perfect agreement of theory and experiment is obtained. This occurs because the growth kinetics can also be controlled by mechanisms other than the atomic transport, for instance by the kinetics of B atoms moving across the phase boundary during growth of the precipitate, or for precipitates at crystal defects where other diffusion mechanisms can operate.

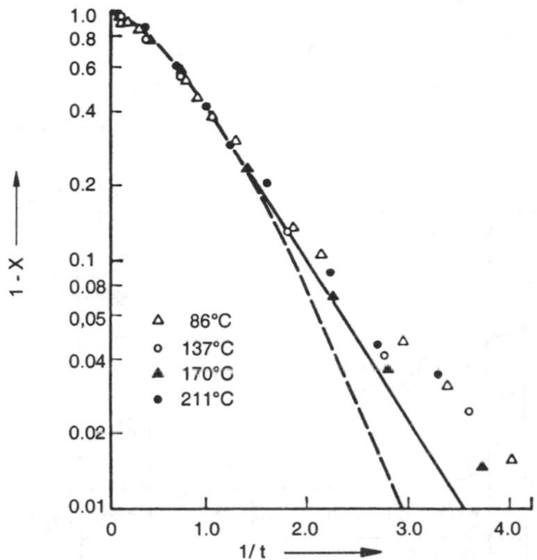

Fig. 9.21. Growth kinetics of C in α-Fe. Solid line - precise theory; dashed line - dependence according to Eq. (9.22).

One is tempted to expect that the growth of the precipitates ceases once the equilibrium composition has been attained. In reality one observes, however, that in the course of annealing small precipitates are dissolved, while large precipitates continue to grow. This "competitive" precipitate coarsening is referred to as Ostwald ripening. The driving force for this process is a decrease of the total interface boundary energy. The total energy would be smallest were there only one huge particle instead of many small precipitates. Whether a precipitate is to dissolve or to grow depends on the chemical potential of atoms in the vicinity of a precipitate. Since equilibrium is attained only if the chemical potential is constant everywhere, there is always a flux of atoms from the higher to lower chemical potential. For spherical particles the chemical potential is determined by the curvature of the surface. Correspondingly there is a chemical potential difference between two particles with different radii r_1 and r_2 (Gibbs-Thomson equation)

$$\Delta\mu = 2\gamma_{\alpha\beta}\Omega\left(\frac{1}{r_1} - \frac{1}{r_2}\right) = kT\frac{\Delta c_B}{\hat{c}_B} \tag{9.24}$$

($\hat{c}_B$ equilibrium concentration of the matrix for a flat interface boundary ($r \to \infty$), $\gamma_{\alpha\beta}$ - interface boundary energy and Ω atomic volume). The chemical potential difference is obviously related to a concentration gradient $\Delta c_B = c(r_1) - c(r_2)$ (Fig. 9.22) which causes a diffusion flux from the small to the large particle so that the large particle grows while the small particle dissolves. The solution of the diffusion problem yields the time dependence of the average particle size $\bar{r}$

$$\bar{r}^3 - \bar{r}_0^3 \sim \gamma_{\alpha\beta} D_B t \tag{9.25}$$

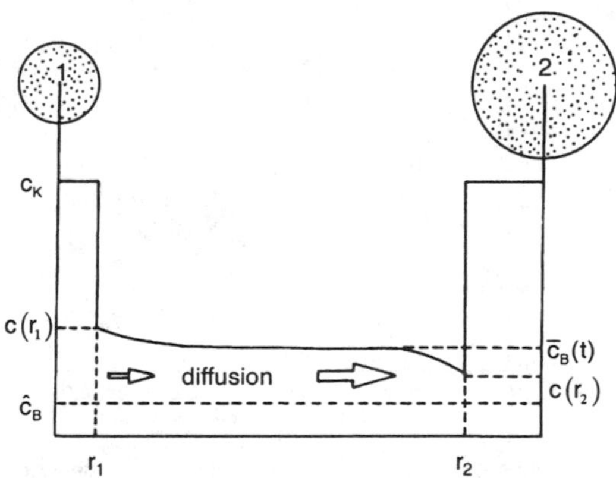

Fig. 9.22. Concentration distribution and diffusion flux between particles of different size.

In this model elastic distortions were not accounted for. Also, if the precipitates are plates or needles, e.g. due to the shape factor in Eq. 9.14, one has to take into account the deviation from a spherical geometry. Irrespectively, the dependency $r \sim t^{1/3}$ is observed for any metric of a particle, for instance for each edge of a cuboidal particle although with different kinetic coefficients ($\gamma_{\alpha\beta} D_B$) for different spatial directions.

Finally, it is noted that particle coarsening is not confined to the equilibrium phase rather it also plays a role for metastable phases. For instance during age hardening the hardness maximum and overaging are due to Ostwald ripening although in most cases the equilibrium phase has not yet been formed.

9.2.1.7 Eutectoid decomposition and discontinuous precipitation

Although precipitation usually proceeds by nucleation and growth of individual nuclei, there are cases where the solid state phase transformation can proceed by motion of a reaction front, for instance in the cases of eutectoid decomposition and discontinuous precipitation.

Eutectoid decomposition is akin to the process of eutectic solidification, except that it occurs in the solid state. One phase decomposes into two different phases: $\alpha \rightarrow \beta + \gamma$. Since β and γ occur simultaneously, but have different composition — possibly even different crystal structure — the morphology of the phase transformation is a lamellar structure because the concurrent precipitation can occur only by exchange of atoms via short-range diffusion. One of the commercially most important examples is the pearlite reaction in steel. In this case the carbon-rich fcc γ-Fe decomposes in carbon-poor α-Fe and cementite (Fe_3C).

In the majority of cases the reaction is initiated at a grain boundary, where nucleation is favored. If the α phase, for instance, forms the first nucleus (Fig. 9.24a) then the necessary enrichment of one element (for instance B) will be provided from its immediate environment, which becomes depleted of this element but enriched with atoms of the other element (A) by which a β nucleus is generated. The depletion of the environment of A atoms in turn promotes nucleation of α and so on. The result is a lamellar microstructure (Fig. 9.24b). Since the thickness of the lamellae is determined by the range of diffusion, the lamellar spacing decreases with increasing transformation rate (Fig. 9.23) (Eq. 8.9).

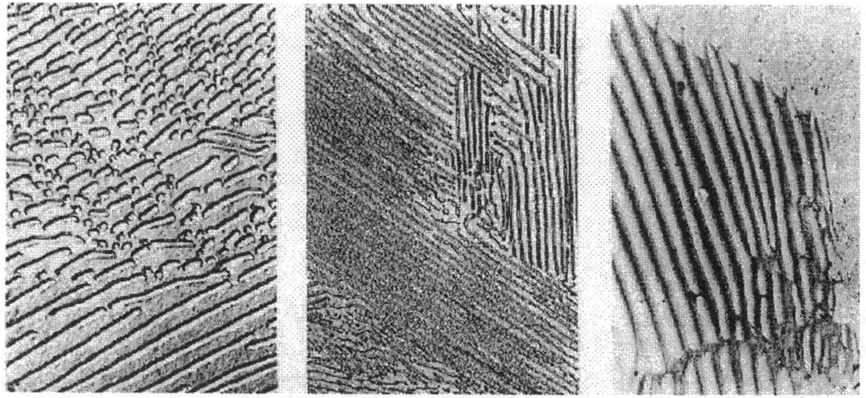

Fig. 9.23. Microstructure of eutectoidally decomposed Fe-0.78%C-0.97%Mn. (a) Cooled slowly (solidification at 680°C) [9.11]; (b) cooled rapidly (solidification at 639%C [9.11]; (c) incipient forming of pearlite with advancing carbide lamella in Fe-0.8%C (carbide lamella are light colored, ferrite lamella dark colored) [9.12].

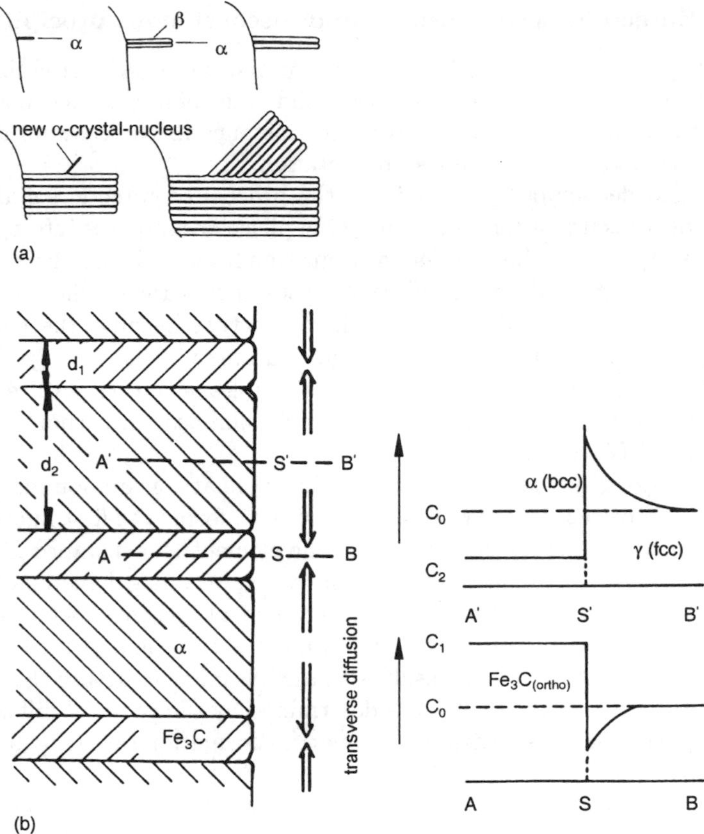

Fig. 9.24. (a) Illustration of the principle of lamella generation during eutectoid decomposition; (b) lamella arrangement and carbon concentration distribution in growth direction during the pearlite reaction.

Another form of transformation by motion of a reaction front is discontinuous precipitation. In this case the reaction at the transformation front is $\alpha \rightarrow \alpha + \beta$, i.e. only one new phase is formed. Usually nucleation also begins at grain boundaries and as in eutectoid decomposition a lamellar microstructure is generated. Since the precipitation is not homogeneous in the grain interior — as in continuous precipitation — rather it occurs at a few preferred locations, mostly at grain boundaries, this process is termed discontinuous precipitation. It is characteristic for discontinuous precipitation that the reaction is connected to grain boundary migration. Therefore, it is also considered as a recrystallization process. Since diffusion proceeds faster in a grain boundary, the decomposition is facilitated and accelerated in grain boundaries. Consequently, behind the moving grain boundary the new phase β is precipitated, which grows due to concurrent grain boundary displace-

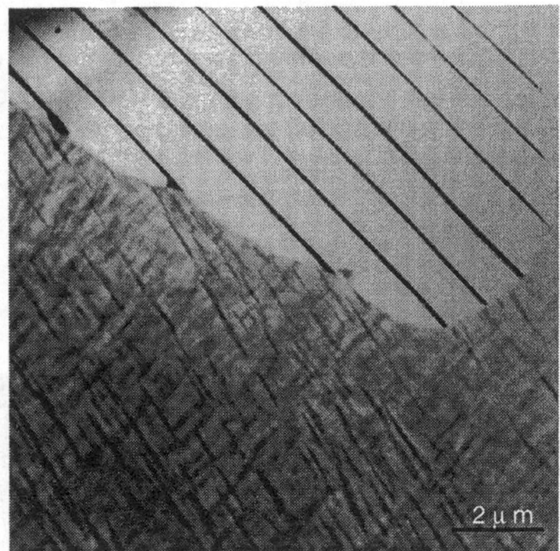

Fig. 9.25. Reaction front during discontinuous precipitation in Al-2.8%Ag-1%Ga.

ment in a lamellar morphology (Fig. 9.25). The grain boundary migration rate is further increased if the supersaturated solid solution is plastically deformed, because the energy of the dislocations serves as driving force for grain boundary migration in addition to the chemical driving force for phase transformation (see Chapter 7). Quite analogous is the reverse process, namely the dissolution of precipitates by moving grain boundaries during the recrystallization of two-phase materials in a microstructure where the homogeneous solid solution is stable.

9.2.2 Martensitic Transformations

In pure metals the new crystal structure forms spontaneously at the transition temperature, and the phase transformation cannot be suppressed even at high cooling rates. The situation is different for alloys, because the formation of a new phase goes along with a change of composition, the rate of which is controlled by diffusion. With increasing cooling rate the diffusivity slows down until for high cooling rates diffusion fails to accomplish the necessary concentration changes. In the latter case the phase transformation is suppressed. With increasing supercooling of an unstable phase the driving forces for transformation rise drastically. If the phase transformation involves a change of crystal structure, as in the system Fe-C from fcc to bcc, the driving forces can become sufficiently large to change the crystal structure spontaneously without change of concentration. These spontaneous phase transformations without concentration change are generally referred to as martensitic transformations in allusion to the commercially most important spontaneous transformation in the system Fe-C when the formation of ferrite is suppressed.

If the spontaneous change of the crystal structure is associated with a volume change or change of shape which may be further complicated by the suppressed change of composition, transformation proceeds by successive displacive transformations of needle or lath shaped regions into the new crystal structure (Fig. 9.26). Consecutively transformed regions will be limited in their spatial extent by prior transformed regions.

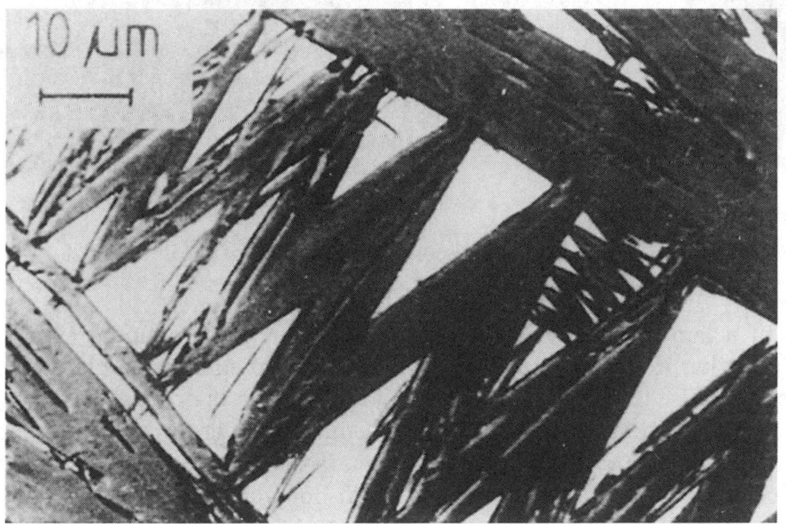

Fig. 9.26. Martensite and retained austenite in an FeNiAl-alloy [9.13].

The volume fraction of the martensite phase usually does not depend on time but only on the temperature to which the material was quenched, and increases with decreasing temperature (Fig. 9.27). Above a certain temperature M_s (martensite start, defined as 1% martentensitic volume fraction in the microstructure) virtually no transformation takes place. Temperatures substantially lower than M_s are needed to obtain complete martensitic transformation (M_f - martensite finish, defined as 99% martensitic volume fraction). This range of existence of martensite depends on composition (Fig. 9.28). With increasing concentration M_s usually decreases. On heating, martensite does not retransform at temperature M_s into the fcc austenite. Superheating to A_s (austenite start) is necessary where $A_s > M_s$ also depends on composition (Fig. 9.28). In the special case of Fe-C, martensite usually decomposes on annealing into ferrite and cementite, instead of retransforming to austenite. By plastic deformation the difference between the transformation start temperatures M_s and A_s can be reduced. It is reasonable to assume that the thermodynamic equilibrium temperature T_0 for both phases (same free energy) is approximately $T_0 = (M_s + A_s)/2$, a result which is supported by experimental results.

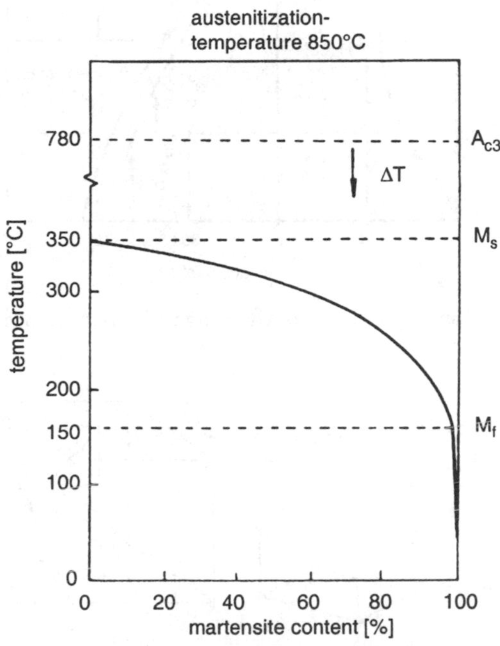

Fig. 9.27. Existence curve of martensite in Fe-0.45%C. A_{c3} is the temperature of beginning transformation to austenite during heating. In practice M_s and M_f are determined by martensite contents of 1% and 99%, respectively.

An explanation for the crystallographic relationships observed during martensitic transformation in the system Fe-C was given by Bain (Fig. 9.29). The transformation consists of a change of the crystal structure from fcc to bcc. The bcc unit cell is commonly observed to be tetragonally distorted in the presence of carbon as will be explained below. The center of two next neighbor fcc unit cells (lattice parameter a_0) contains a tetragonal body-centered unit cell (bct) with the lattice parameters $a = a_0/\sqrt{2}$ and $c = a_0$. To obtain a bcc unit cell from a bct unit cell the latter has to be compressed along the c direction and stretched in the plane perpendicular to the c direction, by which distortion the volume will change by about 3 to 5%. This model complies

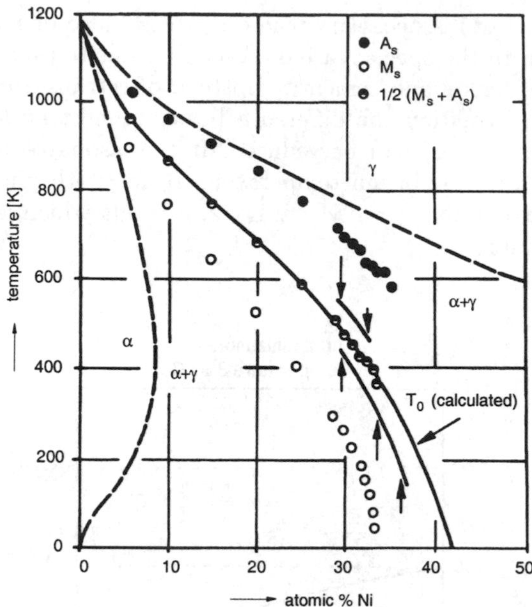

Fig. 9.28. Concentration dependence of the range of existence of martensite in FeNi (after [9.14].

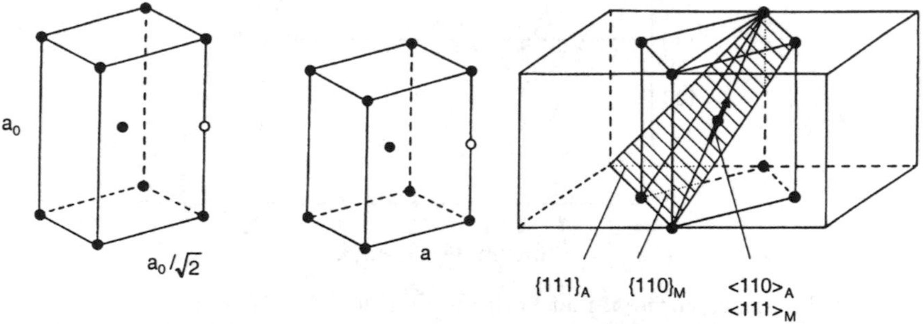

Fig. 9.29. The Bain-model of martensite formation in the FeC system and the corresponding orientation correlation of austenite and martensite after Kurdjumov-Sachs; open circles indicate possible positions of carbon.

with the constraint that during martensite transformation the atomic positions should not change much, and that next neighbors should remain next neighbors also after phase transformation. The Bain correspondence is substantiated by X-ray crystallography investigations that confirm an orientation relationship Fe-C $\{111\}_\gamma \parallel \{110\}_\alpha$ and $\langle 1\bar{1}0\rangle_\gamma \parallel \langle 1\bar{1}1\rangle_\alpha$ (Kurdjumov-Sachs relationship).

Instead of a transition from fcc to bcc, as occurs in pure iron, or for austenite/ferrite the martensitic transformation of Fe-C is a lattice transformation from fcc to bct. This does not contradict the Bain correspondence, rather it is caused by the carbon in solution. The c/a ratio in martensite increases with increasing carbon concentration, whereas the magnitude of the a axis is virtually independent of the carbon content (or decreases slightly) (Fig. 9.30). In contrast, the lattice parameter a_0 increases with increasing carbon content in the γ phase. These observations can be explained within the Bain model. Apparently, carbon in austenite is located on the octahedral interstitial sites of the iron lattice, i.e. in the center of the cube or on the center of its edges. According to the Bain correspondence, after transformation the carbon atoms are located only on the c axis. Since the carbon atoms are larger than the octahedral interstitial sites this leads to a slight increase of the lattice parameter of austenite in the c direction. However, because of the arrangement of all carbon atoms on the c-axis in martensite the lattice parameter in the c-direction increases conspicuously, i.e. tetragonal martensite is formed. If carbon is allowed to diffuse upon heating it will become partitioned to all lattice sites or it will precipitate, for instance during tempering of martensite, where bcc martensite or ferrite is obtained.

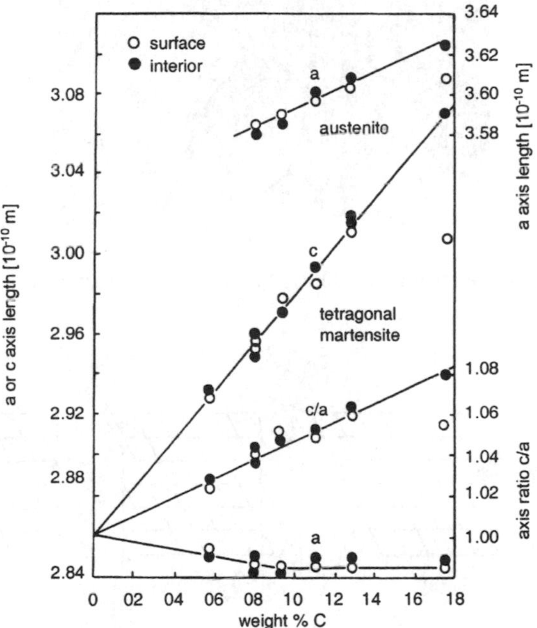

Fig. 9.30. Dependence of the lattice parameters of austenite and martensite on carbon content (after [9.15]).

Carbon plays a particular role for the special mechanical properties of martensite. On the one hand the solubility of carbon in γ-iron is much larger than in α-iron (larger octahedral interstitial sites in the fcc lattice), so that the bct martensite represents a strongly supersaturated solid solution, on the other hand the tetragonal distortion reduces the dislocation mobility (increased Peierls stress, see section 6.3.1).

The Bain correspondence proves that the martensitic transformation causes a substantial plastic deformation besides the change of lattice structure. The compression and stretching of the unit cell to comply with the cubic (respectively tetragonal) shape proceeds by shear deformation. This results in a shape change of the transformed region, which becomes apparent in form of reliefs on the surface (Fig. 9.31). These shape changes would entail large elastic compatibility strains in the immediate environment of the martensite plates, which can be reduced by plastic deformation within the martensite via glide and twinning (Figs. 9.31 and 9.32). The overall deformation corresponds to a shear parallel to an undistorted plane (for instance $\{111\}_{fcc} = (0001)_{hex}$ during the cobalt transformation). This plane is referred to as the habit plane, and frequently (e.g. for Fe-C), it is of irrational crystallographic orientation.

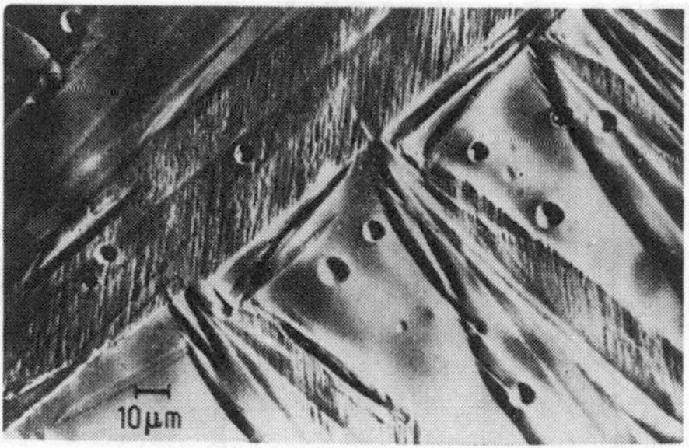

Fig. 9.31. Plate martensite in Fe-33.2%Ni with internal twinning [9.16].

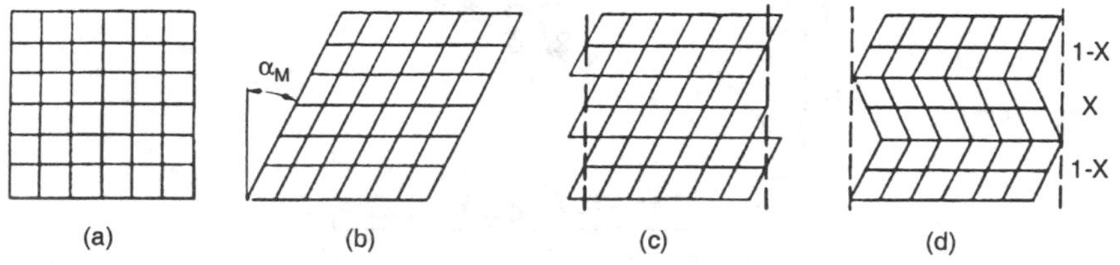

Fig. 9.32. The formation of martensite changes the shape of a crystal (a) by shear (b). Slip (c) or twinning (d) in martensite can largely compensate for this shape change.

9.2.3 Applications

9.2.3.1 TTT diagrams

Phase transformations are among the most important processes for imparting the necessary mechanical properties to structural materials. The character and distribution of particles of a second phase have a strong influence on the mechanical properties, i.e. strength, ductility, toughness and fracture behavior. Phase transformations are the foundation of thermomechanical treatment of metallic materials. In this context we want to address a few important examples that elucidate the underlying principles.

The volume fraction of the precipitated phase and its dependence on annealing temperature and annealing time is of prime importance for the annealing treatment of systems that undergo phase transformation. Although the final volume fraction of the respective phases in equilibrium can be determined from the phase diagram according to the lever rule, for commercial processes the temporal evolution of phase transformations is also of considerable interest. Such information may be derived from the TTT diagram, (Time-Temperature-Transformation diagrams). In such diagrams iso-transformation lines are plotted in a field defined by the annealing time and annealing temperature. In particular, start, finish or a defined fraction of precipitated phase is plotted (Fig. 9.33). Since these diagrams relate to the transformation kinetics, the characteristics of kinetics ought to be reflected in such a diagram. The reciprocal value of the nucleation rate $1/\dot{N}$ can be interpreted as the time for nucleation t_N. The temperature dependence of $\dot{N}$ as plotted in Fig. 8.5 corresponds to a plot of T vs t_N (that is $1/\dot{N}$ vs. T, with the axes exchanged) and explains the "nose" shape of the TTT diagram for the start of precipitation as shown in Fig. 9.33.

Nucleation below the transition temperature depends much more strongly on diffusion than in the case of solidification, therefore, there are only narrow temperature ranges where a phase can occur in technically relevant times. Correspondingly, there are also temperature regimes where transformation does not take place in such times, for instance in Fig. 9.33 in the range between pearlite and Bainite for alloy steels. Because nucleation needs a finite incubation time, the time for the start of precipitation is finite, so that by adequate cooling conditions a transformation can be avoided or obtained to a desired degree. This is important, for instance, if special material properties are to be achieved by martensitic transformation. In this case the cooling rate must be high enough to avoid the pearlite and Bainite reaction (Fig. 9.33).

9.2.3.2 Technological importance of martensite transformations: some examples

The martensitic transformation in the system Fe-C is important primarily for the improvement and control of the strength of steels. As indicated previously,

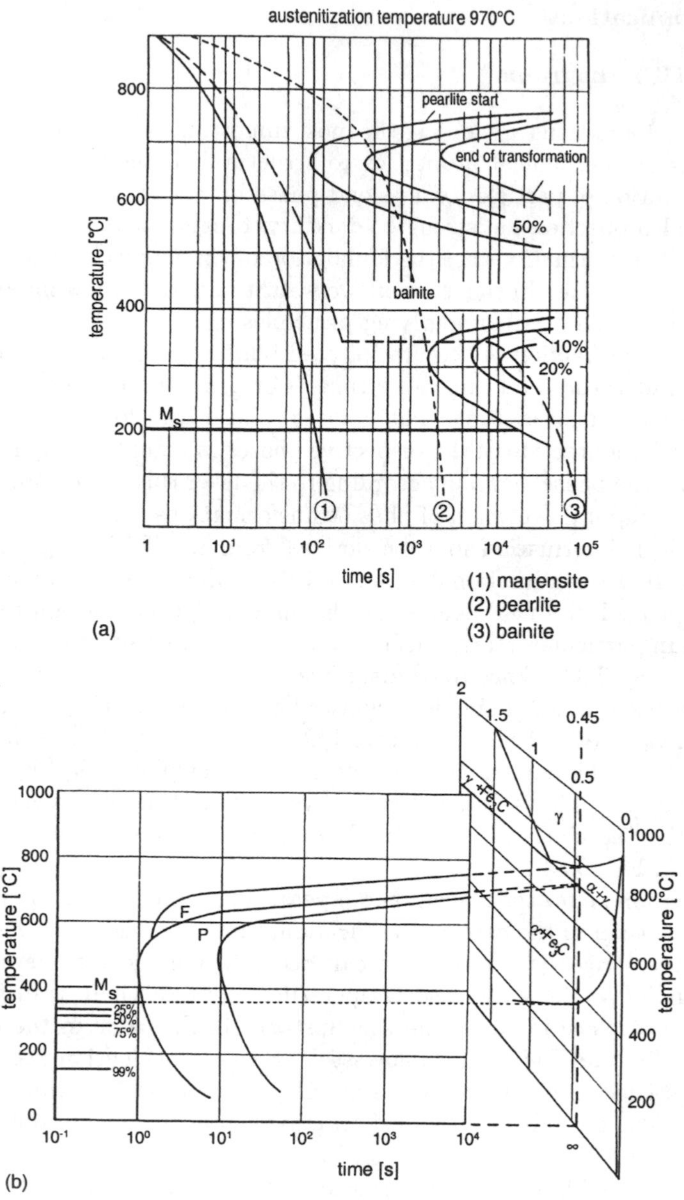

Fig. 9.33. (a) TTT-diagram of an alloy steel (Fe-C-Cr) with cooling curves. Different cooling rates yield martensite (1), pearlite (2) or bainite (3). (b) Relationship of TTT-diagram and phase diagram for Fe-C. Only for very low cooling rates occurs the transformation at equilibrium temperature (F - ferrite; P - pearlite; M_s - martensite start) (after [9.17]).

substantial contributions to the increase of strength are caused by solid solution hardening of the supersaturated bct α-crystals, as well as by the elastic distortion and plastic deformation caused by the martensitic transformation. Moreover, the martensite plates limit the slip length of moving dislocations that further increases work hardening. Inasmuch as the volume fraction of martensite is determined only by the temperature to which the material was quenched, strength, ductility, and toughness of the material may be varied at desire. An abundance of processes has been developed to obtain the optimum properties of steels with and without martensitic transformation. These efforts include grain refinement prior to pearlite transformation (for instance "ausforming") and the age hardening of soft (nickel-rich and carbon-poor) martensite ("Maraging").

A special variant are the so-called TRIP steels (TRansformation Induced Plasticity). Their properties are caused by the fact that the martensitic transformation is strongly favored by deformation. By appropriate chemical composition the transformation temperature in the presence of deformation M_d is adjusted to lie slightly above the service temperature of the material. If the loading conditions in service are such that plastic deformation occurs the material will undergo martensitic transformation. The shear deformation associated with the martensitic transformation contributes to plastic deformation and besides, causes an additional increase of strength and suppression of crack formation and crack propagation. Therefore, these materials show excellent strength, ductility and high toughness. To optimize the strength of the material the steel will be processed thermomechanically prior to service.

The martensitic transformation is not confined to the system Fe-C. It can be found in many other alloy systems where the crystal structure changes as function of temperature. A good example are the shape memory alloys. They are famous for their property to regain the shape they had prior to a deformation by a respective heat treatment. The physical cause for this effect is the shear deformation that is associated with the martensitic transformation in combination with the variety of crystallographically equivalent variants of the martensitic phase (Fig. 9.34). On cooling of the shape memory material through the transformation temperature M_f the crystallographically equivalent variants are adjusted in such a way that the shape of the specimen remains unchanged. During deformation below M_f the shape change is not accomplished by dislocation motion, rather the crystallographically more favorably oriented martensitic variants grow at the expense of the other variants to cause the imposed shape change. If afterwards the material is heated to a temperature A_f, martensite disappears and the original shape is restored. This effect was first discovered in InTl alloy but it became commercially important only through the development of NiTi alloys (Nitinol). Today there is a large number of shape memory alloys where usually the high temperature phase has a disordered bcc structure, whereas the martensitic phase has a long-range ordered bcc or orthorhombic structure. There is an abundance of applications for this effect from medicine to aerospace.

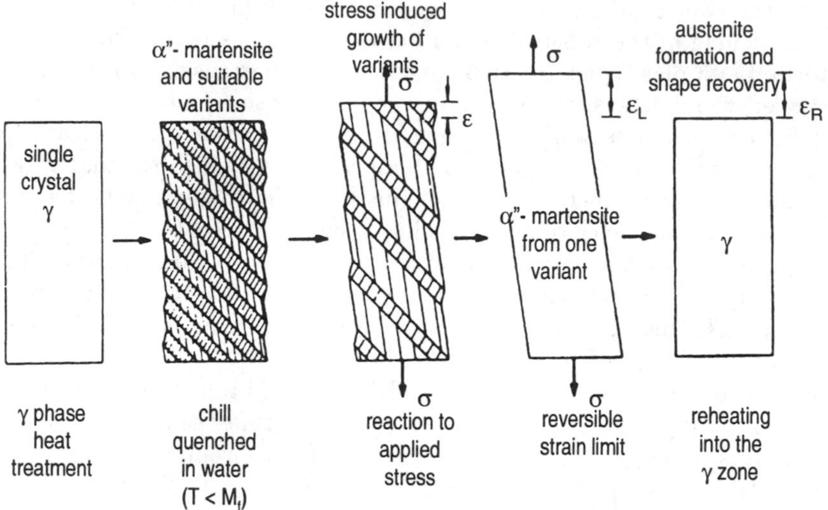

Fig. 9.34. Schematic illustration of how the shape-memory-effect works in a single crystal.

A similar effect causes so-called pseudo elasticity. In this case the martensitic phase is introduced by an applied mechanical stress. This is accompanied by a large although non-linear strain that completely disappears upon unloading. Such materials have excellent damping properties owing to the easily introduced martensite transformation, because the energy of mechanical impact will be consumed by the martensitic transformation rather than by the emission of elastic waves (sound waves). Another popular commercial application is the use of superelastic materials for eyeglass frames which survive virtually any misuse.

10

Physical Properties

10.1 Fundamentals of Electron Theory

The properties of a solid mainly reflect the properties of the electronic structure of its atoms. The existence of a solid state as the stable low-temperature configuration of the atoms proves that there are attractive forces between the atoms. If two atoms are far apart, this attractive force is small and caused by the dipole moment of their electron structure. A dipole moment results from the fact that the centers of gravity of the positive charge (nucleus) and the negative charge (electrons) are never identical, because of fluctuations of the electron density distribution. A locally separated positive and negative charge forms an electrical dipole. The interaction among dipoles always is attractive and, therefore, causes atoms to approach each other (Fig. 10.1). The attractive force increases rapidly with decreasing distance between dipoles. If the atoms reach a distance where their outermost electrons interact, various outcomes can occur depending on the different types of chemical bonding as covered in Chapter 2. A further decrease of the atomic spacing causes the electron orbitals to overlap. According to the Pauli principle, two electrons cannot share the same energy state. Correspondingly, an overlap forces some electrons to change to "free" states, which are of higher energy. The corresponding rapid increase of electron energy results in a strong repulsive interatomic force. The sum of the attractive and repulsive forces constitutes the total interaction force between the atoms (Fig. 10.2). At the equilibrium spacing, repulsive and attractive forces balance, i.e., the sum of forces is zero. This simple concept of the formation of a two-atomic molecule can be generalized to a solid consisting of many atoms, where the same interactions determine the next neighbor arrangement, qualitatively.

The periodic arrangement of atoms in a crystalline solid engenders a special electronic structure: although free electrons can assume any state of energy, electrons in crystals are allowed to occupy only certain ranges of energy that are separated by "forbidden" zones. This situation is captured by the band model of electron theory. In simple terms one can describe the sit-

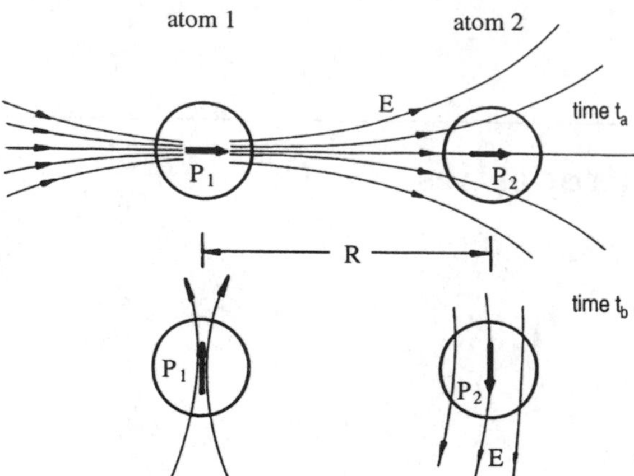

Fig. 10.1. The origin of dipole interaction in the classical model: The dipole moment P_1 of atom 1 causes the alignment of dipole moment P_2 of atom 2. Due to fluctuations the dipole moments are not invariable with time, so that alignment of the moments can be parallel or antiparallel. Regardless of the type of alignment, interaction always causes an attractive force between the atoms.

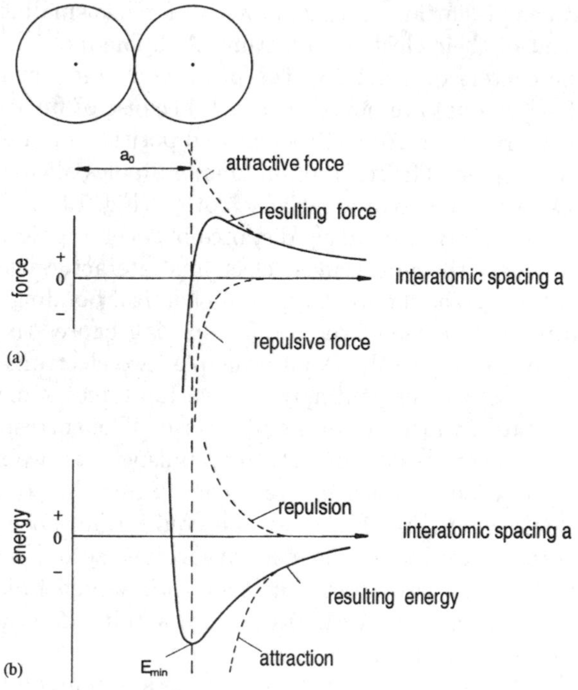

Fig. 10.2. (a) Force between two atoms as a function of their distance. (b) Energy of two atoms plotted versus interatomic spacing.

uation as follows: in an isolated atom the electrons assume discrete energy values. These energies are the same for all atoms of the same element, as long as these atoms are completely separated from each other. If two atoms with the same electron structure approach, at least one electron must change its energy state, i.e. the energy state splits into two energy levels (Fig. 10.3). For N atoms, correspondingly, the energy state splits into N energy levels. The number of atoms in a solid is large and, therefore, N is large ($\approx 10^{23}/\mathrm{cm}^3$). The N energy levels form a quasi continuous energy band. The extent of splitting of energy levels into bands depends on the spacing of the atoms and is usually confined to the outermost shells of the electron structure, because the inner electrons are strongly bound and practically unaffected by the overlap of the electron orbitals. In the band theory of solids, two bands are most important, namely, the energetically highest, completely filled band, the "valence" band, and the next higher band that can be partially filled, or completely empty, called the "conduction" band.

The physical reason for this somewhat complicated hierarchical energy structure is the wave character of the electrons. The state of electrons can be expressed in terms of their wave function, $\psi(\mathbf{r}, t)$, which is the solution of the Schroedinger equation

$$-\frac{\hbar^2}{2m}\nabla^2\psi + V\psi = i\hbar\dot{\psi} \tag{10.1}$$

where m is the electronic mass, h - Planck's constant $= 6.63 \cdot 10^{-20}$Js, $\hbar = h/2\pi$ and V the potential energy. The time independent part $\varphi(\mathbf{r})$ of the solution $\psi(\mathbf{r}, t) = \varphi(\mathbf{r}) \cdot \exp(-i[E/\hbar]) \cdot t)$, where E is the energy, then obeys the partial differential equation

$$\nabla^2\varphi + \frac{2m}{\hbar^2}(E - V)\varphi = 0 \tag{10.2}$$

For a given atomic potential and specified boundary conditions one can calculate the wave function φ of the electrons from Eq. (10.2).

For a crystal lattice the atomic potential must be periodic, and in the simplest approach can be represented by a periodic potential well of height V_0 (Kronig-Penney potential, Fig. 10.4). The solution of Eq. (10.2) in the one dimensional case can be written as

$$\varphi(x) = u(x) \cdot e^{ikx} \tag{10.3}$$

where $u(x)$ is a periodic function, $k = 2\pi/\lambda$, the wave number, and e^{ikx} represents a linear wave. The solution for a one-dimensional crystal requires

$$P\frac{\sin \alpha a}{\alpha a} + \cos \alpha a = \cos ka \tag{10.4}$$

where a is the spacing of the potential wells (Fig. 10.4), $P = maV_0W/\hbar^2$, and $\alpha = \sqrt{2mE}/\hbar$. Depending on the potential height, V_0, different limiting

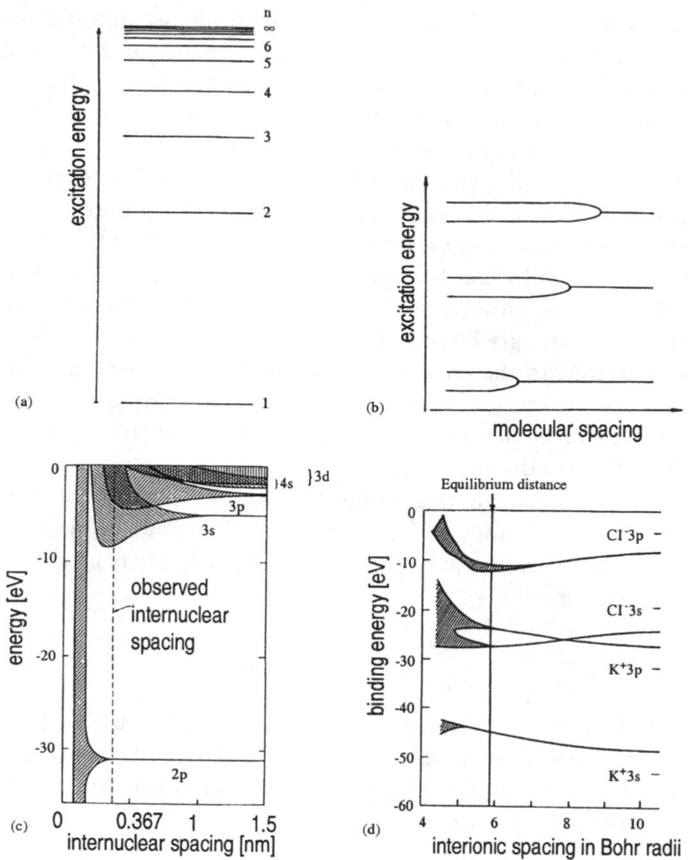

Fig. 10.3. (a) Schematic diagram of the energy levels of an electron in an atom; (b) splitting of energy levels when a molecule forms; (c) energy levels in a solid (e.g. sodium). At small interatomic spacing the energy levels of the sodium atom become bands. In solid sodium the interatomic spacing is 0.367 nm (after [10.1]). (d) Energy levels of a compound (KCl). The four highest occupied energy bands of KCl: Calculated dependence on interionic separation in Bohr radii $(a_0 = 5.29 \cdot 10^{-9} cm)$ (after [10.2]).

situations can be distinguished (Fig. 10.5).

For a low height of the periodic potential V_0, i.e. $P \approx 0$, one obtains the limit of "free" electrons, with $\alpha = k$ and $E = \hbar^2 k^2/2\text{m}$, for any arbitrary energy state (Fig. 10.5b). In case of a deep potential well, a solution exists only for sin $\alpha a \approx 0$, because $|\text{cos}ka| \leq 1$, i.e., $\alpha = n \cdot \pi/a$ with n = 1,2,3,.. which represents discrete energy states. This solution portrays the state of strongly bound electrons, such as the innermost electrons (Fig. 10.5a).

For intermediate values of V_0 not all values of α yield a solution because

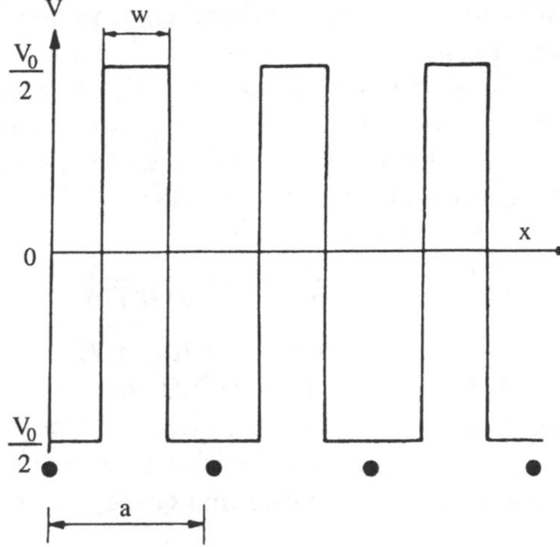

Fig. 10.4. The Kroning-Penney-potential. The filled circles indicate the positions of the ion cores.

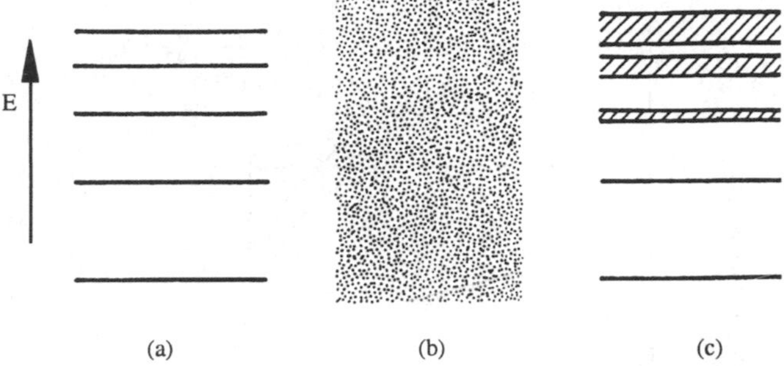

Fig. 10.5. Allowed energy levels for (a) bound electrons, (b) free electrons, and (c) electrons in a solid.

$|\cos ka| \leq 1$. In other words, there are "forbidden" energy zones, or band gaps that separate the allowed energy zones, i.e. the bands (Fig. 10.5c). This reflects the electronic state of crystalline solids and is the foundation of the electronic band structure model of solids.

Elementary particles with a spin $s = \pm 1/2$, for instance electrons, obey the Pauli principle, i.e. two particles cannot share the same energy state. Such particles are called Fermions. In systems with many particles, for instance free electrons in a solid, the electrons have to assume different and, therefore, successively higher energy states. Without thermal activation, i.e. for $T = 0K$, all energy levels below an energy ε_F, the Fermi-energy, are occupied. The Fermi-energy is a material constant. At higher temperatures the electrons can move into higher energy states by thermal activation. The thermal energy kT is small compared to the Fermi-energy, however. Thus, only the relatively few electrons that have an energy near to the Fermi-energy can change to a higher as yet unoccupied energy state. By contrast, electrons far below the Fermi-level, remain completely unaffected by the crystal's thermal energy because they can not move into higher energy states, which are already occupied. The temperature dependence of the probability that a state of energy E is occupied is given by the Fermi-Dirac-distribution $f(E)$ (Eq. (10.5)).

$$f(E) = \frac{1}{1 + \exp\left((E - \varepsilon_F)/kT\right)} \tag{10.5}$$

At the absolute zero of temperature ($T = 0K$), $f(E) = 1$ for all electron energies $E < \varepsilon_F$ and $f(E) = 0$ for $E > \varepsilon_F$. At any non-zero temperature, (T>0), $f(E) = 1$ for $E \ll \varepsilon_F$, but for $E > \varepsilon_F$, $f(E) > 0$ (Fig. 10.6). Formally, the Fermi-energy can be associated with a temperature T_F, the Fermi-temperature, where the corresponding thermal energy corresponds to ε_F

$$kT_F = \varepsilon_F \tag{10.6}$$

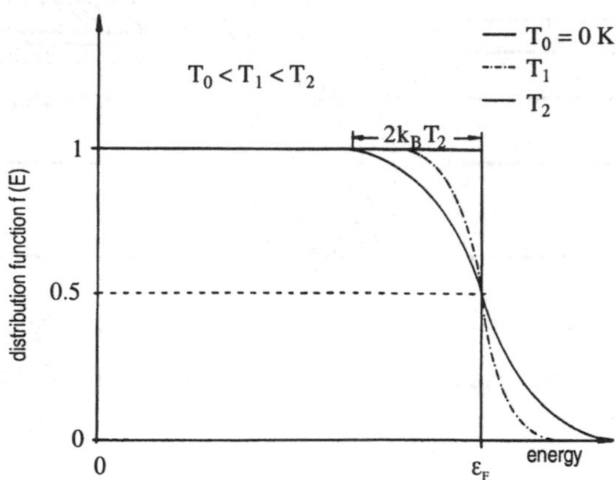

Fig. 10.6. Fermi-Dirac-distribution at different temperatures.

Since ε_F is large, T_F usually is far higher than the melting temperature of a solid (Table 10.1). The Fermi-energy can also be expressed in terms of a Fermi-velocity, according to the electron's kinetic energy

$$\frac{1}{2}mv_F^2 = \varepsilon_F \tag{10.7}$$

Fermi-temperature and Fermi-velocity are characteristics of electrons with states near the Fermi-energy. Since $T_F \gg T_m$ (T_m - melting temperature) thermal activation scarcely affects the properties of electrons in a solid. For

Table 10.1. Calculated values of Fermi-energy and Fermi-temperature for free electrons in metals at room temperature. (For Na, K, Rb, and Cs at 5 K; for Li at 78 K).

Valency	Metal	Fermi-energy in $[eV]$	Fermi-temperature $T_F = \varepsilon_F/k$ in $[10^4 K]$
1	Li	4.72	5.48
	Na	3.23	3.75
	K	2.12	2.46
	Rb	1.85	2.15
	Cs	1.58	1.83
	Cu	7.00	8.12
	Ag	5.48	6.36
	Au	5.51	6.39
2	Be	14.14	16.41
	Mg	7.13	8.27
	Ca	4.68	5.43
	Sr	3.95	4.58
	Ba	3.65	4.24
	Zn	9.39	10.90
	Cd	7.46	8.66
3	Al	11.63	13.49
	Ga	10.35	12.01
	In	8.60	9.98
4	Pb	9.37	10.87
	Sn	10.03	11.64

instance, only the electrons near the Fermi-level can contribute to the electrical and thermal conductivity of a solid.

Particles with integer spin numbers, for instance protons, neutrons, photons (light particles) or phonons (vibration particles) are referred to as Bosons. They need not comply with the Pauli principle and, therefore, any number of Bosons can share the same energy state, in particular the ground state $E = 0$. The probability that a bosonic state is occupied at $T = 0\text{K}$, is $f(E) = 1$ for $E = 0$ and $f(E) = 0$ for $E \neq 0$. For $T > 0\text{K}$ higher energy levels will be assumed by thermal activation, and all energy levels are possible states for all Bosons. In this case the probability for the occupation of an energy state is given by the Bose-Einstein distribution

$$f(E) = \frac{1}{\exp(E/kT) - 1} \tag{10.8}$$

This distribution is important in solids in the context of lattice vibrations and superconductivity (see sections 10.2 and 10.4).

10.2 Mechanical and Thermal Properties

The atoms in a solid are arranged in such a way that their (free) energy has a minimum, i.e. the sum of all forces is zero. This is represented by the minimum of the interatomic potential, which is schematically plotted in Fig. 10.2b. A closed form calculation of interatomic potentials is difficult to achieve. Usually interatomic potentials are approximated by functions that qualitatively follow the shape of the curves in Fig. 10.2 and their adjustable parameters are fitted to measurable material data. The most common radially symmetric potential functions are the Morse potentials (exponential approach)

$$V = D \cdot \left(e^{[-2\alpha(r-r_0)]} - 2\, e^{[-\alpha(r-r_0)]} \right) \tag{10.9}$$

and the Lennard-Jones potential (approximation by a power law)

$$V = \frac{A}{r^{12}} - \frac{B}{r^6} \tag{10.10}$$

where A, B, and D and α are constants, and r_0 is the equilibrium spacing of the nuclei.

These potentials, however, capture only the pairwise radial interaction between next neighbor atoms. Frequently, long-range interactions are also of importance. In this case, embedded atom potentials can be used that account for the influence of other atoms.

Mechanical equilibrium is attained when the atoms assume their equilibrium spacing, which is given by the minimum of the interatomic potential. The equilibrium spacing determines the lattice parameter of a crystal, and its molar volume (Fig. 10.2). Since thermal expansion is not considered in this approach, one obtains the lattice parameter at $T = 0K$. If solute atoms are added to a pure element, the lattice parameter changes because the atomic volume of the solute is different from the matrix atom. It is found for alloys with complete solubility in the solid state that the lattice parameter approximately changes linearly with concentration (Vegard's rule, Fig. 10.7).

The slope of the potential function defines the force or the mechanical stress, if normalized by the area, that is required to change the spacing (Fig. 10.2a). At the equilibrium spacing the stress is zero and changes in proportion to small displacements of the atoms. The proportionality constant is the slope of the curve $\sigma = f(a)$ for $a = a_0$, hence it is the second derivative of the interatomic potential at a_0. Macroscopically this corresponds to the resistance against a length change, i.e., to Young's modulus, and the proportionality between stress and strain $\varepsilon = \Delta a/a_0$ reflects Hooke's law. For larger displacements, the stress does not change in proportion to the spacing, i.e., elasticity becomes nonlinear in this case. Equivalently, instead of a length change one can also consider the relative volume change, $\Delta V/V$, for a given hydrostatic stress p. It yields a similar behavior as that shown in Fig. 10.2a. The slope of $p(V)$ for $\Delta V/V_0 = 0$ corresponds to the compressibility of the material.

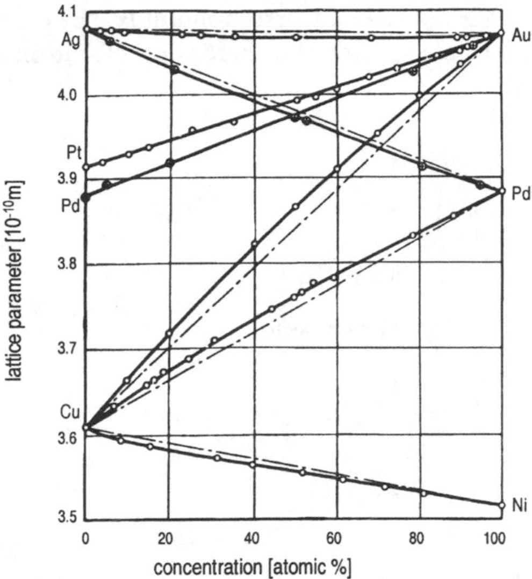

Fig. 10.7. Composition dependence of the lattice parameter for several completely miscible alloys. The dashed lines correspond to a linear dependence, the open circles are measured values.

The proportionality of stress and strain is also found in the behavior of a spring, as the energy of such a spring increases with the square of its length change. If we consider a crystal to be composed of hard sphere atoms connected by springs (see Fig. 6.2), its interatomic potential (Fig. 10.2b), would have the shape of a parabola (harmonic approximation). This cannot be exactly true, however, because a harmonic potential leads to no thermal expansion. On the other hand, the ball-spring-model is useful in providing an explanation of the thermal properties of solids as will be shown below. If a solid is heated its atoms begin to vibrate, which we experience as increase of its temperature. The larger the amount of absorbed heat, the higher the temperature and the larger the vibration amplitude of the atoms. If the potential were symmetrical with regard to its energy minimum (harmonic approximation) the centroid of the vibrations would remain independent of the vibration amplitude, i.e. the atom spacing would be independent of temperature, at the equilibrium spacing for $T = 0K$. This behavior is in contrast to observations; however, rather one observes that for sufficiently high temperatures ($T > \Theta_D/2$, Θ_D is the Debye temperature, see below) that thermal expansion, $\Delta\ell/\ell$, progresses linearly with temperature (Fig. 10.8b). Thus, the coefficient of thermal expansion, $\alpha \equiv (1/\ell)d\ell/dT$, is constant. The experimentally observed fact of thermal expansion implies that the interatomic potential is asymmetric with regard to the energy minimum. More specifically, the interatomic potential must be steeper as the atoms approach each other than if they separate. The same amplitude of vibration causes a larger separation of the atoms than their approach. Correspondingly, the centroid of the vibrations shift to larger atomic spacings for increasing vibration amplitude (higher temperature) (Fig. 10.8a).

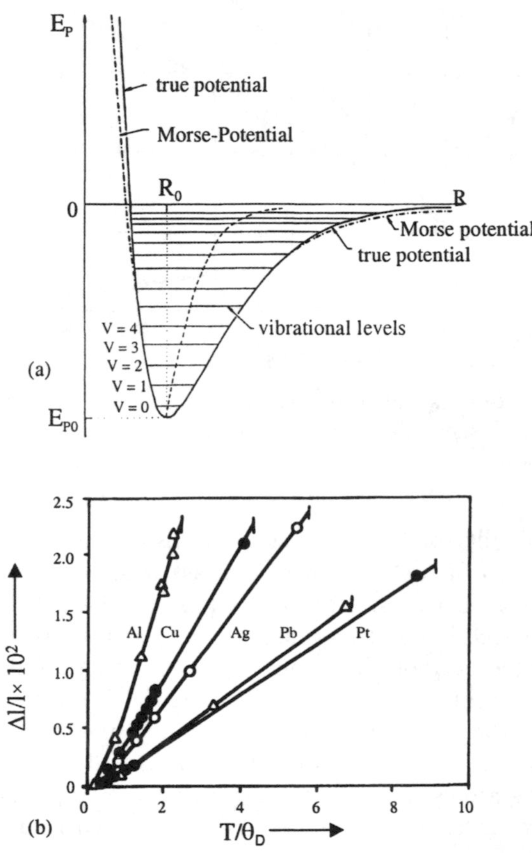

Fig. 10.8. (a) Schematic potential energy curve of a diatomic molecule plotted versus interatomic spacing. The dashed line indicates a shift of the centroid at higher vibrational levels, i.e. at higher temperature. (b) Thermal expansion of some metals as a function of temperature over Debye temperature (after [10.4]).

The change of the average atomic spacing, in turn, affects the mechanical properties. The elastic constants (Young's modulus, compressibility, etc.) are related to the second derivative of the interatomic potential at the location of the equilibrium spacing. The equilibrium spacing at $T = 0K$ is defined by the energy minimum. With increasing temperature the spacing is changed to larger values from the atomic vibrations. At larger atomic spacing, however, the interatomic potential rises less steeply, and, correspondingly, its second derivative becomes smaller (Fig. 10.2b). Therefore, the elastic constants decrease slowly with increasing temperature. For temperatures not too low they change approximately linear with the temperature (Fig. 10.9). As a rule of thumb, Young's modulus decreases by about 50% from 0K to the melting temperature.

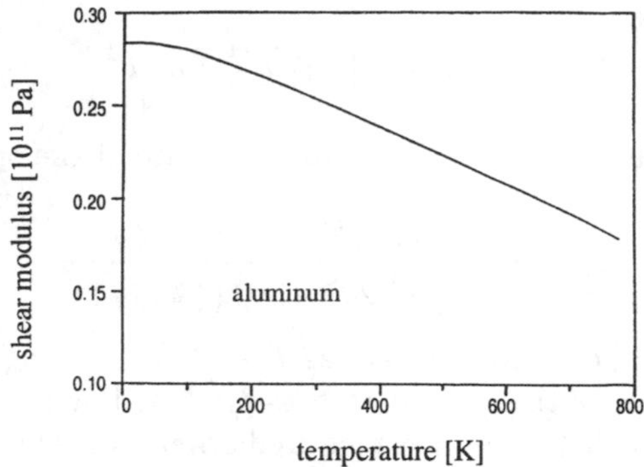

Fig. 10.9. Temperature dependence of the shear modulus of aluminum.

The energy of lattice vibrations determines the crystal's heat content, U, and, hence, may be related to the specific heat. The specific heat at constant volume V is

$$c_V = \left.\frac{\partial U}{\partial T}\right|_V \tag{10.11}$$

The heat content of a solid is the total vibrational energy of all its atoms. At high temperatures the heat content of N atoms increases according to classical thermodynamics linearly with temperature: $U = 3 \cdot N \cdot kT$. The specific heat is, therefore, independent of temperature according to Eq. (10.11) with $c_v = 3Nk = 3R \approx 6$ cal/(mole ·K). This is the law of Dulong-Petit. At low temperatures the specific heat of solids is observed to decrease with falling temperature. This deviation from classical behavior was first explained by Einstein. He described the vibrations of the atoms as the vibration of harmonic oscillators, which, however, have to comply with the laws of quantum mechanics; i.e., the oscillators assume discrete energy states. The elementary quantum of lattice vibrations is a quasi-particle called a phonon, which one can visualize as the elementary quantum of a vibration. If one assumes that all atoms vibrate independently of each other with the same frequency, ν, then the specific heat can be calculated easily. For a vibration frequency ν all possible energy states of an oscillator are given by $E_n = h\nu(n + 1/2)$, where n is the quantum number, $n = 0, 1, \ldots$.

The frequency distribution of the quantum states, n, for an oscillator with vibration frequency ν is given by the Bose-Einstein distribution

$$\langle n \rangle = \frac{1}{e^{+\frac{h\nu}{kT}} - 1} \tag{10.12}$$

For N atoms and vibrations occurring in all three spatial directions the total energy of a solid may be expressed as

$$U = 3 \cdot N \cdot \left(\langle n \rangle + \frac{1}{2} \right) h \cdot \nu \tag{10.13}$$

and with Eq. (10.11) we obtain Einstein's estimate of the specific heat

$$c_v = 3Nk \left(\frac{h\nu}{kT} \right)^2 \cdot \frac{\exp\left(\frac{h\nu}{kT}\right)}{\left(\exp\left(\frac{h\nu}{kT}\right) - 1\right)^2} \tag{10.14}$$

Indeed, the specific heat drops to zero as $T \to 0$. The strong exponential temperature dependence of the specific heat, as predicted by Eq. (10.14), however, is not observed in reality. This incorrect prediction occurs because the model is too simple in the sense that not all the atoms are vibrating independently with the same frequency. Rather (according to a proposal of Debye) they vibrate as coupled oscillators with a large number of different frequencies, namely from the shortest possible wave length corresponding to twice the atomic spacing, up to the largest possible wave length that is twice the specimen length. To calculate the heat content properly the energy contributions of all the different vibrations have to be summed. The frequency spectrum is practically continuous, so one obtains by integration Debye's estimate of the energy content of a crystal of coupled atomic oscillators,

$$U = \int_0^{\nu_D} D(\nu) \left(n(\nu) + \frac{1}{2} \right) h\nu \, d\nu \tag{10.15}$$

Here ν_D is the highest possible frequency (the Debye "cut-off" frequency), $n(\nu)$ is the probability of states with quantum number n, and $D(\nu)$ represents the number of vibration states in a frequency interval between ν and $d\nu$ in a cube of length L. This so-called density of states is given by

$$D(\nu) = \frac{2\nu^2 \cdot L^3}{V_S^3} \tag{10.16}$$

(V_S = speed of sound).

If we define the Debye temperature $\Theta_D \equiv (h\nu_D/k)$ we obtain as an estimate for the specific heat for $T \ll \Theta_D$

$$c_v \cong 234 \; Nk \left(\frac{T}{\Theta_D} \right)^3 \tag{10.17}$$

The low temperature dependence of the specific heat on the third power of the temperature has been confirmed for very many different solids (Fig. 10.10).

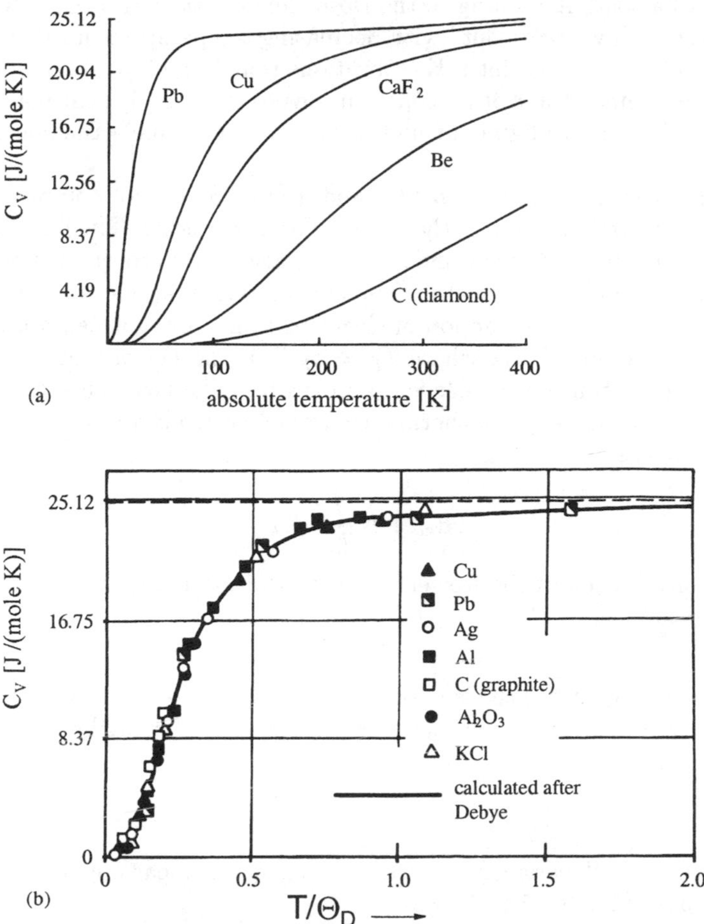

Fig. 10.10. (a) Molar heat of different substances versus absolute temperature; (b) specific heat of different substances versus temperature over Debye temperature Θ. The dashed line indicates the constant Dulong and Petit value 25.12 J/(mol · K), reached at high temperature (after [10.5]).

For $T \gg \Theta_D$, Debye's formula approaches the classical limit of Dulong-Petit

$$c_v \cong 3Nk \tag{10.18}$$

In homogeneous alloys or multiphase systems the molar specific heat follows closely a linear rule of mixtures of the specific heats of the components (Neumann-Kopp's rule).

The energy of lattice vibrations contributes the largest amount to the specific heat of a solid. According to the Bose-Einstein statistics (Eq. (10.8)) the vibrations steadily "freeze out" with decreasing temperatures until eventually only the so-called zero-point (0K) vibrations remain ($\langle n \rangle = 0$). Hence, at extremely low temperatures other contributions to the specific heat may become important that are negligible at high temperatures, namely the specific heat of free electrons.

Electrons are Fermions (spin 1/2) and, therefore, do not comply with the Bose-Einstein statistics, but rather follow Fermi-statistics. The only electrons that can contribute to the specific heat must also absorb thermal energy. This condition holds only for free electrons and, therefore, applies only in metals. At a temperature T, the fraction of electrons in metals that can absorb thermal energy is about T/T_F, where $T_F = \varepsilon_F/k$ is the Fermi-temperature (ε_F - Fermi-energy). Since each electron absorbs in a classical sense the thermal energy of magnitude kT, the thermal energy of electrons for N atoms may be approximated as

$$E_{el} \cong N\frac{T}{T_F} \cdot kT \tag{10.19}$$

and the contribution of the free electrons to the specific heat is

$$c_v^{el} \sim T \tag{10.20}$$

The total specific heat is the sum of the contributions of the lattice vibrations and of the free electrons, i.e., at low temperatures these contributions combine as

$$c_v = AT^3 + BT \tag{10.21}$$

Indeed, a plot c_v/T versus T^2 reveals the suggested linear dependence at low temperatures (Fig. 10.11).

10.3 Thermal Conductivity

Temperature gradients in a solid produce a heat flux

$$\dot{q} = -\lambda\frac{dT}{dx} \tag{10.22}$$

where $\dot{q}$ is the heat flux density (thermal energy per unit area and time), T is the temperature, and λ is the thermal conductivity. The magnitude of λ determines whether a material is a good or a poor thermal conductor. Usually excellent electrical conductors, i.e. metals, are also good thermal conductors,

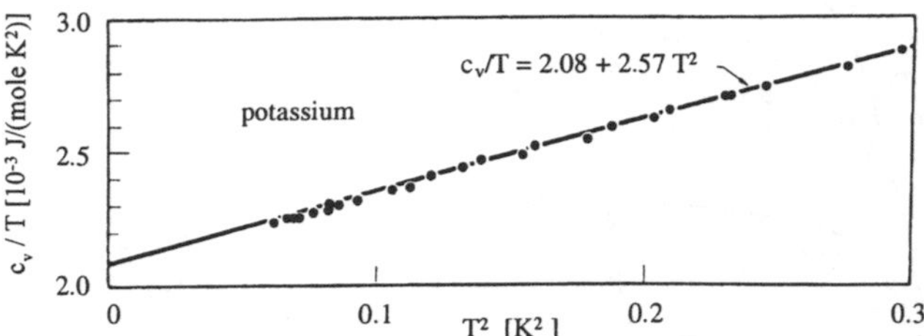

Fig. 10.11. Experimentally determined values of the molar heat c_v of potassium plotted as c_v/T versus T^2 [10.6].

and vice versa, electrical insulators usually also have a low thermal conductivity (Table 10.2). However, the thermal conductivity depends on temperature (Fig. 10.12). At low temperatures some insulators, such as diamond, also have excellent thermal conductivity.

Table 10.2. Thermal conductivity λ in $J/(cm \cdot s \cdot K)$ at room temperature.

Al	Cu	Na	Ag	NaCl	KCl	Cr-Al-alloy
2.26	3.94	1.38	4.19	0.071	0.071	0.019

The kinetic theory of an ideal gas yields the following relation

$$\lambda = \frac{1}{3} C \cdot v \cdot \ell \tag{10.23}$$

where C is the specific heat at constant volume, v the average particle velocity, and ℓ the average mean free-path between two successive collisions of a particle. The temperature in a solid is a measure of the intensity of the lattice vibrations. The heat flux is caused by a transmission of lattice vibrations to less excited regions. This transmission can occur principally via two different mechanisms: either through lattice vibrations or through free electrons. In insulators there are few freely moving charges (electrons), therefore, the thermal conductivity proceeds almost exclusively via phonons through the crystal lattice. In metals the number of free electrons is very high, therefore the thermal conductivity is mainly via the free electrons.

The thermal conductivity through the crystal lattice proceeds by exchange of high energy vibrations among atoms. This can be visualized most easily, if the lattice vibrations are considered as quasi-particles, or so-called phonons. More specifically, a phonon is the energy quantum of a vibration. The model and terminology are analogous to electromagnetic radiation. An electromagnetic energy quantum (for instance a light wave) is called a photon, which can

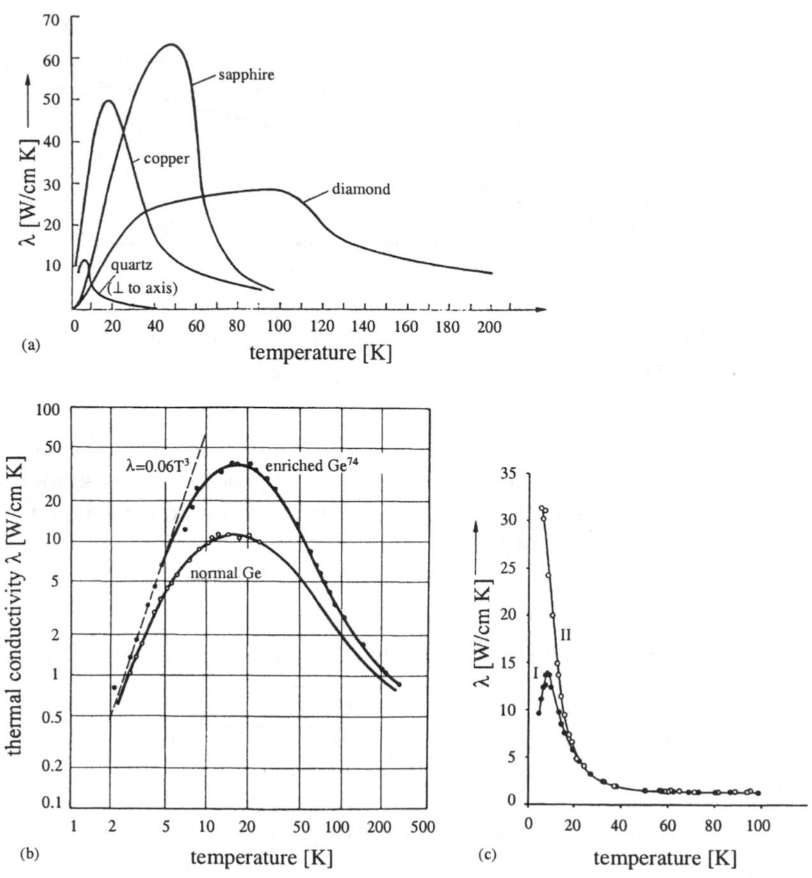

Fig. 10.12. (a) Thermal conductivity of copper, quartz, synthetic sapphire and diamond; (b) the influence of isotopic composition on thermal conductivity in germanium. Regardless of its maximum value, thermal conductivity is proportional to T^3 at low temperature. (c) Thermal conductivity of a high purity sodium crystal (II) and impure sodium (I) (after [10.7]).

also be identified with a particle to simplify understanding. In analogy with electromagnetic waves, elastic waves of atoms in a solid also occur in discrete quanta, the unit of which is referred to as a phonon. The transmission of a lattice vibration can be visualized in this model by a collision of one phonon with another phonon, by which energy is exchanged, namely heat. The quantities C, v, and ℓ in Eq. (10.23) correspond to the properties of the phonons with regard to heat conduction in a solid. The specific heat of the phonons is the specific heat of the lattice, v is the velocity of propagation of elastic waves, which is the temperature independent speed of sound. At high temperatures C is constant and ℓ approximately proportional to $1/T$. The number of phonons increases with T, and the average mean-free path decreases inversely proportional to the number of the phonons, all of which are potential collision partners. At low temperatures ℓ becomes eventually as large as the specimen dimensions, in which case the temperature dependence of the thermal conductivity is given by the specific heat that decreases proportionally to T^3 (Fig. 10.12). The overall dependency yields a maximum of thermal conductivity at low temperatures, as experimentally observed.

Perturbations of the perfect crystal, like point defects, dislocations or impurities cause phonon scattering and, therefore, reduce the mean free-path. Hence, the thermal conductivity always is degraded by the introduction of lattice defects. Even isotopes in an otherwise ideal crystal will lower λ for the same reason (Fig. 10.12b).

In metals the heat flux is carried by free electrons in addition to the phonons. The contribution of electrons in pure metals, however, is much larger than the phonon contribution. In alloys both contributions can be of comparable magnitude (Table 10.2).

The thermal conduction of electrons can be visualized by collisions of electrons with atoms by which they absorb energy that is lost in subsequent collisions. The atoms transform the absorbed energy of collision into more intense vibrations that are felt as a temperature increase. The specific heat of electrons changes in proportion to the temperature. The speed of electrons is given by the Fermi-energy, respectively the Fermi-velocity, both of which are independent of temperature since $\varepsilon_F = 1/2mv_F^2$ is a material constant. The mean free-path, ℓ, is determined by the scattering of electrons by phonons and lattice defects. The maximum of λ is caused by the competing effects of a decreasing specific heat at lower temperatures and increasing mean free-path. The maximum free path is usually not determined by the specimen dimensions rather than by the average spacing of the lattice defects, in particular imposed by the impurities, hence λ increases with higher purity (Fig. 10.12c).

In metals the free electrons carry both thermal conductivity, λ, and electrical conductivity, σ, so both quantities are correlated. The correlation between λ and σ is given by the Wiedemann-Franz law

$$\frac{\lambda}{\sigma} = L \cdot T \qquad (10.24)$$

where L is the Lorenz number which has a constant value for all metals $L = 2.45 \cdot 10^{-8} W\Omega/K^2$.

10.4 Electrical Properties

10.4.1 Conductors, semi-conductors and insulators

Whether a solid is an electrical conductor or insulator is determined by its electronic structure or, more precisely, by its band structure (Fig. 10.13a). The valence band which is completely filled is separated from the conduction band by an energy gap, E_g. If the conduction band is completely empty, the material is an insulator, for instance ceramic materials. If the conduction band is partially filled, the solid is an electrical conductor. Metals are excellent electrical conductors. In case that electrons can be easily lifted from the valence band to the conduction band the material is a semi-conductor (Fig. 10.13b). The frequency, f, for this process is given by the probability that an electron assumes sufficient thermal energy to surmount the energy gap, E_g, which is given by the Boltzmann factor

$$f \sim e^{\left(-\frac{E_g}{kT}\right)} \tag{10.25}$$

(Fig. 10.13c).

The number of electrons N_e in the conduction band of an insulator or semiconductor is $N_e \sim f$ and, therefore, depends exponentially on the temperature. If E_g is large (for instance 5.33 eV for diamond) then at all temperatures below the melting temperature the number of conduction electrons remains negligibly small, the material is always an excellent insulator. If E_g is small, however, (for instance 0.67 eV for Ge) then the number of thermally activated conduction electrons is of considerable magnitude at ambient temperature. Semiconductors are characterized by a strong decrease of the electrical resistivity with increasing temperature (Fig. 10.14). The elements germanium and silicon and the compound GaAs are well known examples of semiconductors. Doping with impurities can substantially decrease the size of the band gap, by which the conductivity is further increased (Fig. 10.15).

If a conduction electron is promoted by thermal activation, the corresponding electron is missing from the valence band. This "hole" also contributes to the electrical conductivity and can be considered as charge carrier with a positive charge. In pure semi-conductors the number of electrons and holes is balanced, and the electrical conduction is termed "intrinsic" conductivity. The addition of impurities disturbs the balance of electrons and holes. So-called "acceptor" atoms scavenge electrons of a semi-conductor and thus create holes. Such semi-conductors are referred to as p-type semi-conductors, because the majority of charge carriers is positive. By analogy, a "donor" atom liberates one of its electrons so the majority of carriers has a negative charge. Such a material is called a n-type semi-conductor. The contribution of impurities to the conductivity is termed "extrinsic" conductivity. At lower temperatures extrinsic conductivity dominates (Fig. 10.16). Controlled doping with impurities and, therefore, control of extrinsic conductivity has made semiconductors technologically important for electronic applications. By combination of p- and n-type semi-conductors diodes, transistors and eventually microprocessors have been created that comprise the elements of modern microelectronics.

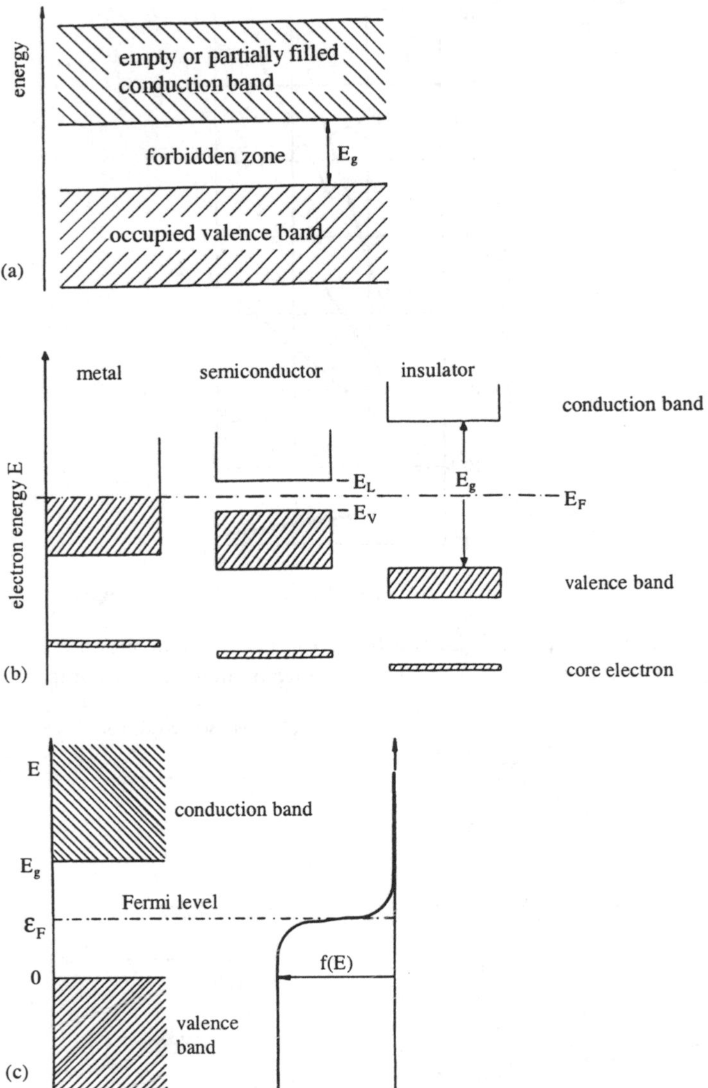

Fig. 10.13. (a) Band structure, E_g is the width of the band gap; (b) schematic representation of the occupation of the allowed energy bands in a metal, a semiconductor and an insulator; (c) band structure of an intrinsic semiconductor. At absolute zero the conduction band is empty, and separated from the occupied valence band by the energy gap E_g. Also plotted is the Fermi-Dirac-distribution for a temperature T with kT> E_g. By thermal activation electrons can change to the conduction band according to the Fermi-distribution. (Note that the Fermi-function only determines the occupation probability. The actual number of electrons is obtained by multiplying the Fermi-function with the density of states, which is zero in the forbidden regions (e.g. the energy gap).)

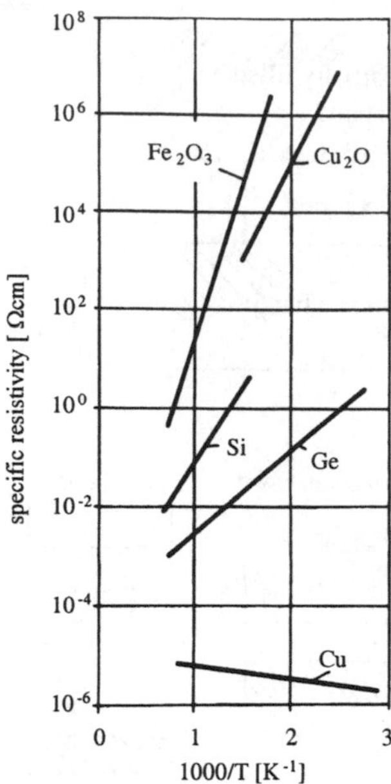

Fig. 10.14. Arrhenius plot of the resistivity of several semiconductors over reciprocal temperature. The resistivity curve of copper is included for comparison.

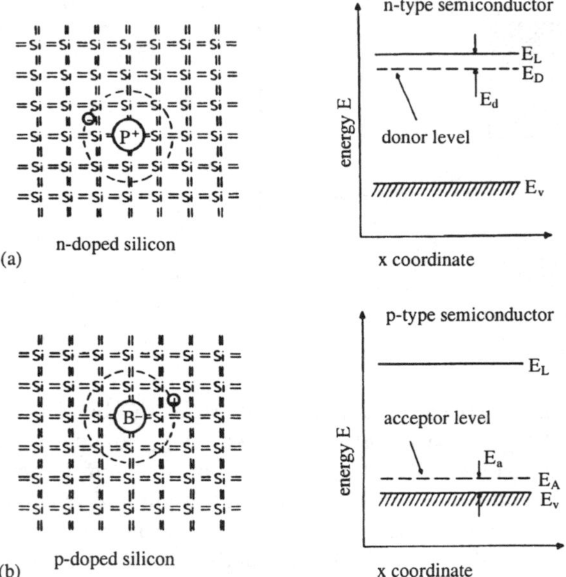

Fig. 10.15. Schematic illustration of how (a) a phosphorus atom acts as a donor, (b) a boron atom acts as an acceptor in a silicon crystal. E_d = donor ionization energy; E_a = acceptor ionization energy.

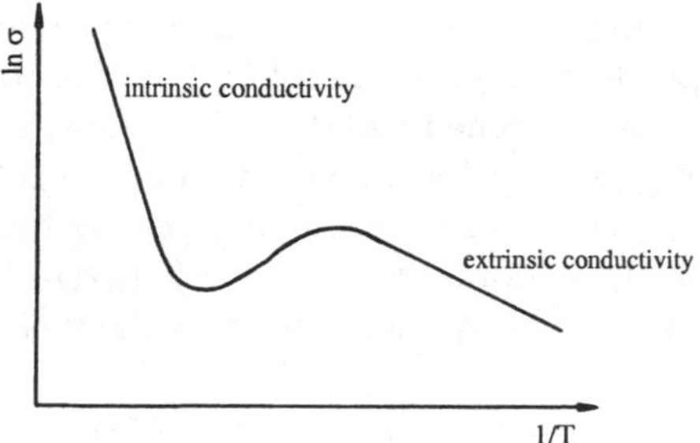

Fig. 10.16. Arrhenius plot of the conductivity of a doped semiconductor (temperature increases from right to left).

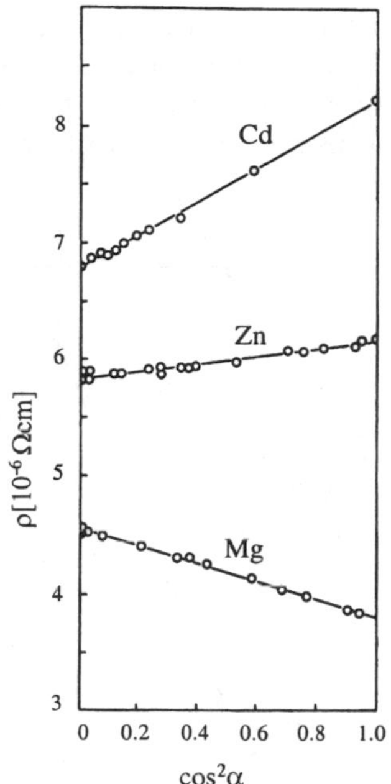

Fig. 10.17. Electrical resistivity of some hexagonal metals as a function of the angle α with the basal plane. Resistivity change is linear with $\cos^2\alpha$ (after [10.8]).

10.4.2 Conductivity of metals

Metals are excellent electrical conductors. The number of charge carriers in metals is practically independent of temperature, but the conductivity depends on the material and is strongly influenced by temperature, purity, lattice defects, and constitution. The specific electrical resistivity ρ is the reciprocal value of the electrical conductivity: $\rho = 1/\sigma$. A typical order of magnitude for metals is $\rho \approx 10^{-7} \Omega m$.

All transport properties, and thus, also the electrical conductivity of crystalline solids are not scalars but rather a symmetrical tensor of rank 2, because the conductivity depends on crystallographic direction. In cubic crystals the conductivity, however, is the same in all directions. Only for this isotropic case is the conductivity represented by a single number, a scalar. For hexagonal (or lower symmetry crystal structure metals) the conductivity, respectively the resistivity, are different along the $c(\rho_\perp)$ and $a(\rho_{||})$ axis (Fig. 10.17). For an arbitrary direction inclined by the angle α to the basal plane

$$\rho_\alpha = \rho_\perp + \left(\rho_{||} - \rho_\perp\right) \cdot \cos^2 \alpha \tag{10.26}$$

For instance for magnesium $\rho_{||} = 3.5 \cdot 10^{-8} \Omega m$ and $\rho_\perp = 4.2 \cdot 10^{-8} \Omega m$; also, the temperature coefficient of conductivity is different for the two directions. For even lower crystal symmetry the electrical conductivity will be different in all three spatial directions, for instance gallium (50.5; 16.1; 7.5) $\cdot 10^{-8} \Omega m$. Engineering materials usually are polycrystalline. In this case the conductivity is an average value over all orientations and, therefore, would be isotropic except if the material has a pronounced crystallographic texture.

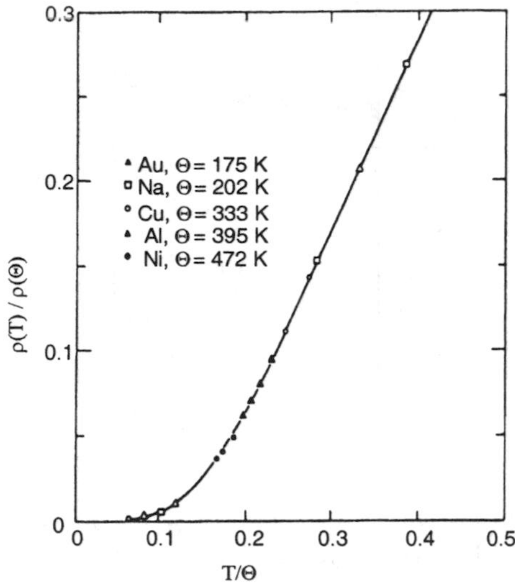

Fig. 10.18. Theoretical temperature dependence of electrical resistivity (after Grüneisen) and measured values for different metals (after [10.9]).

The electrical resistivity also depends on temperature, but in contrast to semiconductors it increases with rising temperature (Fig. 10.18 - Fig. 10.19). Above a temperature Θ, which depends on the material, the resistivity increases linearly with temperature (Fig. 10.19). At low temperatures the resistivity deviates from a linear dependency. Close to absolute zero the resistivity remains practically constant. The corresponding value as $T \to 0$ is referred to as the residual resistivity. With increasing purity the residual resistivity decreases (Fig. 10.20) and, theoretically, an absolutely pure metals, free of lattice defects, would have zero resistivity at absolute zero temperature.

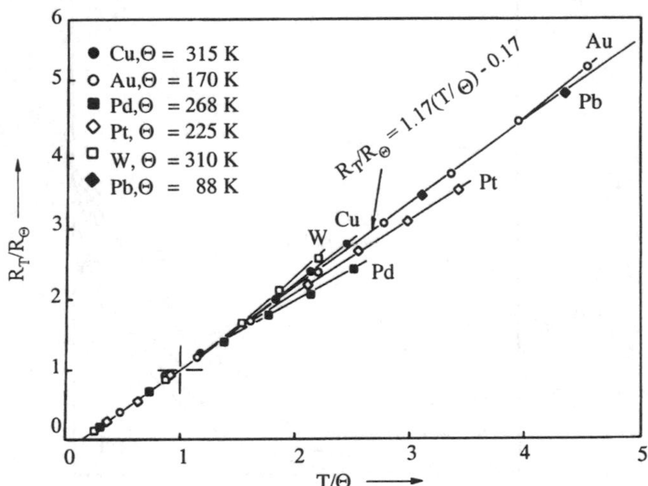

Fig. 10.19. Electrical resistivity of several metals as a function of normalized temperature (Θ = Debye temperature) (after [10.10]).

With increasing temperature the resistivity first increases with a temperature dependence of about T^5 until eventually it changes to a linear temperature dependency above some characteristic temperature (Fig. 10.21).

The resistivity also depends on composition. With increasing concentration of solute the resistivity increases (Fig. 10.22a). In a binary alloy the resistivity will increase for solid solutions of both elements that compose the alloy, and hence, the resistivity will pass through a maximum at some intermediate composition, not necessarily at the concentration of 50% as evident for Ag-Pd (Fig. 10.22b).

Although the absolute value of the resistivity changes considerably with increasing concentration, the temperature coefficient of the resistivity $d\rho/dT$ is practically independent of concentration, at least for dilute alloys (Fig. 10.23a). This is known as Matthiessen's rule. From accurate measurements one knows that Matthiessen's rule does not always hold exactly, and for some alloying elements even strong deviations can be observed, for instance Cr in Au (Fig. 10.23b).

The absolute change of the resistivity depends on the alloying element. According to Norbury's rule the electrical resistivity increases in proportion to the square of the valence difference of the partners (Fig. 10.24). A totally different behavior of the resistivity is observed if intermetallic phases appear

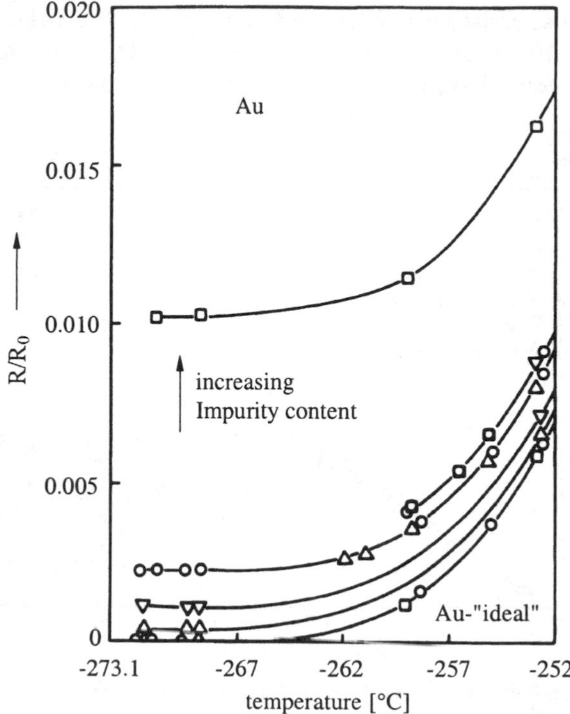

Fig. 10.20. Resistivity of gold at very low temperatures. The residual resistivity increases with increasing number of crystalline defects, i.e. impurities (after [10.11]).

in the phase diagram, with long range ordered atomic arrangements. Since the electrical resistivity is an effect of the wave nature of electrons, it is caused by deviations from the strictly periodic crystal structure rather than by scattering of individual atoms. The resistivity will notably decrease upon transition from a disordered to an ordered state (Fig. 10.25).

In heterogeneous alloys the resistivity depends on the geometrical arrangement of the phase mixture. In case the different phases are arranged in layers, the resistivities of the layers would add up, if the electrical current were to flow perpendicular to those layers (Fig. 10.26a). If the current is parallel to the layers the conductivities will add to the total conductivity (Fig. 10.26b). The latter case represents much better the morphology of real materials (Fig. 10.26c), when the second phase is completely embedded in the mother phase. Correspondingly, the measured resistivity reflects a parallel circuit of the phases (Fig. 10.26b, 10.27).

10.4.3 Modeling of the Electrical Conductivity

Let us consider the charge carriers as free particles that experience a force in an electrical field. Although this is a simple model, it can explain many

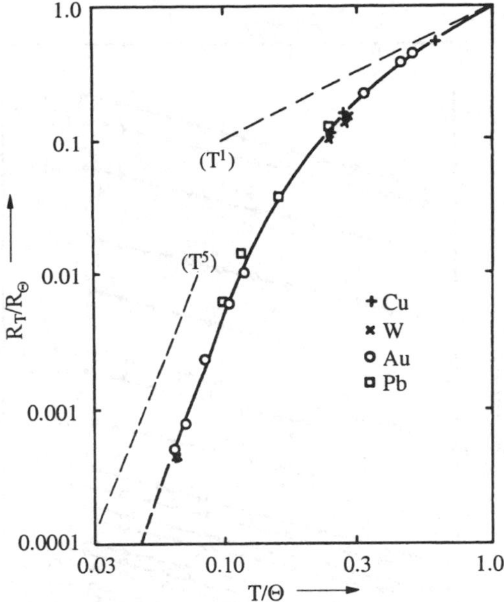

Fig. 10.21. A normalized log-log-plot of the resistivity of different metals (after [10.12]).

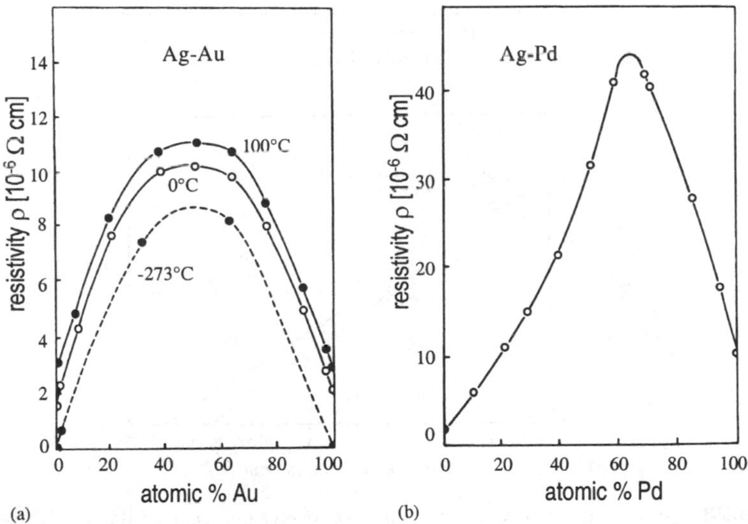

Fig. 10.22. Electrical resistivity at room temperature as a function of alloy composition for the completely miscible alloys AgAu (a) and AgPd (b) (after [10.13]).

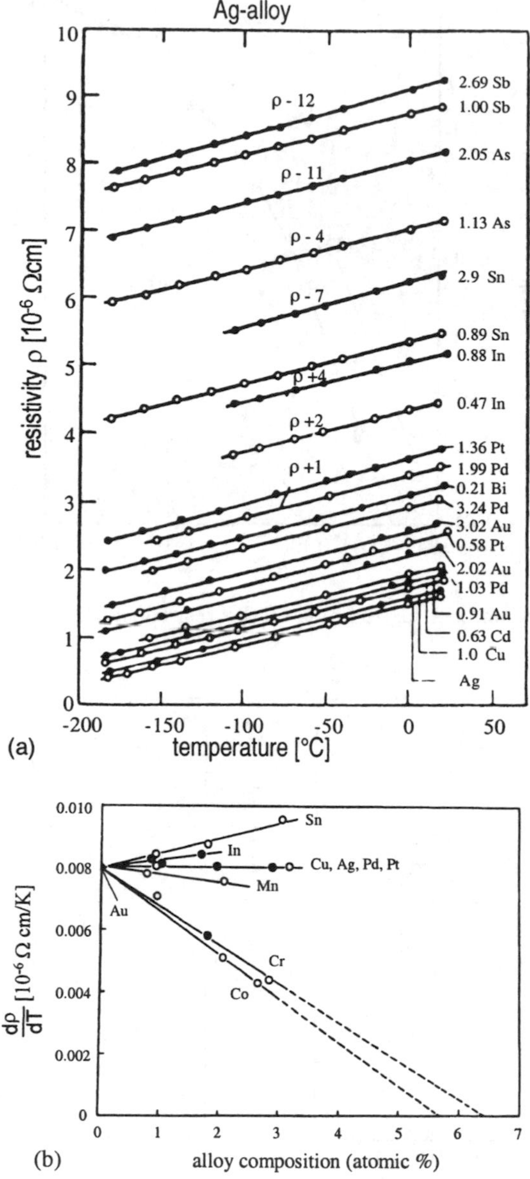

Fig. 10.23. (a) Temperature dependence of electrical resistivity in dilute silver alloys. In a first order approximation the temperature dependence of resistivity is constant (Matthiessen's rule). (b) Temperature derivative of electrical resistivity for different alloys of gold (after [10.14]).

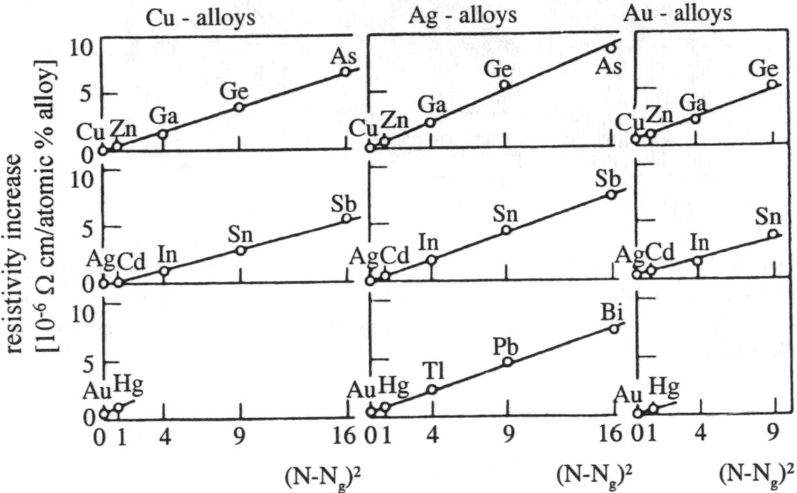

Fig. 10.24. Increase in atomistic resistivity of Cu, Ag, and Au due to dissolved metal as a function of valence difference. Resistivity increase is proportional to the square of valence (Norbury rule) (after [10.15]).

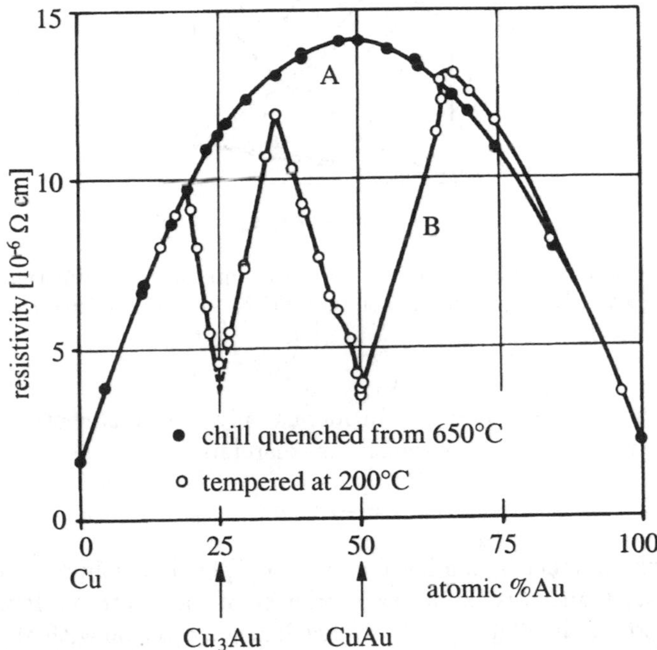

Fig. 10.25. When the ordered phases Cu_3Au and CuAu are established, this always leads to a decrease of resistivity (after [10.16]).

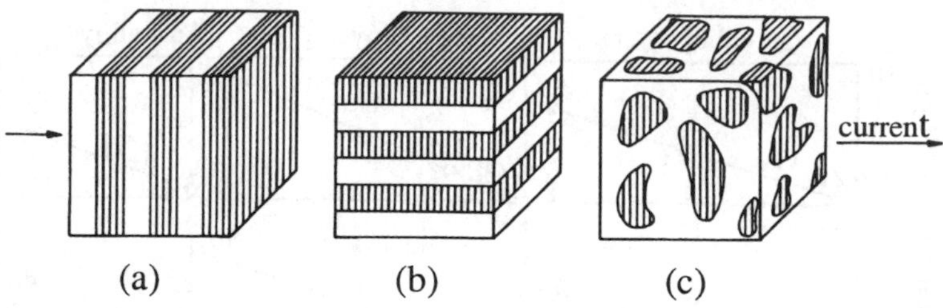

Fig. 10.26. Dependence of the resistivity of heterogeneous alloys on the geometrical arrangement of the phases. (a) Sum of the resistivity; (b) sum of the conductivities of the two phases; (c) realistic distribution: one phase is not contiguous; its behavior is similiar to (b).

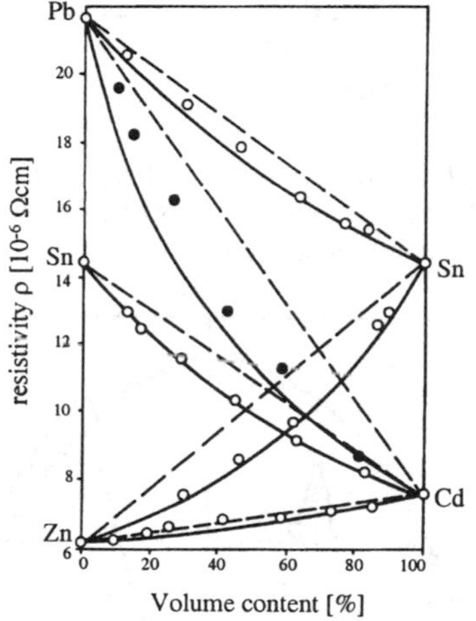

Fig. 10.27. Electrical resistivity of several phase mixtures as a function of volume content of the two phases (— sum of conductivities, - - - sum of resistivities) (after [10.17]).

aspects of electrical conductivity. In metals only electrons contribute to the electrical current. The current density is, therefore,

$$j = -nv \cdot e \tag{10.27}$$

where n is the number of conduction electrons per unit volume, e is the elementary charge, and v is the average velocity of the electrons. If we assume that an electron loses all its kinetic energy during a collision with an atom and will be accelerated during the time τ between successive collisions we obtain for an electrical field E the force F

$$F = -eE = \frac{d}{dt}(mv)$$

$$v_{\max} = \int_0^\tau \frac{F}{m} dt = \int_0^\tau \frac{-eE}{m} dt = \frac{-eE}{m}\tau \tag{10.28}$$

Since the velocity rises in proportion to time, the average velocity (drift velocity) is given by $v = v_{max}/2$ and

$$j = n\frac{e^2E}{2m} \cdot \tau \tag{10.29}$$

Comparison with Ohm's law

$$j = \sigma E \tag{10.30}$$

yields the electrical conductivity

$$\sigma = \frac{ne^2\tau}{2m} \tag{10.31}$$

and if we introduce the mobility

$$\mu = \frac{v}{E} = \frac{e\tau}{2m} \tag{10.32}$$

$$\sigma = ne\mu \tag{10.33a}$$

In semi-conductors the electrical current is made up by the contributions of electrons and holes. Since both charge carriers have different densities (n and p) and mobilities (μ_n and μ_p) but the same absolute value of charge, we obtain

$$\sigma = e\left(n\mu_n + p\mu_p\right) \tag{10.33b}$$

Note that in metals only the mobility depends on temperature, but in semiconductors the density of the charge carriers also depends on temperature. The mobility is only affected via the time τ between collisions, which decreases with increasing numbers of scattering events. Such scattering centers are mainly lattice defects (impurity atoms) and lattice vibrations (phonons). The influence of impurities can be probed as residual resistivity, because it dominates at low temperatures where the lattice vibrations are frozen. The residual resistivity increases with increasing concentration of solute atoms, because τ decreases with increasing concentration. At high temperatures the contribution of phonons dominates. Since the amplitude of lattice vibrations increases with rising temperature, the probability of collisions grows proportionally to temperature and, therefore, $\tau \sim 1/T$. Correspondingly, the resistivity in metals rises in proportion to temperature. Since the lattice vibrations in dilute alloys are little influenced by solute atoms, this also explains

Matthiessen's rule. With falling temperatures the lattice vibrations progressively decay. According to Debye the lattice vibrations decrease in proportion to T^3. The observed dependency, proportional to T^5 is explained by the fact that the spectrum of exited phonons consists mainly of long-wave phonons, the momentum of which is so small that the electrons are scattered only through small angles, which is described by another T^2 dependency.

For semiconductors the electrical conductivity is markedly different from zero only at high temperatures. Again, solute atoms and phonons are the main scattering centers for the charge carriers. In contrast to the free electrons in metals, in semi-conductors both the number of charge carriers and their mobility depend on temperature. With $\tau(T) = (Tv)^{-1}$ and $v = \sqrt{2E/m} = \sqrt{2kT/m}$ we obtain $\tau \sim T^{-3/2}$. Moreover, $n \sim \exp(-E_g/kT)$. The same holds for holes.

Apparently, the temperature dependency of the mobility can be neglected compared to the temperature dependency of the density of charge carriers, therefore, the temperature dependence of the conductivity of semi-conductors essentially follows an Arrhenius dependency (see Fig. 10.14).

However, not all phenomena of conductivity can be understood with this simple model of free electrons. The contribution of solute atoms or of crystal defects to the resistivity is difficult to calculate. The same is true for the decrease of the resistivity in ordered structures. In such cases the wave nature of charge carriers has to be considered to yield quantitative results, which poses serious difficulties even to date.

10.4.4 Superconductivity

In some metals and some oxide ceramics the electrical resistivity completely vanishes at a characteristic temperature above $0K$. This temperature is referred to as the critical temperature (Fig. 10.28). The respective materials change to a superconducting state. The state of normal conduction is regained if either the temperature, or the current density, or an external magnetic field exceed certain critical values, which are different for each material. The superconducting state, therefore, is confined to a certain range of imposed conditions (Fig. 10.29). Besides the complete loss of electrical resistivity the superconducting state also shows another remarkable property, namely the complete expulsion of a magnetic field from a material's interior (Fig. 10.30). The superconductor behaves like an ideal diamagnet, the magnetic induction in its interior is zero (see section 10.5.1). This effect is referred to as Meissner-Ochsenfeld effect. It is not a consequence of the loss of resistivity. If an ideal metal would be cooled in a magnetic field to $0K$ its electrical resistivity would completely disappear, however, the magnetic field would remain in its interior, in contrast to a superconductor.

Superconductivity is found in many metallic systems, in elements, alloys, and intermetallic compounds (Table 10.3 and Table 10.4). In metallic systems the highest critical temperature observed to date is $T_c = 23.1\ K$ in Nb_3Ge.

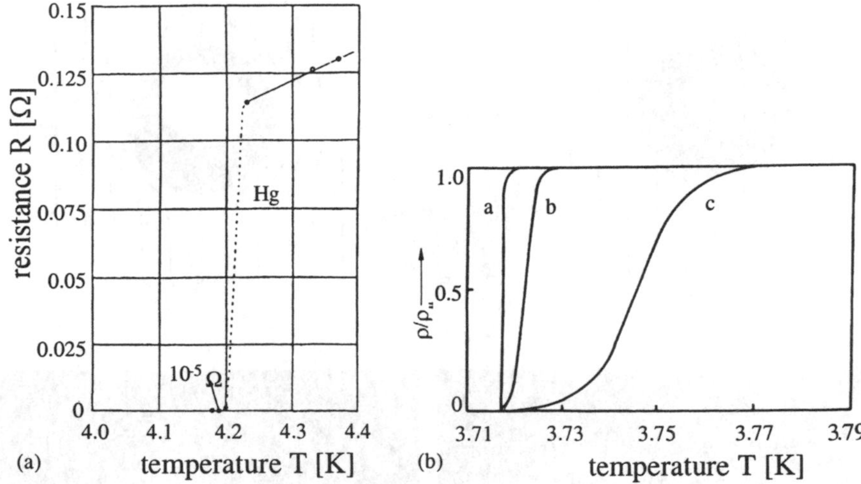

Fig. 10.28. (a) Dependence of the resistance (in Ω) of a mercury sample on absolute temperature.This graph by Kammerling Onnes marked the discovery of superconductivity (after [10.18]). (b) Typical low temperature resistivity curve of a superconducting material, in this example tin: a - single crystal; b - polycrystal; c - impure material.

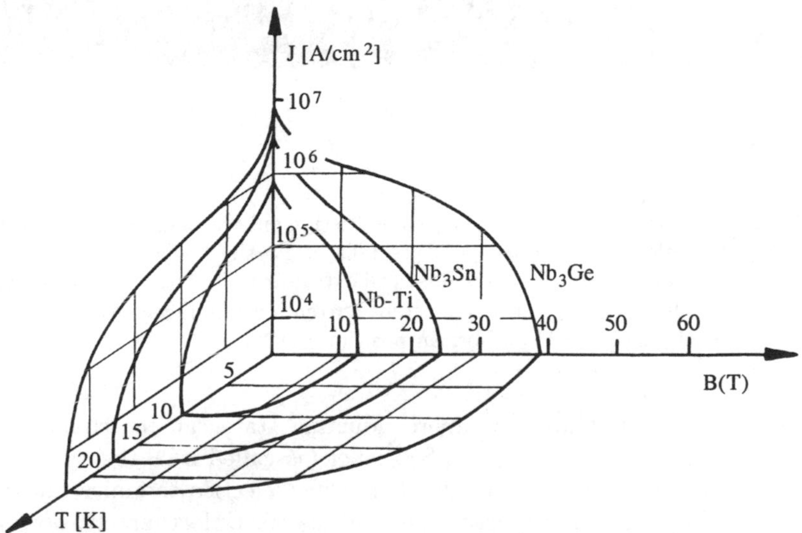

Fig. 10.29. Graph of the critical values of current density J, magnetic field B and temperature T for some superconductors (after [10.19]).

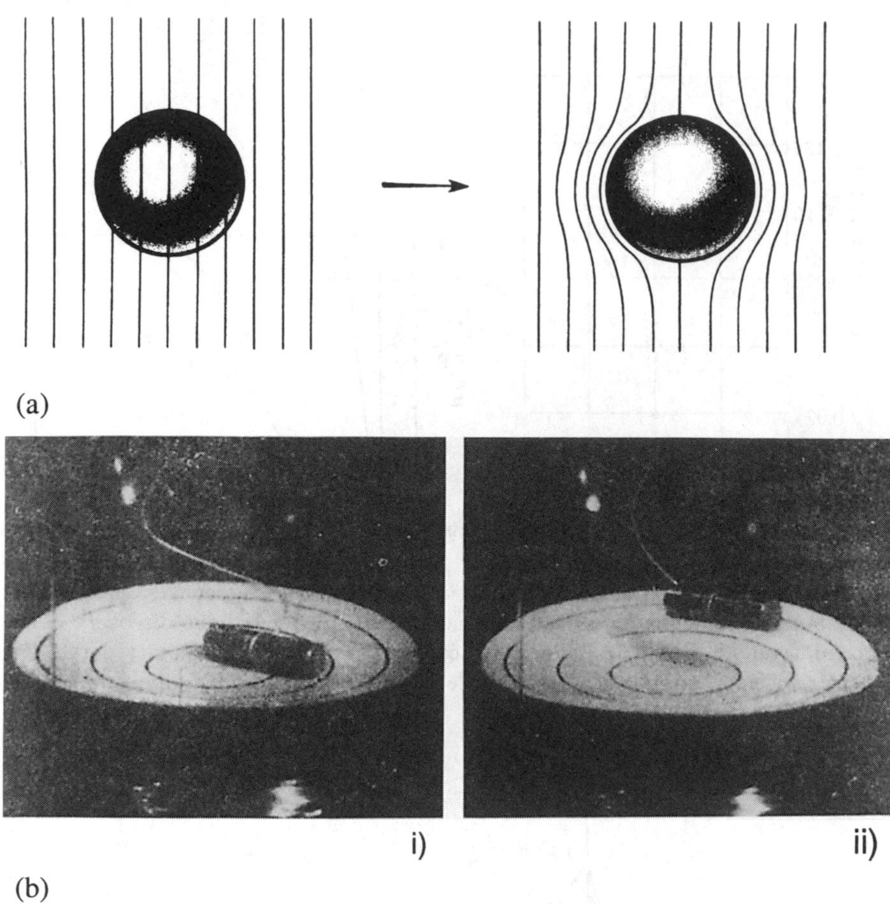

Fig. 10.30. (a) Meissner effect in a superconducting sphere being cooled in a constant magnetic field; when the temperature falls below the transition temperature magnetic flux lines are expelled (schematic illustration); (b) "the levitating magnet" illustrates the Meissner effect; persistent currents are induced by lowering the magnet: i) initial position; ii) equilibrium position [10.20].

For these metallic systems the superconducting state can be explained by the theory of Bardeen, Cooper, and Schrieffer (so-called BCS theory). In the year 1986 superconductivity was also found in dielectric ceramics but with much higher critical temperatures, $80K$ and above. Unfortunately, however, the critical current of ceramic superconductors is much smaller than in metals which is caused by grain boundaries in the oxide superconductors. Despite all efforts there is not yet a satisfactory theory to explain such high critical temperatures. In the following we confine ourselves to metallic systems.

Table 10.3. Critical temperature of selected compounds.

Compound	T_c in $[K]$	Compound	T_c in
Nb_3Ge	23.1	Nb_3Au	11.5
Nb_3Sn	18.05	La_3In	10.4
Nb_3Al	17.5	Ti_2Co	3.44
V_3Si	17.1	Nb_6Sn_5	2.07
NbN	16.0	InSn#)	1.9
MoN	12.0	$(SN)_x$polymer	0.26
$YBa_2Cu_3O_7$	80	$HgBa_2Ca_3CuO_{8+x}$	134

#) metallic phase

Table 10.4. Positioning of the superconductors in the periodic table of elements. Upper number: critical temperature in $[K]$; lower number: (where available) critical field at $T = 0K$ in 10^{-4} *Tesla*; *) = superconducting only under high pressure or in thin films.

H																	He
Li	Be 0.026											B	C	N	O	F	Ne
Na	Mg											Al 1.140 105	Si* 6.7	P* 4.6-6.1	S*	Cl	Ar
K	Ca	Sc	Ti 0.39 100	V 5.38 1420	Cr	Mn	Fe	Co	Ni	Cu	Zn 0.875 53	Ga 1.091 51	Ge* 5.4	As* 0.5	Se* 6.9	Br	Kr
Rb	Sr	Y* 1.5-2.7	Zr 0.546 47	Nb 9.5 1980	Mo 0.92 95	Tc 7.77 1410	Ru 0.51 70	Rh 0.0003 0.049	Pd	Ag	Cd 0.56 30	In 3.4035 293	Sn 3.722 309	Sb* 3.6	Te* 4.5	I	Xe
Cs* 1.5	Ba* 1.8-5.1	*la.)*	Hf 0.12	Ta 4.483 830	W 0.012 1.07	Re 1.4 1.98	Os 0.655 65	Ir 0.14 19	Pt	Au	Hg 4.153 412	Tl 2.39 171	Pb 7.193 803	Bi* 3.9-8.5	Po	At	Rn
Fr	Ra	*ac.)*															

la.) - lanthanoids	La 6.00 1100	Ce* 1.7	Pr	Nd	Pm	Sm	Eu	Gd	Tb	Dy	Ho	Er	Tm	Yb	Lu 0.1
ac.) - actinoids	Ac	Th 1.368 1.62	Pa 1.4	U 0.2	Np	Pu	Am	Cm	Bk	Cf	Es	Fm	Md	No	Lr

e distinguish superconductors of type I (soft superconductors) and su-onductors of type II (hard superconductors) (Fig. 10.31). Hard supercon-ctors allow the magnetic field to penetrate above a critical field strength $_{C1}$ into a thin surface layer of the material, but the material remains su-perconducting. Only for $H > H_{C2}$ the material becomes a normal conductor. The field H_{C2} can be two orders of magnitude larger than the critical field of a soft superconductor. Therefore, hard superconductors are interesting for commercial applications, for instance for high-field superconducting magnets.

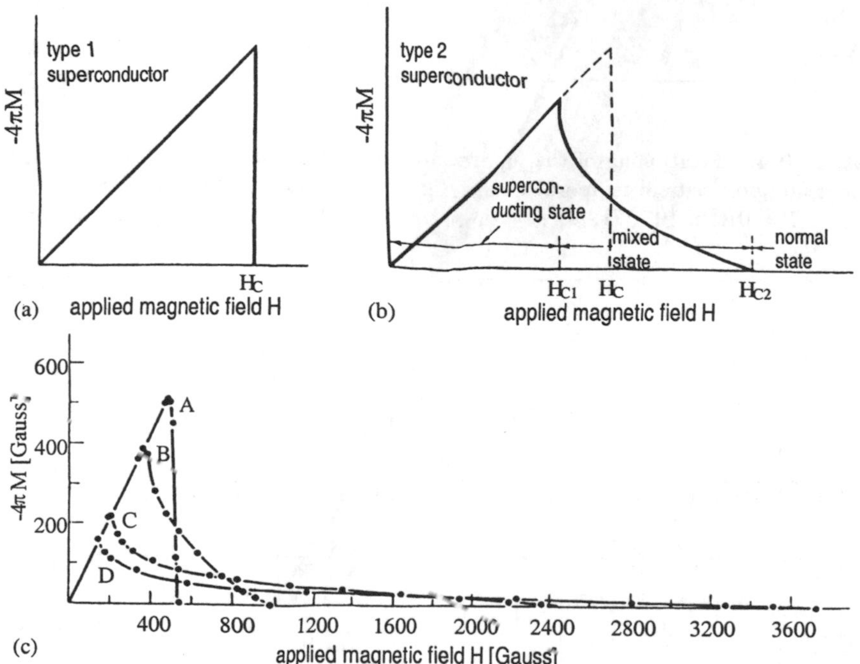

Fig. 10.31. (a) Magnetization curve of a type I superconductor (perfect diamagnet); (b) magnetization curve of a type II superconductor. At field strength H_{C1} flux begins to penetrate into the sample; (c) magnetization curve of tempered polycrystalline lead-indium alloys at 4.2K; (A) lead; (B) lead 2.8 w/o indium; (C) lead 8.32 w/o indium; (D) lead 20.4 w/o indium (after [10.21]).

In a strongly simplified model superconductivity can be explained by the formation of pairs of two electrons with antiparallel momentum and antiparallel spin at low temperatures. The attractive interaction between the electrons occurs by the exchange of so-called virtual phonons through the crystal lattice (Fig. 10.32). Such a Cooper pair forms a new particle $(\mathbf{p}\uparrow, -\mathbf{p}\downarrow)$ with zero total momentum and zero total spin. A particle with a total spin zero, how-

ever, is a Boson rather than a Fermion and, therefore, is not cons
Fermi-statistics. Instead, all Cooper pairs can occupy the same quant
Since all Cooper pairs occupy the same quantum state they move as
in phase with zero-point lattice vibrations. In the presence of an electr
the entity of all Cooper pairs assumes a slightly higher energy state, in w
all pairs share the same momentum, which corresponds to a superconduct
current. The superconducting state requires the same quantum mechanic
state for all pairs. Therefore, a single pair cannot interact with the crystal lattice, i.e. to exchange momentum, without leaving the superconducting state, for which a finite energy has to be expended, namely the interaction energy between Cooper pairs. Therefore, below T_c only the absorption of large energies in electrical or magnetic fields will eventually destroy the superconducting property.

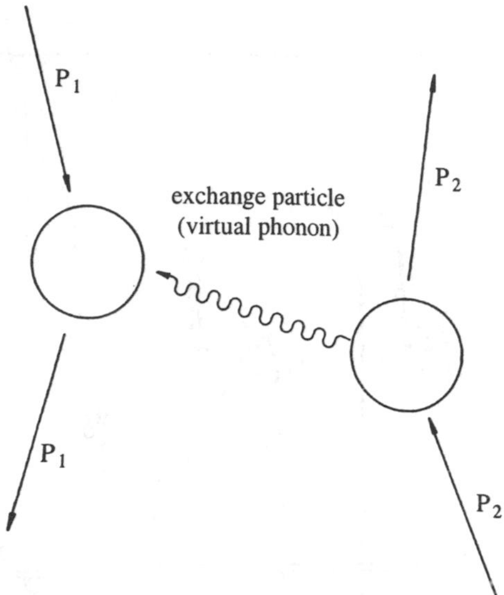

Fig. 10.32. Electron-electron interaction by phonon exchange.

10.5 Magnetic Properties

10.5.1 Dia- and paramagnetism

In solids three kinds of magnetisms are known, namely diamagnetism, paramagnetism, and ferromagnetism. Every material is diamagnetic, however, the effect is so small that it is by far overcompensated in paramagnetic and ferromagnetic materials.

agnetism is caused by magnetic induction. An external magnetic field es a current in the electron shell of an atom, the magnetic field of which racts the external field. The induced magnetic moment M weakens the nal magnetic field H, i.e. the susceptibility χ_D

$$M = \chi_D H \tag{10.34}$$

is negative. Diamagnetism is of little importance for solids except for special effects in solid state physics (magnetic resonances) which will not be addressed here. As mentioned in the previous section superconductors are ideal diamagnets since they completely expel the magnetic field from their interior. In this case χ_D = -1. In other materials χ_D is very small, of the order of 10^{-8} and independent of temperature.

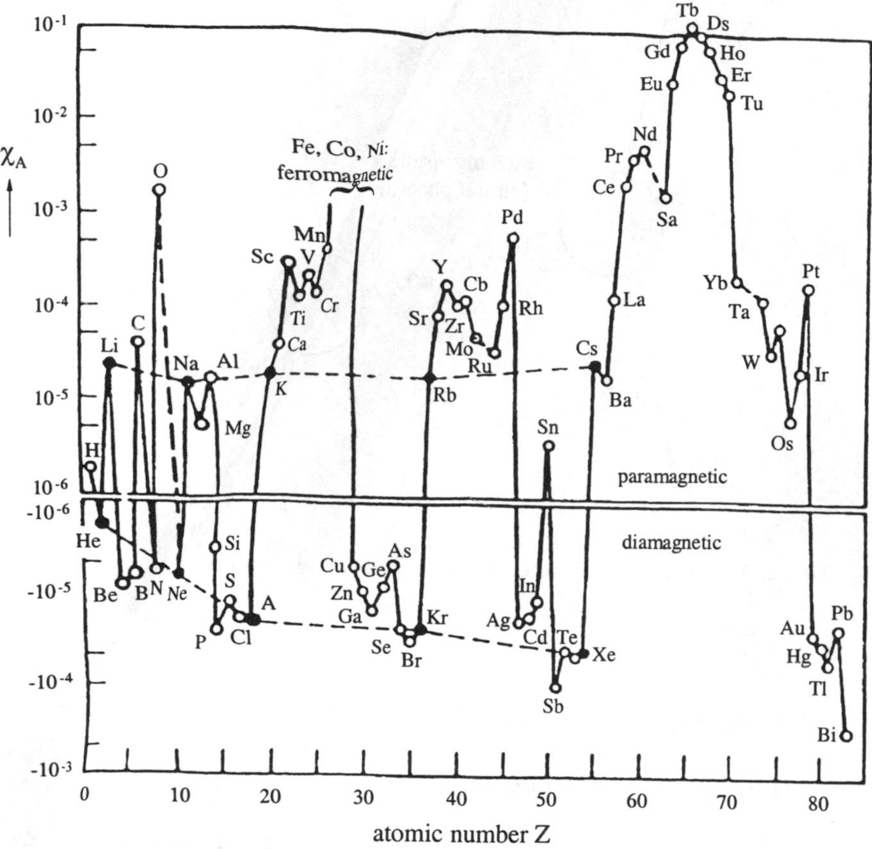

Fig. 10.33. Dependence of the atomic susceptibility on the atomic number (after [10.22]).

Paramagnetic materials contain elements where the electronic shell has a permanent magnetic moment, which is always the case if an electronic shell is filled incompletely (Fig. 10.33). Since the magnetic moments are statistically distributed in space the magnetization of a paramagnetic solid vanishes on average. However, if an external field is applied the magnetic moments become aligned by precessing around the direction of the external field. Magnetic moments that are oriented opposite to the external magnetic field have a higher energy by $\mu_z H$(μ_z-component of the magnetic moment antiparallel to the field). Therefore, they tend to reorient in the positive field direction with increasing external field. In this manner the solid gains a magnetization that is proportional to the external field

$$M = \chi_P H \tag{10.35}$$

for small fields, and $\chi_P > 0$. If the field strength is high all moments will align with the field direction and magnetic saturation is obtained (Fig. 10.34). This alignment is degraded by thermal activation so that χ_P decreases with increasing temperature. According to Curie's law of paramagnetism

$$\chi_P = \frac{C}{T} \tag{10.36}$$

where C is a constant (Fig. 10.35).

Electrons have an intrinsic angular momentum, a spin, that is either parallel or antiparallel to the magnetic field. Free electrons in a conductor, notably in metals, therefore, contribute to paramagnetism. This contribution is referred to as Pauli-paramagnetism. Pauli-paramagnetism is a small effect because only electrons close to the Fermi-level can absorb magnetic energy (to attain higher energy levels). On the other hand a magnetic field also induces an electric current in free electrons which causes diamagnetic behavior (Landau diamagnetism). This diamagnetic (Landau) susceptibility, however, is even much smaller than the paramagnetic (Pauli) susceptibility.

$$\chi_{\text{Landau}} = -\frac{1}{3}\chi_{\text{Pauli}} \tag{10.37}$$

The Pauli-paramagnetism diminished by the diamagnetic (Landau) contribution causes the difference between the magnetism of free atoms and the same atoms in a solid. Free atoms of noble metals like Cu, Ag, or Au are diamagnetic.

10.5.2 Ferromagnetism

Ferromagnetism is the commercially most important kind of magnetism in solids. In contrast to paramagnetic solids, ferromagnets have a spontaneous magnetic moment. This spontaneous magnetization results from the alignment of the electron spins in a material. This alignment is attributed to an internal

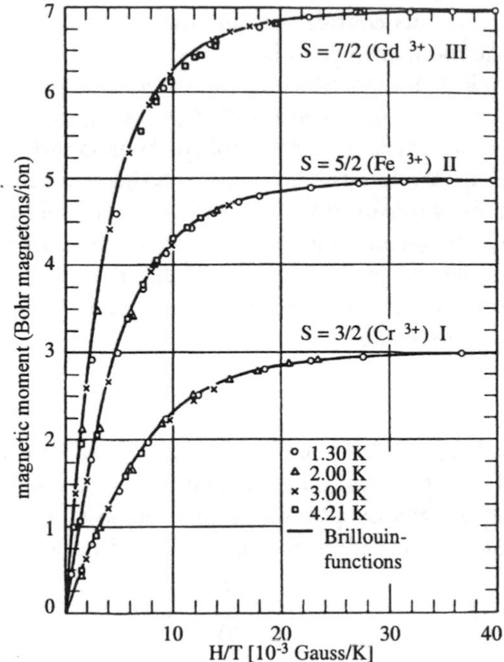

Fig. 10.34. Dependence of the magnetic moment on H/T for spherical samples of (I) potassium-chromium-alum, (II) iron-III-alum and (III) gadolinium-sulfate-octahydrate (after [10.23]).

magnetic field H_E which is caused by spin interaction and is referred to as exchange field or molecular field. It is extremely large, attaining 10^7 Gauss ($= 10^3$ Tesla). The magnetization M is proportional to the molecular field strength

$$H_E = \lambda M \tag{10.38}$$

where λ is a material constant independent of temperature. If all spins are aligned the magnet is in a state of maximum magnetization, i.e. it is saturated. The internal field H_E is so large that practically all spins are aligned. However, temperature counteracts this alignment. Therefore, the saturation magnetization decreases with increasing temperature, and at a critical temperature T_C, the Curie temperature, it vanishes (Fig. 10.36). At T_C the spontaneous magnetization disappears, and for $T > T_C$ the material behaves as a paramagnet. The material constants T_C and λ are related. For an externally applied field H and $T > T_C$

$$M = \chi \cdot (H + H_E) \tag{10.39}$$

and $\chi = C/T$. Therefore,

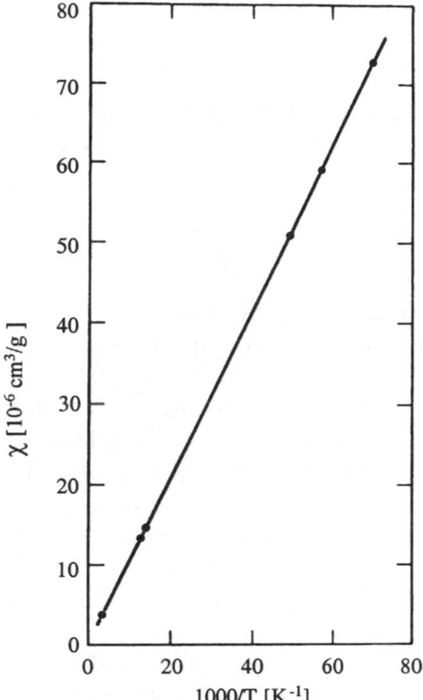

Fig. 10.35. Dependence of susceptibility on reciprocal temperature for powderized $CuSO_4 \cdot K_2SO_4 \cdot 6H_2O$. Curve shape corresponds to Curie's law (after [10.24]).

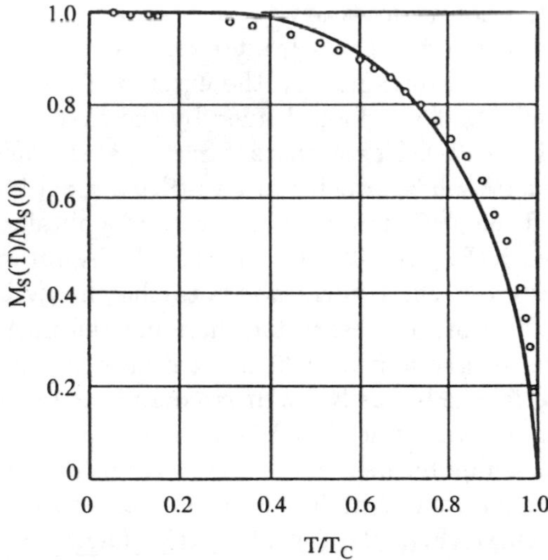

Fig. 10.36. Saturation magnetization of nickel as a function of temperature; the theoretical curve was derived from mean (or molecular) field theory (after [10.25]).

$$\chi = \frac{C}{T - C\lambda} \tag{10.40}$$

χ has a singularity for $T = C\lambda$; below this temperature there exists a spontaneous magnetization, that is $T_C = C\lambda$. For $T > T_C$ we obtain the Curie-Weiss law

$$\chi = \frac{C}{T - T_C} \tag{10.41}$$

In a commercial ferromagnet the magnetization is usually much smaller than the saturation magnetization M_S, it can even be zero. This is caused by the microstructure of a ferromagnetic solid. It consists of regions (domains) where the spins are perfectly aligned, but the alignment in the different domains is different (Fig. 10.37). In the extreme case the alignment of the domains can be such that the magnetization is zero (Fig. 10.38). The interfaces between domains are called Bloch walls. If an external magnetic field is applied to a ferromagnetic solid the Bloch walls are displaced in such a way that domains that are favorably aligned with regard to the field grow, and the magnetization increases (Fig. 10.39). Saturation M_S is obtained if the specimen is composed of only a single domain, the spins of which are all aligned in the field direction. To obtain full saturation, besides displacements of Bloch walls, rotations of the direction of internal magnetization become necessary, which can only be achieved by the application of high magnetic fields. If the external magnetic field is turned off, a permanent magnetization, the remanence M_R, remains. An external field H_C (coercive force) opposite to the direction of internal magnetization is necessary to demagnetize a material. The magnetization curve, also referred to as hysteresis, describes the dependence of the magnetization of a ferromagnet on the external field strength. The hysteresis ($M(H)$curve) is a closed loop defined by the characteristic values M_S, M_R, and H_C (Fig. 10.40a). These characteristic values can vary widely for different ferromagnetic solids, which can be utilized for a broad spectrum of applications. In soft magnetic materials saturation is obtained already at low external field strength (Fig. 10.40b). In this case the hysteresis is narrow. The area inside the hysteresis curve corresponds to the energy loss per magnetization cycle, i.e., the work necessary for changing the magnetization by an alternating external magnetic field, which eventually is converted to heat. A soft magnetic material can be easily magnetized and the magnetization can be reversed without large losses. Such materials are used as magnetic storage media, for instance as computer memory, or for cores of electrical transformers. Hard magnetic materials have a high saturation magnetization and, therefore, a broad hysteresis curve (Fig. 10.40b). They are difficult to demagnetize and, therefore, can be used as permanent magnets, for instance in loudspeakers.

The magnetization in crystals depends on crystal orientation and so does the hysteresis curve. In iron the $\langle 100\rangle$ direction is magnetically the softest, in nickel it is the $\langle 111\rangle$ direction (Fig. 10.41). This magnetic anisotropy is caused

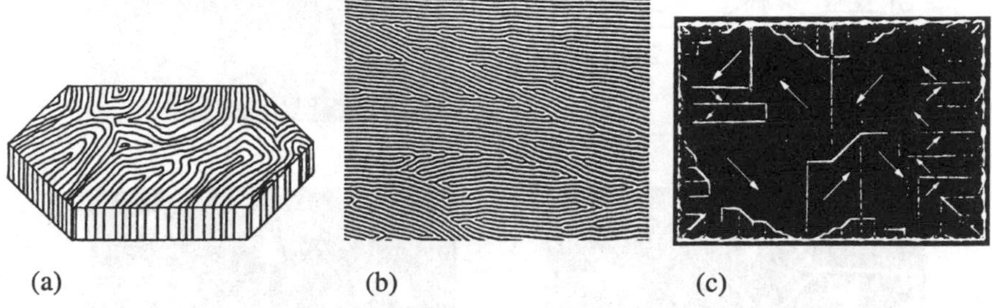

Fig. 10.37. (a) Drawing of the Bloch (or domain) walls in a $BaFe_{12}O_{19}$ crystal; (b) magnetic domains in a $Gd_{0.94}Tb_{0.75}Er_{1.31}Al_{0.5}Fe_{4.5}O_{12}$-garnet. The black and white regions represent domains with different magnetization direction. (c) Ferromagnetic domains at the surface of a monocrystalline nickel lamina. The domain boundaries have been made visible by the Bitter method (after [10.26]).

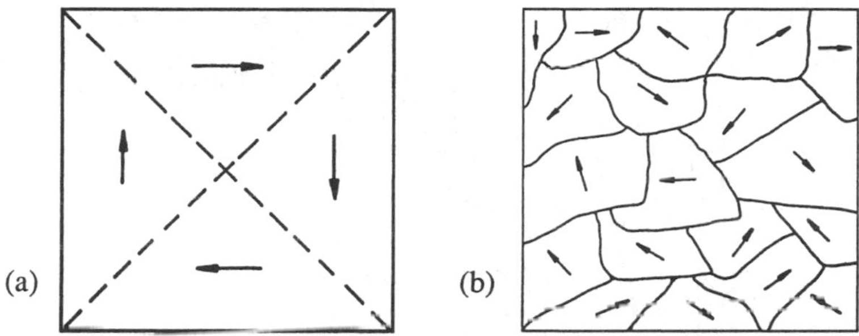

Fig. 10.38. Schematic illustration of domain ordering in (a) a single crystal and (b) a polycrystal. The resulting magnetic moment vanishes. Generally domain boundaries and grain boundaries do not coincide.

by the overlap of the electron density distribution of next neighbor atoms, which results in a non-isotropic charge distribution. In polycrystals with random orientation distribution the average magnetization is zero. If the material, however, has a pronounced crystallographic texture, the anisotropy becomes felt externally (Fig. 10.40c). This is why grain oriented electrical steels have a Goss texture $\{011\}\langle 100\rangle$ where the magnetically soft $\langle 100\rangle$ direction is parallel to the rolling direction and, therefore, parallel to the direction of alternating magnetization. The Goss texture is obtained by secondary recrystallization (see Chapter 7). In amorphous materials like metallic glasses there is no orientation dependence by definition. Moreover, Bloch walls cannot be hindered by lattice defects. Therefore, these materials are very soft magnetically.

In some crystal structures the spin/spin coupling can cause next neighbor spins to be antiparallel rather than parallel. In this case the material is antiferromagnetic, because in saturation the magnetization is zero. In analogy to ferromagnetism also antiferromagnets have a temperature T_N, the Néel

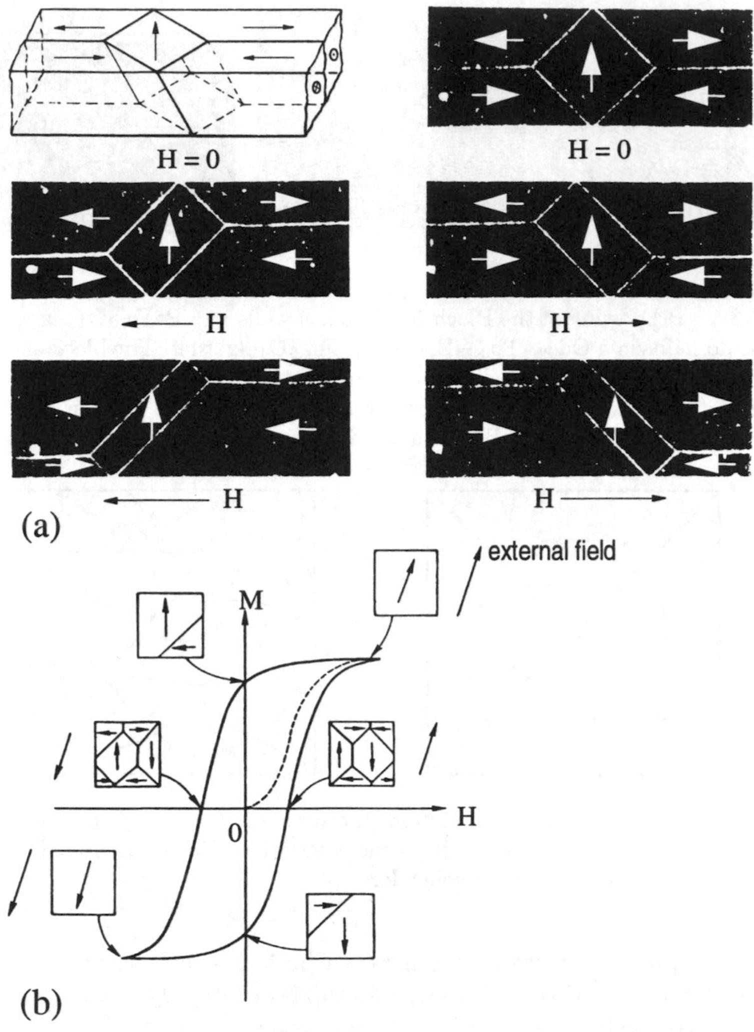

Fig. 10.39. (a) Continuous and reversible motion of a domain wall in an iron crystal. The domains aligned with the applied field grow at the expense of the others [10.27]. (b) Changing domain microstructure during a magnetic hysteresis cycle [10.28].

temperature, where antiferromagnetism disappears, and the material becomes paramagnetic for $T > T_N$.

Akin to antiferromagnetism is ferrimagnetism. In such materials specific lattice sites are occupied with parallel or antiparallel spins (Fig. 10.42a), however, the number of parallel and antiparallel spins need not be equal. Therefore, in saturation the magnetization is not zero but much smaller than the magnetization saturation of a ferromagnet. The term ferrimagnetism has historical reasons, since it was first discovered in ferrites, which are the oxides of

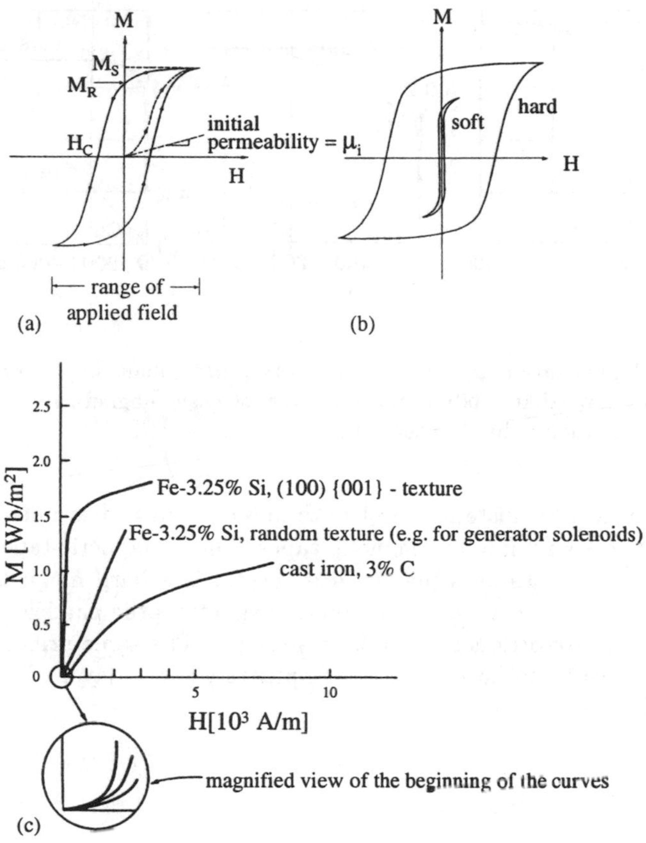

Fig. 10.40. (a) Hysteresis of a ferromagnetic material (- - - initial curve; M_S = saturation magnetization; M_R = remanence, H_C = coercive field). (b) Schematic diagram of hysteresis in hard and soft magnetic materials. (c) Comparison of the initial curve of three iron alloys.

iron, for instance magnetite $FeO \cdot Fe_2O_3$. Ferrimagnetics have a spinel crystal structure ($MgAl_2O_4$ structure, Fig. 10.42b). In magnetite the iron atoms are arranged on the octahedral and tetrahedral sites, but with antiparallel spins, of a cubic most densely packed oxygen lattice.

Ferrites are the commercially most important ceramic magnets. Typically they have low electrical conductivity. This is advantageous for applications at high frequencies. In alternating fields an electrical current is induced in an electrical conductor (Eddy current), which eventually is converted to heat. For very high frequency alternating fields the losses by Eddy currents would be high in metals. Besides, this is also the reason why the silicon content in FeSi sheets of electrical transformers is as high as possible, to minimize the electrical conductivity and, therefore, the Eddy currents. In the nonconducting ferrites, however, the Eddy currents are of minor concern. Typical applica-

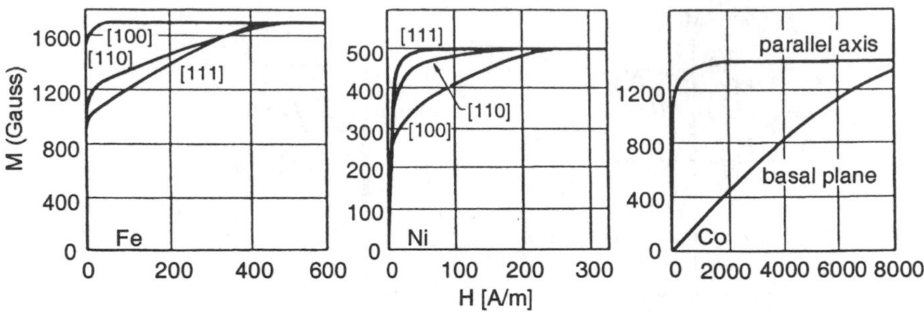

Fig. 10.41. Magnetization curves for iron, nickel, and cobalt single crystal. The curves for iron show that ⟨100⟩ is the direction of easy magnetization, in ⟨111⟩ direction magnetization is hard (after [10.29]).

tions of ferrites are for instance the ferrite antennas in radios but also electrical transformers and, finally, magnetic tapes. These magnetic tapes consist of fine γ-Fe_2O_3 particles on a plastic tape. Fe_2O_3 is a hard magnetic material. The electrical field caused by the sound magnetizes the particles, and the magnetization is proportional to the field strength. The same principle holds also for floppy discs and hard discs of computers where a layer of iron oxide is deposited on a plastic or metal disc.

10.6 Optical Properties

10.6.1 Light

Optics is the science of light. Light consists of electromagnetic waves where the magnetic field oscillates perpendicular to the electrical field. It is characterized by the wavelength, λ, which is related to its frequency ν by the relation

$$\nu\lambda = c \tag{10.42}$$

where c is the velocity of light, which amounts to $3{\cdot}10^{10}$m/s in vacuum, but which is smaller in solids. Light consists of elementary quanta, the photons, which do not have mass but have energy

$$E = h\nu = hc/\lambda \tag{10.43}$$

Visible light is electromagnetic radiation with a wave length in the range of $0.34\mu m \leq \lambda \leq 0.74\mu m$. Ultraviolet light has a shorter wave length and infrared radiation (heat radiation) microwaves and radiowaves have larger wave lengths than visible light. Optical properties of solids concern interaction phenomena of a solid with electromagnetic radiation especially in the visible spectrum.

Generally, a solid can be characterized optically according to its color and

(a)

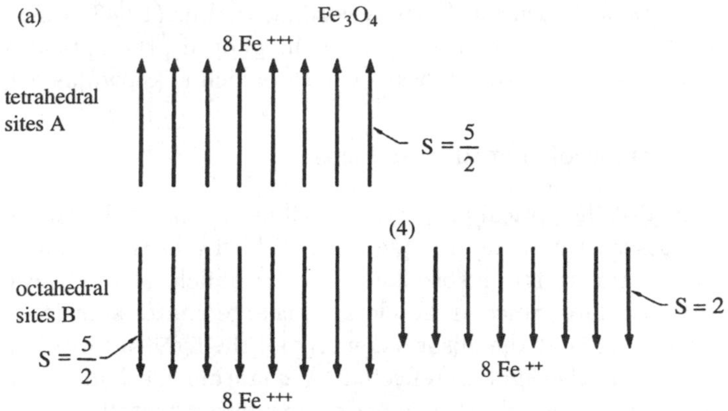

(b)

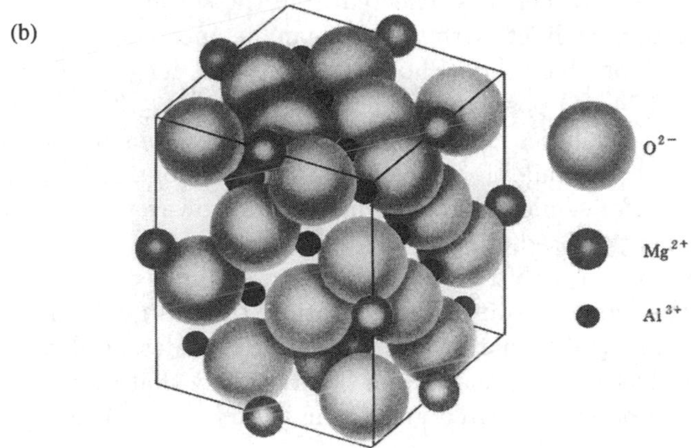

Fig. 10.42. (a) Schematic spin orientation in magnetite, $FeO{\cdot}Fe_2O_3$. The moments of the Fe^{3+} ions cancel each other out, only the moments of the Fe^{++} ions remain. (b) Crystal lattice of normal spinel $MgAl_2O_4$; the Mg^{++} ions are located at the corners of a tetrahedron, surrounded by four oxygen atoms each; the Al^{3+} ions occupy the corners of an octahedron and are surrounded by six oxygen ions each.

transparency. Transparency quantifies the capability of light to penetrate a solid. Many electrical insulators are transparent. If a solid is not transparent, it absorbs the light, either completely, then it appears black, or only partially, then it looks colored, or it reflects light. Most metals completely reflect light and, therefore, appear as silver-white or reflect light partially, then they appear colored, such as the red color of copper or the yellow color of gold. White light is a mixture of all visible wave lengths. Light of a single wave length is referred to as monochromatic light; it has a defined color, which is only determined by its wave length. Since according to Eq. (10.43) different wave lengths correspond to different energies of the photons, the optical properties of a solid usually depend on the wave length, which is known as dispersion.

10.6.2 Reflection of metallic surfaces

More specifically, the optical properties of solids are caused by the interaction of electromagnetic waves, i.e., the photons, with the electrons of a solid. Metals contain free electrons. The oscillating electrical field of a light wave causes them also to oscillate. Since an accelerated charge emits radiation, the oscillating electron will emit the absorbed energy of the light in form of radiation, which means that the light is reflected. A quantum mechanical calculation of the penetration of an electromagnetic wave into a metallic surface by application of Eq. (10.1) shows the amplitude of the light wave to attenuate exponentially with penetration depth, which simply means that it does not penetrate the metal, rather it is reflected from the surface.

The interaction of light with a solid can be most easily illustrated in the band model of the solid. Photons of a wave length of visible light ($0.34\mu m \leq \lambda \leq 0.74\mu m$) possess an energy $1.7\text{eV} \leq E \leq 3.5\text{eV}$ according to Eq. (10.43). These energies can be transferred to electrons in the conduction band or valence band and, therefore, move the electrons into excited states of higher energy. For semiconductor or insulators, however, this requires that the observed energy of radiation is at least as large as the band gap E_g. By contrast, for metals any arbitrary excitation energy is possible. If the electrons fall back into their ground state after excitation, the energy difference is emitted as radiation. If specific energy levels are excited preferentially, the emitted (reflected) light has a color according to the frequency that corresponds to the energy difference between excited and ground state (Fig. 10.43).

10.6.3 Insulators

10.6.3.1 Color

Electrical insulators exhibit a wide variety of optical properties. Pure insulators are usually completely transparent, i.e. they do not have a color. They can become colored by addition of impurities even in very small concentrations, for instance, the red ruby or the blue sapphire. The red color of ruby is caused by impurities of about 0.5% Cr^{3+} ions. The blue color of the sapphire results from dissolved Ti^{3+} ions. The excitation states of the impurity atoms determine the wave length of the reflected light and, therefore, the color of the crystal.

If the band gap between the valence and conduction band is smaller than the energy of the incident light, the light energy will cause - in particular for high energy (blue) light - an electron-hole pair and, therefore, the light becomes absorbed in the crystal. The color of the crystal is in this case the color

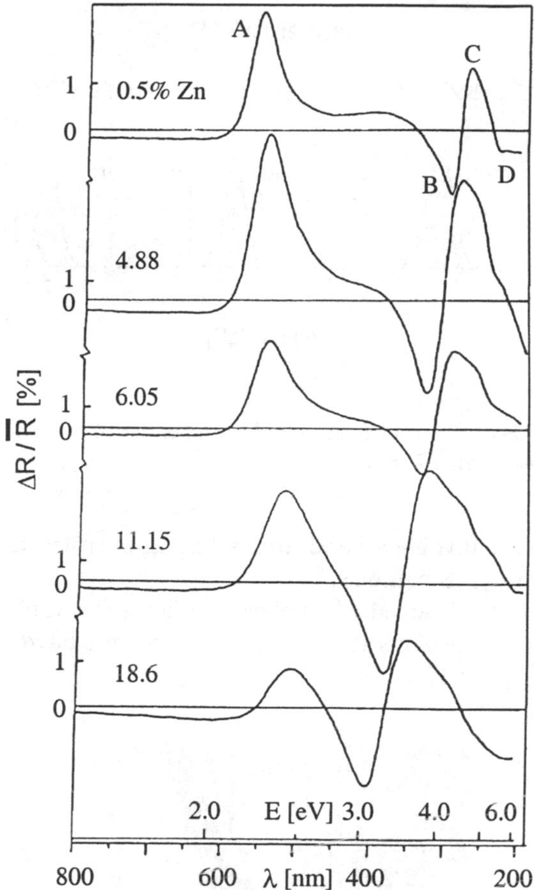

Fig. 10.43. Experimentally determined reflectograms for various Cu-Zn alloys. The indicated parameter is the average zinc concentration in atomic percent (after [10.30]).

of the remaining transmitted light. For instance, CdS appears yellow-orange, because the high energy blue light is absorbed. Transition elements frequently have atomic excitation states in the visible range. Crystals that contain such transition elements then appear colored even if the band gap is not in the visible range.

Uncolored crystals can become colored after exposure to an X-ray beam or by irradiation with elementary particles. The reason for this phenomenon is the formation of color centers which consist of electrons localized in irradiation induced anionic vacancies. If electrons have excitation states with energies of visible light this results in absorption bands and, therefore, in a change of crystal color. The F-bands of some (otherwise transparent) alkali halides are shown in Fig. 10.44. There exist also more complicated structures

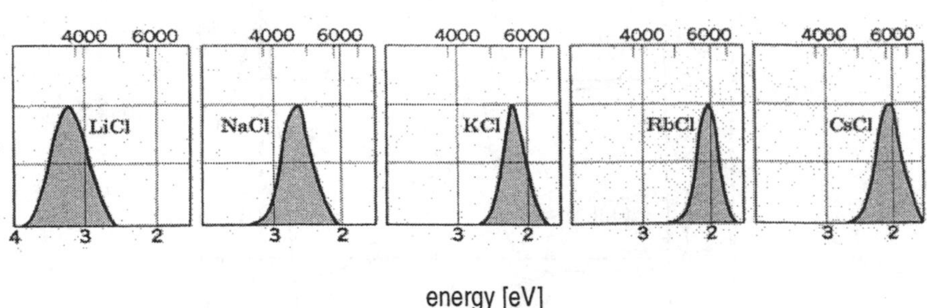

Fig. 10.44. F-bands of different alkali halides; dependence of optical absorption on wavelength of crystals with F-centers.

of color centers that involve several atoms and impurities. In such case also several absorption bands can occur.

The same effect is obtained when electrical fields are applied to an insulator. The electrons in the generated space charge regime have excitation levels of visible light (Fig. 10.45).

Fig. 10.45. Color centers (dark (blue) region) in NaCl, caused by the high electric field at the tip of the (upper) electrode.

10.6.3.2 Absorption

If an electron is moved from the valence band to the conduction band by absorption of radiation with an energy larger than the band gap, a hole is generated in the valence band. If the energy input is slightly less than the band gap energy, the formation of an electron hole pair is still possible, however, they remain correlated as a pair. Such a bound pair is called exciton (Fig. 10.46). The crystal reveals absorption in the respective energy range, as Fig. 10.47 illustrates for GaAs. Alkali halides show very typical exciton absorption in the ultraviolet spectrum, as this corresponds to the location of the electron hole pair that has lower excitation levels than the positive alkali zones.

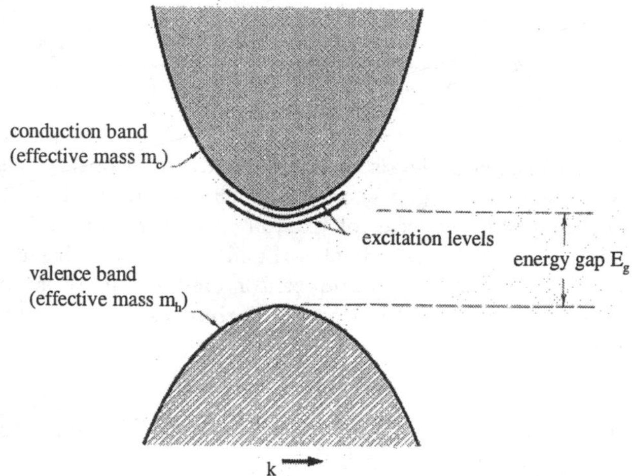

Fig. 10.46. Position of the exciton levels relative to the conduction band minimum in a simple band structure: The conduction band minimum and valence band maximum occur at $k = 0$. An exciton can have (kinetic) translational energy. However, if its translational energy exceeds its binding energy, it is metastable against decomposition into a free (unlocalized) electron and hole. All excitons are potentially unstable against radiation recombination, where the electron falls back into a hole in the valence band, emitting a photon or phonon.

10.6.3.3 Photoconductivity

If the electrical conductivity of a crystalline insulator is increased by incident electromagnetic radiation this is referred to as photoconductivity. The reason for photoconductivity is the increase of the charge carrier concentration due to electron-hole formation, provided that the photon energy is sufficiently high. If such pairs are homogeneously formed in the entire crystal the photoconductivity can be easily calculated. Let A be the absorption rate of photons and

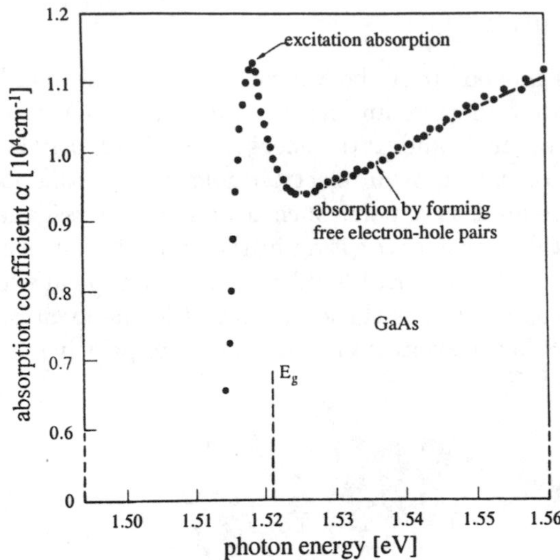

Fig. 10.47. Effect of exciton levels on a semiconductor's optical absorption with photon energy close to band gap energy E_g. The diagram shows the beginning of optical absorption and the peak of exciton absorption in gallium arsenide at 21K (after [10.31]). The ordinate indicates the absorption coefficient α, defined by $I(x) = I_0 \cdot \exp(-\alpha x)$. The gap energy and exciton binding energy can be determined from the absorption curve: the band gap is derived as 1.521eV, exciton binding energy 0.0034eV.

R the rate for recombination of holes and electrons. The temporal change of the charge carrier concentration n is

$$dn/dt = A - Rn^2 \tag{10.44}$$

In steady state $\dot{n} = 0$ and we obtain the steady state photoelectron concentration

$$n_0 = \sqrt{\frac{A}{R}} \tag{10.45}$$

and the photoconductivity according to Eq. (10.33a)

$$\sigma = ne\mu = \sqrt{\frac{A}{R}} e\mu \tag{10.46}$$

Commercial applications are crystal detectors and exposure meters in photocameras.

10.6.3.4 Luminescence

Solids show a variety of optical phenomena which are subsumed under the term luminescence. For instance fluorecense and phosphorescence. Luminescence denotes, in general, the glowing effects owing to absorbed energy that may be caused by incident light, mechanical impact, chemical reactions, or heat input. If light emission occurs during excitation within 10^{-8}s the phenomenon is referred to as fluorescence. Phosphorescence or after-glowing denotes the case that the light is emitted only subsequent to excitation. The delay time may be milliseconds or hours. Crystalline solids which show luminescence are generally referred to as phosphors. Luminescence is caused by the excitation of electrons of so-called activators, i.e., impurity atoms in crystals, upon exposure to energy and the emission of visible light upon return to the ground state. Commonly it is observed that the absorbed energy has a certain range, which is explained by the fact that part of the energy is also imparted to vibrations of the respective atoms (Fig. 10.48).

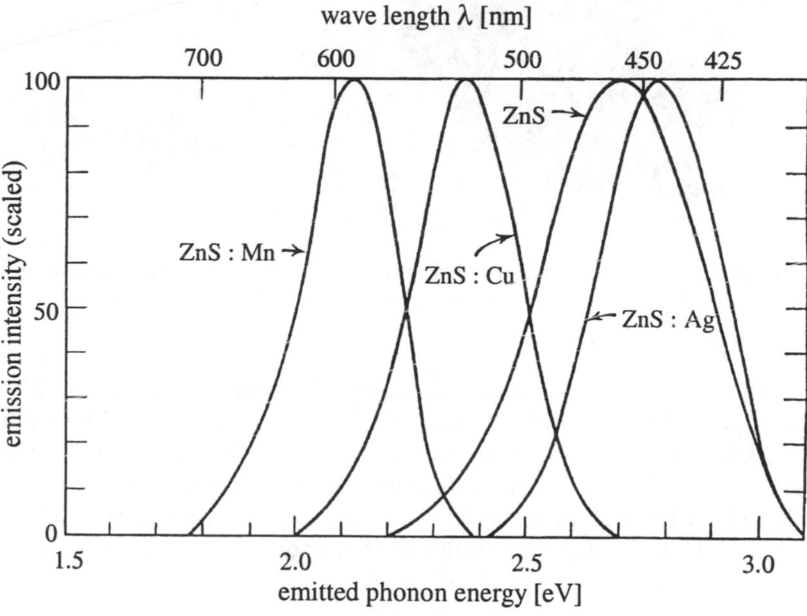

Fig. 10.48. Luminenscence emission spectra of ZnS, ZnS:Ag, ZnS:Cu, and ZnS:Mn (after [10.32]).

10.6.4 Applications

Optical properties of glasses and crystals have major commercial importance today. Since long the known refraction of light in glasses has been used for lenses in optical instruments. An important example for large scale commercial applications of absorption and reflection is controlled heat insulation. By appropriate coatings the behavior of a material is tuned in such a way that it absorbs heat, i.e. infrared radiation, or reflects it, while is transmits the visible light. Window panes in air conditioned buildings, therefore, are usually coated and have a characteristic color. Widely used are also self-coloring glasses, where the coloring depends on the light intensity, in particular during ultraviolet radiation. In such glasses specific absorption bands are exited by the incident ultraviolet radiation (mostly photochemically, Fig. 10.49).

Fig. 10.49. Photochromic glass is primarily used for self-toning sunglasses. Here one of the two lenses has been selectively exposed to ultraviolet light.

Applications of luminescence are found in fluorescent lights and displays of oscilloscopes. Photoconductivity is utilized in exposure meters and crystal detectors. Also metallic surfaces are increasingly utilized for decorative purposes, for instance anodically anodized aluminum. In this context the reflectivity of metals and the light of the reflected radiation plays a dominant role. The same principle controls the function of precision mirrors of metallized surfaces.

References

[0.1] Agricola, De Re Metallica, VDI-Verlag, Düsseldorf, 1961

[1.1] Institut für Eisenhüttenkunde, RWTH Aachen, unpublished
[1.2] G. Schmitz, P. Haasen, Acta metall. mater. 40 (1992), p. 2209-2217

[2.1] C.J. McHargue, L.K. Jetter, J.C. Ogle, in C. Barrett, T. B. Massalski, Structure of Metals, 1980, p. 546
[2.2] E. Hornbogen, Werkstoffe, Springer-Verlag, Berlin, 1979, p. 7
[2.3] D. Raabe, Archive MPI Eisenforschung, Düsseldorf
[2.4] H. Esokrig, J. Fink, L. Schultz, Physik-Journal 1 (2002) p. 45

[3.1] P. Haasen, Physikalische Metallkunde, Springer-Verlag, Berlin, 1984, p. 198
[3.2] P. Haasen, Physikalische Metallkunde, Springer-Verlag, Berlin, 1984, p. 195
[3.3] T. Schröder, Max Planck Forschung 4/2002, p. 33
[3.4] G. Cox, Forschungszentrum Jülich, Annual Report 1992, p. 11
[3.5] G. Hosson et al. in: The Nature and Behaviour of Grain Boundaries, TMS-AIME, H. Hu (ed.), p. 13
[3.6] W. Bollmann, Crystal Defects and Crystalline Interfaces, Springer-Verlag, Berlin, 1970, p. 125
[3.7] Progress in Materials Science 3, B. Chalmers (ed.), Pergamon Press, London, 1952, p. 293-319
[3.8] W. Bollmann, Crystal Defects and Crystalline Interfaces, Springer-Verlag, Berlin, 1970, p. 187
[3.9] V. Vitek et al. in: Grain Boundary Structure and Kinetics, TMS-AIME, 1979, p. 115-148
[3.10] W. Krakow, J. Materials Research 5 (1990), p. 2660
[3.11] A. Ourmazd, MRS Bulletin 15 (1990), p. 58-64
[3.12] F.S. Shieu, S.L. Sass, Acta. metall. mater. 38 (1990), p. 1653
[3.13] A.G. Evans, M. Rühle, MRS Bulletin 15 (1990), p. 46-50
[3.14] Archives, Institut für Metallkunde und Metallphysik, RWTH Aachen
[3.15] Archives, Institut für Metallkunde und Metallphysik, RWTH Aachen

[4.1] T.B. Massalski, Binary Alloy Phase Diagrams, ASM Int., 1992
[4.2] Archives, Institut für Metallkunde und Metallphysik, RWTH Aachen

[4.3] W. Dahl, K. Lücke, Archiv für Eisenhüttenwesen 25 (1954), p. 241-250
[4.4] Archives, Institut für Metallkunde und Metallphysik, RWTH Aachen
[4.5] G.E.R. Schulze, Metallphysik, Berlin, Akademie-Verlag 1967, J9, J13, J15
[4.6] Progress in Metal Physics, B. Chalmers (ed.), Pergamon Press, Oxford, 1952, p. 42-75
[4.7] K. Lücke, H. Haas, H.A. Schulze, J. Phys. Chem. Solids 37 (1976), p. 979
[4.8] J.F. Shackelford, Introduction to Materials Science for Engineers, New York, London, 1985

[5.1] W.M. Franklin in Norwick, Burton, Diffusion in Solids, Academic Press, New York, 1975, p. 2
[5.2] W. Seith, Diffusion in Metallen, Berlin, Göttingen, Heidelberg, Springer-Verlag, 1955, p. 39
[5.3] D.R. Askeland, The Science and Engineering of Materials, Boston PWS-KENT 1989, Fig. 5-7
[5.4] L.H. Van Vlack, Elements of Materials Science and Engineering, Reading-Massachusetts, Addison-Wesley, 1985, Fig. 4-7
[5.5] D.R. Askeland, The Science and Engineering of Materials, Boston, PWS-KENT 1989
[5.6] W. Seith, Diffusion in Metallen, Springer-Verlag, Berlin, 1955, p. 107
[5.7] W. Seith, Diffusion in Metallen, Springer-Verlag, Berlin, 1955, p. 99
[5.8] W. Seith, Diffusion in Metallen, Springer-Verlag, Berlin, 1955, p. 106
[5.9] W. Seith, Diffusion in Metallen, Springer-Verlag, Berlin, 1955, p. 130
[5.10] W. Seith, Diffusion in Metallen, Springer-Verlag, Berlin, 1955, p. 140
[5.11] W. Seith, Diffusion in Metallen, Springer-Verlag, Berlin, 1955, p. 186
[5.12] W. Seith, Diffusion in Metallen, Springer-Verlag, Berlin, 1955, p. 192
[5.13] N.L. Peterson, Grain Boundary Structure and Kinetics, ASM Int., Ohio, 1979, p. 219
[5.14] N.L. Peterson, Grain Boundary Structure and Kinetics, ASM Int., Ohio, 1979, p. 216
[5.15] P. Shewmon, Diffusion in Solids, TMS, 1989, p. 157
[5.16] P. Shewmon, Diffusion in Solids, TMS, 1989, p. 161

[6.1] J. Marin, Mechnical Behavior of Engineering Materials, Prentice Hall Inc., 1962, p. 24
[6.2] G. Masing, Grundlagen der Metallkunde, Springer-Verlag, Berlin, 1955, p. 104
[6.3] G. Masing, Grundlagen der Metallkunde, Springer-Verlag, Berlin, 1955, p. 108
[6.4] G. Masing, Grundlagen der Metallkunde, Springer-Verlag, Berlin, 1955, p. 106
[6.5] G. Masing, Grundlagen der Metallkunde, Springer-Verlag, Berlin, 1955, p. 106
[6.6] W. Boas, E. Schmid, Z. Phys. 54 (1929), p. 16
[6.7] G.E. Dieter, Mechanical Metallurgy, 3nd ed., McGraw-Hill Book Company, 1986, p. 124
[6.8] G.E. Dieter, Mechanical Metallurgy, 3nd ed., McGraw-Hill Book Company, 1986, p. 302
[6.9] P. Haasen, Physikalische Metallkunde, Springer-Verlag, Berlin, 1984, p. 239

[6.10] P. Haasen, Physikalische Metallkunde, Springer-Verlag, Berlin, 1984, p. 232
[6.11] J.D. Livingston, Acta metall. 10 (1962), p. 229
[6.12] T.H. Courtney, Mechanical Behavior of Materials, McGraw-Hill Publishing Company, New York, 1990, p. 127
[6.13] F. Kirch, Doctoral thesis, RWTH Aachen, 1970
[6.14] Werkstoffkunde Eisen und Stahl Bd 1, Verlag Stahleisen mbH, Düsseldorf, 1983, p. 58
[6.15] T.E. Mitchell, R.A. Foxall, P.B. Hirsch, Phil. Mag. 8 (1963), p. 1895
[6.16] M.J. Whelan, P.B. Hirsch, R.W. Horne, W. Bollmann, Proc. Roy. Soc. A204 (1957), p. 524
[6.17] J. Guerland, Stereology and Qualitative Metallography ASTM STP 504, 1972, p. 108
[6.18] E. Schmid, Phys. Zeitschrift 31 (1930) p. 892
[6.19] G. Sachs, J. Weets, Z. Phys. 62 (1930), p. 473
[6.20] J.O. Linde, S. Edwards, Arkiv Fysik 8 (1954), p. 511 and R.L. Fleischer, Acta metall. 11 (1963), p. 203, und T.J. Koppenal, M.E. Fine, Tranp. TMS-AIME 224 (1962), p. 347 and C. Wert, Tranp. TMS-AIME 188 (1950), p. 1242 und P.R.V. Evans, J. Less Common Metals 4 (1962), p. 78 and A.G. Evans, T. Langdon, Progress in Materials Science 21 (1976), p. 11
[6.21] R.L. Fleischer, Acta metall. 11 (1963), p. 203
[6.22] A.H. Cottrell, An Introduction to Metallurgy, Edward Arnold Ltd., London (1967)
[6.23] P.B. Hirsch, F.J. Humphreys, Physics of Strength and Plasticity, A. Argon (ed.), M.I.T. Press, Cambridge, 1969
[6.24] M.F. Ashby, Z. Metallkunde 55 (1964), p. 5
[6.25] P. Haasen, Physikalische Metallkunde, Springer-Verlag, Berlin, 1984, p. 295
[6.26] T.H. Courtney, Mechanical Behavior of Materials, McGraw-Hill Publishing Company, New York, 1990, p. 190
[6.27] R.E. Reed-Hill, Physical Metallurgy Principles, 2nd ed., D. Van Nostrand Company, New York, 1973, p. 842
[6.28] D.A. Holt, W.A. Backofen, Trans. Quart. ASM 61 (1968), p. 329
[6.29] N. Furushiro, S. Hori, Superplasticity in Metals, Ceramics and Intermetallics, M.J. Mayo, M. Kobayashi, J. Wadsworth (ed.), 1990, MRS, p. 252
[6.30] O.D. Sherby, J. Wadsworth, Superplasticity in Metals, Ceramics and Intermetallics, M.J. Mayo; M. Kobayashi; J. Wadsworth (ed.), 1990, MRS, p. 9
[6.31] W. Blum, B. Ilschner, Phys. Stat. Sol. 20 (1967), p. 629 and E.C. Norman, S.A. Duran, Acta metall. 18 (1970), p. 723 and C.Y. Cheng, A. Karim, T.G. Langdorn, J.E. Dorn, Trans. Met. Soc. AIME 242 (1968), p. 584
[6.32] R.E. Reed-Hill, Physical Metallurgy Principles, D. van Nostrand Company, New York, 1973, p. 854
[6.33] L.E. Poteat, C.S. Yust, Ceramic Microstructure, R.M. Fulrath, J.A. Pask (ed.), Wiley, New York, 1968, p. 649
[6.34] T.H. Courtney, Mechanical Behavior of Materials, McGraw-Hill Publishing Company, New York, 1990, p. 287
[6.35] Archives, Institut für Metallkunde und Metallphysik, RWTH Aachen
[6.36] T.B. King, R.W. Cahn, B. Chalmers, Nature, London, 1948, p. 682
[6.37] T.S. Kê, Phys. Rev. LXXI (1947), p. 533
[6.38] T.S. Kê, Phys. Rev. LXXII (1947), p. 41
[6.39] T.S. Kê, Phys. Rev. LXXI (1947), p. 533

[6.40] H. Domininghaus, Die Kunststoffe und ihre Eigenschaften, VDI-Verlag, Düsseldorf, 1992, p. 187
[6.41] R.W.K. Honeycombe, The Plastic Deformation of Metals, Edward Arnold, 1984, p. 146

[7.1] G. Gottstein, Rekristallisation metallischer Werkstoffe, DGM, Oberursel, 1984, p. 29
[7.2] F.L. Vogel jr., Acta metall. 3 (1955) p. 245
[7.3] S. Mader in: Moderne Probleme der Metallphysik, A. Seeger (ed.), Vol. 1, Springer-Verlag, Berlin, 1965, p. 203
[7.4] W.R. Hibbard, C.Dunn, Acta metall. 4 (1956), p. 311
[7.5] L.M. Clarebrough, M.E. Hargreaves, M.H. Loretto, Recovery and Recrystallization of Metals, L. Himmel (ed.), Interscience, N.Y. (1963) p. 63
[7.6] Hayendy in: Grundlagen der Wärmebehandlung von Stahl, Verlag Stahl-Eisen
[7.7] R.D. Doherty, R.W. Cahn, J. Less Common Metals Vol. 28 (1972), p. 279
[7.8] R.A. Vandermeer, P. Gordon, Trans AIME 215 (1959), p. 577
[7.9] E. Hornbogen, U. Köster in: Recrystallization of Metallic Materials, F. Haessner (ed.), Dr. Riederer-Verlag, Stuttgart, 1978, p. 159-194
[7.10] B.B. Rath, H.Hu, Trans. TMS AIME 245 (1969), p. 1243-52 and 1577-85
[7.11] B. Liebmann, K. Lücke, G. Masing, Z. Metallkunde 47 (1956), p. 57
[7.12] G. Gottstein, H.C. Murmann, G. Renner, C. Simpson, K. Lücke, Textures of Materials Vol. 1, Springer-Verlag, Berlin, 1978, p. 530
[7.13] D.W. Demianczuk, K.T. Aust, Acta metall. 23 (1975), p. 1149 und E.M. Friedman, C.V. Kopezky, L.S. Shvindlerman, Z. Metallk. 66 (1975), p. 533
[7.14] W.A. Anderson, R.F. Mehl, Trans. AIME 161 (1945), p.140
[7.15] F.W. Rosenbaum, Doctoral thesis, RWTH Aachen (1972)
[7.16] E. Hornbogen, Werkstoffe, Springer-Verlag, Berlin, 1979, p. 85
[7.17] W.A. Anderson, R.F. Mehl, Trans. AIME 161 (1945), p.140
[7.18] O. Dahl, F. Pawlek, Z. Metallk. 28 (1936), p. 266
[7.19] K. Detert, K. Lücke, Report No. AFOSR - TN - 56 -103 AD - 82016, Brown Univ. (1956)
[7.20] P. Gordon, R.A.Vandermeer, Recrystallisation, Grain Growth and Textures, ASM Metals Park, Ohio, 1956, p. 205
[7.21] C. Frois, O. Dimitrov, Mem. Sci. Rev. Met. 59 (1962), p. 643
[7.22] W. Grünwald, F. Haessner, Acta metall. 18 (1970), p. 217
[7.23] W.C. Leslie, J.T. Michalak, F.W. Aul, Iron and its Solid Solutions, Interscience Publishers, 1963, p. 119
[7.24] N. Hansen, H.R. Jones, Recovery and Recrystallization of Particle Containing Materials, 24 colloque de metallurgie, Sacley, 1981, p. 95
[7.25] G.E. Burke, D. Turnbull, Progress in Metal Physics 3 (1952), p. 274
[7.26] T. Grey, J. Higgins, Acta metall. 21 (1973), p. 310
[7.27] P.A. Beck, M.L. Holzwerth, P.R. Sperry, Trans. AIME 180 (1949), p. 163
[7.28] C. Rossard, Metaux 35 (1960), p. 102, 140, 190
[7.29] R.A. Petkovic, PhD thesis, McGill University Montreal (1975)
[7.30] C.M. Sellars, W.J.McG. Tegart, Mem. Sci. Rev. Met. 63 (1966), p. 731
[7.31] R. Bromley, C.M. Sellars, Proc. Int. Conf. Strength of Metals and Alloys 3, Cambridge, 1973, Vol.1, p. 380
[7.32] M.J. Luton, C.M. Sellars, Acta metall.17 (1969) p. 1033

[7.33] C.M. Sellars, J.A. Whiteman, Met. Sci. 13 (1979), p. 187

[7.34] J. Poirier, M. Nicholson, J. Geol. 83 (1975)

[8.1] F. Sauerwald, Lehrbuch der Metallkunde des Eisens und der Nichteisenmetalle, Springer-Verlag, Berlin, 1929

[8.2] P. Debye, H. Menke, Ergebn. techn. Röntgenkunde 2 (1938), p. 18

[8.3] G. Tammann, Aggregatzustände, Leipzig, p. 223

[8.4] E. Scheil, Z. Metallkunde 32 (1940), p. 171

[8.5] L. Horn, G. Masing, Z. Elektrochemie 46 (1940), p. 109

[8.6] L.E. Murr, Interfacial Phenomena in Metals and Alloys, Addison Publishing Company, London, 1975, p. 8

[8.7] Archives, Institut für Metallkunde und Metallphysik, RWTH Aachen

[8.8] Archives, Institut für Metallkunde und Metallphysik, RWTH Aachen

[8.9] W. Schatt, Einführung in die Werkstoffwissenschaften, VEB Verlag für Grundstoffindustrie, Leipzig, 1981, p. 121

[8.10] Esaka, Straunke, Kurz, Columnar Dendrite Growth in SCN-Acetone (Videobänder), EPFL-Lausanne (1985)

[8.11] P.C. Huang, M.E. Glicksman, Acta metall. 29 (1981), p. 717

[8.12] W. Schatt, Einführung in die Werkstoffwissenschaften, VEB Verlag für Grundstoffindustrie, Leipzig, 1981, p. 122

[8.13] T.B. Massalski, Binary Alloy Phase Diagrams, ASM Int., 1992 und G. Masing, Grundlagen der Metallkunde, Springer-Verlag, Berlin, 1955, p. 30, 31

[8.14] T. Donomoto, N. Miura, K. Funatani, N. Miyake, Ceramic Fiber Reinforced Piston for High Performance Diesel Engine, SAE Tech. Paper No. 83052, Detroit, MI, 1983

[8.15] A. Sieverts, Z. Metallkunde 21 (1929), p. 37

[8.16] Archives,Institut für Eisenhüttenkunde, RWTH Aachen

[8.17] G. Masing, Lehrbuch der Allgemeinen Metallkunde, Springer-Verlag, Berlin, 1950, p. 235

[8.18] G. Masing, Lehrbuch der Allgemeinen Metallkunde, Springer-Verlag, Berlin, 1950, p. 230, 231

[8.19] O. Greis, Nachr. Chem. Lab. 38 (1990), p. 1346-50

[9.1] G. Masing, Lehrbuch der Allgemeinen Metallkunde, Springer-Verlag, Berlin, 1950, p. 479

[9.2] T.B. Massalski, Binary Alloy Phase Diagrams, ASM International (1990)

[9.3] J.E. Hilliard in Phase Transformation, ASM Metals Park, Ohio, 1970

[9.4] P. Haasen, Physikalische Metallkunde, Springer-Verlag, Berlin, 1984, p. 173

[9.5] T.B. Massalski, Binary Alloy Phase Diagrams, ASM International (1990)

[9.6] J.M. Silcock, J. Inst. Metals 89 (1960), p. 203-210

[9.7] E. Hornbogen, Aluminium 43 (1967), p. 41

[9.8] R.B. Nicholson et al, J. Inst. Metals 87 (1958), p. 431

[9.9] R.B. Nicholson, T. Nutting, J. Inst. Met. 87 (1958), p. 34

[9.10] H.K. Hardy, T.J. Heal, in Progress in Metal Physics 5 (1954), Pergamon Press, p. 177

[9.11] P.G. Shewmon, Transformations in Metals, McGraw-Hill Book Company, New York, 1969, p. 227

[9.12] D. Horstmann, Das Zustandsdiagramm Fe-C, Verlag Stahleisen, Düsseldorf, 1985

[9.13] E. Hornbogen in: Advanced Structural and Functional Material, W.G.J. Bunk (ed.), Springer-Verlag, 1991, p. 140
[9.14] P.G. Shewmon, Transformations in Metals, McGraw-Hill Book Company, New York, 1969, p. 328
[9.15] E. Houdremont, Handbuch der Sonderstahlkunde, 3. Auflage, Band 1, Springer-Verlag, Berlin, 1956
[9.16] P. Haasen, Physikalische Metallkunde, Springer-Verlag, Berlin, 1974, p. 267
[9.17] H.P. Hougardy in: Werkstoffkunde Stahl Bd. 1, Verlag Stahleisen, Düsseldorf, 1984, p. 198-231

[10.1] J.C. Slater, Phys. Rev. 45 (1934), p. 794
[10.2] H. Ibach, H. Lüth, Festkörperphysik, Springer-Verlag, Berlin, 1989, p. 119
[10.3] G. Masing, Lehrbuch der Allgemeinen Metallkunde, Springer-Verlag, Berlin, 1950, p. 262
[10.4] G. Masing, Lehrbuch der Allgemeinen Metallkunde, Springer-Verlag, Berlin, 1950, p. 263
[10.5] G.E.R. Schulze, Metallphysik, Berlin, Akademie-Verlag, 1967, p. 150
[10.6] C. Kittel, Einführung in die Festkörperphysik, R. Oldenbourg-Verlag, München, 1968, p. 262
[10.7] C. Kittel, Einführung in die Festkörperphysik, R. Oldenbourg-Verlag, München, 1968, p. 240, 241
[10.8] G. Masing, Lehrbuch der Allgemeinen Metallkunde, Springer-Verlag, Berlin, 1950, p. 266
[10.9] C. Kittel, Einführung in die Festkörperphysik, R. Oldenbourg-Verlag, München, 1968, p. 2
[10.10] G. Masing, Lehrbuch der Allgemeinen Metallkunde, Springer-Verlag, Berlin, 1950, p. 267
[10.11] H.K. Hardy, T.J. Heal in: Progress in Metal Physics 5 (1954), Pergamon Press, Oxford, p. 177
[10.12] Archives, Institut für Metallkunde und Metallphysik, RWTH Aachen
[10.13] G. Masing, Lehrbuch der Allgemeinen Metallkunde, Springer-Verlag, Berlin, 1950, p. 271
[10.14] G. Masing, Lehrbuch der Allgemeinen Metallkunde, Springer-Verlag, Berlin, 1950, p. 270
[10.15] G. Masing, Lehrbuch der Allgemeinen Metallkunde, Springer-Verlag, Berlin, 1950, p. 270
[10.16] Progress in Metal Physics, B. Chalmers (ed.), Pergamon Press, Oxford, 1952, p. 42-75
[10.17] G. Masing, Lehrbuch der Allgemeinen Metallkunde, Springer-Verlag Berlin, 1950, p. 269
[10.18] C. Kittel, Einführung in die Festkörperphysik, R. Oldenbourg-Verlag, München, 1968, p. 398
[10.19] B. Stritzker, Anwendungen der Supraleitung I, Vortrag 24 in Vorlesungsmanuskripte des 19. IFF-Ferienkurses in der KFA Jülich, KFA Jülich (ed.), 1988, Jülich
[10.20] W. Buckel, Supraleitung, Physik-Verlag, 1977, p. 12
[10.21] C. Kittel, Einführung in die Festkörperphysik, R. Oldenbourg-Verlag, München, 1968, p. 399
[10.22] K.M. Koch, W. Jellinghaus, Einführung in die Physik der magnetischen Werkstoffe, Franz Deuticke, Wien, 1957

[10.23] C. Kittel, Einführung in die Festkörperphysik, R. Oldenbourg-Verlag, München, 1968, p. 504
[10.24] C. Kittel, Einführung in die Festkörperphysik, R. Oldenbourg-Verlag, München, 1968, p. 507
[10.25] C. Kittel, Einführung in die Festkörperphysik, R. Oldenbourg-Verlag, München, 1968, p. 532
[10.26] C. Kittel, Einführung in die Festkörperphysik, R. Oldenbourg-Verlag, München, 1968, p. 565
[10.27] C. Kittel, Einführung in die Festkörperphysik, R. Oldenbourg-Verlag, München, 1968, p. 566
[10.28] J.F. Shackelford, Introduction to Materials Science for Engineers, Macmillan Publishing Company, New York, 1988
[10.29] C. Kittel, Einführung in die Festkörperphysik, R. Oldenbourg-Verlag, München, 1968, p. 568
[10.30] R.J. Nastasi, Andrews, R.E. Hummel, Phys. Rev. B16 (1977), p. 4314
[10.31] J. Sturge, Phys. Rev. 127 (1962), p. 768
[10.32] R.H. Bube, Photoconductivity of Solids, Wiley, New York, 1960

Specific Literature

Chap.1:

– J. Guerland — Stereology and Quantitative Metallography ASTM STP 504 (1972)

Chap.2:

– C.S. Barrett, T.B. Massalski — Structure of Metals (Pergamon Press 1980)

– B.D. Cullity — Elements of X-Ray Diffraction (Addison-Wesley 1978)

Chap.3:

– W. Bollmann — Crystal Defects and Crystalline Interfaces (Springer-Verlag 1970)

– D. Hull, D.J. Bacon — Introduction to Dislocations (Pergamon Press 1989)

– A.P. Sutton, R.W. Balluffi — Interfaces in Crystalline Materials (Clarendon Press Oxford, 1995)

– J. Weertmann, J.R. Weertmann — Elementary Dislocation Theory (Oxford University Press 1992)

Chap.4:

– A.H. Cottrell — Theoretical Structural Metallurgy (Edward Arnold Verlag 1955)

– P. Haasen — Physical Metallurgy (Springer Verlag 1984)

Chap.5:

– R.J. Borg, G.J. Dienes — An Introduction to Solid State Diffusion (Academic Press 1991)

– J. Crank — Mathematics of Diffusion (Oxford University Press 1975)

– P.G. Shewmon — Diffusion in Solids (TMS 1989)

– M.E. Glicksman — Diffusion in Solids State Principles (Wiley Interscience 2000)

Chap.6:

– T.H. Courtney — Mechanical Behavior of Materials (MacGraw-Hill 1990)

– G.E. Dieter — Mechanical Metallurgy (MacGraw-Hill 1986)

– J. Friedel — Dislocations (Pergamon Press 1967)

– J.P. Hirth, J. Lothe — Theory of Dislocations (Krieger Publishing Company 1992)

– R.W.K. Honeycombe — The Plastic Deformation of Metals (Edward Arnold Verlag 1984)

– D. Hull, D.J. Bacon — Introduction to Dislocations (Pergamon Press 1989)

– M.A. Meyers, K.K. Chawla — Mechanical Metallurgy: Principles and Applications (Prentice Hall 1984)

– F.R.N. Nabarro — Dislocations in Solids (North-Holland Publ. 1979ff, Vol. 1-8)

– J. Weertmann, J.R. Weertmann — Elementary Dislocation Theory (Oxford University Press 1992)

– C. Zener — Elasticity and Anelasticity of Metals (The University of Chicago Press 1965)

Chap.7:

– P. Cotterill, P.R. Mould — Recrystallization and Grain Growth in Metals (Krieger Publishing Company 1976)

– F. Haessner — Recrystallization of Metallic Materials (Dr. Riederer Verlag 1978)

–M. Hatherly, F.J. Humphreys — Recrystallization and Related Annealing Phenomena (Pergamon 1995)

– G. Gottstein, L.S. Shvindlerman — Grain Boundary Migration in Metals (CRC Press, 1999)

Chap.8:

- B. Chalmers — Principles of Solidification (J. Wiley 1964)
- A.H. Cottrell — Theoretical Structural Metallurgy (Edward Arnold Verlag 1955)
- E. Murr — Interfacial Phenomena in Metals and Alloys (Techbooks 1975)
- D.A. Porter, K.E. Easterling — Phase Transformations in Metals (Van Nostrand Reinhold 1992)
- R.E. Reed-Hill — Physical Metallurgy Principles (PWS Publishers 1992)
- R.E. Smallmann — Modern Physical Metallurgy (Butterworths 1985)

Chap.9:

- J.W. Christian — Transformation in Metals and Alloys (Pergamon Press 1981)
- D.A. Porter, K.E. Easterling — Phase Transformations in Metals (Van Nostrand Reinhold 1992)

Chap.10:

- R.H. Bube — Electrons in Solids (Academic Press 1992)
- R.E. Hummel — Electronic Properties of Materials (Springer-Verlag 1992)
- C. Kittel — Introduction to Solid State Physics (Oldenbourg 1991)
- L. Solymar, D. Walsh — Lectures on Electrical Properties of Materials (Oxford Science Publications 1993)

Index

D

I

J

K

L

Q

T

Dr. Günter Gottstein is full professor and, since 1989, head of the Institute of Physical Metallurgy and Metal Physics at RWTH Aachen University, Germany. After his doctoral and post-doctoral studies and research at RWTH, he continued his career in the USA and became full professor of materials science at Michigan State University. In addition to research on interfaces, crystallographic textures, crystal plasticity and materials simulation, Prof. Gottstein has a long teaching experience on topics ranging from solid-state physics to physical metallurgy and materials engineering.

In this vivid and comprehensible introduction to materials science, the author expands the modern concepts of metal physics to formulate basic theory applicable to other engineering materials, such as ceramics and polymers. Written for engineering students and working engineers with little previous knowledge of solid-state physics, this textbook enables the reader to study more specialized and fundamental literature of materials science. Dozens of illustrative photographs, many of them Transmission Electron Microscopy images, plus line drawings, aid developing a firm appreciation of this complex topic. Hard-to-grasp terms such as "textures" are lucidly explained – not only the phenomenon itself, but also its consequences for the material properties. This excellent book makes materials science more transparent.

ISBN 3-540-40139-3
9 783540 401391

› springeronline.com